NAWCWD TP 8347

I0041853

**NAVAL AIR WARFARE CENTER WEAPONS DIVISION
POINT MUGU, CALIFORNIA**

NAV AIR

ELECTRONIC WARFARE AND RADAR SYSTEMS ENGINEERING HANDBOOK

NAVAIR Electronic Warfare/Combat Systems

Avionics Department 4.5

The Pulse of Naval Aviation

1 April 1997
w/Rev 4 of 1 Jun 2012

Approved for public release: Distribution is unlimited

Naval Air Systems Command
Naval Air Warfare Center Weapons Division

FOREWORD

This handbook is designed to aid electronic warfare and radar systems engineers in making general estimations regarding capabilities of systems. This handbook is sponsored by the NAVAIR Director of Electronic Warfare / Combat Systems.

Approved by:
MARK SCHALLHEIM,
Chief Engineer
Electromagnetic Spectrum Dominance
Naval Air Systems Command

Approved by:
Dr. RONALD SMILEY
Director Avionics Department
Naval Air Warfare Center Weapons Division
And
Director, Electronic Warfare/Combat Systems
Naval Air Systems Command

Under authority of:
MATHIAS W. WINTER.
RADM U.S. Navy
Commander

Jun 1, 2012

Released for publication by:
SCOTT O'NEIL
Executive Director and Director for Research and Engineering

NAWCWD Technical Publication 8347 - Rev 4

Published by .... Technical Information Division
Supersedes .... NAWCWD TP 8347 Rev 3
Collation .... Cover, 163 leaves (Rev 4), + Tabs
Third Printing (Rev 4)...50 copies
Second Printing (Rev 3).. 100 copies
First Printing... 2,000 copies
Previously published as NAWCWPNS TS 92-78 (1st through 4th Edition)................................. 4,040 copies

CONTENTS

ABBREVIATIONS and ACRONYMS

a	Acceleration or atto (10^{-18} multiplier)
A	Ampere, Area, Altitude, Angstrom (Å), Antenna Aperture, or Aerial
A-799	No evidence of failure report
A/A, A-A, AA	Air-to-Air or Anti-Aircraft
AA-()	Air-to-Air missile number ()
AAA	Anti-Aircraft Artillery
AAAA	Army Aviation Association of America
AAED	Advanced Airborne Expendable Decoy
AAM	Air-to-Air Missile
AARGM	Advanced Anti-Radiation Guided Missile
AAW	Anti-Air Warfare
A-BIT	Automatic Built-in-Test
ABM	Air Breathing Missile or Anti-ballistic Missile
A/C	Aircraft (also acft.)
AC	Alternating Current
ACA	Associate Contractor Agreement or Airspace Coordination Area
ACAT	Acquisition Category
ACC	Air Combat Command
ACCB	Aircraft Configuration Control Board
Acft	Aircraft (also A/C)
ACLS	Aircraft Carrier Landing System
ACM	Advanced Cruise Missile or Air Combat Maneuvering
ACQ	Acquisition
ACS	Antenna Coupler Set
ACTD	Advanced Concept Technology Demonstration
A/D	Analog to Digital
ADM	Advanced Development Model
ADP	Automatic Data Processing or Advanced Development Program
AEA	Airborne Electronic Attack
AEC	Aviation Electronic Combat (Army)
AEGIS	Automatic Electronic Guided Intercept System
AEL	Accessible Emission Limit
AESA	Active Electronically Scanned Array
AEW	Airborne Early Warning
AF	Antenna Factor, Air Force, or Audio Frequency
AFB	Air Force Base or Airframe Bulletin
AFC	Automatic Frequency Control or Airframe Change
AFIPS	Automated Financial Information Processing System

AFOTEC	Air Force Operational T&E Center
A/G	Air-to-Ground
AGB	Autonomous Guided Bomb
AGC	Automatic Gain Control
AGI	Auxiliary General Intelligence (Intelligence-gathering Ship)
AGL	Above Ground Level
AGM	Air-to-Ground Missile
AGS	Angle Gate Stealer
AHWS	Advanced Helicopter Weapons System
AI	Artificial Intelligence, Air Intercept, or Airborne Interceptor
AIAA	American Institute of Aeronautics and Astronautics
AIC	Air Intercept Control
AIM	Air Intercept Missile
AIRLANT	Commander, U.S. Naval Air Forces, Atlantic Fleet
AIRPAC	Commander, U.S. Naval Air Forces, Pacific Fleet
AJ	Anti-jamming or Anti-Jam
A-Kit	Aircraft wiring kit for a system (includes cabling, racks, etc. excluding WRAs)
ALC	Air Logistics Center
AM	Amplitude Modulation
AMD	Aircraft Maintenance Department
AMES	Advanced Multiple Environment Simulator
AMLV	Advanced Memory Loader/Verifier
Amp	Amplifier
AMRAAM	Advanced, Medium-Range, Air-to-Air Missile
ANSI	American National Standards Institute
ANT	Antenna
Ao	Operational Availability
AO	Acousto-Optical
AOA	Angle of Arrival, Angle of Attack, or Analysis of Alternatives (similar to COEA)
AOC	Association of Old Crows (Professional EW Society) or Award of Contract
AOT	Angle Only Track, Angle Off Tail, or Acquisition-on-Target
APC	Amphenol Precision Connector or Armored Personnel Carrier
APN	Aircraft Procurement, Navy
APO	Armed Forces (or Army or Air) Post Office, Acquisition Program Office
APU	Auxiliary Power Unit

AR	Anti-reflection or Aspect Ratio	B	Bandwidth (also BW) or Magnetic inductance
ARM	Anti-radiation Missile		
ARO	After Receipt of Order	BAFO	Best and Final Offer
A/S, A-S, AS	Air-to-Surface	BAU	Bus Adapter Unit
ASC	Air Systems Command	BC	Bus Controller
ASCM	Anti-ship Cruise Missile	BDA	Battle Damage Assessment
ASE	Aircraft Survivability (or Survival) Equipment, Allowable Steering Error, or Automatic Support Equipment	BDI	Battle Damage Indication
		BFO	Beat Frequency Oscillator
		BI	Background Investigation
ASIC	Application Specific Integrated Circuit	BIFF	Battlefield Identification, Friend, or Foe
ASK	Amplitude Shift Keying	BIT	Built-in-Test, Binary Digit or Battlefield Information Technology
ASM	Air-to-Surface Missile		
ASO	Aviation Supply Office		
A-Spec	System Specification	BITE	Built-in-Test Equipment
ASPJ	Airborne Self-Protection Jammer	BIU	Bus Interface Unit
ASPO	Avionics Support (also Systems) Project Office (also Officer)	B-Kit	Avionics "Black Box" WRAs
		B/N	Bombardier/Navigator
ASR	Advanced Special Receiver or Airport/Airborne Surveillance Radar	BNC	Bayonet Navy Connector
		BOA	Basic Ordering Agreement
ASRAAM	Advanced Short Range Air-to-Air Missile	BOL	Swedish chaff dispenser in a launcher
		BPF	Band Pass Filter
ASTE	Advanced Strategic and Tactical Expendables	BPS	Bits Per Second
		BUMED	Bureau of Medicine (Navy)
ASW	Anti-submarine Warfare	BUNO	Bureau Number (aircraft)
ATA	Advanced Tactical Aircraft	BUR	Bottom Up Review
ATARS	Advanced Tactical Air Reconnaissance System	BVR	Beyond Visual Range
		BW	Beamwidth (referring to an antenna) or sometimes Bandwidth
ATC	Air Traffic Control		
ATD	Advanced Technology Demonstration	BWA	Backward Wave Amplifier
ATE	Automatic Test Equipment	BWO	Backward Wave Oscillator
ATEDS	Advanced Technology Expendables and Dispenser Systems		
ATF	Advanced Tactical Fighter (F-22)	c	Speed of Light = 3×10^8 meters/sec = 1.8×10^{12} furlongs per fortnight or 1.8 terafurlongs per fortnight, or centi (10^{-2}) multiplier
ATIMS	Airborne Turret Infrared Measurement System or Airborne Tactical Information Management System		
ATIRCM	Advanced Threat Infrared Countermeasures	C	Electron Charge, Coulomb, Capacitance, Celsius, Centigrade, Confidential, Roman numeral for 100, or a programming language (also C+ and C++)
ATP	Acceptance Test Procedure		
ATR	Autonomous Target Recognition, Airborne Transportable Rack, Atlantic Test Range		
		C^2	Command and Control
ATRJ	Advanced Threat Radar Jammer	C^3 (C^4)	Command, Control, Communications (and Computers)
AUTODIN	Automatic Digital Network		
AUX	Auxiliary	C^3I (C^4I)	Command, Control, Communications, (Computers) and Intelligence
avdp.	Avoirdupois (system of measures)		
Avg	Average	CAD	Computer-Aided Design
AWACS	Airborne Warning and Control System	CAE	Computer-Aided Engineering
AZ	Azimuth (also Az)	CAG	Carrier Air Group
		CAGE	Commercial and Government Entry
		CAIV	Cost as an Independent Variable
		CAL	Calibration
		CAM	Computer-Aided Manufacturing or Constant Addressable Memory

CAO	Competency Aligned Organization or Contract Administrative Officer	CM	Countermeasures or Configuration Management
CAP	Combat Air Patrol	CMC	Command Mission Computer or Commandant Marine Corps
CAS	Close Air Support or Calibrated Airspeed	CMDS	Countermeasure Dispensing System
CASS	Consolidated Automated Support System	CMOS	Complementary Metal-Oxide Semiconductor
CAT	Catapult or Cockpit Automation Technology	CMP	Configuration Management Plan
CB	Citizens Band (also see Seabee)	CMWS	Common Missile Warning System
CBD	Commerce Business Daily	CNAF	Commander, Naval Air Forces
CBIT	Continuous Built-in-Test	CNAL	Commander, Naval Air Forces Atlantic (also COMNAVAIRLANT)
CBO	Congressional Budget Office	CNAP	Commander, Naval Air Forces Pacific (also COMNAVAIRPAC)
CCA	Circuit Card Assembly		
CCB	Configuration Control Board	CNI	Communications, Navigation, and Identification
CCD	Charge Coupled Device		
CCM	Counter-Countermeasures	CO	Commanding Officer, Contracting Officer, Change Order, or Carbon Monoxide
CCN	Contract Change Number or Configuration Change Notice		
CCU	Cockpit Control Unit	COB	Close of Business
cd	Candela (SI unit of luminous intensity)	COCOM	Combatant Command
CD	Compact Disk or Control and Display	COEA	Cost and Operational Effectiveness Analysis
CDC	Combat Direction Center		
CDR	Critical Design Review	COG	Center of Gravity or Cognizant
CDRL	Contract Data Requirements List	COMM	Communications
CE	Conducted Emission	COMSEC	Communications Security
CECOM	Communications and Electronics Command (Army)	CONSCAN	Conical Scanning Radar
		CONUS	Continental United States
CEESIM	Combat Electromagnetic Environment Simulator	CO-OP	Cooperative (countermeasures)
		COR	Contracting Officers Representative
CEP	Circular Error Probability	CORPORAL	Collaborative On-Line Reconnaissance Provider/Operationally Responsive Attack Link
CFA	Cognizant Field Activity		
CFAR	Constant False Alarm Rate		
CFE	Contractor Furnished Equipment	Cos	Cosine
CG	Center of Gravity, Commanding General, Command Guidance, or Cruiser	COSRO	Conical-Scan on Receive Only
		COTS	Commercial Off-The-Shelf (hardware/software)
CI	Configuration Item	CP	Circularly Polarized (antenna), Central Processor, or Command Post
CIA	Central Intelligence Agency		
CIC	Combat Information Center (now called CDC)	CPS	Computer or Control Power Supply (depends on application)
CID	Combat Identification or Charge Injection Device	CPU	Central Processing Unit
		CRC	Originally Chemical Rubber Company, now published reference books by CRC Press
CILOP	Conversion in Lieu of Procurement		
CINC	Commander in Chief		
CIP	Capital Improvement Program	CRFM	Coherent RF Memory
CIS	Commonwealth of Independent States (11 of 15 former Soviet Union territories except Estonia, Georgia, Latvia, and Lithuania)	CRISD	Computer Resources Integrated Support Document
		CRLCMP	Computer Resources Life Cycle Management Plan
CIWS	Close-In Weapon System	CRO	Countermeasures Response Optimization
CJ	Coherent Jamming		
CLC	Command Launch Computer	CRT	Cathode Ray Tube or Combat Rated Thrust (afterburner)
cm	Centimeter		

CSAR	Combat Search and Rescue		deg	Degree
Crypto	Cryptographic		DEMVAL	Demonstration Validation (also DEM/VAL)
CS	Conducted Susceptibility			
CSC	Commodity Software Change		DET	Detachment
CSCI	Computer Software Configuration Item		DF	Direction Finding
			DFT	Discrete Fourier Transform
C-Spec	Product Specification		DI	Data Item
CSS	Contractor Support Services		DIA	Defense Intelligence Agency or Diameter
CTR	Chesapeake Test Range			
CV	Aircraft Carrier		DID	Data Item Description
CVN	Nuclear Powered Aircraft Carrier		DIRCM	Directed Infrared Countermeasures
CVR	Crystal Video Receiver		DJ	Deceptive Jamming
CW	Continuous Wave or Chemical Warfare		D-Level	Depot Level Maintenance
			DM	Data Management (also manager)
CWBS	Contract Work Breakdown Structure		DMA	Direct Memory Address or Defense Mapping Agency
CWI	Continuous Wave Illuminator			
CY	Calendar Year		DME	Distance Measuring Equipment
			DNA	Defense Nuclear Agency, Does Not Apply, or Deoxyribonucleic Acid
			DOA	Direction of Arrival
d	Distance, Diameter, or deci (10^{-1} multiplier)		DOD or DoD	Department of Defense
			DoDISS	DoD Index of Specifications and Standards
D	Distance, Diameter, Electron displacement, Detectivity, Doppler, Density, or Roman numeral for 500			
			DOM	Depth of Modulation
			DON	Department of the Navy
da	deca (10^0 multiplier)		DOS	Disk Operating System
D/A	Digital-to-Analog		DOT&E	Director, Operational Test & Evaluation
DAB	Defense Acquisition Board			
DAC	Digital to Analog Converter or Dept of Army Civilian		DPRO	Defense Plant Representative Office
			DRB	Defense Review Board
DAR	Defense Acquisition Regulation		DRFM	Digital RF Memory
DARPA	Defense Advanced Research Projects Agency		DSARC	Defense Systems Acquisition (and) Review Council
DB	Database		DSN	Defense Switching Network
dB	Decibel		DSO	Dielectrically Stabilized Oscillator
dBc	dB referenced to the Carrier Signal		DSP	Digital Signal Processor
dBi	Decibel antenna gain referenced to an isotropic antenna		D-Spec	Process Specification
			DT (&E)	Development or Developmental Test (and Evaluation)
dBm	Decibel referenced to the power of one milliwatt			
			DTC	Design to Cost
DBOF	Defense Business Operations Fund		DTE	Data Terminal Equipment
dBsm	Decibel value of radar cross section referenced to a square meter		DTO	Digitally Tuned Oscillator or Defense Technology Objectives
			DTRMC	Defense (or DoD) Test Recourse Management Center
dBW	Decibel referenced to the power of one watt			
DC	Direct Current, Discrete Circuit, or District of Columbia			
DCE	Data Communication Equipment			
DCS	Direct Commercial Sales or Distributed Control System		e	Electron charge or base of natural logarithms (2.71828...)
			E	Electric Field Intensity or Strength, Energy, East, or Exa (10^{18} multiplier)
DDI	Digital Display Indicator			
DDS	Direct Digital Synthesizers			
DECM	Deceptive Electronic Countermeasures (also Defensive ECM)		E^3	Electromagnetic Environmental Effects

EA	Electronic Attack (similar to older term of ECM)	EOCM	Electro-Optic Countermeasures
EC	Electronic Combat	EOF	Electro-Optical Frequency (300 to 3 x 10^7 GHz)
ECAC	Electromagnetic Compatibility Analysis Center (DOD), now Joint Spectrum Center	EP	Electronic Protection (similar to older terms of ECCM and DECM)
ECCM	Electronic Counter-Countermeasures (similar to newer term of EP)	EPA	Environmental Protection Agency
		EPROM	Electrically Programmable Read-only Memory
ECL	Emitter Coupled Logic	ERAM	Electronic Counter-Countermeasures (also Protection) Requirements and Assessment Manual
ECM	Electronic Countermeasures (similar to newer term of EA)		
ECN	Engineering Change Notice	ERP	Effective Radiated Power
ECO	Engineering Change Order	ES	Electronic Surveillance (similar to older term of ESM)
ECP	Engineering Change Proposal or Egress Control Point	ESD	Electrostatic Discharge
ECR	Electronic Combat Range (China Lake) or Electronic Combat & Reconnaissance	ESM	Electronic Support Measures (similar to newer term of ES)
		ESSM	Evolved Sea Sparrow Missile
ECS	Environmental Control System	ET	Electronics Technician
ECSEL	Electronic Combat Simulation and Evaluation Laboratory	ETI	Elapsed Time Indicator
		ETIRMS	EW Tactical Information and Report Management System
ECU	Electronic Control Unit		
EDM	Engineering Development Model	ETR	Estimated Time to Repair
EED	Electro-Explosive Device	EVM	Earned Value Management
EEPROM	Electrically Erasable/Programmable Read-only Memory	EW	Electronic Warfare , Early Warning, or Expeditionary Warfare
EHF	Extremely High Frequency [30 to 300 GHz]	EWBM	EW Battle Management
		EWDS	EW Data Systems
EIA	Electronic Industries Associates	EWIA	EW Intelligence Analysis
EID	Emitter Identification Data	EWIR	Electronic Warfare Integration & Reprogramming (USAF database)
EIRP	Effective Isotropic Radiated power		
EL	Elevation (also El)	EWMP	Electronic Warfare Master Plan
ELF	Extremely Low Frequency [3 Hz to 3 kHz]	EWO	Electronic Warfare Officer
		EWRL	Electronic Warfare Reprogrammable Library (USN)
ELINT	Electronics Intelligence		
ELNOT	Emitter Library Notation	EWSI	EW Systems Integration
EM	Electromagnetic	EWSSA	EW Software Support Activity
E-Mail	Electronic Mail	EXP	Expendable Countermeasure
EMC	Electromagnetic Compatibility		
EMCAB	EMC Advisory Board		
EMCON	Emission Control	f	femto (10^{-15} multiplier), Frequency (also F), or lens f number
EMD	Engineering and Manufacturing Development		
		F	Frequency (also f), Force, Farad, Faraday Constant, Female, Fahrenheit, Noise Figure, Noise Factor or "Friendly" on RWR display
EME	Electromagnetic Environment		
EMI	Electromagnetic Interference		
EMP	Electromagnetic Pulse		
EMR	Electromagnetic Radiation	F2T2EA	Find, Fix, Track, Target, Engage, Assess (targeting of hostile forces)
EMS	Electromagnetic Susceptibility, Electromagnetic Spectrum		
		F/A	Fighter/Attack
EMV	Electromagnetic Vulnerability	FAA	Federal Aviation Administration
EO	Electro-Optic, Electro-Optical, or Engineering Order	FAC	Forward Air Controller
		FAR	Federal Acquisition Regulations or False Alarm Rate
EOB	Electronic Order of Battle or Expense Operating Budget		
		f_c	Footcandle (unit of illuminance)

FCA	Functional Configuration Audit	gal	Gallon
FCR	Fire Control Radar	GAO	General Accounting Office
FDR	Frequency Domain Reflectometry	GBU	Guided Bomb Unit
FEBA	Forward Edge of the Battle Area	GCA	Ground Controlled Approach
FET	Field-Effect Transistor	GCI	Ground Control Intercept
FEWC	Fleet EW Center	GENSER	General Service
FFT	Fast Fourier Transform	GEN-X	Generic Expendable
FIFO	First In / First Out	GFE	Government Furnished Equipment
FIPR	Federal Information Processing Resources	GHz	GigaHertz
fl	fluid	GI	Government Issue
FLAK	AAA Shrapnel, from the German "Flieger Abwher Kanone" (AAA gun that fires fast and furiously)	GIDEP	Government Industry Data Exchange Program
		GIG	Global Information Grid
		GIGO	Garbage In / Garbage Out
FLIR	Forward Looking Infrared	GOCO	Government Owned Contract Operated
FLPS	Flightline Payload Simulator		
FLT	Flight	GOFO	General Officer / Flag Officer
FM	Frequency Modulation or Failure Mode	GP	General Purpose
		GPI	Ground Plane Interference
FME	Foreign Material Exploitation	GPIB	General Purpose Interface Bus
FMEA	Failure Mode and Effects Analysis	GPS	Global Positioning System
FMS	Foreign Military Sale(s)	GSE	Ground Support Equipment
FOC	Full Operational Capability		
FOD	Foreign Object Damage		
FORCECAP	Force Combat Air Patrol	h	hours, hecto (10^2 multiplier), Plank's constant, or height (also H)
FOT&E	Follow-On Test and Evaluation		
FOTD	Fiber Optic Towed Device	H	Height (also h), Henry (Inductance), or Irradiance
FOUO	For Official Use Only		
FOV	Field of View	HARM	High-speed Anti-Radiation Missile
FPA	Focal Plane Array	HAWK	Homing All the Way Killer
fps	feet per second	HDBK	Handbook
FRACAS	Failure, Reporting, Analysis, and Corrective Actions System	HDF	High Duty Factor
		HE	High Explosive
FRB	Failure Review Board	HEF	High Energy Frequency (3×10^7 to 3×10^{14} GHz)
FRD	Functional Requirements Document		
FSD	Full Scale Development	HEL	High Energy Laser
FSED	Full Scale Engineering Development	HELO	Helicopter
FSK	Frequency Shift Keying	HERF	Hazards of Electromagnetic Radiation to Fuel
FSU	Former Soviet Union		
ft	Feet or Foot	HERO	Hazards of Electromagnetic Radiation to Ordnance
FTC	Fast Time Constant		
FTD	Foreign Technology Division (USAF)	HERP	Hazards of Electromagnetic Radiation to Personnel
FWD	Forward		
FY	Fiscal Year	hex	hexadecimal
		HF	High Frequency [3 - 30 MHz]
		HIL or HITL	Hardware-in-the-Loop
g	Gravity (also G)	HOJ	Home-On-Jam
G	Universal Gravitational Constant (also K), Giga (10^9 multiplier), Conductance, or Gain	HOL	Higher Order Language
		HPF	High-Pass Filter
		HP-IB	Hewlett-Packard Interface Bus
		HP-IL	Hewlett-Packard Interface Loop
G&A	General and Administrative (expense)	HPM	High Powered Microwave
GaAs	Gallium Arsenide	HPRF	High Pulse Repetition Frequency
GACIAC	Guidance and Control Information Analysis Center (DoD)	hr	hour

HSDB	High Speed Data Bus		INT	Intensity
HUD	Heads-Up Display		IO	Information Operations
HV	High Voltage		I/O	Input/Output
H/W	Hardware		IOC	Initial Operational (also Operating) Capability
HWCI	Hardware Configuration Item			
HWIL	Hardware-in-the-loop		IOT&E	Initial Operational Test and Evaluation
Hz	Hertz (Cycles per second)		IPO	International Projects (Program) Office
			IPR	In-Progress/Process Review
			IPT	Integrated Product (also Program) Team
i	current (also I)			
I	Current (also i), Intensity, Irradiance, Intermediate, or Roman Numeral for One		IR	Infrared
			IR&D	Independent Research and Development
IA	Information Assurance		IRCM	Infrared Countermeasures
IADS	Integrated Air Defense System		IRDS	Infrared Detecting System
I&Q	In-Phase and Quadrature		IREXP	IR Expendables
IAS	Indicated Airspeed		IRIG-B	Inter-range Instrumentation Group B
IAW	In Accordance With		IRLS	Infrared Line Scanner
IBIT	Initiated Built-in-Test		IRS	Interface Requirements Specification, IR Suppression or Internal Revenue Service
IBU	Interference Blanker Unit			
IC	Integrated Circuit			
ICD	Interface Control Document Initial Capabilities Document		IRST	Infrared Search and Track
			ISAR	Inverse Synthetic Aperture Radar
ICMD	Improved Countermeasure Dispenser		ISO	Derived from the Greek "isos" meaning "equal", the official title is International Organization for Standardization
ICNIA	Integrated Communication, Navigation, Identification Avionics			
ICS	Inverse Conical Scan or Intercommunications System (aircraft)			
			ISP	Integrated Support Plan Intelligence Support Plan
ICW	In Compliance With			
ID	Identification		ISR	Interference to Signal Ratio (also I/S)
IDA	Institute For Defense Analysis		ITU	International Telecommunications Union
IDECM	Integrated Defensive Electronic Countermeasures			
			IV&V	Independent Validation and Verification
IEEE	Institute of Electrical and Electronic Engineers			
			IW	Information Warfare
IF	Intermediate Frequency			
IFF	Identification Friend-or-Foe			
IFM	Instantaneous Frequency Measurement			
IFR	Instrument Flight Rules		J	Jamming, Radiance, Current Density, or Joules
IG	Inspector General			
IIR	Imaging Infrared		JAAS	Joint Architecture for Aircraft Survivability
I-Level	Intermediate Level of Repair (also "I" Level)			
			JAFF	Jammer (illuminating) Chaff
ILS	Integrated Logistic Support, Instrument Landing System, or Inertial Locator System		JAG	Judge Advocate General
			JAMS	Jamming Analysis Measurement System
			JARS	Jamming Aircraft & Radar Simulation
ILSMT	Integrated Logistic Support Management Team		JASSM	Joint Air-to-Surface Standoff Missile
IM	Intermodulation or Item Manager		JAST	Joint Advanced Strike Technology
IMA	Intermediate Maintenance Activity		JATO	Jet Assisted Takeoff or JAmmer Technique Optimization
in	Inch			
INEWS	Integrated Electronic Warfare System		JC2WC	Joint Command and Control Warfare Center (now JIOWC)
INS	Inertial Navigation System			

JCTD	Joint Concept Technology Demonstration	KIAS	Knots Indicated Air Speed
JCS	Joint Chiefs of Staff or Joint Spectrum Center (formerly ECAC)	km	Kilometer
		KSLOC	Thousand Source Lines of Code (software)
JDAM	Joint Direct Attack Munition	kt	Knot (nautical miles per hour)
JEACO	Joint Electronic Attack and Compatibility Office	kW	Kilowatt
JED	Journal of Electronic Defense (Published by the Association of Old Crows)	l	length (also L) or liter
JEM	Jet Engine Modulation	L	Length (also l), Loss, inductance, Luminance, or Roman Numeral for fifty
JETS	Joint Emitter Targeting System		
JEWC	Joint EW Conference or Joint EW Center (then JC2WC & now JIOWC)	LADAR	Laser Detection and Ranging (i.e., laser radar)
JEWEL	Joint Electronic Warfare Effects Laboratory	LAN	Local Area Network
JIOWC	Joint Information Operations Warfare Command	LANTIRN	Low Altitude Navigation & Targeting Infrared for Night
JMEM	Joint Munitions Effectiveness Manual	LASER	Light Amplification by Stimulated Emission of Radiation
JMR	Jammer	LAT	Latitude (0-90° N or S from equator)
JOVIAL	Julius' Own Version of International Algorithmic Language (Air Force computer programming language)	lbs	pounds
		LCC	Life Cycle Cost(s)
JPATS	Joint Primary Aircraft Training System	LCD	Liquid Crystal Display or Lowest Common Denominator
J/S	Jamming to Signal Ratio		
JSF	Joint Strike Fighter	LCP or LHCP	Left-hand Circular Polarization
JSGCC	Joint Services Guidance and Control Committee	LDF	Low Duty Factor
		LDS	Laser Detecting Set
JSIR	Joint Spectrum Interference Resolution (signal interference portion of MIJI)	LED	Light-Emitting Diode
		LEX	Leading Edge Extension
JSOW	Joint Stand-Off Weapon (AGM-154A)	LGB	Laser Guided Bomb
JSTARS	Joint Surveillance Target Attack Radar System	LF	Low Frequency [30 - 300 kHz]
		LIC	Low Intensity Combat or Laser Intercept Capability
JTA	Jammer Threat Analysis		
JTAT	JATO Techniques Analysis and Tactics	LISP	List Processing (A programming language used in artificial intelligence)
JTCG/AS	Joint Technical Coordinating Group for Aircraft Survivability	LLL	Low Light Level (as in LLL TV)
		lm	lumen (SI unit of luminous flux)
JTCG/ME	Joint Technical Coordinating Group for Munitions Effectiveness	ln	Natural Logarithm
		LO	Local Oscillator or Low Observable
JTIDS	Joint Tactical Information Distribution System	LOA	Letter of Agreement (or Acceptance)
		LOB	Line of Bearing (see also AOA)
JV or J/V	Joint Venture	LOG	Logarithm to the base 10 (also log) or Logistician
		LONG	Longitude (0-180° E or W from Greenwich, U.K.)
k	kilo (10^3 multiplier) or Boltzmann Constant		
		LOR	Level of Repair
K	Kelvin, Cathode, Universal gravitational constant (also G), or Luminous efficacy	LORA	Level of Repair Analysis
		LORAN	Long Range Navigation
		LORO	Lobe on Receive Only
KCAS	Knots Calibrated Airspeed	LOS	Line-of-Sight
kg	kilogram	LPAR	Large Phased-Array Radar
kHz	KiloHertz	LPD	Low Probability of Detection
KIA	Killed in Action		

LPF	Low Pass Filter	MDU	Multipurpose Display Unit
LPI or LPOI	Low Probability of Intercept	MF	Medium Frequency (300 kHz to 3 MHz)
LPRF	Low Pulse Repetition Frequency		
LR	Lethal Range	MFD	Multifunction (video) Display
LRA	Line Replaceable Assembly	MG	Missile Guidance
LRF	Laser Rangefinder	MHz	MegaHertz (10^6 Hz)
LRIP	Low Rate Initial Production	MIA	Missing in Action
LRU	Line Replaceable Unit	MIC	Microwave Integrated Circuit or Management Information Center
LSA	Logistic Support Analysis		
LSAR	Logistic Support Analysis Record	MICRON	10^{-6} meter
LSB	Least Significant Bit	MiG	Mikoyan-Gurevich (Soviet aircraft manufacturer)
LSI	Large Scale Integration		
LSO	Landing Signal Officer	MIGCAP	MiG Combat Air Patrol
LSSO	Laser System Safety Officer	MIJI	Meaconing, Intrusion, Jamming, & Interference (also see JSIR)
LTBB	Look Through Blanking Bus		
LWIR	Long Wave Infrared	mil	One-thousandth of an inch
LWR	Laser Warning Receiver	MIL	Military power (100%, no afterburner) or Military
lx	Lux (SI unit of illuminance)		
LZ	Landing Zone	MILCON	Military Construction
		MILSPEC	Military Specification
		MILSTRIP	Military Standard Requisitioning and Issue Procedure(s)
m	milli (10^{-3} multiplier), meter, or electron mass	MIMIC	Microwave Monolithic Integrated Circuit (also MMIC)
M	Mega (10^6 multiplier), Male, Mach number, or Roman numeral for 1,000	MIN	Minimum
		Mincon	Minimal Construction
MA	Missile Alert or Missile Active	MIPPLE	RWR display switching between ambiguous emitters
MAD	Magnetic Anomaly Detection (also Detector)		
		MIPS	Millions of (Mega) Instructions Per Second
MADD	Microwave Acoustic Delay Device		
MAF	Maintenance Action Form	ML	Missile Launch
MAG	Marine Aircraft Group or Magnetic	MLC	Main Lobe Clutter
MAGTF	Marine Air-Ground Task Force	MLV	Memory Loader Verifier
MANPADS	Man-portable Air Defense System	MLVS	Memory Loader Verifier Set
M&S	Modeling and Simulation	mm	Millimeter
MASER	Microwave Amplification by Simulated Emission of Radiation	MM	Man Month
		MMIC	Microwave Monolithic Integrated Circuit (also MIMIC)
MATE	Modular Automatic Test Equipment		
MAW	Missile Approach Warning system (also MAWS) or Marine Aircraft Wing	MMW	Millimeter Wave (40 GHz or higher per IEEE, but commonly used down to 30 GHz)
MAX	Maximum or Maximum aircraft power (afterburner)	MOA	Memorandum of Agreement
		MOE	Measure of Effectiveness
MBFN	Multiple Beam Forming Network	MOM	Methods of Moments (also MoM) or Metal-Oxide-Metal
MC	Mission Computer		
MCIOC	Marine Corps Information Operations Center	MOP	Modulation on Pulse or Measure of Performance
		MOPS	Million Operations Per Second
MCP	Micro-Channel Plate	MOS	Minimum Operational Sensitivity, Military Occupational Specialty, Metal-Oxide Semiconductor, or Measure of Suitability
MDF	Mission Data File		
MDI	Multiple Display Indicator or Miss Distance Indicator		
MDG	Mission Data Generator		
MDS	Minimum Discernible Signal or Minimum Detectable Signal	MOSAIC	Modeling System for Advanced Investigation of Countermeasures

MOU	Memorandum of Understanding	NAWCWD	Naval Air Warfare Center Weapons Division
MPD	Multi-Purpose Display or Microwave Power Device	NBC	Nuclear, Biological, Chemical
MPE	Maximum Permissible Exposure	NCTR	Non-Cooperative Target Recognition
mph	Miles per Hour	NDI	Non-Developmental Item or Non Destructive Inspection
MPLC	Multi-Platform Launch Controller		
MPM	Microwave Power Module	NEI	Noise Equivalent Power
MPPS	Million Pulses Per Second	NEMP	Nuclear Electromagnetic Pulse
MPRF	Medium Pulse Repetition Frequency	NEOF	No Evidence of Failure
mr or mrad	Milliradian	NEP	Noise Equivalent Power
MRC	Maintenance Requirement Card or Medium Range CAP	NF	Noise Figure or Noise Factor (also F)
		NFO	Naval Flight Officer
MRE's	Meals Ready to Eat	NGJ	Next Generation Jammer
ms	Milliseconds	NIOC	Navy Information Operations Command
MSB	Most Significant Bit		
MSI	Multi-Sensor (also Source) Integration, Management Support Issues, or Medium Scale Integration	NIPO	Navy International Program Office
		NIR	Near Infrared
		nm	nanometer or Nautical Mile (also NM or NMI)
MSIC	Missile and Space Intelligence Center		
MSL	Mean Sea Level (altitude) or Missile	NM or NMI	Nautical Mile (also nm)
MTBF	Mean Time Between Failures	NMCI	Navy Marine Corps Intranet
MTI	Moving Target Indicator (or Indication)	NNWC	Naval Network Warfare Command
		NOHD	Nominal Ocular Hazard Distance
MTTR	Mean Time To Repair	NORAD	North American Air Defense Command
MUXBUS	Multiplex Bus		
MVS	Minimum Visible Signal	NPG or NPGS	Naval Post Graduate School
mw	Microwave	NRE	Non-Recurring Engineering
mW	Milliwatt	NRL	Naval Research Laboratory
MWIR	Mid Wave Infrared	NRZ	Non Return to Zero
MWS	Missile Warning Set	NSA	National Security Agency
MY	Man Year	nsec or ns	Nanosecond
		NSN	National Stock Number
		NSWC	Naval Surface Weapons Center
		nt	Nit (SI unit of luminance)
n	nano (10^{-9} multiplier) or number of elements	NUWC	Naval Undersea Warfare Center
		NVG	Night Vision Goggles
N	Noise, Newton (force), Radiance, North, or No	NWIP	Naval Warfare Information Publication
		NWP	Naval Warfare Publication
n/a	Not Applicable (also N/A)		
NA	Numerical Aperture		
NADEP	Naval Aviation Depot		
NASA	National Aeronautics and Space Administration	O	Optical
		OADR	Originating Agency's Determination Required
NATO	North Atlantic Treaty Organization		
NATOPS	Naval Air Training and Operating Procedures Standardization	OAG	Operational Advisory Group
		O&MN	Operations and Maintenance, Navy (also O&M,N)
NAV	Navigation		
NAVAIR	Naval Air Systems Command (also NAVAIRSYSCOM)	OBE	Overtaken (Overcome) By Events
		OCA	Offensive Counter Air
NavMPS	Naval Mission Planning System	OEWTPS	Organizational Electronic Warfare Test Program Set
NAVSEA	Naval Sea Systems Command (also NAVSEASYSCOM)		
		OFP	Operational Flight Program
NAWCAD	Naval Air Warfare Center Aircraft Division	OJT	On-the-Job Training

O-Level	Organizational Level of Repair (also "O" Level)	PEL	Personnel Exposure Limits
OMA	Organizational Maintenance Activity	PEM	Photoelectromagnetic
OMB	Office of Management and Budget	PEO	Program Executive Officer
OMEGA	Optimized Method for Estimating Guidance Accuracy (VLF Navigation System)	pf	Power Factor or Pico Farads
		PFA	Probability of False Alarm
		PGM	Precision Guided Munition
ONR	Office of Naval Research	ph	Phot (unit of illuminance)
OOK	On-Off Keying	P_h	Probability of Hit
OPEVAL	Operational Evaluation	pi	Greek letter π
OPM	Office of Personnel Management	P_i	Probability of Intercept (also POI)
OPSEC	Operational Security	PID	Positive Identification
OPTEVFOR	Operational Test and Evaluation Force	PIN	Personal Identification Number
OR	Operational Requirement or Operationally Ready	PIP	Product Improvement Plan or Predicted Intercept Point
ORD	Operational Requirements Document	Pixel	Picture Element
OSD	Office of the Secretary of Defense	P_k	Probability of Kill or Peak
OSHA	Occupational Safety and Health Act	PLSS	Precision Location Strike System
OSIP	Operational Safety Improvement Program	PM	Phase Modulation or Program Manager
OSM	Operating System Memory or SMA connector made by Omni-Spectra	PMA	Program (also Project) Manager, Air
		PMAWS	Passive Missile Approach Warning System
OSRB	Operational Software Review Board	PMS	Program Manager, Ship
OT (&E)	Operational Test (and Evaluation)	PMT	Photomultiplier Tube
OTD	Operational Test Director	PMW	Pragram Manager, Warfare
OTH	Over the Horizon	P-N	Positive to Negative Junction (also p-n)
OTH-B	Over-the-Horizon Backscatter		
OTH-R	Over-the-Horizon Radar	PN or P/N	Part Number
OTH-T	Over-the-Horizon Targeting	POC	Point of Contact
OTRR	Operational Test Readiness Review	POET	Primed Oscillator Expendable Transponder
OUSD	Office of the Under Secretary of Defense	POI	Probability of Intercept (also PI)
		POL	Polarization
oz	ounce	POM	Program Objective Memorandum
		POP	Pulse-on-Pulse or Product Optimization Program
p	pico (10^{-12} multiplier) or page		
P	Power, Pressure, or Peta (10^{15} multiplier)	POST	Passive Optical Seeker Technology (Stinger missile)
P^3I	Pre-Planned Product Improvement	PPI	Plan Position Indicator
Pa	Pascal (pressure)	PPS	Pulses Per Second
PA	Public Address or Program Analyst	PRF	Pulse Repetition Frequency
PBIT	Periodic Built-in-Test	PRI	Priority or Pulse Repetition Interval
PC	Pulse Compression, Personal Computer, or Photoconductive	PROM	Programmable Read-only Memory
		PRR	Production Readiness Review or Pulse Repetition Rate
PCA	Physical Configuration Audit		
PCM	Pulse Code Modulation	PRT	Pulse Repetition Time
P_d	Probability of Detection	P_s	Probability of Survival
PD	Pulse Doppler	P's & Q's	Pints and Quarts (small details)
PDI	PD Illuminator or Post Detection Integration	PSK	Phase-shift Keying
		PUPS	Portable Universal Programming System
PDP	Plasma Display Panel		
PDQ	Pretty Darn [sic] Quick	PV	Photovoltaic
PDR	Preliminary Design Review	pw or PW	Pulse Width
PDW	Pulse Descriptor Word	PWB	Printed Wiring Board

q	electron charge	RIO	Radar Intercept Officer
Q	Quantity Factor (figure of merit), Quadrature, aerodynamic pressure, or Charge (coulomb)	RM	Radar Mile
		rms or RMS	Root Mean Square
		RNG	Range
QA	Quality Assurance	ROC	Required Operational Capability
QC	Quality Control	ROE	Rules of Engagement
QED	Quod Erat Demonstradum (end of proof)(Satirically "quite easily done")	ROI	Return on Investment
		ROM	Read-only Memory or Rough Order of Magnitude
QML	Qualified Manufacturer Listing	ROR	Range Only Radar or Rate of Return (financial)
QPL	Qualified Parts List		
QRC	Quick-Reaction Capability	ROT	Rate of Turn
QRD	Quick Reaction Demonstration	ROWG	Response Optimization Working Group
QRT	Quick-Reaction test		
		RPG	Receiver Processor Group
		RPM	Revolutions per Minute
		RPT	Repeat
r or R	Radius or Range	RPV	Remotely Piloted Vehicle
R	Resistance, Reliability, or Roentgen	RRT	Rapid Reprogramming Terminal (a type of MLVS)
rad	Radian		
R&D	Research and Development	RS	Radiated Susceptibility or Remote Station
RADAR	Radio Detection and Ranging		
RADHAZ	Radiation Hazard	RSDS	Radar Signal Detecting Set
RAM	Random Access Memory, Radar Absorbing Material, Rolling Airframe Missile, or Reliability, Availability, and Maintainability	RSO	Range Safety Officer or Receiver, Set-on
		RST	Receiver Shadow Time
R&M	Reliability and Maintainability	RT	Remote Terminal, Termination Resistance, or Receiver/Transmitter (also R/T)
RAT	Ram Air Turbine		
RBOC	Rapid Blooming Offboard Chaff	RUG	Radar Upgrade
RCP or RHCP	Right-hand Circular Polarization	RWR	Radar Warning Receiver
RCS	Radar Cross Section	Rx	Receive
RCVR	Receiver		
RDT&E	Research, Development, Test, & Evaluation	s, S, or sec	seconds
RDY	Ready	S	Signal Power, Surface Area, Secret, Electrical conductance (siemens), South, Scattering (as in S-parameters), or Seconds
RE	Radiated Emissions		
REC	Receive		
RET	Return		
RF	Radio Frequency	SA	Situational Awareness, Semi-Active, Spectrum Analyzer, or Surface-to-Air (also S/A or S-A)
RFEXP	RF Expendables		
RFI	Radio Frequency Interference, Ready-For-Issue, or Request for Information		
		SA-()	Surface-to-Air missile number ()
RFP	Request for Proposal	SAE	Society of Automotive Engineers
RFQ	Request for Quotation	SAM	Surface-to-Air Missile
RFSS	Radio Frequency Simulation System (Army)	SA-N-()	Naval Surface-to-Air missile number
		SAR	Synthetic Aperture Radar, Special Access Required, Semi-Active Radar, Search and Rescue, or Specific Absorption Rate
RGPO	Range Gate Pull Off		
RGS	Range Gate Stealer		
RGWO	Range Gate Walk Off (see RGPO)		
RHAW	Radar Homing and Warning Receiver or Radar Homing All the Way	SATS	Semi-Active Test System
		SAW	Surface Acoustic Wave
RHAWS	Radar Homing and Warning System	SBIR	Small Business Innovative Research
RINT	Radiation Intelligence	SCI	Sensitive Compartmented Information

SCIF	Sensitive Compartmented Information Facility	SNORT	Supersonic Naval Ordnance Research Track
SCN	Specification Change Notice	SNTK	Special Need to Know
SCR	Software Change Request	SOF	Safety of Flight
SCP	Software Change Proposal	SOJ	Stand-off Jammer
SCRB	Software Configuration Review Board	SONAR	Sound Navigation and Ranging
SCUD	Soviet short-range surface-to-surface missile	SOO	Statement of Objectives (replacing SOW)
SE	Support Equipment	SOP	Standard Operating Procedures
SDLM	Standard Depot Level Maintenance	SORO	Scan-on-Receive Only
SDI	Strategic Defense Initiative	SOS	"Save Our Ship" (distress call with easy Morse code, i.e. ● ● ● – – – ● ● ●)
SEAD	Suppression of Enemy Air Defense (pronounced "seed" or "C add")		
		SOW	Statement of Work (being replaced by SOO)
SEAL	Sea-Air-Land (Navy special forces)		
sec	seconds (also S or s)	SPAWAR	Space and Naval Warfare Systems Command
SECDEF	Secretary of Defense		
SEI	Specific Emitter Identification	SPEC	Specification
SEMA	Special Electronic Mission Aircraft	SPIRITS	Spectral Infrared Imaging of Targets and Scenes
SERD	Support Equipment Recommendation Data		
		SPO	System Program Office
SHAPE	Supreme Headquarters Allied Powers Europe (NATO military command)	SPY	Radar on an AEGIS ship
		sq	Square
SHF	Super High Frequency (3 to 30 GHz)	sr	Steradian
SI	Special Intelligence or System International (Units)	SRA	Shop Replaceable Assembly
		SRAM	Static Random Access Memory
SIF	Selective Identification Feature	SRB	Software Review Board
SIGINT	Signals Intelligence	SRBOC	Super Rapid Blooming Offboard Chaff
SIJ	Stand-In Jamming (also S/J)	SRD	Systems Requirements Document
SIM	Simulation	SRS	Software Requirements Specification
sin	Sine	SRU	Shop Replaceable Unit
SINCGARS	Single Channel Ground and Airborne Radio System	SSA	Software (also Special or System) Support Activity, Source Selection Activity, or Solid State Amplifier
SIRCM	Suite of IR Countermeasures		
SIRFC	Suite of Integrated RF Countermeasures	SSB	Single Side Band
		SSI	Small Scale Integration
SJ	Support Jamming	SSJ	Self Screening Jamming
S/J	Stand-In Jamming or Signal to Jamming Ratio	SSM	Surface-to-Surface Missile
		SSRO	Sector Scan Receive Only
SL	Side lobe or Sea Level (also S.L.)	SSW	Swept Square Wave
SLAM	Standoff Land Attack Missile	S&T	Science and Technology
SLAR	Side-Looking Airborne Radar	STANAG	Standardization Agreement (NATO)
SLC	Side Lobe Clutter	STAR	System Threat Assessment Report
SLOC	Source Lines of Code or Sea Lines of Communication	stat	Statute
		STBY	Standby
SM	Statute Mile (also sm) or Standard Missile	STC	Sensitivity Time Control, Short Time Constant or SHAPE Technical Center
SMA	Scheduled Maintenance Action or Sub-Miniature A connector	STD	Software Test Description, Standard, or Sexually Transmitted Disease
SMC	Sub-Miniature C connector	STE	Standard Test Equipment
SME	Subject Matter Expert	STOVL	Short Takeoff and Vertical Landing
SML	Support Material List	STP	Software Test Plan, or Standard Temperature and Pressure (0°C at 1 atmosphere)
SMS	Stores Management Set or Status Monitoring System		
S/N or SNR	Signal-to-Noise Ratio		

STR	Software (also System) Trouble Report	TENA	Training Enabling Architecture
STT	Single Target Track	TERPES	Tactical Electronic Reconnaissance
STU	Secure Telephone Unit		Processing and Evaluation System
SUBSAM	Subsurface-to-Air Missile	TGT	Target
SUT	System Under Test	TIM	Technical Interchange Meeting
S/W	Software (also SW)	TM	Telemetry, Transverse Magnetic, or
SWAP	Size, Weight, And Power		Technical Manual
SWC	Scan With Compensation	TMD	Theater Missile Defense
SWM	Swept Wave Modulation	TNC	Threaded Navy Connector
SYSCOM	Systems Command	TOA	Time of Arrival
		TOJ	Track on Jam
		TOO	Target of Opportunity (HARM
t	Time (also T)		operating mode)
T	Time (also t), tera (10^{12} multiplier),	TOR	Tentative (also Tactical) Operational
	Temperature, or Telsa		Requirement or Time of Receipt
TA	Target Acquisition or Terrain	TOS	Time on Station
	Avoidance	TOT	Time on Target
TAAF	Test, Analyze, and Fix	TOW	Tube-Launched, Optically-Tracked,
TAC	Tactical Air Command (now ACC)		Wire-guided
TACAIR	Tactical Aircraft	TPI	Test Program Instruction
TACAMO	Take Charge and Move Out (airborne	TPS	Test Program Set or Test Pilot School
	strategic VLF communications relay	TPWG	Test Plan Working Group
	system)	TQM	Total Quality Management
TACAN	Tactical Air Navigation	T/R	Transmit / Receive
TACDS	Threat Adaptive Countermeasures	TRB	Technical Review Board
	Dispensing System	TRD	Test Requirements Document
TACTS	Tactical Aircrew Combat Training	TREE	Transient Radiation Effects on
	System		Electronics
TAD	Threat Adaptive Dispensing,	TRF	Tuned Radio Frequency
	Temporary Additional (also Active)	TRR	Test Readiness Review
	Duty, or Tactical Air Direction	TS	Top Secret
T&E	Test & Evaluation	TSS	Tangential Sensitivity
TALD	Tactical Air Launched Decoy	TSSAM	Tri-Service Standoff Attack Weapon
TAMPS	Tactical Automated Mission Planning	TT	Target Track
	System	TTI	Time To Impact/Intercept
TAR	Target Acquisition Radar or Training	TTG	Time-to-Go
	Administrative Reserve	TTL	Transistor-Transistor Logic
TAS	True Airspeed	TTR	Target Tracking Radar
TBA	To Be Announced	TV	Television
TBD	To Be Determined	TVC	Thrust Vector Control
TBMD	Theater Ballistic Missile Defense	TWS	Track While Scan or Tail Warning
TD	Technical Directive (also Director)		System
TDD	Target Detection Device	TWSRO	Track While Scan on Receive Only
TDM	Time Division Multiplexing	TWT	Traveling Wave Tube
TE	Transverse Electric	TWTA	Traveling Wave Tube Amplifier
TEA	Technology Exchange Agreement	Tx	Transmit
TEAMS	Tactical EA-6B Mission Support	TYCOM	Type Commander
TECHEVAL	Technical Evaluation		
TEL	Transporter Erector Launcher		
TEM	Transverse Electromagnetic		
TEMP	Test and Evaluation Master Plan	u or μ	micron / micro (10^{-6} multiplier)
TEMPEST	Not an acronym. Certification of	U	Unclassified, Unit, or Unknown (on
	reduced electromagnetic radiation for		RWR display)
	security considerations		

UAS	Unmanned Aerial System	VMAQ	Prefix for Marine Tactical EW Squadron
UAV	Unmanned (also uninhabited) Air (or Aerial) Vehicle	VP	Prefix for Navy patrol squadron
UCAV	Uninhabited Combat Air Vehicle (new USAF term for UAV)	VQ	Prefix for Navy special mission (usually reconnaissance) squadron
UDF	User Data File	VRAM	Video Random Access Memory
UDFG	User Data File Generator	VS or vs	Velocity Search or Versus (also vs.)
UDM	User Data Module	V/STOL	Vertical/Short Take-off and Landing (also VSTOL)
UHF	Ultra High Frequency (300 MHz to 3 GHz)	vt	Velocity (also V or v)
ULF	Ultra Low Frequency (3 to 30 Hz)	VTOL	Vertical Takeoff and Landing
μm	Micrometer	VSWR	Voltage Standing Wave Ratio
UN	United Nations	VVA	Voltage Variable Attenuator
UNK	Unknown (also U)		
UPC	Unique Planning Component		
UPS	Uninterruptable Power Supply		
us or μs	Microseconds		
U.S.	United States	W	Watts, Weight, or West
USA	United States of America or United States Army	W&T	Warning & Targeting
		WARM	Wartime Reserve Mode
USAF	United States Air Force	wb	Weber (magnetic flux)
USMC	United States Marine Corps	WBS	Work Breakdown Structure
USN	United States Navy	WC	Waveguide, circular
UTA	Uninhabited Tactical Aircraft	WFT	Windowed Fourier Transform
UUT	Unit Under Test	WGIRB	Working Group on Infrared Background
UV	Ultraviolet		
		WORM	Write Once Read Many [times] (Refers to optical disks)
		WOW	Weight on/off Wheels (also WonW or WoffW)
v	Volts (also V), Velocity (also V or v_t)	WPN	Weapons Procurement, Navy or Weapon
V	Volts (also v), Velocity (also v or v_t), Volume, or Roman Numeral for five	WR	Waveguide, rectangular
		WRA	Weapon Replaceable Assembly
VA	Veterans Administration, Volt-Amperes	WRD	Waveguide, rectangular double ridged
VAQ	Prefix for Navy tactical EW squadron	WSSA	Weapons System Support Activity
V&V	Validation and Verification	WVR	Within Visual Range
VCO	Voltage Controlled Oscillator		
Vdc or VDC	Volts Direct Current		
VDT	Video Display Terminal		
VECP	Value Engineering Change Proposal	x	Multiplication symbol
VF	Prefix for Navy fighter squadron	X	Reactance, Experimental, Extraordinary, Roman Numeral for ten, or X axis
VFO	Variable Frequency Oscillator		
VFR	Visual Flight Rules		
VGPO	Velocity Gate Pull Off	X-EYE	Cross Eye
VGS	Velocity Gate Stealer	X-POL	Cross Polarization
VGWO	Velocity Gate Walk Off	XMIT	Transmit
VHF	Very High Frequency (30 - 300 MHz)	XMTR	Transmitter
VHSIC	Very High Speed Integrated Circuit		
VID	Visual Identification		
VLF	Very Low Frequency (3 to 30 kHz)	Y	Yes or Y-Axis
VLSI	Very Large Scale Integration	YAG	Yttrium-Aluminum Garnet
VLSIC	Very Large Scale Integrated Circuit	yd	Yard
		YIG	Yttrium-Iron Garnet

Z	Impedance, Zenith, or Z-Axis
1xLR, 2xLR	One (or two or three etc.) Times Lethal Range
1v1 or 1-v-1	One versus One (Aerial engagement)
2D	Two Dimension
3D	Three Dimension
3M	Navy Maintenance and Material Management System

FUNDAMENTALS

CONSTANTS, CONVERSIONS, and CHARACTERS

DECIMAL MULTIPLIER PREFIXES

Prefix	Symbol	Multiplier
exa	E	10^{18}
peta	P	10^{15}
tera	T	10^{12}
giga	G	10^{9}
mega	M	10^{6}
kilo	k	10^{3}
hecto	h	10^{2}
deka	da	10^{1}
deci	d	10^{-1}
centi	c	10^{-2}
milli	m	10^{-3}
micro	μ	10^{-6}
nano	n	10^{-9}
pico	p	10^{-12}
femto	f	10^{-15}
atto	a	10^{-18}

EQUIVALENCY SYMBOLS

Symbol	Meaning
$\propto$	Proportional
$\sim$	Roughly equivalent
$\approx$	Approximately
$\cong$	Nearly equal
$=$	Equal
$\equiv$	Identical to, defined as
$\neq$	Not equal
$\gg$	Much greater than
$>$	Greater than
$\geq$	Greater than or equal to
$\ll$	Much less than
$<$	Less than
$\leq$	Less than or equal to
$\therefore$	Therefore
$\circ$	Degrees
$'$	Minutes or feet
$"$	Seconds or inches

UNITS OF LENGTH

1 inch (in)	=	2.54 centimeters (cm)
1 foot (ft)	=	30.48 cm = 0.3048 m
1 yard (yd)	$\cong$	0.9144 meter
1 meter (m)	$\cong$	39.37 inches
1 kilometer (km)	$\cong$	0.54 nautical mile
	$\cong$	0.62 statute mile
	$\cong$	1093.6 yards
	$\cong$	3280.8 feet
1 statute mile	$\cong$	0.87 nautical mile
(sm or stat. mile)	$\cong$	1.61 kilometers
	=	1760 yards
	=	5280 feet
1 nautical mile	$\cong$	1.15 statute miles
(nm or naut. mile)	$\cong$	1.852 kilometers
	$\cong$	2025 yards
	$\cong$	6076 feet
1 furlong	=	1/8 mi (220 yds)

UNITS OF SPEED

1 foot/sec (fps)	$\cong$	0.59 knot (kt)*
	$\cong$	0.68 stat. mph
	$\cong$	1.1 kilometers/hr
1000 fps	$\approx$	600 knots
1 kilometer/hr	$\cong$	0.54 knot
(km/hr)	$\cong$	0.62 stat. mph
	$\cong$	0.91 ft/sec
1 mile/hr (stat.)	$\cong$	0.87 knot
(mph)	$\cong$	1.61 kilometers/hr
	$\cong$	1.47 ft/sec
1 knot*	$\cong$	1.15 stat. mph
	$\cong$	1.69 feet/sec
	$\cong$	1.85 kilometer/hr
	$\cong$	0.515 m/sec

*A knot is 1 nautical mile per hour.

UNITS OF VOLUME

1 gallon	$\cong$	3.78 liters
	$\cong$	231 cubic inches
	$\cong$	0.1335 cubic ft
	$\cong$	4 quarts
	$\cong$	8 pints

1 fl ounce $\cong$ 29.57 cubic centimeter (cc) or milliliters (ml)

$1\ in^3$ $\cong$ 16.387 cc

UNITS OF AREA

1 sq meter $\cong$ 10.76 sq ft

1 sq in	$\cong$	645 sq millimeters (mm)
	=	1,000,000 sq mil
1 mil	=	0.001 inch
1 acre	=	43,560 sq ft

UNITS OF WEIGHT

1 kilogram (kg)	$\cong$	2.2 pounds (lbs)
1 pound	$\cong$	0.45 Kg
	=	16 ounce (oz)
1 oz	=	437.5 grains
1 carat	$\cong$	200 mg
1 stone (U.K.)	$\cong$	6.36 kg

NOTE: These are the U.S. customary (avoirdupois) equivalents, the troy or apothecary system of equivalents, which differ markedly, was used long ago by pharmacists.

UNITS OF POWER / ENERGY

1 H.P.	=	33,000 ft-lbs/min
	=	550 ft-lbs/sec
	$\cong$	746 Watts
	$\cong$	2,545 BTU/hr

(BTU = British Thermal Unit)

1 BTU	$\cong$	1055 Joules
	$\cong$	778 ft-lbs
	$\cong$	0.293 Watt-hrs

SCALES

OCTAVES

"N" Octaves = Freq to Freq x 2^N
i.e. One octave would be 2 to 4 GHz
Two Octaves would be 2 to 8 GHz
Three octaves would be 2 to 16 GHz

DECADES

"N" Decades = Freq to Freq x 10^N
i.e. One decade would be 1 to 10 MHz
Two decades would be 1 to 100 MHz
Three decades would be 1 to 1000 MHz

TEMPERATURE CONVERSIONS

$°F = (9/5)\ °C + 32$
$°C = (5/9)(\ °F - 32)$
$°K = °C + 273.16$
$°F = (9/5)(\ °K - 273) + 32$
$°C = °K - 273.16$
$°K = (5/9)(\ °F - 32) + 273$

UNITS OF TIME

1 year	=	365.2 days
1 fortnight	=	14 nights
		(2 weeks)
1 century	=	100 years
1 millennium	=	1,000 years

NUMBERS

1 decade	=	10
1 Score	=	20
1 Billion	=	1×10^9 (U.S.)
		(thousand million)
	=	1×10^{12} (U.K.)
		(million million)

RULE OF THUMB FOR ESTIMATING DISTANCE TO LIGHTNING / EXPLOSION:

km - Divide 3 into the number of seconds which have elapsed between seeing the flash and hearing the noise.
miles - Multiply 0.2 times the number of seconds which have elapsed between seeing the flash and hearing the noise.
Note: Sound vibrations cause a change of density and pressure within a media, while electromagnetic waves do not. An audio tone won't travel through a vacuum but can travel at 1100 ft/sec through air. When picked up by a microphone and used to modulate an EM signal, the modulation will travel at the speed of light.

Physical Constant	Quoted Value	S*	SI unit	Symbol
Avogadro constant	6.0221367×10^{23}	36	mol^{-1}	N_A
Bohr magneton	$9.2740154 \times 10^{-24}$	31	$J \bullet T^{-1}$	μ_B
Boltzmann constant	1.380658×10^{-23}	12	$J \bullet K^{-1}$	$k(=R\ N_A)$
Electron charge	$1.602177\ 33 \times 10^{-19}$	49	C	$-e$
Electron specific charge	$-1.758819\ 62 \times 10^{11}$	53	$C \bullet kg^{-1}$	$-e/m_e$
Electron rest mass	$9.1093897 \times 10^{-31}$	54	kg	m_e
Faraday constant	9.6485309×10^{4}	29	$C \bullet mol^{-1}$	F
Gravity (Standard Acceleration)	9.80665 or 32.174	0	m/sec^2 ft/sec^2	g
Josephson frequency to voltage ratio	4.8359767×10^{14}	0	$Hz \bullet V^{-1}$	2e/hg
Magnetic flux quantum	$2.06783461 \times 10^{-15}$	61	Wb	φ_o
Molar gas constant	8.314510	70	$J \bullet mol^{-1} \bullet K^{-1}$	R
Natural logarithm base	$\cong 2.71828$	-	dimensionless	e
Newtonian gravitational constant	6.67259×10^{-11}	85	$m^3 \bullet kg^{-1} \bullet s^{-2}$	G or K
Permeability of vacuum	$4\pi \times 10^{-7}$	d	H/m	μ_o
Permittivity of vacuum	$\cong 8.8541878 \times 10^{-12}$	d	F/m	ε_o
Pi	$\cong 3.141592654$		dimensionless	π
Planck constant Planck constant/2π	6.62659×10^{-34} $1.05457266 \times 10^{-34}$	40 63	$J \bullet s$ $J \bullet s$	h $h(=h2\pi)$
Quantum of circulation	$3.63694807 \times 10^{-4}$	33	$J \bullet s \bullet kg^{-1}$	$h/2m_e$
Radius of earth (Equatorial)	6.378×10^{6} or 3963		m miles	
Rydberg constant	$1.0973731534 \times 10^{7}$	13	m^{-1}	R_χ
Speed of light	2.9979246×10^{8}	1	$m \bullet s^{-1}$	c
Speed of sound (dry air @ std press & temp)	331.4	-	$m \bullet s^{-1}$	-
Standard volume of ideal gas	22.41410×10^{-3}	19	$m^3 \bullet mol^{-1}$	V_m
Stefan-Boltzmann constant	5.67051×10^{-8}	19	$W \bullet K^{-4} \bullet m^{-2}$	σ

* S is the one-standard-deviation uncertainty in the last units of the value, d is a defined value.
(A standard deviation is the square root of the mean of the sum of the squares of the possible deviations)

THE SPEED OF LIGHT			
ACTUAL	UNITS	RULE OF THUMB	UNITS
$\cong 2.9979246 \times 10^8$	m/sec	$\cong 3 \times 10^8$	m/sec
$\cong 299.79$	m/μsec	$\cong 300$	m/μsec
$\cong 3.27857 \times 10^8$	yd/sec	$\cong 3.28 \times 10^8$	yd/sec
$\cong 5.8275 \times 10^8$	NM/hr	$\cong 5.8 \times 10^8$	NM/hr
$\cong 1.61875 \times 10^5$	NM/sec	$\cong 1.62 \times 10^5$	NM/sec
$\cong 9.8357105 \times 10^8$	ft/sec	$\cong 1 \times 10^9$	ft/sec
$\cong 9.8357105 \times 10^2$	ft/μsec	$\cong 1 \times 10^3$	ft/μsec

SPEED OF LIGHT IN VARIOUS MEDIUMS

The speed of EM radiation through a substance such as cables is defined by the following formula:

$$V = c/(\mu_r \varepsilon_r)^{1/2}$$

Where: μ_r = relative permeability

ε_r = relative permittivity

The real component of ε_r = dielectric constant of medium.

EM propagation speed in a typical cable might be 65-90% of the speed of light in a vacuum.

APPROXIMATE SPEED OF SOUND (MACH 1)

Sea Level (CAS/TAS)		36,000 ft* (TAS)	(CAS)
1230 km/hr	Decreases	1062 km/hr	630 km/hr
765 mph	Linearly	660 mph	391 mph
665 kts	To $\Rightarrow$	573 kts	340 kts

* The speed remains constant until 82,000 ft, when it increases linearly to 1215 km/hr (755 mph, 656 kts) at 154,000 ft. Also see section 8-2 for discussion of Calibrated Air Speed (CAS) and True Airspeed (TAS) and a plot of the speed of sound vs altitude.

SPEED OF SOUND IN VARIOUS MEDIUMS

Substance	Speed (ft/sec)
Vacuum	Zero
Air	1,100
Fresh Water	4,700
Salt Water	4,900
Glass	14,800
Steel	20,000

DECIMAL / BINARY / HEX CONVERSION TABLE

Decimal	Binary	Hex	Decimal	Binary	Hex	Decimal	Binary	Hex
1	00001	01h	11	01011	0Bh	21	10101	15h
2	00010	02h	12	01100	0Ch	22	10110	16h
3	00011	03h	13	01101	0Dh	23	10111	17h
4	00100	04h	14	01110	0Eh	24	11000	18h
5	00101	05h	15	01111	0Fh	25	11001	19h
6	00110	06h	16	10000	10h	26	11010	1Ah
7	00111	07h	17	10001	11h	27	11011	1Bh
8	01000	08h	18	10010	12h	28	11100	1Ch
9	01001	09h	19	10011	13h	29	11101	1Dh
10	01010	0Ah	20	10100	14h	30	11110	1Eh

When using hex numbers it is always a good idea to use "h" as a suffix to avoid confusion with decimal numbers.

DECIMAL TO HEX CONVERSION

Both the following methods must use long division. Method one computes the digits from right to left while method two works from left to right.

Method one: To convert a decimal number above 16 to hex, divide the number by 16, then record the integer resultant and the remainder. Convert the remainder to hex and write this down - this will become the far right digit of the final hex number. Divide the integer you obtained by 16, and again record the new integer result and new remainder. Convert the remainder to hex and write it just to the left of the first decoded number. Keep repeating this process until dividing results in only a remainder. This will become the left-most character in the hex number. i.e. to convert 60 (decimal) to hex we have 60/16 = 3 with 12 remainder. 12 is C (hex) - this becomes the right most character. Then 3/16=0 with 3 remainder. 3 is 3 (hex). This becomes the next (and final) character to the left in the hex number, so the answer is 3C.

Method two: Use table of powers to work the digits from left to right:

For Example: Here is your Decimal Number - 9379

Step 1 - Set up your chart:

65536	4096	256	16	1
16^4	16^3	16^2	16^1	16^0

Step 2 - Look in the table for the highest divisible number in the chart.
   9379 / 4096 = 2 (the left-most Hex digit)
   Must use long division to calculate the remainder.(1187)

Step 3 - Divide the remainder with its highest divisible number in the chart:
   1187 / 256 = 4 (the next digit to the right)
   Must use long division to calculate the remainder (163)

Step 4 - Divide the remainder with its highest divisible number in the chart:
   163 / 16 = 10 (or "A" from table L-1) (the next digit to the right)
   Must use long division to calculate the remainder (3)

Step 5 - The remainder will not divide: remainder = 3 (the right-most Hex digit)

65536	4096	256	16	1
16^4	16^3	16^2	16^1	16^0
	2	4	A	3

HEX TO DECIMAL CONVERSION

To convert a hex number to decimal, multiply each hex digit converted to decimal by the decimal equivalent of the hex power represented and add the results.

For example: Here is your Hex Number - 24A3

Step 1 - Set up your chart:

4096	256	16	1
16^3	16^2	16^1	16^0

Step 2 – Place the numbers in a table:

4096	256	16	1
16^3	16^2	16^1	16^0
2	4	A	3

Step 3 - Multiply the Hex number times the power value:
   2 x 4096   =   8192
   4 x 256   =   1024
   A(10) x 16   =   160
   3 x 1   =   3

Step four - add up your values:
   Decimal value is 9379

GREEK ALPHABET

Case		Greek Alphabet Name	English Equivalent
Upper	Lower		
A	α	alpha	a
B	β	beta	b
Γ	γ	gamma	g
Δ	δ	delta	d
E	ε	epsilon	e
Z	ζ	zeta	z
H	η	eta	e
Θ	θ , '	theta	th
I	ι	iota	i
K	κ	kappa	k
Λ	λ	lambda	l
M	μ	mu	m

Case		Greek Alphabet Name	English Equivalent
Upper	Lower		
N	ν	nu	n
Ξ	ξ	xi	x
O	o	omicron	o
Π	π	pi	p
P	ρ	rho	r
Σ	σ	sigma	s
T	τ	tau	t
Y	υ	upsilon	u
Φ	φ	phi	ph
X	χ	chi	ch
Ψ	ψ	psi	ps
Ω	ω	omega	o

LETTERS FROM THE GREEK ALPHABET COMMONLY USED AS SYMBOLS

Symbol	Name	Use
α	alpha	space loss, angular acceleration, or absorptance
β	beta	3 dB bandwidth or angular field of view [radians]
Γ	Gamma	reflection coefficient
γ	gamma	electric conductivity, surface tension, missile velocity vector angle, or gamma ray
Δ	Delta	small change or difference
δ	delta	delay, control forces and moments applied to missile, or phase angle
ε	epsilon	emissivity [dielectric constant] or permittivity [farads/meter]
η	eta	efficiency or antenna aperture efficiency
Θ	Theta	angle of lead or lag between current and voltage
θ or '	theta	azimuth angle, bank angle, or angular displacement
Λ	Lambda	acoustic wavelength or rate of energy loss from a thermocouple
λ	lambda	wavelength or Poisson Load Factor
μ	mu	micro 10^{-6} [micron], permeability [henrys/meter], or extinction coefficient [optical region]
ν	nu	frequency
π	pi	3.141592654+
ρ	rho	charge/mass density, resistivity [ohm-meter], VSWR, or reflectance
Σ	Sigma	algebraic sum
σ	sigma	radar cross section [RCS], Conductivity [1/ohm-meter], or Stefan-Boltzmann constant
T	Tau	VSWR reflection coefficient
τ	tau	pulse width, atmospheric transmission, or torque
Φ	Phi	magnetic/electrical flux, radiant power [optical], or Wavelet's smooth function [low pass filter]
φ	phi	phase angle, angle of bank, or beam divergence [optical region]
Ψ	Psi	time-dependent wave function or Wavelet's detail function [high pass filter]
ψ	psi	time-independent wave function, phase change, or flux linkage [weber]
Ω	Omega	Ohms [resistance] or solid angle [optical region]. Note: inverted symbol is conductance [mhos]
ω	omega	carrier frequency in radians per second

MORSE CODE and PHONETIC ALPHABET

A - alpha	• —	J - juliett	• — —	S - sierra	• • •	1	• — — — —
B - bravo	— • • •	K - kilo	— • —	T - tango	—	2	• • — — —
C - charlie	— • — •	L - lima	• — • •	U - uniform	• • —	3	• • • — —
D - delta	— • •	M - mike	— —	V - victor	• • • —	4	• • • • —
E - echo	•	N - november	— •	W - whiskey	• — —	5	• • • • •
F - foxtrot	• • — •	O - oscar	— — —	X - x-ray	— • • —	6	— • • • •
G - golf	— — •	P - papa	• — — •	Y - yankee	— • — —	7	— — • • •
H - hotel	• • • •	Q - quebec	— — • —	Z - zulu	— — • •	8	— — — • •
I - india	• •	R - romeo	• — •	0	— — — — —	9	— — — — •

Note: The International Maritime Organization agreed to officially stop Morse code use by February 1999, however use may continue by ground based amateur radio operators (The U.S. Coast Guard discontinued its use in 1995).

BASIC MATH / GEOMETRY REVIEW

EXPONENTS

$$a^x\, a^y = a^{x+y}$$

$$a^x / a^y = a^{x-y}$$

$$(a^x)^y = a^{xy}$$

$$a^0 = 1$$

Example:

$$\frac{x}{\sqrt{x}} = x \bullet x^{-\frac{1}{2}} = x^{(1-\frac{1}{2})} = x^{\frac{1}{2}} = \sqrt{x}$$

LOGARITHMS

$$\log (xy) = \log x + \log y$$

$$\log (x/y) = \log x - \log y$$

$$\log (x^N) = N \log x$$

If $z = \log x$ then $x = 10^z$

Examples: $\log 1 = 0$
$\log 1.26 = 0.1$; $\log 10 = 1$

if $10 \log N = dB\#$,
then $10^{(dB\#/10)} = N$

TRIGONOMETRIC FUNCTIONS

$$\sin x = \cos (x-90°)$$

$$\cos x = -\sin (x-90°)$$

$$\tan x = \sin x / \cos x = 1 / \cot x$$

$$\sin^2 x + \cos^2 x = 1$$

A radian is the angular measurement of an arc which has an arc length equal to the radius of the given circle, therefore there are 2π radians in a circle. One radian = $360°/2\pi$ = 57.296....°

ELLIPSE
Area = $\pi\, a\, b$
Approx circumference $= 2\pi \sqrt{\dfrac{a^2 + b^2}{2}}$

RECTANGLULAR SOLID
Area = lw
Volume = lwh

CYLINDER
Volume — $\pi r2h$
Lateral surface area — $2\pi rh$

ANGLES
r^2- x^2 + y^2
Sin θ = y/r   Cos θ = x/r
Tan θ = y/x

TRIANGLES

Angles: $A + B + C = 180°$

$c^2 = a^2 + b^2 - 2ab \cos C$

Area $= 1/2\ hc = 1/2\ ab \sin C$

$b = \sqrt{e^2 + h^2}$

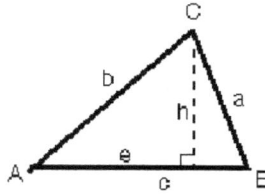

SPHERE

Surface area $= 4\pi r^2$

Volume $= 4/3\ \pi r^3$

Cross Section (circle)
Area $= \pi r^2$

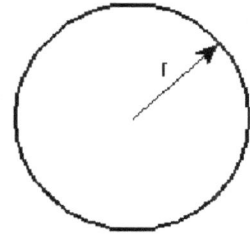

DERIVATIVES

Assume: a = fixed real #; u, v & w are functions of x

$$d(a)/dx = 0 \ ; \ d(\sin u)/dx = du(\cos u)/dx$$

$$d(x)/dx = 1 \ ; \ d(\cos v)/dx = -dv(\sin v)/dx$$

$$d(uvw)/dx = uvdw/dx + vwdu/dx + uwdv/dx +...etc$$

INTEGRALS

Note: All integrals should have an arbitrary constant of integration added, which is left off for clarity

Assume: a = fixed real #; u, & v are functions of x

$$\int a\,dx = ax \quad and \quad \int a\ f(x)dx = a\int f(x)dx$$

$$\int (u + v)dx = \int u\,dx + \int v\,dx \ ; \ \int e^x dx = e^x$$

$$\int (\sin ax)dx = -(\cos ax)/a \ ; \ \int (\cos ax)dx = (\sin ax)/a$$

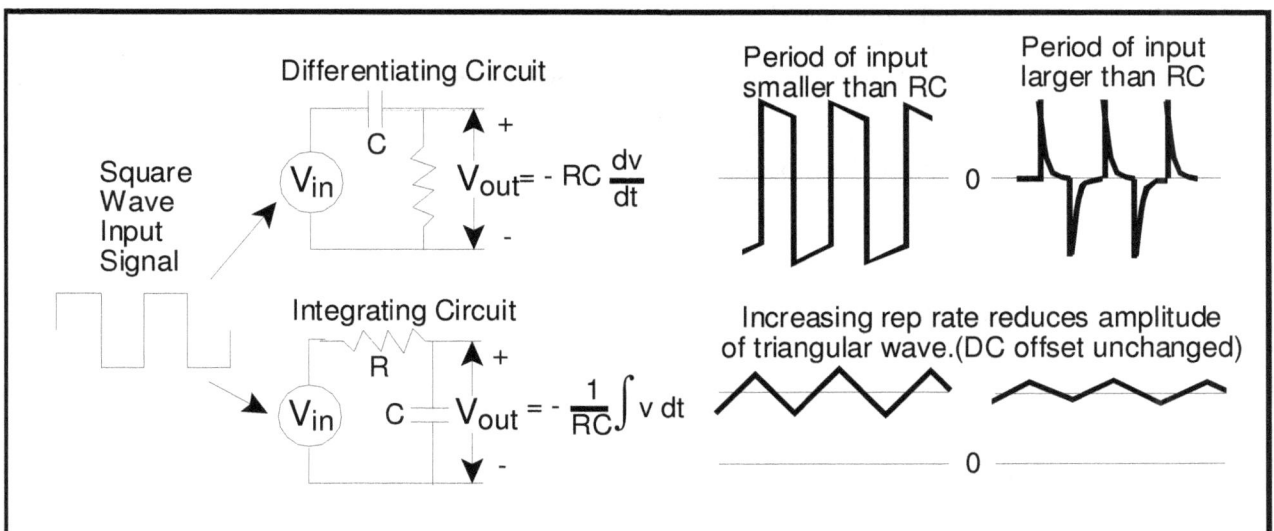

MATHEMATICAL NOTATION

The radar and EW communities generally accept some commonly used notation for the various parameters used in radar and EW calculations. For instance, "P" is almost always power and "G" is almost always gain. Textbooks and reference handbooks will usually use this notation in formulae and equations.

A significant exception is the use of "α" for space loss. Most textbooks don't develop the radar equation to its most usable form as does this reference handbook, therefore the concept of "α" just isn't covered.

Subscripts are a different matter. Subscripts are often whatever seems to make sense in the context of the particular formula or equation. For instance, power may be "P", "P_T", "P_t", or maybe "P_1". In the following list, generally accepted notation is given in the left hand column with no subscripts. Subscripted notation in the indented columns is the notation used in this handbook and the notation often used in the EW community.

α = Space loss
   α_1 = One way space loss, transmitter to receiver
   α_2 = Two way space loss, xmtr to target (including radar cross section) and back to the rcvr
   α_{1t} = One way space loss, radar transmitter to target, bistatic
   α_{1r} = One way space loss, target to radar receiver, bistatic

Other notation such as α_{tm} may be used to clarify specific losses, in this case the space loss between a target and missile seeker, which could also be identified as α_{1r} .

A = Antenna aperture (capture area)
   A_e = Effective antenna aperture
   Å = Angstrom

B = Bandwidth (to 3dB points)
   B_{IF} = 3 dB IF bandwidth of the receiver (pre-detection)
   B_J = Bandwidth of the jamming spectrum
   B_{MHz} = 3 dB bandwidth in MHz
   B_N = Equivalent noise bandwidth, a.k.a. B
   B_V = 3 dB video bandwidth of the receiver (post-detection) (Subscript V stands for video)

BF = Bandwidth reduction factor (jamming spectrum wider than the receiver bandwidth)
BW = Beamwidth (to 3 dB points)

c = Speed of Light

f = Frequency (radio frequency)
   f_c = Footcandle (SI unit of illuminance)
   f_D = Doppler frequency
   f_R = Received frequency
   f_T = Transmitted frequency

G = Gain
   G_t = Gain of the transmitter antenna
   G_r = Gain of the receiver antenna
   G_{tr} = Gain of the transmitter/receiver antenna (monostatic radar)
   G_J = Gain of the jammer
   G_{JA} = Gain of the jammer antenna
   G_{JT} = Gain of the jammer transmitter antenna

G_{JR}	=	Gain of the jammer receiver antenna
G_σ	=	Gain of reflected radar signal due to radar cross section
h	=	Height or Planks constant
h_{radar}	=	Height of radar
h_{target}	=	Height of target
J	=	Jamming signal (receiver input)
J_1	=	Jamming signal (constant gain jammer)
J_2	=	Jamming signal (constant power jammer)
J/S	=	Jamming to signal ratio (receiver input)
k	=	Boltzmann constant
$K_{1,2,3,4}$	=	Proportionality constants, see Sections 4-3, 4-4, 4-5, and 4-1 respectively.
λ	=	Lambda, Wavelength or Poisson factor
L	=	Loss (due to transmission lines or circuit elements)
N	=	Receiver equivalent noise input (kT_oB)
NF	=	Noise figure
P	=	Power
P_d	=	Probability of detection
P_D	=	Power density
P_J	=	Power of a jammer transmitter
P_n	=	Probability of false alarm
P_r	=	Power received
P_t	=	Power of a transmitter
R	=	Range (straight line distance)
R_1	=	Bistatic radar transmitter to target range
R_2	=	Bistatic radar target to receiver range
R_J	=	Range of jammer to receiver (when separate from the target)
R_{NM}	=	Range in nautical miles
σ	=	Sigma, radar cross section (RCS)
S	=	Signal (receiver input)
S_R	=	Radar signal received by the jammer
S_{min}	=	Minimum receiver sensitivity
t	=	Time
t_{int}	=	Integration time
t_r	=	Pulse Rise Time
τ	=	Pulse Width
V	=	Velocity
V_r	=	Radial velocity

FREQUENCY SPECTRUM

Figure 1, which follows, depicts the electromagnetic radiation spectrum and some of the commonly used or known areas. Figure 2 depicts the more common uses of the microwave spectrum. Figure 3 shows areas of the spectrum which are frequently referred to by band designations rather than by frequency.

Section 7-1 provides an additional breakdown of the EO/IR spectrum.

To convert from frequency (f) to wavelength (λ) and vice versa, recall that f = c/λ, or λ = c/f; where c = speed of light.

$$\lambda_{meter} = \frac{3x10^8}{f_{Hz}} = \frac{3x10^5}{f_{kHz}} = \frac{300}{f_{MHz}} = \frac{0.3}{f_{GHz}} \qquad \text{or} \qquad f_{Hz} = \frac{3x10^8}{\lambda_{meter}} \quad f_{kHz} = \frac{3x10^5}{\lambda_{meter}} \quad f_{MHz} = \frac{300}{\lambda_{meter}} \quad f_{GHz} = \frac{0.3}{\lambda_{meter}}$$

Some quick rules of thumb follow:

Metric:

      Wavelength in cm = 30 / frequency in GHz

      For example: at 10 GHz, the wavelength = 30/10 = 3 cm

English:

      Wavelength in ft = 1 / frequency in GHz

      For example: at 10 GHz, the wavelength = 1/10 = 0.1 ft

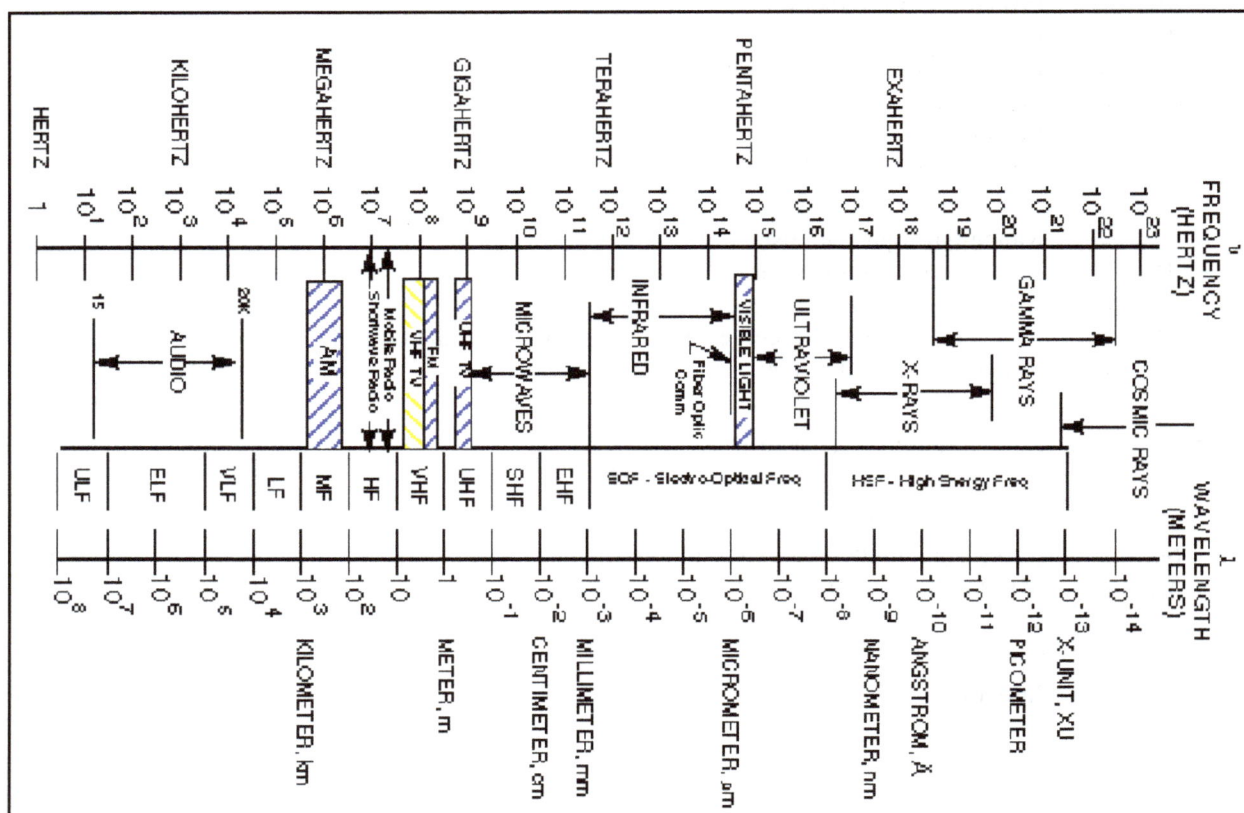

Figure 1. Electromagnetic Radiation Spectrum

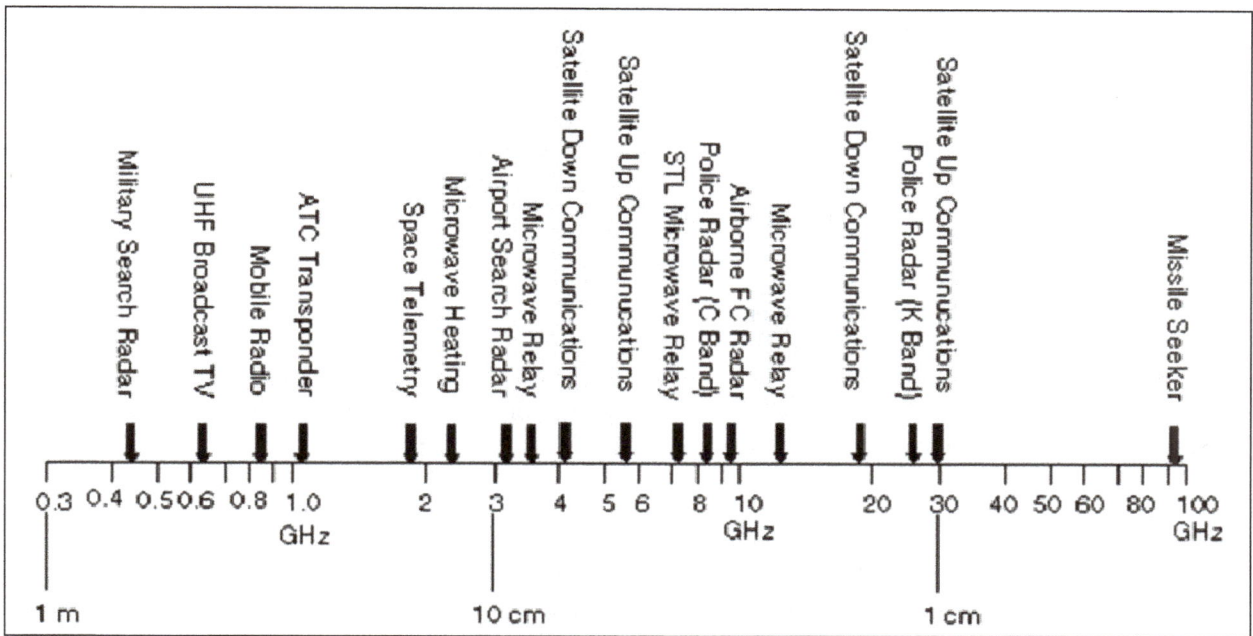

Figure 2. The Microwave Spectrum

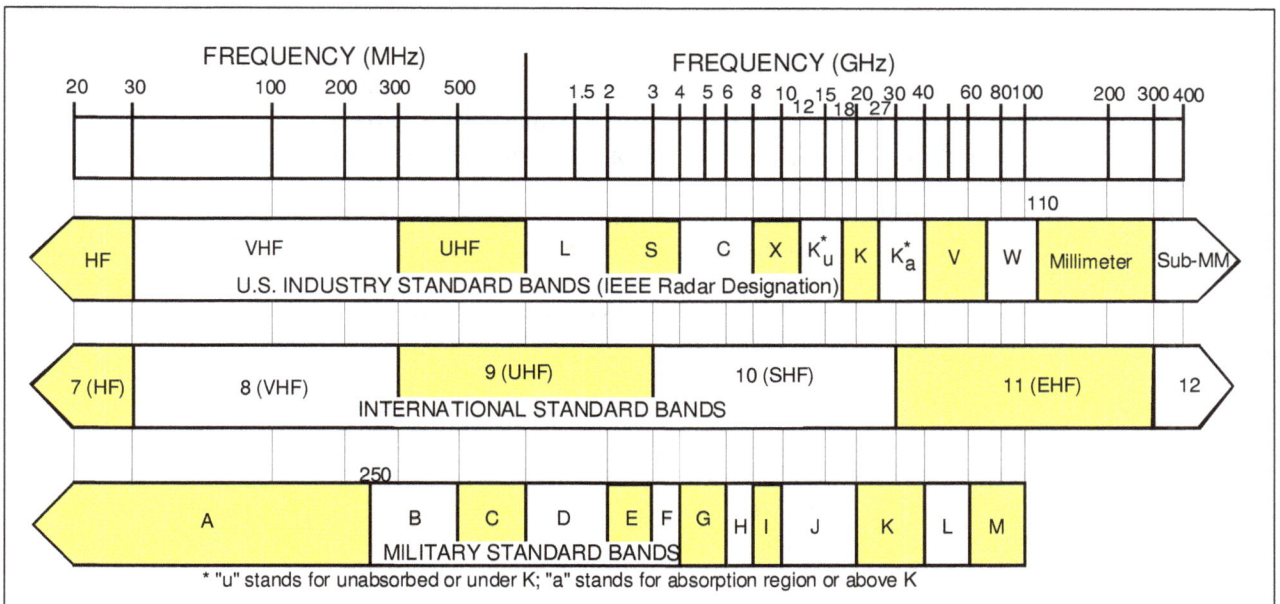

Figure 3. Frequency Band Designations

See Section 7, Figure 1 for a more detailed depiction of the UV and IR spectrum.

DECIBEL (dB)

The Decibel is a subunit of a larger unit called the bel. As originally used, the bel represented the power ratio of 10 to 1 between the strength or intensity i.e., power, of two sounds, and was named after Alexander Graham Bell. Thus a power ratio of 10:1 = 1 bel, 100:1 = 2 bels, and 1000:1 = 3 bels. It is readily seen that the concept of bels represents a logarithmic relationship since the logarithm of 100 to the base 10 is 2 (corresponding to 2 bels), the logarithm of 1000 to the base 10 is 3 (corresponding to 3 bels), etc. The exact relationship is given by the formula

$$\text{Bels} = \log(P_2/P_1) \qquad\qquad [1]$$

where P_2/P_1 represents the power ratio.

Since the bel is a rather large unit, its use may prove inconvenient. Usually a smaller unit, the Decibel or dB, is used. 10 decibels make one bel. A 10:1 power ratio, 1 bel, is 10 dB; a 100:1 ratio, 2 bels, is 20 dB. Thus the formula becomes

$$\text{Decibels (dB)} = 10 \log(P_2/P_1) \qquad\qquad [2]$$

The power ratio need not be greater than unity as shown in the previous examples. In equations [1] and [2], P_1 is usually the reference power. If P_2 is less than P_1, the ratio is less then 1.0 and the resultant bels or decibels are negative. For example, if P_2 is one-tenth P_1, we have

$$\text{bels} = \log(0.1/1) = -1.0 \text{ bels}$$
and $\qquad$ $$\text{dB} = 10 \log(0.1/1) = -10 \text{ dB}.$$

It should be clearly understood that the term decibel does not in itself indicate power, but rather is a ratio or comparison between two power values. It is often desirable to express power levels in decibels by using a fixed power as a reference. The most common references in the world of electronics are the milliwatt (mW) and the watt. The abbreviation dBm indicates dB referenced to 1.0 milliwatt. One milliwatt is then zero dBm. Thus P_1 in equations [1] or [2] becomes 1.0 mW. Similarly, The abbreviation dBW indicates dB referenced to 1.0 watt, with P_2 being 1.0 watt, thus one watt in dBW is zero dBW or 30 dBm or 60 dBμW. For antenna gain, the reference is the linearly polarized isotropic radiator, dBLI. Usually the "L" and/or "I" is understood and left out.

dBc is the power of one signal referenced to a carrier signal, i.e. if a second harmonic signal at 10 GHz is 3 dB lower than a fundamental signal at 5 GHz, then the signal at 10 GHz is -3 dBc.

THE DECIBEL, ITS USE IN ELECTRONICS

The logarithmic characteristic of the dB makes it very convenient for expressing electrical power and power ratios. Consider an amplifier with an output of 100 watts when the input is 0.1 watts (100 milliwatts); it has an amplification factor of
$$P_2/P_1 = 100/0.1 = 1000$$
or a gain of:
$$10 \log(P_2/P_1) = 10 \log(100/0.1) = 30 \text{ dB}.$$
(notice the 3 in 30 dB corresponds to the number of zeros in the power ratio)

The ability of an antenna to intercept or transmit a signal is expressed in dB referenced to an isotropic antenna rather than as a ratio. Instead of saying an antenna has an effective gain ratio of 7.5, it has a gain of 8.8 dB (10 log 7.5).

A ratio of less than 1.0 is a loss, a negative gain, or attenuation. For instance, if 10 watts of power is fed into a cable but only 8.5 watts are measured at the output, the signal has been decreased by a factor of

8.5/10 = .85

or   10 log(.85) = -0.7 dB.

This piece of cable at the frequency of the measurement has a gain of -0.7 dB. This is generally referred to as a loss or attenuation of 0.7 dB, where the terms "loss" and "attenuation" imply the negative sign. An attenuator which reduces its input power by factor of 0.001 has an attenuation of 30 dB. The utility of the dB is very evident when speaking of signal loss due to radiation through the atmosphere. It is much easier to work with a loss of 137 dB rather than the equivalent factor of 2×10^{-14}.

Instead of multiplying gain or loss factors as ratios we can add them as positive or negative dB. Suppose we have a microwave system with a 10 watt transmitter, and a cable with 0.7 dB loss connected to a 13 dB gain transmit antenna. The signal loss through the atmosphere is 137 dB to a receive antenna with a 11 dB gain connected by a cable with 1.4 dB loss to a receiver. How much power is at the receiver? First, we must convert the 10 watts to milliwatts and then to dBm:

10 watts = 10,000 milliwatts

and

10 log (10,000/1) = 40 dBm

Then

40 dBm - 0.7 dB + 13 dB - 137 dB + 11 dB - 1.4 dB = -75.1 dBm.

-75.1 dBm may be converted back to milliwatts by solving the formula:
$$mW = 10^{(dBm/10)}$$

giving: $10^{(-75.1/10)} = 0.00000003$ mW

Voltage and current ratios can also be expressed in terms of decibels, provided the resistance remains constant. First we substitute for P in terms of either voltage, V, or current, I. Since P=VI and V=IR we have:
$$P = I^2R = V^2/R$$

Thus for a voltage ratio we have: dB $= 10 \log[(V_2^2/R)/(V_1^2/R)] = 10 \log(V_2^2/V_1^2)$
$= 10 \log(V_2/V_1)^2 \qquad = 20 \log(V_2/V_1)$

Like power, voltage can be expressed relative to fixed units, so one volt is equal to 0 dBV or 120 dBμV.

Similarly for current ratio: $dB = 20 \log(I_2/I_1)$

Like power, amperage can be expressed relative to fixed units, so one amp is equal to 0 dBA or 120 dBμA.

Decibel Formulas (where Z is the general form of R, including inductance and capacitance)

When impedances are equal: $dB = 10 \log \dfrac{P2}{P1} = 20 \log \dfrac{E2}{E1} = 20 \log \dfrac{I2}{I1}$

When impedances are unequal: $dB = 10 \log \dfrac{P2}{P1} = 20 \log \dfrac{E2\sqrt{Z1}}{E1\sqrt{Z2}} = 20 \log \dfrac{I2\sqrt{Z2}}{I1\sqrt{Z1}}$

SOLUTIONS WITHOUT A CALCULATOR

Solution of radar and EW problems requires the determination of logarithms (base 10) to calculate some of the formulae. Common "four function" calculators don't usually have a log capability (or exponential or fourth root functions either). Without a scientific calculator (or math tables or a Log-Log slide rule) it is difficult to calculate any of the radar equations, simplified or "textbook". The following gives some tips to calculate a close approximation without a calculator.

DECIBEL TABLE

DB	Power Ratio	Voltage or Current Ratio	DB	Power Ratio	Voltage or Current Ratio
0	1.00	1.00	10	10.0	3.16
0.5	1.12	1.06	15	31.6	5.62
1.0	1.26	1.12	20	100	10
1.5	1.41	1.19	25	316	17.78
2.0	1.58	1.26	30	1,000	31.6
3.0	2.00	1.41	40	10,000	100
4.0	2.51	1.58	50	10^5	316
5.0	3.16	1.78	60	10^6	1,000
6.0	3.98	2.00	70	10^7	3,162
7.0	5.01	2.24	80	10^8	10,000
8.0	6.31	2.51	90	10^9	31,620
9.0	7.94	2.82	100	10^{10}	10^5

For dB numbers which are a multiple of 10

An easy way to remember how to convert dB values that are a multiple of 10 to the absolute magnitude of the <u>power</u> ratio is to place a number of zeros equal to that multiple value to the right of the value 1.
i.e. 40 dB = 10,000 : 1   (for Power)

Minus dB moves the decimal point that many places to the left of 1.
i.e. -40 dB = 0.0001 : 1 (for Power)

For <u>voltage or current</u> ratios, if the multiple of 10 is even, then divide the multiple by 2, and apply the above rules. i.e.   40 dB = 100 : 1   (for Voltage)
            -40 dB = 0.01 : 1

If the power in question is not a multiple of ten, then some estimation is required. The following tabulation lists some approximations. Some would be useful to memorize.

DB RULES OF THUMB

Multiply Current / Voltage By			Multiply Power By:	
if +dB	if -dB	**dB**	if +dB	if -dB
1	1	0	1	1
1.12	0.89	1	1.26	0.8
1.26	0.79	2	1.58	0.63
1.4	0.707	3	2	0.5
2.0	0.5	6	4	0.25
2.8	0.35	9	8	0.125
3.16	0.316	10	10	0.1
4.47	0.22	13	20	0.05
10	0.1	20	100	0.01
100	0.01	40	10,000	0.0001

You can see that the list has a repeating pattern, so by remembering just three basic values such as one, three, and 10 dB, the others can easily be obtained without a calculator by addition and subtraction of dB values and multiplication of corresponding ratios.

Example 1:
A 7 dB increase in power (3+3+1) dB is an increase of (2 x 2 x 1.26) = 5 times whereas
A 7 dB decrease in power (-3-3-1) dB is a decrease of (0.5 x 0.5 x 0.8) = 0.2.

Example 2:   Assume you know that the ratio for 10 dB is 10, and that the ratio for 20 dB is 100 (doubling the dB increases the power ratio by a factor of ten), and that we want to find some intermediate value.

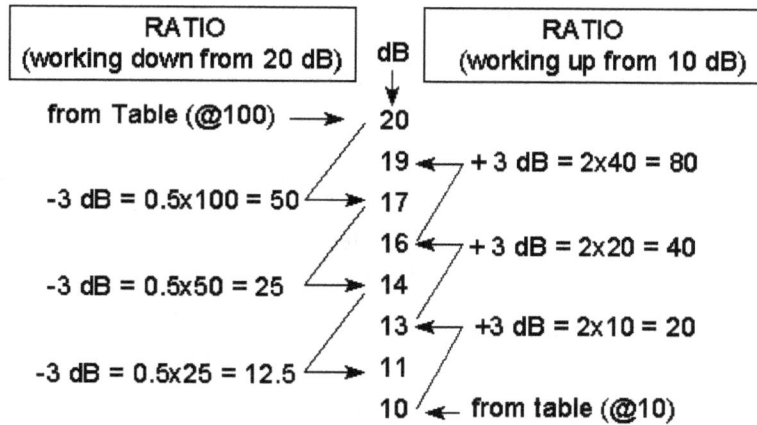

RATIO (working down from 20 dB)	dB	RATIO (working up from 10 dB)

from Table (@100) ⟶ 20

19 ⟵ + 3 dB = 2x40 = 80

-3 dB = 0.5x100 = 50 ⟶ 17

16 ⟵ + 3 dB = 2x20 = 40

-3 dB = 0.5x50 = 25 ⟶ 14

13 ⟵ +3 dB = 2x10 = 20

-3 dB = 0.5x25 = 12.5 ⟶ 11

10 ⟵ from table (@10)

We can get more intermediate dB values by adding or subtracting one to the above, for example, to find the ratio at 12 dB we can:

work up from the bottom; 12 = 1+11 so we have 1.26 (from table) x 12.5 = 15.75

alternately, working down the top 12 = 13-1 so we have 20 x 0.8 (from table) = 16

The resultant numbers are not an exact match (as they should be) because the numbers in the table are rounded off. We can use the same practice to find any ratio at any other given value of dB (or the reverse).

dB AS ABSOLUTE UNITS

Power in absolute units can be expressed by using 1 Watt (or 1 milliwatt) as the reference power in the denominator of the equation for dB. We then call it dBW or dBm. We can then build a table such as the adjoining one.

From the above, any intermediate value can be found using the same dB rules and memorizing several dB values i.e. for determining the absolute power, given 48 dBm power output, we determine that 48 dBm = 50 dBm - 2 dB so we take the value at 50 dB which is 100W and <u>divide</u> by the value 1.58 (ratio of 2 dB) to get:

100 watts/1.58 = 63 W or 63,291 mW.

dB AS ABSOLUTE UNITS			
dBµW	dBm	POWER	dBW
120	90	1 MW	60
90	60	1 kW	30
80	50	100 W	20
70	40	10 W	10
60	30	1 W (1000 mW)	0
50	20	100 mW	-10
40	10	10 mW	-20
33	3	2 mW	-27
32	2	1.58 mW	-28
31	1	1.26 mw	-29
30	0	1 mW	-30

Because dBW is referenced to one watt, the Log of the power in watts times 10 is dBW. The Logarithm of 10 raised by any exponent is simply that exponent. That is: $Log(10)^4 = 4$. Therefore, a power that can be expressed as any exponent of 10 can also be expressed in dBW as that exponent times 10. For example, 100 kW can be written 100,000 watts or 10^5 watts. 100 kW is then +50 dBW. Another way to remember this conversion is that dBW is the number of zeros in the power written in watts times 10. If the transmitter power in question is conveniently a multiple of ten (it often is) the conversion to dBW is easy and accurate.

DUTY CYCLE

Duty cycle (or duty factor) is a measure of the fraction of the time a radar is transmitting. It is important because it relates to peak and average power in the determination of total energy output. This, in turn, ultimately affects the strength of the reflected signal as well as the required power supply capacity and cooling requirements of the transmitter.

Although there are exceptions, most radio frequency (RF) measurements are either continuous wave (CW) or pulsed RF. CW RF is uninterrupted RF such as from an oscillator. Amplitude modulated (AM), frequency modulated (FM), and phase modulated (PM) RF are considered CW since the RF is continuously present. The power may vary with time due to modulation, but RF is always present. Pulsed RF, on the other hand, is bursts (pulses) of RF with no RF present between bursts. The most general case of pulsed RF consists of pulses of a fixed pulse width (PW) which come at a fixed time interval, or period, (T). For clarity and ease of this discussion, it is assumed that all RF pulses in a pulse train have the same amplitude. Pulses at a fixed interval of time arrive at a rate or frequency referred to as the pulse repetition frequency (PRF) of so many pulse per second. Pulse repetition interval (PRI) and PRF are reciprocals of each other.

$$PRF = 1/T = 1/PRI \qquad \qquad [1]$$

Power measurements are classified as either peak pulse power, P_p, or average power, P_{ave}. The actual power in pulsed RF occurs during the pulses, but most power measurement methods measure the heating effects of the RF energy to obtain an average value of the power. It is correct to use either value for reference so long as one or the other is consistently used. Frequently it is necessary to convert from P_p to P_{ave}, or vice versa; therefore the relationship between the two must be understood. Figure 1 shows the comparison between P_p and P_{ave}.

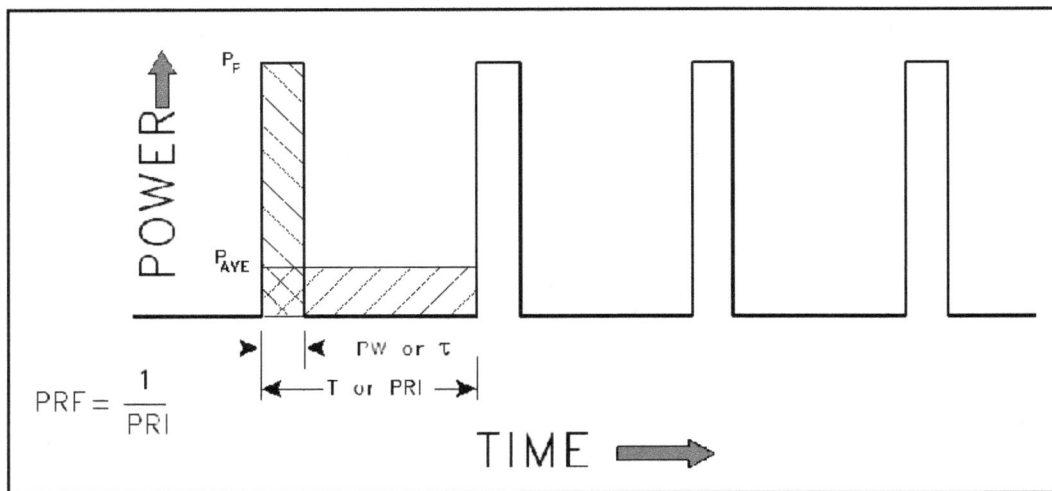

Figure 1. RF Pulse Train

The average value is defined as that level where the pulse area above the average is equal to area below average between pulses. If the pulses are evened off in such a way as to fill in the area between pulses, the level obtained is the average value, as shown in Figure 1 where the shaded area of the pulse is used to fill in the area between pulses. The area of the pulse is the pulse width multiplied by the peak pulse power. The average area is equal to the average value of power multiplied by the pulse period.

Since the two values are equal:

$$P_{ave} \times T = P_p \times PW \qquad\qquad [2]$$

or

$$P_{ave}/P_p = PW/T \qquad\qquad [3]$$

Using [1]

$$P_{ave}/P_p = PW/T = PW \times PRF = PW/PRI = \text{duty cycle} \qquad\qquad [4]$$

(note that the symbol τ represents pulse width (PW) in most reference books)

The ratio of the average power to the peak pulse power is the duty cycle and represents the percentage of time the power is present. In the case of a square wave the duty cycle is 0.5 (50%) since the pulses are present 1/2 the time, the definition of a square wave.

For Figure 1, the pulse width is 1 unit of time and the period is 10 units. In this case the duty cycle is:

$$PW/T = 1/10 = 0.1 \ (10\%).$$

A more typical case would be a PRF of 1,000 and a pulse width of 1.0 microseconds. Using [4], the duty cycle is $0.000001 \times 1,000 = 0.001$. The RF power is present one-thousandth of the time and the average power is 0.001 times the peak power. Conversely, if the power were measured with a power meter which responds to average power, the peak power would be 1,000 time the average reading.

Besides expressing duty cycle as a ratio as obtained in equation [4], it is commonly expressed as either a percentage or in decibels (dB). To express the duty cycle of equation [4] as a percentage, multiply the value obtained by 100 and add the percent symbol. Thus a duty cycle of 0.001 is also 0.1%.

The duty cycle can be expressed logarithmically (dB) so it can be added to or subtracted from power measured in dBm/dBW rather than converting to, and using absolute units.

$$\text{Duty cycle (dB)} = 10 \log(\text{duty cycle ratio}) \qquad\qquad [5]$$

For the example of the 0.001 duty cycle, this would be $10 \log(0.001) = -30$ dB. Thus the average power would be 30 dB less than the peak power. Conversely, the peak power is 30 dB higher than the average power.

For pulse radars operating in the PRF range of 0.25-10 kHz and PD radars operating in the PRF range of 10-500 kHz, typical duty cycles would be:

Pulse	~	0.1 - 3%	=	0.001 - .03	=	-30 to -15 dB
Pulse Doppler	~	5 - 50%	=	0.05 - .5	=	-13 to -3 dB
Continuous Wave	~	100%	=	1	=	0 dB

Intermediate Frequency Bandwidths of typical signals are:

Pulse	1 to 10 MHz
Chirp or Phase coded pulse	0.1 to 10 MHz
CW or PD	0.1 to 5 kHz

PRF is usually subdivided into the following categories: Low 0.25-4 kHz; Medium 8-40 kHz; High 50-300 kHz.

DOPPLER SHIFT

Doppler is the apparent change in wavelength (or frequency) of an electromagnetic or acoustic wave when there is relative movement between the transmitter (or frequency source) and the receiver.

Summary RF Equation for the Two-Way (radar) case

$$f_{Rec} = f_{Xmt} + f_D = f_{Xmt} + \frac{2(V_{Xmtr} + V_{Tgt})\, f_{Xmt}}{c}$$

Summary RF Equation for the One-Way (ESM) case

$$f_{Rec} = f_{Xmt} + f_D = f_{Xmt} + \frac{V_{Xmtr\ or\ Rec}\, f_{Xmt}}{c}$$

Rules of Thumb for two-way signal travel
(divide in half for one-way ESM signal measurements)

At 10 GHz, $f_D \cong$
35 Hz per Knot
19 Hz per km/Hr
67 Hz per m/sec
61 Hz per yd/sec
20 Hz per ft/sec

To estimate f_D at other frequencies, multiply these by:

$$\left[\frac{f_{Xmt}(GHz)}{10} \right]$$

The Doppler effect is shown in Figure 1. In everyday life this effect is commonly noticeable when a whistling train or police siren passes you. Audio Doppler is depicted, however Doppler can also affect the frequency of a radar carrier wave, the PRF of a pulse radar signal, or even light waves causing a shift of color to the observer.

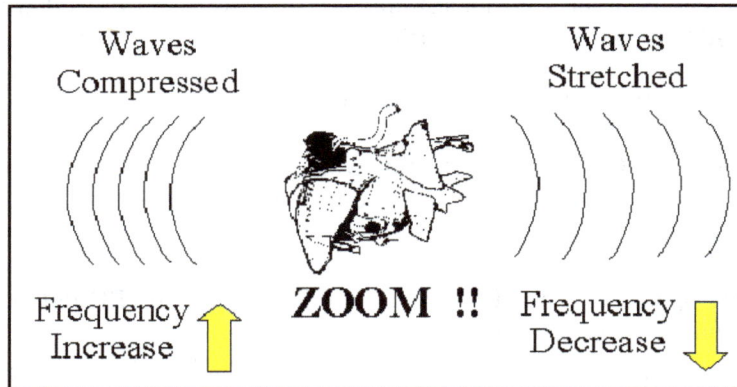

Figure 1. Doppler Frequency Creation From Aircraft Engine Noise

How do we know the universe is expanding?

Answer: The color of light from distant stars is shifted to red (see Section 7-1: higher λ or lower frequency means Doppler shift is stretched, i.e. expanding).

A memory aid might be that the lights from a car (going away) at night are red (tail lights)!

Doppler frequency shift is directly proportional to velocity and a radar system can therefore be calibrated to measure velocity instead of (or along with) range. This is done by measuring the shift in frequency of a wave caused by an object in motion (Figure 2).

    * Transmitter in motion
    * Reflector in motion
    * Receiver in motion
    * All three

For a closing relative velocity:
    * Wave is compressed
    * Frequency is increased

For an opening relative velocity:
    * Wave is stretched
    * Frequency is decreased

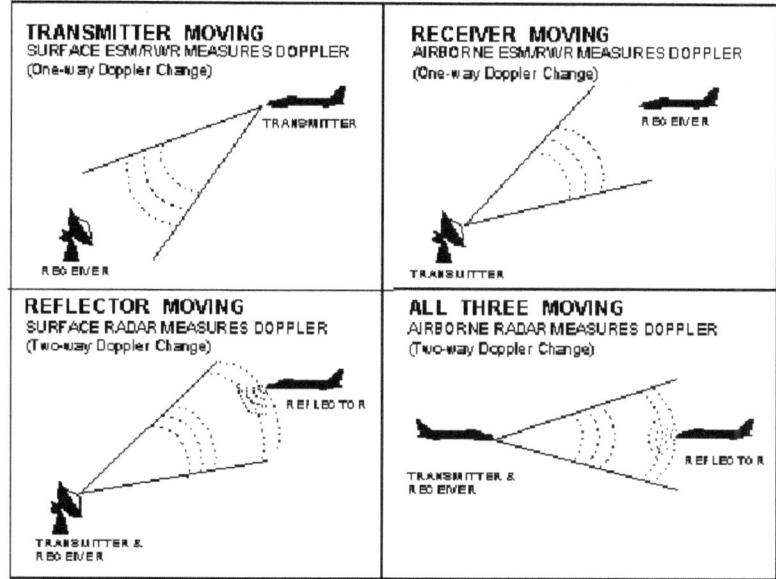

Figure 2. Methods of Doppler Creation

To compute Doppler frequency we note that velocity is range rate; $V = dr/dt$

For the <u>reflector in motion case</u>, You can see the wave compression effect in Figure 3 when the transmitted wave peaks are one wavelength apart. When the first peak reaches the target, they are still one wavelength apart (point a).

When the 2nd peak reaches the target, the target has advanced according to its velocity (vt) (point b), and the first reflected peak has traveled toward the radar by an amount that is less than the original wavelength by the same amount (vt) (point c).

As the 2nd peak is reflected, the wavelength of the reflected wave is 2(vt) less than the original wavelength (point d).

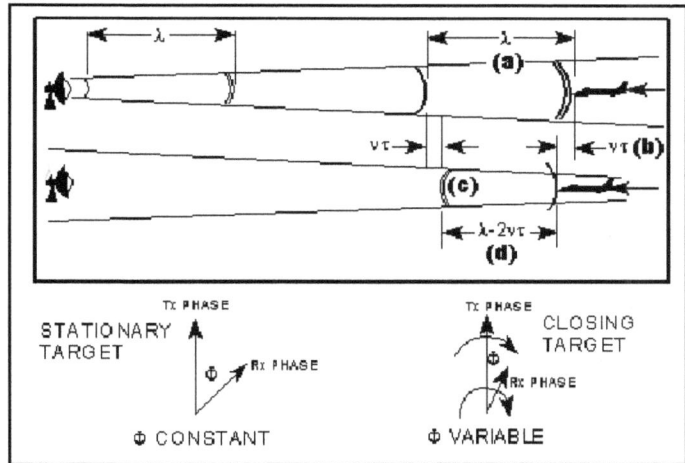

Figure 3. Doppler Compression Equivalent to Variable Phase Shift

The distance the wave travels is twice the target range. The reflected phase lags transmitted phase by 2x the round trip time. For a fixed target the received phase will differ from the transmitted phase by a constant phase shift. For a moving target the received phase will differ by a changing phase shift.

For the closing target shown in Figure 3, the received phase is advancing with respect to the transmitted phase and appears as a higher frequency.

Doppler is dependent upon closing velocity, not actual radar or target velocity as shown in Figure 4.

For the following equations (except radar mapping), we assume the radar and target are moving directly toward one another in order to simplify calculations (if this is not the case, use the velocity component of one in the direction of the other in the formulas).

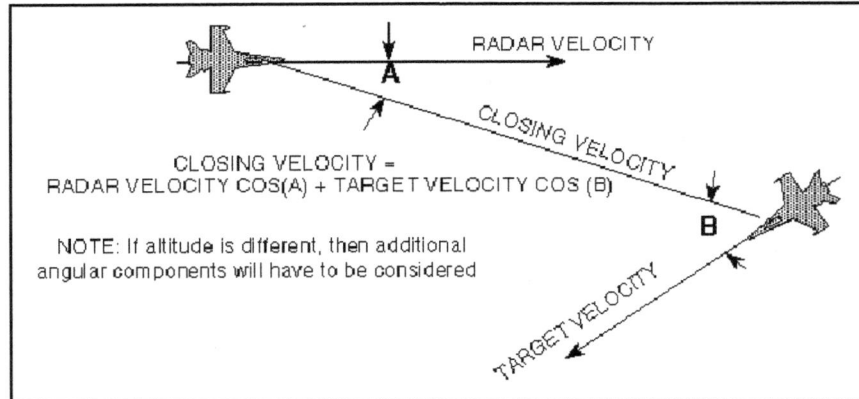

CLOSING VELOCITY = RADAR VELOCITY COS(A) + TARGET VELOCITY COS (B)

NOTE: If altitude is different, then additional angular components will have to be considered

Figure 4. Doppler Depends Upon Closing Velocity

For the case of a moving reflector, doppler frequency is proportional to 2x the transmitted frequency: Higher rf = higher doppler shift

$$f_D = (2 \times V_{Target})(f/c)$$

Likewise, it can be shown that for other cases, the following relationships hold:

For an airplane radar with an airplane target (The "all three moving" case, i.e. aircraft radar transmitter, target, and aircraft radar receiver)

$$f_D = 2(V_{Radar} + V_{Target})(f/c)$$

Speed of Light Conversions
* * *
$c \cong 2.9979 \times 10^8$ m/sec
$c \cong 5.8275 \times 10^8$ nm/hr (knots)
$c \cong 9.8357 \times 10^8$ ft/sec

For the case of a semi-active missile receiving signals (Also "all three moving")

$$f_D = (V_{Radar} + 2V_{Target} + V_{Missile})(f/c)$$

For the airplane radar with a ground target (radar mapping) or vice versa.

$$f_D = 2(V_{Radar} \, \text{Cos}\theta \, \text{Cos}\varphi)(f/c), \quad \text{Where } \theta \text{ and } \varphi \text{ are the radar scan azimuth and depression angles.}$$

For a ground based radar with airborne target - same as previous using target track crossing angle and ground radar elevation angle.

For the ES/ESM/RWR case where only the target or receiver is moving (One-way Doppler measurements)

$$f_D = V_{Receiver \, or \, Target} \, (f/c)$$

Note: See Figure 4 if radar and target are not moving directly towards or away from one another.

Figure 5 depicts the results of a plot of the above equation for a moving reflector such as might be measured with a ground radar station illuminating a moving aircraft.

It can be used for the aircraft-to-aircraft case, if the total net closing rate of the two aircraft is used for the speed entry in the figure.

It can also be used for the ES/ESM case (one-way doppler measurements) if the speed of the aircraft is used and the results are divided by two.

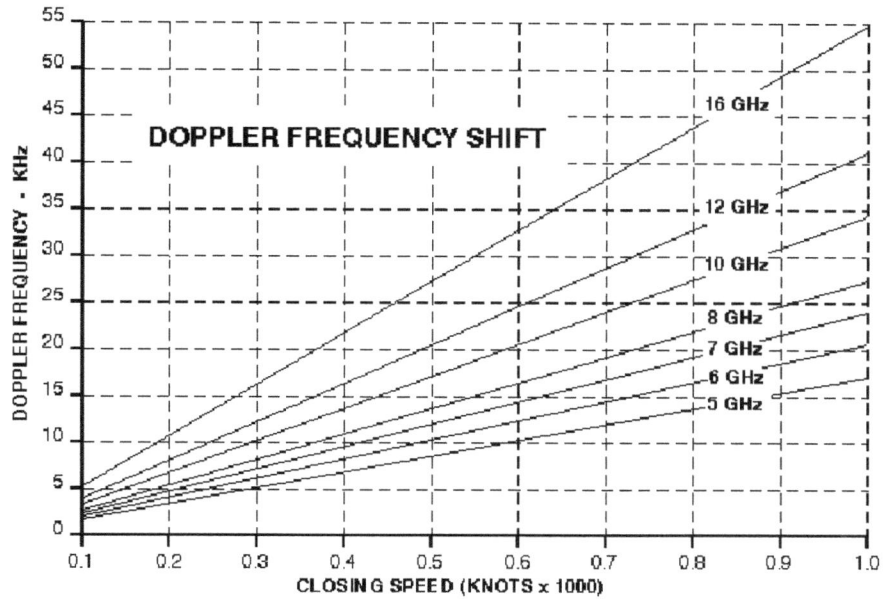

DOPPLER FREQUENCY SHIFT

Figure 5. Two-Way Doppler Frequency Shift

SAMPLE PROBLEMS:

(1) If a ground radar operating at 10 GHz is tracking an airplane flying at a speed of 500 km/hr tangential to it (crossing pattern) at a distance of 10 km, what is the Doppler shift of the returning signal?

Answer: Since the closing velocity is zero, the Doppler is also zero.

(2) If the same aircraft turns directly toward the ground radar, what is the Doppler shift of the returning signal?

Answer: 500 km/hr = 270 kts from Section 2-1. From Figure 4 we see that the Doppler frequency is about 9.2 KHz.

(3) Given that a ground radar operating at 7 GHz is Doppler tracking an aircraft 20 km away (slant range) which is flying directly toward it at an altitude of 20,000 ft and a speed of 800 ft/sec, what amount of VGPO switch would be required of the aircraft jammer to deceive (pull) the radar to a zero Doppler return?

Answer: We use the second equation from the bottom of page 2-6.3 which is essentially the same for this application except a ground radar is tracking an airplane target (vs an airplane during ground mapping), so for our application we use a positive elevation angle instead of a negative (depression) angle.

$f_D = 2(V_r \cos \theta \cos \varphi)(f/c)$, where θ is the aircraft track crossing angle and φ is the radar elevation angle.

Since the aircraft is flying directly at the radar, $\theta = 0°$; the aircraft altitude = 20,000 ft = 6,096 meters.

Using the angle equation in Section 2-1, sin φ = x/r = altitude / slant range, so:
$\varphi = \sin^{-1}$ (altitude/slant range) = $\sin^{-1}$ (6,096 m / 20,000 m) = 17.7°

$F_D = 2(800 \text{ ft/sec} \cos 0° \cos 17.7°)(7x10^9 \text{ Hz} / 9.8357 \times 10^9 \text{ ft/sec}) = \underline{10,845 \text{ Hz}}$

ELECTRONIC FORMULAS

Ohm's Law Formulas for D-C Circuits. $E = IR = \dfrac{P}{I} = \sqrt{PR}$ $P = I^2 R = EI = \dfrac{E^2}{R}$

Ohm's Law Formulas for A-C Circuits and Power Factor.

$E = IZ = \dfrac{P}{I \cos \Theta} = \sqrt{\dfrac{P\,Z}{\cos \Theta}}$ $P = I^2 Z \cos \Theta = IE \cos \Theta = \dfrac{E^2 \cos \Theta}{Z}$

In the above formulas Θ is the angle of lead or lag between current and voltage and cos Θ = P/EI = power factor or *pf*.

$pf = \dfrac{Active\ power\ (in\ watts)}{Apparent\ power\ (in\ volt\text{-}amps)} = P\,or\,E\,I$ $pf = \dfrac{R}{Z}$

Note: Active power is the "resistive" power and equals the equivalent heating effect on water.

Voltage/Current Phase Rule of Thumb   Remember "ELI the ICE man"

ELI: Voltage (E) comes before (leads) current (I) in an inductor (L)
ICE: Current (I) comes before (leads) Voltage (E) in a capacitor (C)

Resistors in Series $R_{total} = R_1 + R_2 = R_3 + ...$

Two Resistors in Parallel $R_t = \dfrac{R_1 R_2}{R_1 + R_2}$

Resistors in Parallel, General Formula

$R_{total} = \dfrac{1}{\dfrac{1}{R_1} + \dfrac{1}{R_2} + \dfrac{1}{R_3} + ...}$

Resonant Frequency Formulas   *Where in the second formula f is in kHz and L and C are in microunits.

$f = \dfrac{1}{2\pi\sqrt{LC}}, \ \ or \ \ f = \dfrac{159.2\,*}{\sqrt{LC}} \ \ \ L = \dfrac{1}{4\pi^2 f^2 C}, \ \ or \ \ L = \dfrac{25,330\,*}{f^2 C} \ \ \ C = \dfrac{1}{4\pi^2 f^2 L}, \ \ or \ \ C = \dfrac{25,330\,*}{f^2 L}$

Conductance $G = \dfrac{1}{R}$ *(for D-C circuit)* $G = \dfrac{R}{R^2 + X^2}$ *(for A-C circuit)*

Reactance Formulas $X_C = \dfrac{1}{2\pi f C}$ $C = \dfrac{1}{2\pi f X_C}$ $X_L = 2\pi fL$ $L = \dfrac{X_L}{2\pi f}$

Impedance Formulas $Z = \sqrt{R^2 + (X_L - X_C)^2}$ *(for series circuit)*

$Z = \dfrac{RX}{\sqrt{R^2 + X^2}}$ *(for R and X in parallel)*

Q or Figure of Merit $Q = \dfrac{X_L}{R} \ \ or \ \ \dfrac{X_C}{R}$

Frequency Response

		Inductor * ⌒⌒⌒	Capacitor * ⊣⊢	Resister ⌁
───────	DC	Pass	Block	Attenuate
∿	Low Freq AC	Attenuate *	Attenuate *	Attenuate
∿∿∿	High Freq	Block	Pass	Attenuate

* Attenuation varies as a function of the value of the each device and the frequency

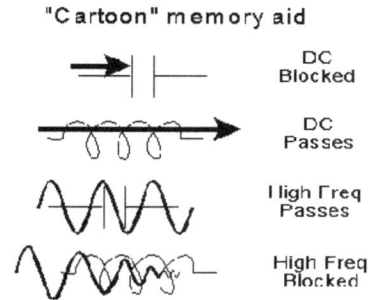

"Cartoon" memory aid

DC Blocked

DC Passes

High Freq Passes

High Freq Blocked

Sinusoidal Voltages and Currents

Effective value   = 0.707 x peak value
[Also known as Root-Mean Square (RMS) value]

Half Cycle Average value = 0.637 x peak

Peak value   = 1.414 x effective value

∴ Effective value   = 1.11 x average value

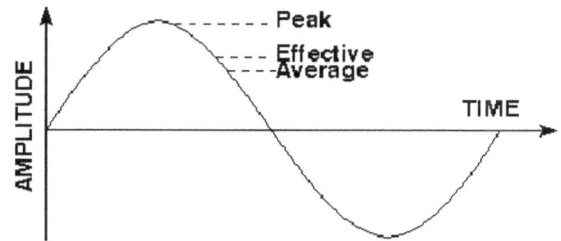

Three-phase AC Configurations

(120° phase difference between each voltage)
If the connection to a three phase AC configuration is miswired, switching any two of the phases will put it back in the proper sequence.
Electric power for ships commonly uses the delta configuration, while commercial electronic and aircraft applications commonly use the wye configuration.

Wye (Y) or Star

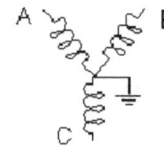

Delta

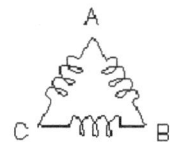

Color Code for House Wiring:	**PURPOSE:**	**Color Code for Chassis Wiring:**
Black or red	HOT	Red
White	NEUTRAL (Return)	White
Green or bare	GROUND	Black

Color Code for Resistors:

First and second band: (and third band # of zeros if not gold/silver)				**Third band** **Multiplier**		**Fourth band** **Tolerance**	
0	Black	5	Green	.1	Gold	5%	Gold
1	Brown	6	Blue	.01	Silver	10%	Silver
2	Red	7	Violet			20%	No color
3	Orange	8	Gray				
4	Yellow	9	White				

The third color band indicates number of zeros to be added after figures given by first two color bands. But if third color band is gold, multiply by 0.1 and if silver multiply by 0.01. Do not confuse with fourth color-band that indicates tolerance. Thus, a resistor marked blue-red-gold-gold has a resistance of 6.2 ohms and a 5% tolerance.

MISSILE AND ELECTRONIC EQUIPMENT DESIGNATIONS

Missiles are designated with three letters from the columns below plus a number (i.e. AIM-7M) Suffixes (M in this case) indicate a modification.

First Letter Launch Environment	Second Letter Mission Symbols	Third Letter Vehicle Type
A Air B Multiple C Coffin H Silo stored L Silo launched M Mobile P Soft Pad R Ship U Underwater	D Decoy E Special electronic G Surface attack I Intercept, aerial Q Drone T Training U Underwater attack W Weather	M Guided Missile N Probe (non-orbital instruments) R Rocket (without installed or remote control guidance)

U.S. military electronic equipment is assigned an identifying alphanumeric designation that is used to uniquely identify it. This system is commonly called the "AN" designation system, although its formal name is the Joint Electronics Type Designation System (JETDS). The letters AN preceding the equipment indicators formerly meant "Army/Navy," but now are a letter set that can only be used to indicate formally designated DOD equipment. The first three letters following the "AN/" indicate Platform Installation, Equipment Type, and Equipment Function, respectively. The appropriate meaning is selected from the lists below. The letters following the AN designation numbers provide added information about equipment. Suffixes (A, B, C, etc.) indicate a modification. The letter (V) indicates that variable configurations are available. The letter (X) indicates a development status. A parenthesis ( ) without a number within it indicates a generic system that has not yet received a formal designation, e.g., AN/ALQ( ). Quite often the () is pronounced "bow legs" since they look like the shape of cowboy legs.

First Letter Platform Installation	Second Letter Equipment Type	Third Letter Function or Purpose

A Piloted aircraft
B Underwater mobile,
   submarine
D Pilotless carrier
F Fixed ground
G General ground use
K Amphibious
M Mobile (ground)
P Portable
S Water
T Ground, transportable
U General utility
V Vehicular (ground)
W Water surface and
   underwater
   combination
Z Piloted-pilotless
   airborne vehicle
   combination
	A Invisible light, heat radiation
	C Carrier
	D Radiac
	F Photographic
	G Telegraph or teletype
	I Interphone and public address
	J Electromechanical or inertial wire
	   covered
	K Telemetering
	L Countermeasures
	M Meteorological
	N Sound in air
	P Radar
	Q Sonar and underwater sound
	R Radio
	S Special or combinations of types
	T Telephone (wire)
	V Visual and visible light
	W Armament
	X Facsimile or television
	Y Data Processing
		B Bombing
		C Communications
		D Direction finder, reconnaissance
		   and/or surveillance
		E Ejection and/or release
		G Fire control or searchlight directing
		H Recording and/or reproducing
		K Computing
		M Maintenance and/or test assemblies
		N Navigation aids
		Q Special or combination of purposes
		R Receiving, passive detecting
		S Detecting and/or range and
		   bearing, search
		T Transmitting
		W Automatic flight or remote control
		X Identification and recognition
		Y Surveillance and control

This Page Blank

RADAR HORIZON / LINE OF SIGHT

There are limits to the reach of radar signals. At the frequencies normally used for radar, radio waves usually travel in a straight line. The waves may be obstructed by weather or shadowing, and interference may come from other aircraft or from reflections from ground objects (Figure 1).

As also shown in Figure 1, an aircraft may not be detected because it is below the radar line which is tangent to the earths surface.

Some rules of thumb are:

Range (to horizon):
$$R_{NM} = 1.23\sqrt{h_{radar}} \quad with\ h\ in\ ft$$

Range (beyond horizon / over earth curvature):
$$R_{NM} = 1.23\left(\sqrt{h_{radar}} + \sqrt{h_{target}}\right) \quad with\ h\ in\ ft$$

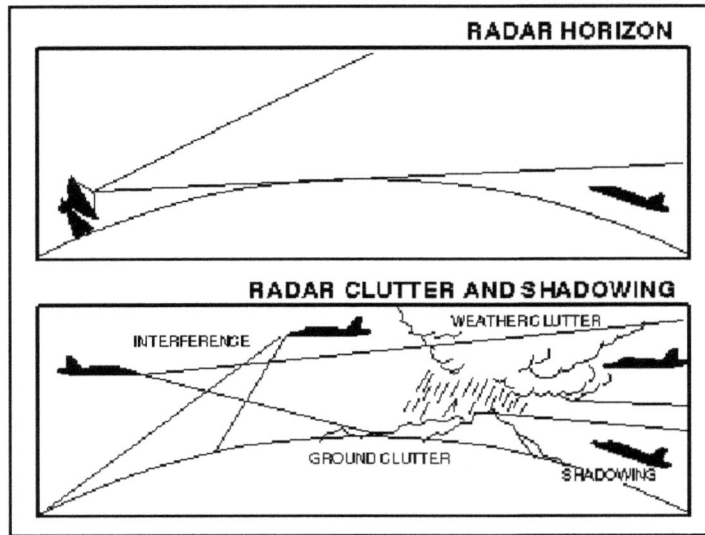

Figure 1. Radar Horizon and Shadowing

In obtaining the radar horizon equations, it is common practice to assume a value for the Earth's radius that is 4/3 times the actual radius. This is done to account for the effect of the atmosphere on radar propagation. For a true line of sight, such as used for optical search and rescue, the constant in the equations changes from 1.23 to 1.06.

A nomograph for determining maximum target range is depicted in Figure 2. Although an aircraft is shown to the left, it could just as well be a ship, with radars on a mast of height "h". Any target of height (or altitude) "H" is depicted on the right side.

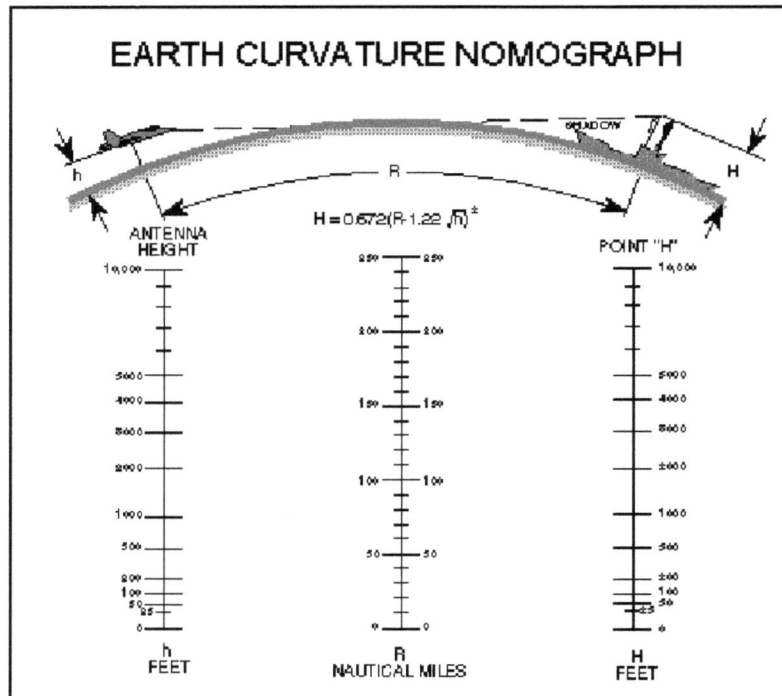

Figure 2. Earth Curvature Nomograph

See also Section 5-1 on ducting and refraction, which may increase range beyond these distances.

This data was expanded in Figure 3 to consider the maximum range one aircraft can detect another aircraft using:

$$R_{NM} = 1.23 \left(\sqrt{h_{radar}} + \sqrt{h_{target}} \right)$$
(with h in feet)

It can be used for surface targets if $H_{target} = 0$. It should be noted that most aircraft radars are limited in power output, and would not detect small or surface objects at the listed ranges.

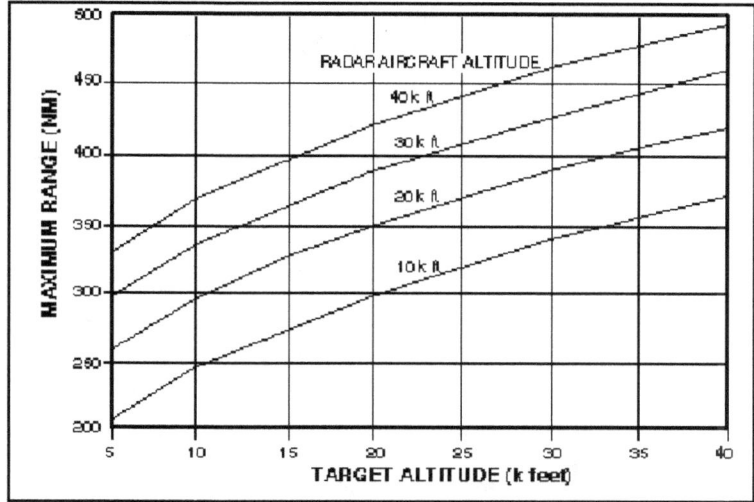

Figure 3. Aircraft Radar vs Aircraft Target Maximum Range

Figure 4 depicts the maximum range that a ship height antenna can detect a zero height object (i.e. rowboat etc).

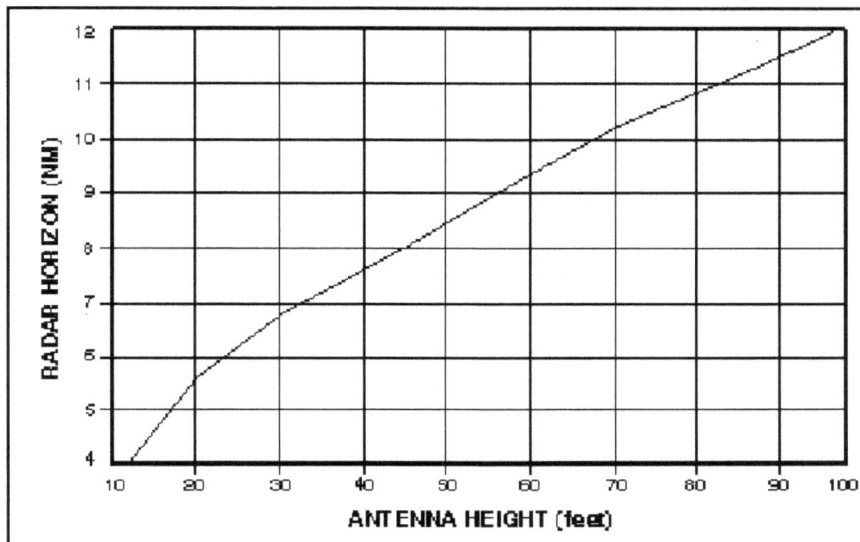

Figure 4. Ships Radar Horizon with Target on the Surface

In this case "H" = 0, and the general equation becomes: $R_{max}(NM) = 1.23 \sqrt{h_r}$

Where h_r is the height of the radar in feet.

Figure 5 depicts the same for aircraft radars. It should be noted that most aircraft radars are limited in power output, and would not detect small or surface objects at the listed ranges.

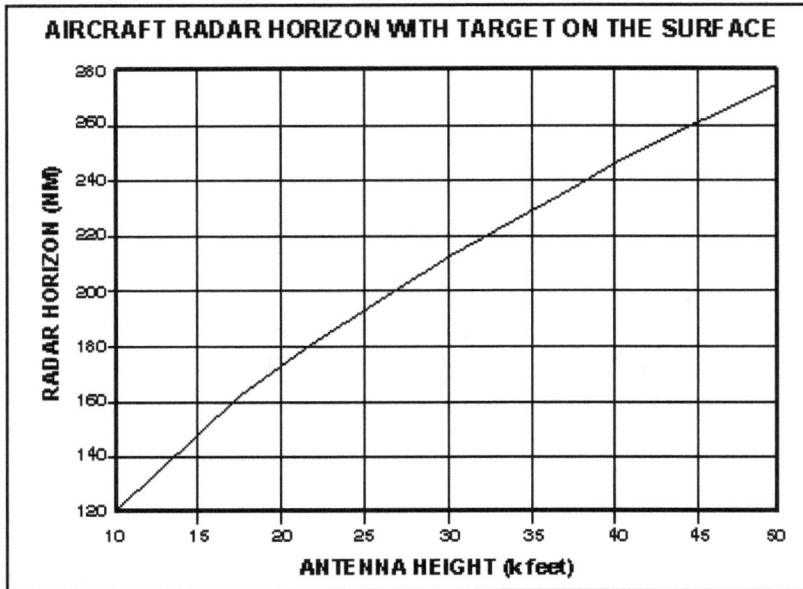

Figure 5. Aircraft Radar Horizon with Target on the Surface

Other general rules of thumb for surface "targets/radars" are:

<table>
<tr><td align="center">**For Visual SAR:**</td><td align="center">**For ESM:**</td></tr>
<tr><td align="center">$R_{Visual}(NM) = 1.05\sqrt{Acft\ Alt\ in\ ft}$</td><td align="center">$R_{ESM}(NM) = 1.5\sqrt{Acft\ Alt\ in\ ft}$</td></tr>
</table>

PROPAGATION TIME / RESOLUTION

1. ROUND TRIP RANGE: $R = \dfrac{c\,t}{2}$ with t = time to reach target

Rules of Thumb

In one µsec round trip time, a wave travels to and from an object at a distance of:

$\cong$ 150 m
$\cong$ 164 yd
$\cong$ 500 ft
$\cong$ 0.08 NM
$\cong$ 0.15 km

The time it takes to travel to and from an object at a distance of:

1 m $\cong$ 0.0067 µsec
1 yd $\cong$ 0.006 µsec
1 ft $\cong$ 0.002 µsec
1 NM $\cong$ 12.35 µsec
1 km $\cong$ 6.7 µsec

2. ONE WAY RANGE: $\boldsymbol{R = ct}$ with t = time to reach target

Time	Distance Traveled		Distance	Time it Takes
1 milli sec (ms)	165 NM		1 NM	6.18 µsec
1 micro sec (µs)	1000 ft		1 km	3.3 µsec
1 nano sec (ns)	1 ft		1 ft	1 nsec

3. UNAMBIGUOUS RANGE

(DISTANCE BETWEEN PULSES): $R = \dfrac{c \bullet PRI}{2}$

Normally a radar measures "distance" to the target by measuring time from the last transmitted pulse. If the inter-pulse period (T) is long enough that isn't a problem as shown in "A" to the right. When the period is shortened, the time to the last previous pulse is shorter than the actual time it took, giving a false (ambiguous) shorter range (figure "B").

Rules of Thumb

RNM $\cong$ 81Pms
RKm $\cong$ 150Pms
Where Pms is PRI in milliseconds

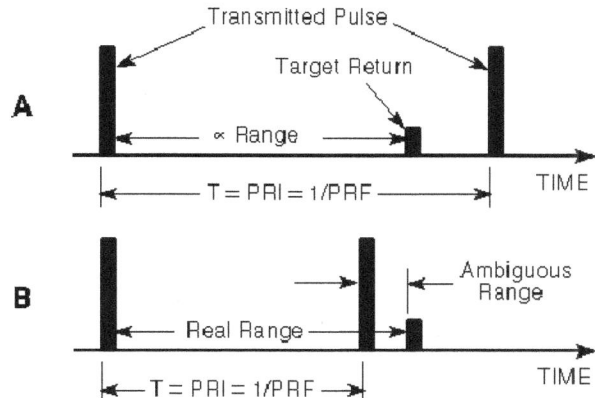

4. RANGE RESOLUTION

Rules of Thumb
500 ft per microsecond of pulse width
500 MHz IF bandwidth provides 1 ft of resolution.

5. BEST CASE PERFORMANCE:

The atmosphere limits the accuracy to 0.1 ft
The natural limit for resolution is one RF cycle.

MODULATION

Modulation is the process whereby some characteristic of one wave is varied in accordance with some characteristic of another wave. The basic types of modulation are angular modulation (including the special cases of phase and frequency modulation) and amplitude modulation. In missile radars, it is common practice to amplitude modulate the transmitted RF carrier wave of tracking and guidance transmitters by using a pulsed wave for modulating, and to frequency modulate the transmitted RF carrier wave of illuminator transmitters by using a sine wave.

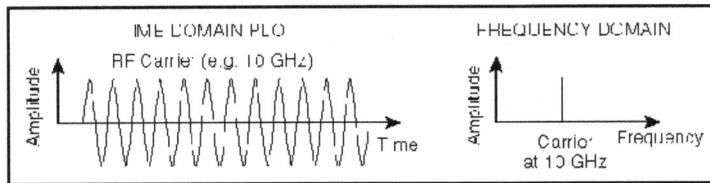

Figure 1. Unmodulated RF Signal

Frequency Modulation (FM) - As shown in Figure 1, an unmodulated RF signal in the time domain has only a single spectral line at the carrier frequency (f_c) in the frequency domain. If the signal is frequency modulated, as shown in Figure 2 (simplified using only two changes), the spectral line will

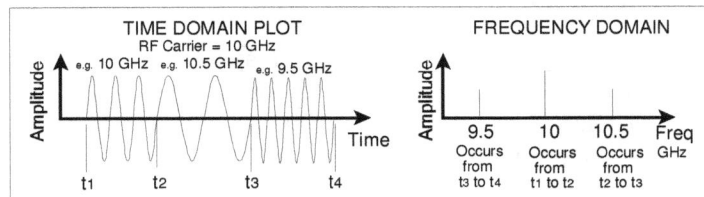

Figure 2. RF Signal With Frequency Modulation.

correspondingly shift in the frequency domain. The bandwidth can be approximated using Carson's rule: BW = 2($\Delta f + fm$), where Δf is the peak deviation of the instantaneous frequency from the carrier and *fm* is the highest frequency present in the modulating signal. There are usually many more "spikes" in the frequency domain than depicted. The number of spikes and shape of the frequency domain envelope (amplitude) are based on the modulation index β. The modulation index is related to the same two factors used in Carson's rule. A high Δf means a higher modulation index with many more "spikes" spread across a wider bandwidth.

Amplitude Modulation (AM) - If the signal in Figure 1 is amplitude modulated by a sinewave as shown in Figure 3, sidebands are produced in the frequency domain at $F_c \pm F_{AM}$. AM other than by a pure sine wave will cause additional sidebands normally at $F_c \pm nF_{AM}$, where n equals 1, 2, 3, 4, etc.

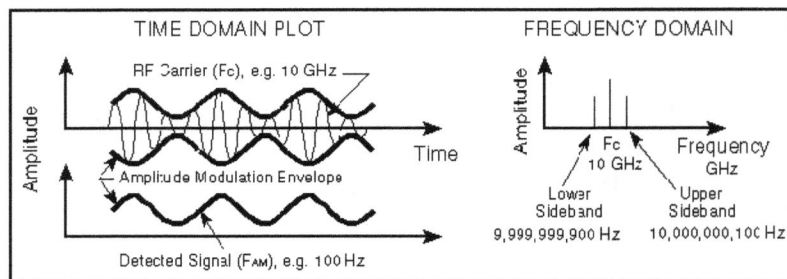

Figure 3. Sinewave Modulated RF Signal

Pulse modulation is a special case of AM wherein the carrier frequency is gated at a pulsed rate. When the reciprocal of the duty cycle of the AM is a whole number, harmonics corresponding to multiples of that whole number will be

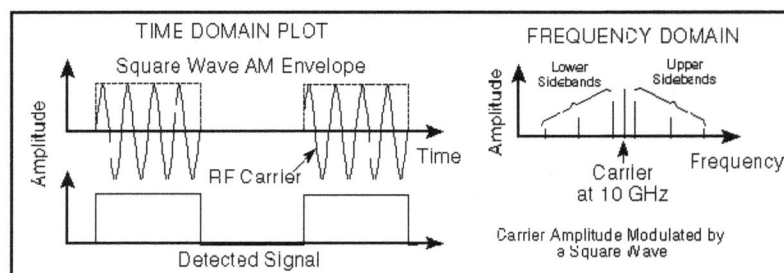

Figure 4. Square Wave Modulated RF Signal (50% Duty Cycle)

missing, e.g. in a 33.33% duty cycle, AM wave will miss the 3rd, 6th, 9th, etc. harmonics, while a square wave or 50% duty cycle triangular wave will miss the 2nd, 4th, 6th, etc. harmonic, as shown in Figure 4. It has sidebands in the frequency domain at $F_c \pm nF_{AM}$, where n = 1, 3, 5, etc. The amplitude of the power level follows a sine x / x type distribution.

Figure 5 shows the pulse width (PW) in the time domain which defines the lobe width in the frequency domain (Figure 6). The width of the main lobe is 2/PW, whereas the width of a side lobe is 1/PW. Figure 5 also shows the pulse repetition interval (PRI) or its reciprocal, pulse repetition frequency (PRF), in the time domain. In the frequency domain, the spectral lines inside the lobes are separated by the PRF or 1/PRI, as shown in Figures 7 and 8. Note that Figures 7 and 8 show actual magnitude of the side lobes, whereas in Figure 4 and 6, the absolute value is shown.

The magnitude of each spectral component for a rectangular pulse can be determined from the following formula:

$$a_n = 2A \frac{\tau}{T} \frac{\sin(n\pi\tau/T)}{n\pi\tau/T} \qquad where: \begin{array}{l} \tau = pulse\ width\ (PW) \\ T = period\ (PRI) \end{array} \qquad and \quad A = Amplitude\ of\ rectangular\ pulse$$

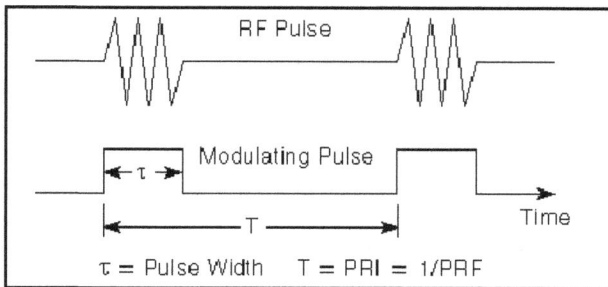

Figure 5. Pulse Width and PRI/PRF Waveforms

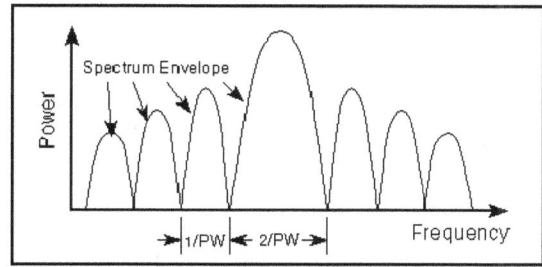

Figure 6. Sidelobes Generated by Pulse Modulation (Absolute Value)

Figure 7 shows the spectral lines for a square wave 50% duty cycle), while Figure 8 shows the spectral lines for a 33.33% duty cycle rectangular wave signal.

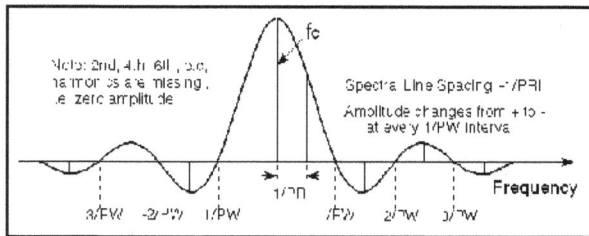

Figure 7. Spectral Times for a Square Wave Modulated Signal

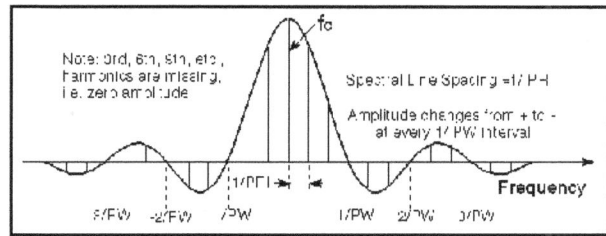

Figure 8. Spectral Lines for a 33.3% Duty Cycle

Figure 9 shows that for square wave AM, a significant portion of the component modulation is contained in the first few harmonics which comprise the wave. There are twice as many sidebands or spectral lines as there are harmonics (one on the plus and one on the minus side of the carrier). Each sideband represents a sine wave at a frequency equal to the difference between the spectral line and f_c.

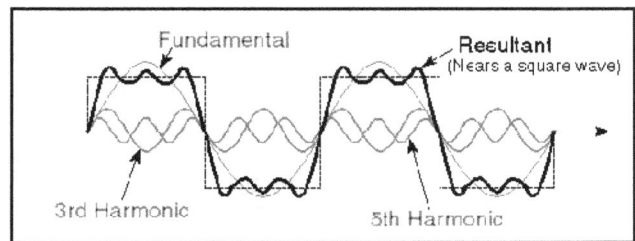

Figure 9. Square Wave Consisting of Sinewave Harmonics

A figure similar to Figure 9 can be created for any rectangular wave. The relative amplitude of the time domain sine wave components are computed using equation [1]. Each is constructed such that at the midpoint of the pulse the sine wave passes through a maximum (or minimum if the coefficient is negative) at the same time. It should be noted that the "first" harmonic created using this formula is <u>NOT</u> the carrier frequency, f_c, of the modulated signal, but at $F_c \pm F_{AM}$.

While equation [1] is for rectangular waves only, similar equations can be constructed using Fourier coefficients for other waveforms, such as triangular, sawtooth, half sine, trapezoidal, and other repetitive geometric shapes.

<u>PRI Effects</u> - If the PW remains constant but PRI increases, the number of sidelobes remains the same, but the number of spectral lines gets denser (move closer together) and vice versa (compare Figure 7 and 8). The spacing between the spectral lines remains constant with constant PRI.

<u>Pulse Width (PW) Effects</u> - If the PRI remains constant, but the PW increases, then the lobe width decreases and vice versa. If the PW approaches PRI, the spectrum will approach "one lobe", i.e., a single spectral line. The spacing of the lobes remains constant with constant PW.

<u>RF Measurements</u> - If the receiver bandwidth is smaller than the PRF, the receiver will respond to one spectral line at a time. If the receiver bandwidth is wider than the PRF but narrower than the reciprocal of the PW, the receiver will respond to one spectral envelope at a time.

Jet Engine Modulation (JEM)

Section 2-6 addresses the Doppler shift in a transmitted radar signal caused by a moving target. The amount of Doppler shift is a function of radar carrier frequency and the speed of the radar and target. Moving or rotating surfaces on the target will have the same Doppler shift as the target, but will also impose AM on the Doppler shifted return (see Figure 10). Reflections off rotating jet engine compressor blades, aircraft propellers, ram air turbine (RAT) propellers used to power aircraft pods, helicopter rotor blades, and protruding surfaces of automobile hubcaps will all provide a chopped reflection of the impinging signal. The reflections are characterized by both positive and negative Doppler sidebands corresponding to the blades moving toward and away from the radar respectively.

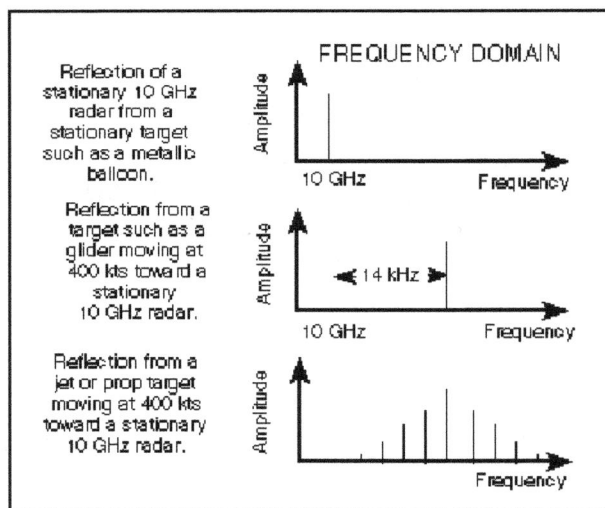

Figure 10. Doppler Return and JEM

Therefore, forward/aft JEM doesn't vary with radar carrier frequency, but the harmonics contained in the sidebands are a function of the PRF of the blade chopping action and its amplitude is target aspect dependent, i.e. blade angle, intake/exhaust internal reflection, and jet engine cowling all effect lateral return from the side. If the aspect angle is too far from head-on or tail-on and the engine cowling provides shielding for the jet engine, there may not be any JEM to detect. On the other hand, JEM increases when you are orthogonal (at a right angle) to the axis of blade rotation. Consequently for a fully exposed blade as in a propeller driven aircraft or helicopter, JEM increases with angle off the boresight axis of the prop/rotor.

This Page Blank

TRANSFORMS / WAVELETS

Transform Analysis

Signal processing using a transform analysis for calculations is a technique used to simplify or accelerate problem solution. For example, instead of dividing two large numbers, we might convert them to logarithms, subtract them, then look-up the anti-log to obtain the result. While this may seem a three-step process as opposed to a one-step division, consider that long-hand division of a four digit number by a three digit number, carried out to four places requires three divisions, 3-4 multiplication's, and three subtractions. Computers process additions or subtractions much faster than multiplications or divisions, so transforms are sought which provide the desired signal processing using these steps.

Fourier Transform

Other types of transforms include the Fourier transform, which is used to decompose or separate a waveform into a sum of sinusoids of different frequencies. It transforms our view of a signal from time based to frequency based. Figure 1 depicts how a square wave is formed by summing certain particular sine waves. The waveform must be continuous, periodic, and almost everywhere differentiable. The Fourier transform of a sequence of rectangular pulses is a series of sinusoids. The envelope of the amplitude of the coefficients of this series is a waveform with a Sin X/X shape. For the special case of a single pulse, the Fourier series has an infinite series of sinusoids that are present for the duration of the pulse.

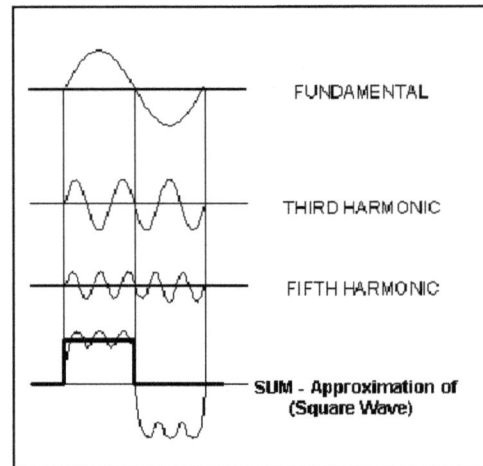

Figure 1. Harmonics

Digital Sampling of Waveforms

In order to process a signal digitally, we need to sample the signal frequently enough to create a complete "picture" of the signal. The discrete Fourier transform (DFT) may be used in this regard. Samples are taken at uniform time intervals as shown in Figure 2 and processed.

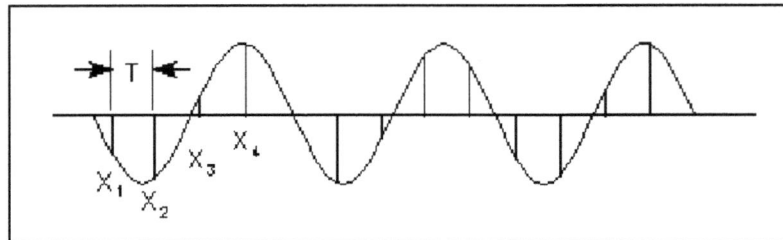

Figure 2. Waveform Sampling

If the digital information is multiplied by the Fourier coefficients, a digital filter is created as shown Figure 3. If the sum of the resultant components is zero, the filter has ignored (notched out) that frequency sample. If the sum is a relatively large number, the filter has passed the signal. With the single sinusoid shown, there should be only one resultant. (Note that being "zero" and relatively large may just mean below or above the filter's cutoff threshold)

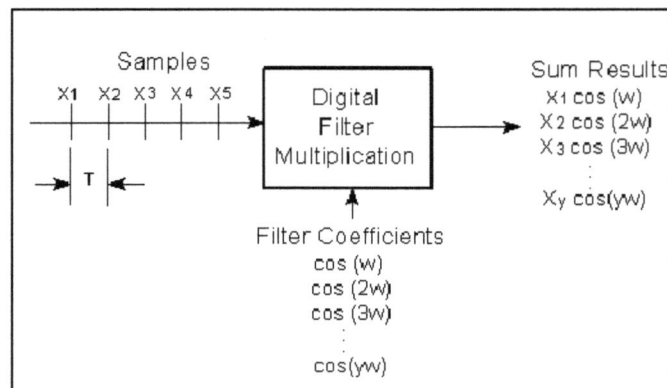

Figure 3. Digital Filtering

Figure 4 depicts the process pictorially: The vectors in the figure just happen to be pointing in a cardinal direction because the strobe frequencies are all multiples of the vector (phasor) rotation rate, but that is not normally the case. Usually the vectors will point in a number of different directions, with a resultant in some direction other than straight up.

In addition, sampling normally has to taken at or above twice the rate of interest (also known as the Nyquist rate),

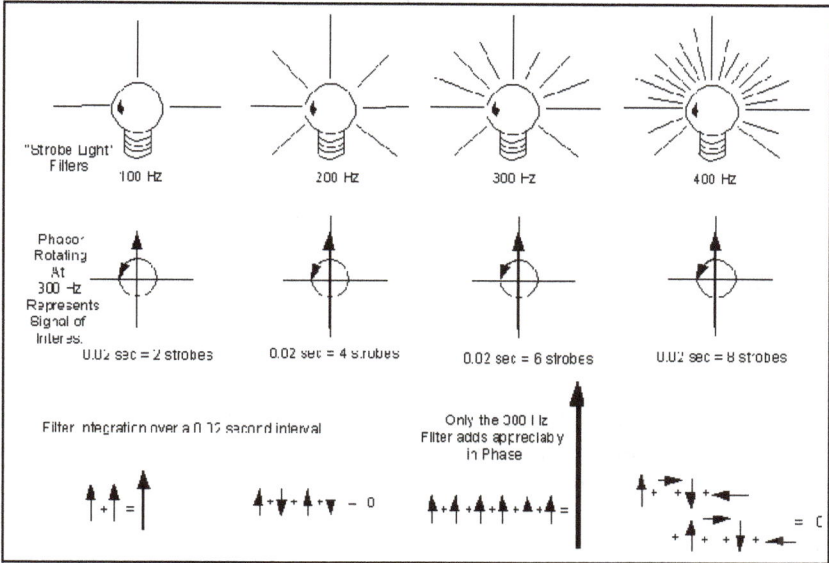

Figure 4. Phasor Representation

otherwise ambiguous results may be obtained. Figure 4 is under-sampled (for clarity) and consequently does not depict typical filtering.

Fast Fourier Transforms

One problem with this type of processing is the large number of additions, subtractions, and multiplications which are required to reconstruct the output waveform. The Fast Fourier transform (FFT) was developed to reduce this problem. It recognizes that because the filter coefficients are sine and cosine waves, they are symmetrical about 90, 180, 270, and 360 degrees. They also have a number of coefficients equal either to one or zero, and duplicate coefficients from filter to filter in a multibank arrangement. By waiting for all of the inputs for the bank to be received, adding together those inputs for which coefficients are the same before performing multiplications, and separately summing those combinations of inputs and products which are common to more than one filter, the required amount of computing may be cut drastically.

- The number of computations for a DFT is on the order of N squared.

- The number of computations for a FFT when N is a power of two is on the order of $N \log_2 N$.

For example, in an eight filter bank, a DFT would require 512 computations, while an FFT would only require 56, significantly speeding up processing time.

Windowed Fourier Transform

The Fourier transform is continuous, so a windowed Fourier transform (WFT) is used to analyze non-periodic

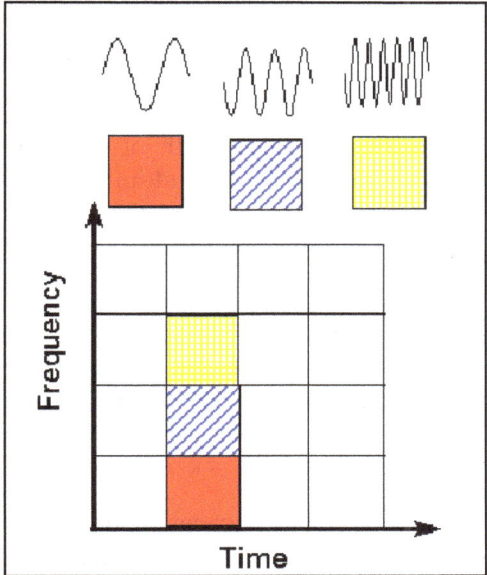

Figure 5. Windowed Fourier Transform

signals as shown in Figure 5. With the WFT, the signal is divided into sections (one such section is shown in Figure 5) and each section is analyzed for frequency content. If the signal has sharp transitions, the input data is windowed so that the sections converge to zero at the endpoints. Because a single window is used for all frequencies in the WFT, the resolution of the analysis is the same (equally spaced) at all locations in the time-frequency domain.

The FFT works well for signals with smooth or uniform frequencies, but it has been found that other transforms work better with signals having pulse type characteristics, time-varying (non-stationary) frequencies, or odd shapes.

The FFT also does not distinguish sequence or timing information. For example, if a signal has two frequencies (a high followed by a low or vice versa), the Fourier transform only reveals the frequencies and relative amplitude, not the order in which they occurred. So Fourier analysis works well with stationary, continuous, periodic, differentiable signals, but other methods are needed to deal with non-periodic or non-stationary signals.

Wavelet Transform

The Wavelet transform has been evolving for some time. Mathematicians theorized its use in the early 1900's. While the Fourier transform deals with transforming the time domain components to frequency domain and frequency analysis, the wavelet transform deals with scale analysis, that is, by creating mathematical structures that provide varying time/frequency/amplitude slices for analysis. This transform is a portion (one or a few cycles) of a complete waveform, hence the term wavelet.

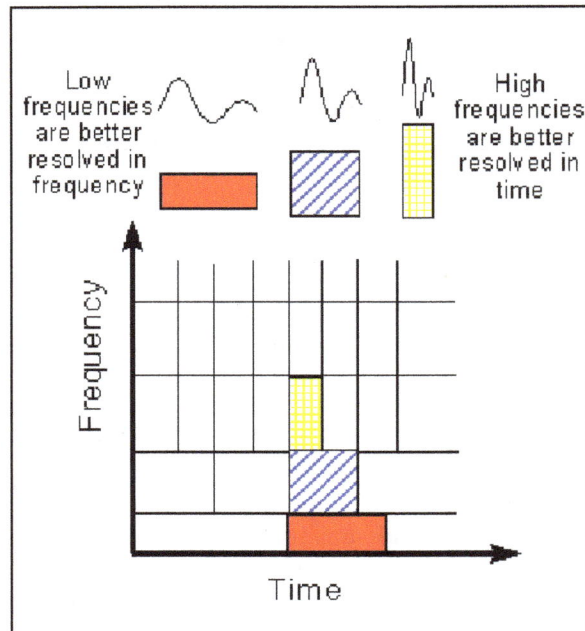

Figure 6. Wavelet Transform

The wavelet transform has the ability to identify frequency (or scale) components, simultaneously with their location(s) in time. Additionally, computations are directly proportional to the length of the input signal. They require only N multiplications (times a small constant) to convert the waveform. For the previous eight filter bank example, this would be about twenty calculations, vice 56 for the FFT.

In wavelet analysis, the scale that one uses in looking at data plays a special role. Wavelet algorithms process data at different scales or resolutions. If we look at a signal with a large "window," we would notice gross features. Similarly, if we look at a signal with a small "window," we would notice small discontinuities as shown in Figure 6. The result in wavelet analysis is to "see the forest and the trees." A way to achieve this is to have short high-frequency fine scale functions and long low-frequency ones. This approach is known as multi-resolution analysis.

For many decades, scientists have wanted more appropriate functions than the sines and cosines (base functions) which comprise Fourier analysis, to approximate choppy signals. (Although Walsh transforms work if the waveform is periodic and stationary). By their definition, sine and cosine functions are non-local (and stretch out to infinity), and therefore do a very poor job in approximating sharp spikes.

But with wavelet analysis, we can use approximating functions that are contained neatly in finite (time/frequency) domains. Wavelets are well-suited for approximating data with sharp discontinuities.

The wavelet analysis procedure is to adopt a wavelet prototype function, called an "analyzing wavelet" or "mother wavelet." Temporal analysis is performed with a contracted, high-frequency version of the prototype wavelet, while frequency analysis is performed with a dilated, low-frequency version of the prototype wavelet. Because the original signal or function can be represented in terms of a wavelet expansion (using coefficients in a linear combination of the wavelet functions), data operations can be performed using just the corresponding wavelet coefficients as shown in Figure 7.

If one further chooses the best wavelets adapted to the data, or truncates the coefficients below some given threshold, the data is sparsely represented. This "sparse coding" makes wavelets an excellent tool in the field of data compression. For instance, the FBI uses wavelet coding to store fingerprints. Hence, the concept of wavelets is to look at a signal at various scales and analyze it with various resolutions.

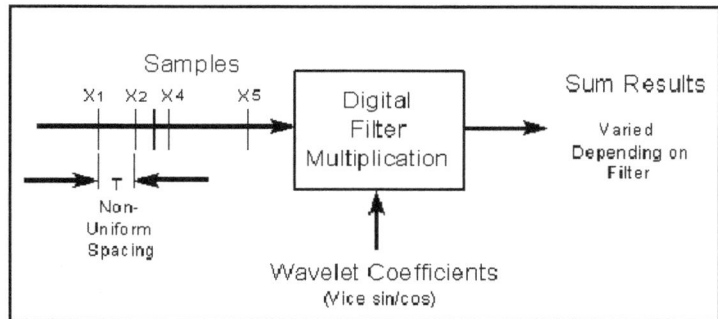

Figure 7. Wavelet Filtering

Analyzing Wavelet Functions

Fourier transforms deal with just two basis functions (sine and cosine), while there are an infinite number of wavelet basis functions. The freedom of the analyzing wavelet is a major difference between the two types of analyses and is important in determining the results of the analysis. The "wrong" wavelet may be no better (or even far worse than) than the Fourier analysis. A successful application presupposes some expertise on the part of the user. Some prior knowledge about the signal must generally be known in order to select the most suitable distribution and adapt the parameters to the

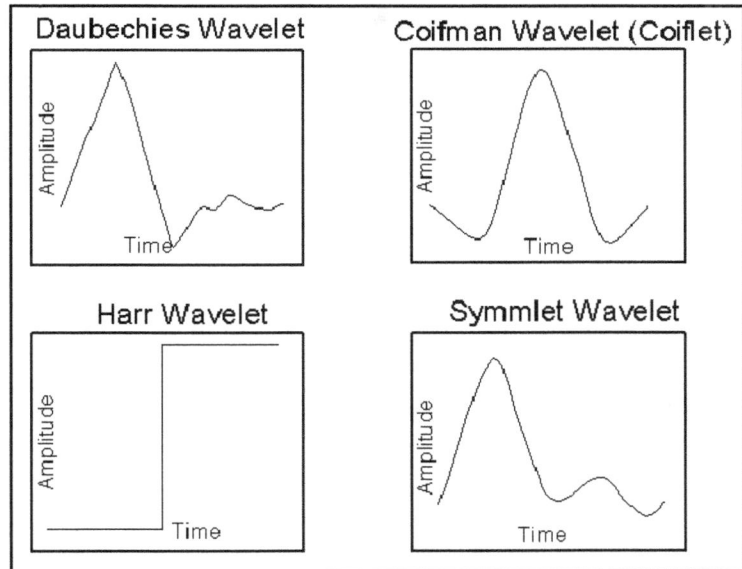

Figure 8. Sample Wavelet Functions

signal. Some of the more common ones are shown in Figure 8. There are several wavelets in each family, and they may look different than those shown. Somewhat longer in duration than these functions, but significantly shorter than infinite sinusoids is the cosine packet shown in Figure 9.

Wavelet Comparison With Fourier Analysis

While a typical Fourier transform provides frequency content information for samples within a given time interval, a perfect wavelet transform records the start of one frequency (or event), then the start of a second event, with amplitude added to or subtracted from, the base event.

Example 1.

Wavelets are especially useful in analyzing transients or time-varying signals. The input signal shown in Figure 9 consists of a sinusoid whose frequency changes in stepped increments over time. The power of the spectrum is also shown. Classical Fourier analysis will resolve the frequencies but cannot provide any information about the times at which each occurs. Wavelets provide an efficient means of analyzing the input signal so that frequencies and the times at which they occur can be resolved. Wavelets have finite duration and must

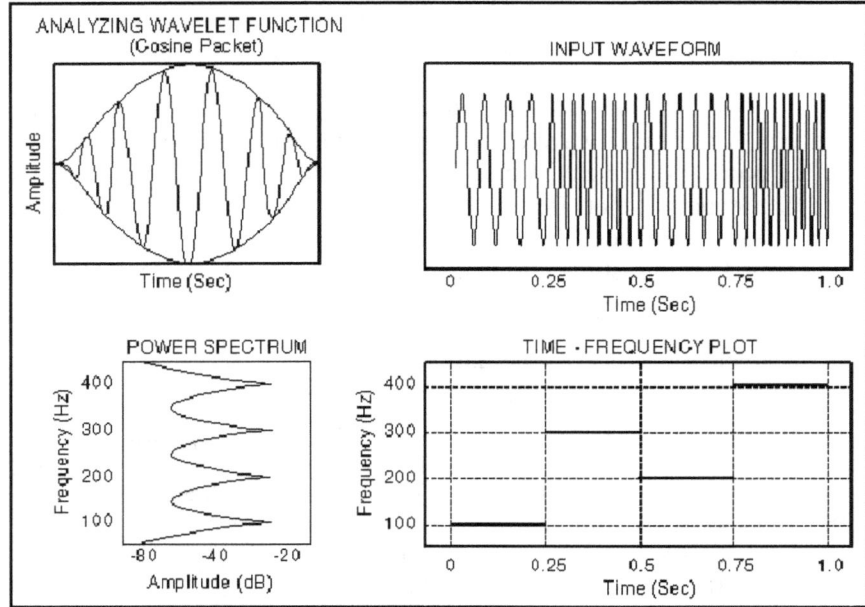

Figure 9. Sample Wavelet Analysis

also satisfy additional properties beyond those normally associated with standard windows used with Fourier analysis. The result after the wavelet transform is applied is the plot shown in the lower right. The wavelet analysis correctly resolves each of the frequencies and the time when it occurs. A series of wavelets is used in example 2.

Example 2. Figure 10 shows the input of a clean signal, and one with noise. It also shows the output of a number of "filters" with each signal. A 6 dB S/N improvement can be seen from the d4 output. (Recall from Section 4.3 that 6 dB corresponds to doubling of detection range.) In the filter cascade, the HPFs and LPFs are the same at each level. The wavelet shape is related to the HPF and LPF in that it is the "impulse response" of an infinite

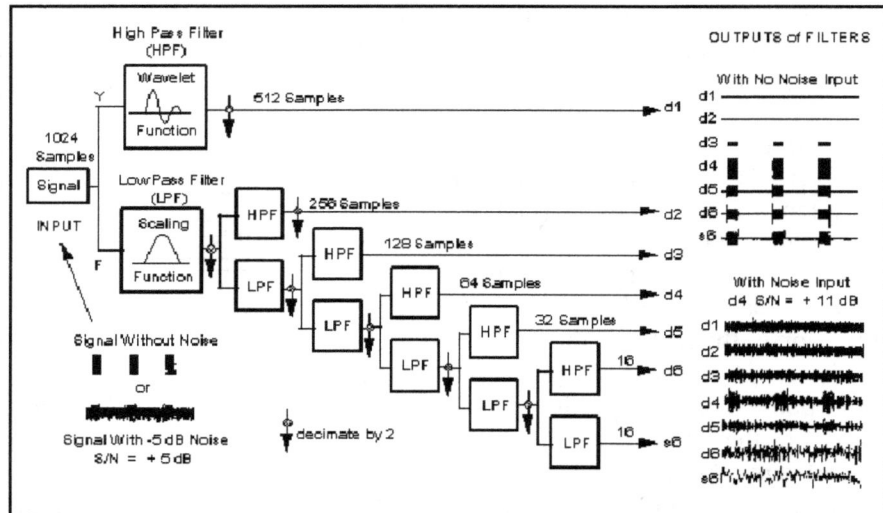

Figure 10. Example 2 Analysis Wavelet

cascade of the HPFs and LPFs. Different wavelets have different HPFs and LPFs. As a result of decimating by 2, the number of output samples equals the number of input samples.

Wavelet Applications   Some fields that are making use of wavelets are: astronomy, acoustics, nuclear engineering, signal and image processing (including fingerprinting), neurophysiology, music, magnetic resonance imaging, speech discrimination, optics, fractals, turbulence, earthquake-prediction, radar, human vision, and pure mathematics applications.

See October 1996 IEEE Spectrum article entitled "Wavelet Analysis", by Bruce, Donoho, and Gao.

ANTENNAS

ANTENNA INTRODUCTION / BASICS

Rules of Thumb:

1. The power gain of an antenna with losses, excluding input impedance mismatch, is given by:

$$G \approx \frac{4\pi\eta A}{\lambda^2} \quad Where \quad \begin{array}{l} \eta = Efficiency \\ A = Physical\ aperture\ area \\ \lambda = wavelength \end{array}$$

Where $BW_{\theta\ and\ \phi}$ *are the elev & az beamwidths in degrees.*

another is :

$$G = \frac{X\ \eta}{BW_\phi\ BW_\theta}$$

For approximating an antenna pattern with :
(1) A rectangle; $X = 41253, \eta_{typical} = 0.7$
(2) An ellipsoid; $X = 52525, \eta_{typical} = 0.55$

2. Directive gain of rectangular X-Band Aperture
   $G = 1.4\ LW$ Where: Length (L) and Width (W) are in cm

3. Power gain of Circular X-Band Aperture
   $G = d^2\eta$ Where: d = antenna diameter in cm
                                      η = aperture efficiency

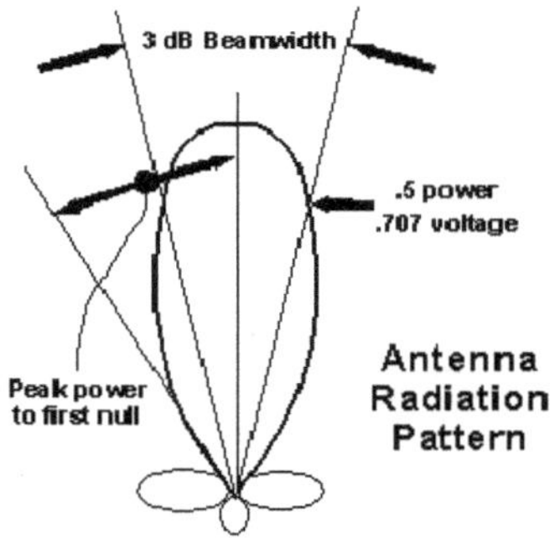

4. Directive gain of an imaginary isotropic antenna radiating in a uniform spherical pattern is one (0 dB).

5. Antenna with a 20 degree beamwidth has approximately a 20 dB directive gain.

Antenna Radiation Pattern

6. 3 dB beamwidth is approximately equal to the angle from the peak of the power to the first null (see figure at right).

7. Parabolic Antenna Beamwidth: $$BW = \frac{70\lambda}{d}$$

Where: BW = antenna beamwidth; λ = wavelength;   d = antenna diameter.

The antenna equations which follow relate to Figure 1 as a typical antenna. In Figure 1, BW_ϕ is the azimuth beamwidth and BW_θ is the elevation beamwidth. Beamwidth is normally measured at the half-power or -3 dB point of the main lobe unless otherwise specified. See Glossary.

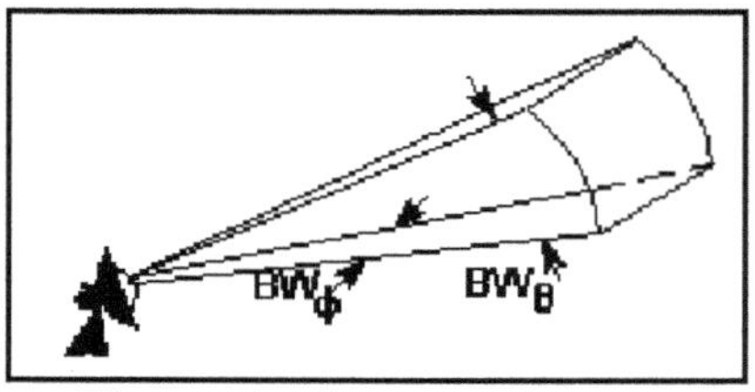

The gain or directivity of an antenna is the ratio of the radiation intensity in a given direction to the radiation intensity averaged over all directions.

Figure 1. Antenna Aperture

Quite often directivity and gain are used interchangeably and it sometimes leads to overly optimistic antenna performance estimations. The difference is that directivity is based solely on antenna pattern shape estimation where antenna losses such as dielectric, ohmic resistance, and polarization mismatch are neglected. If these losses are included in the antenna gain calculations, the antenna gain is then referred to as the power gain. Moreover, if additional impedance mismatch or VSWR losses are included in the antenna system gain estimation, the antenna gain calculations are then referred to as the realized gain. However, using directive gain (or directivity) calculations is very convenient in practice for a first order idealized antenna performance estimation.

Normalizing a radiation pattern by the integrated total power yields the directivity of the antenna. This concept in shown in equation form by:

$$D(\theta,\phi) = 10 \, Log \left[\frac{4\pi P(\theta,\phi)}{\iint P_{in}(\theta,\phi) \, Sin\,\theta \, d\theta \, d\phi} \right] \quad \begin{array}{l} 0 < \phi \leq 360° \\ 0 < \theta \leq 180° \end{array} \qquad [1]$$

Where $D(\theta,\phi)$ is the directivity in dB, and the radiation pattern power in a specific direction is $P_d(\theta,\phi)$, which is normalized by the total integrated radiated power. Another important concept is that when the angle in which the radiation is constrained is reduced, the directive gain goes up. For example, using an isotropic radiating source, the gain would be 0 dB by definition (Figure 2(a)) and the power density (P_d) at any given point would be the power in (P_{in}) divided by the surface area of the imaginary sphere at a distance R from the source. If the spacial angle was decreased to one hemisphere (Figure 2(b)), the power radiated, P_{in}, would be the same but the

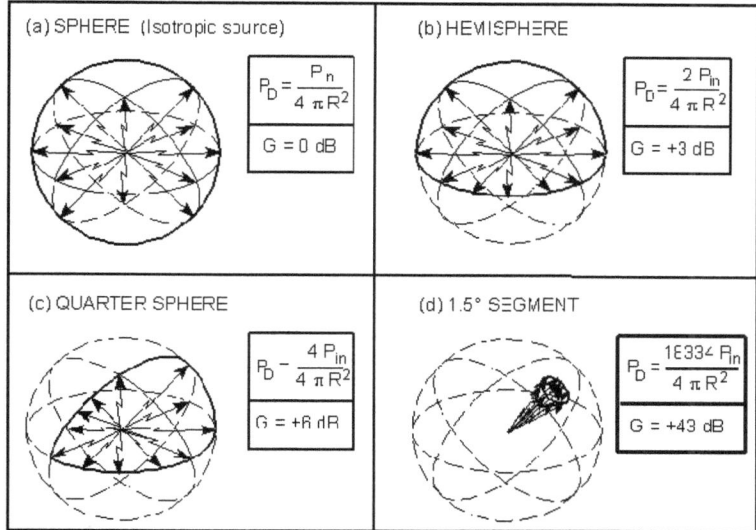

(a) SPHERE (Isotropic source)
$$P_D = \frac{P_n}{4\pi R^2}$$
$$G = 0 \, dB$$

(b) HEMISPHERE
$$P_D = \frac{2\,P_{in}}{4\pi R^2}$$
$$G = +3 \, dB$$

(c) QUARTER SPHERE
$$P_D = \frac{4\,P_{in}}{4\pi R^2}$$
$$G = +6 \, dB$$

(d) 1.5° SEGMENT
$$P_D = \frac{1E334\,P_{in}}{4\pi R^2}$$
$$G = +43 \, dB$$

Figure 2. Notional Representation of Directive Antenna Gain

area would be half as much, so the gain would double to 3 dB. Likewise if the angle is a quarter sphere, (Figure 2(c)), the gain would be 6 dB. Figure 2(d) shows a pencil beam. The gain is independent of actual power output and radius (distance) at which measurements are taken.

Real antennas are different, however, and do not have an ideal radiation distribution. Energy varies with angular displacement and losses occur due to sidelobes. However, if we can measure the pattern, and determine the beamwidth we can use two (or more) ideal antenna models to approximate a real antenna pattern as shown in Figure 3.

Assuming the antenna pattern is uniform, the gain is equal to the area of the isotropic sphere ($4\pi r^2$) divided by the sector (cross section) area.

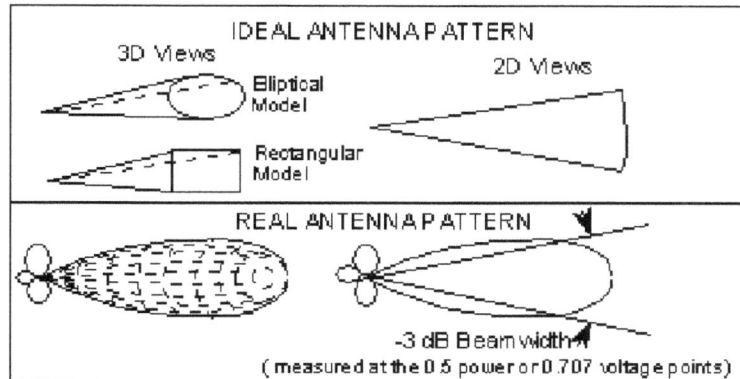

IDEAL ANTENNA PATTERN
3D Views   2D Views
Elliptical Model
Rectangular Model

REAL ANTENNA PATTERN
-3 dB Beamwidth
(measured at the 0.5 power or 0.707 voltage points)

Figure 3. Antenna Beamwidth

$$G = \frac{Area \ of \ Sphere}{Area \ of \ Antenna \ pattern} \qquad [2]$$

It can be shown that:

$$G \approx \frac{4\pi}{BW_{\phi \, az} \, BW_{\theta \, el}} \quad or \quad \frac{4\pi}{\phi\theta \ (radians)} \quad where: \begin{array}{l} BW_{\phi\,az} = Azmith \ beamwidth \ in \ radians \\ BW_{\theta\,el} = Elevation \ beamwidth \ in \ radians \end{array} \qquad [3]$$

From this point, two different models are presented:

(1) Approximating an antenna pattern using an elliptical area, and
(2) Approximating an antenna pattern using a rectangular area.

Approximating the antenna pattern as an **elliptical** area:

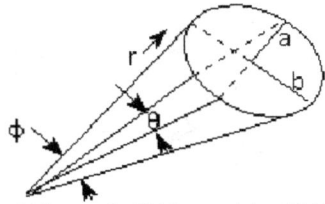

Area of ellipse $= \pi\, a\, b = \pi[\ (r \sin\theta)/2\ ][\ (r \sin\varphi)/2\ ] = (\pi\, r^2 \sin\theta \sin\varphi)/4$

$$G = \frac{Area\ of\ Sphere}{Area\ of\ Antenna\ pattern} = (4\,\pi\,r^2)\left(\frac{4}{\pi\,r^2 \sin\theta \sin\phi}\right) = \frac{16}{\sin\theta \sin\phi}$$

Where $\theta = BW_\theta$, and $\phi = BW_\phi$

For small angles, $\sin\varphi = \varphi$ in radians, so:

$$G = \frac{16}{\sin\phi \sin\theta} \approx \frac{16}{\phi\,\theta\,(radians)} = \frac{16}{\phi\,\theta}\left(\frac{360°}{2\,\pi}\frac{360°}{2\,\pi}\right) = \frac{52525}{\phi\,\theta\,(degrees)} \ or\ \frac{52525}{BW_\phi\,BW_\theta\,(degrees)} \qquad [4]$$

The second term in the equation above is very close to equation [3].

For a very directional radar dish with a beamwidth of 1° and an average efficiency of 55%:

Ideally: G = 52525, or in dB form: 10 log G =10 log 52525 = 47.2 dB

With efficiency taken into account, G = 0.55(52525) = 28888, or in log form: 10 log G = 44.6 dB

Approximating the antenna pattern as a **rectangular** area:

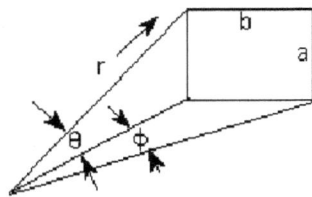

$a = r \sin\theta$, $b = r \sin\varphi$, area $= ab = r^2 \sin\theta \sin\varphi$

$$G = \frac{Area\ of\ Sphere}{Area\ of\ Antenna\ pattern} = \frac{4\,\pi\,r^2}{r^2 \sin\theta \sin\phi} = \frac{4\,\pi}{\sin\theta \sin\phi}$$

Where $\theta = BW_\theta$, and $\phi = BW_\phi$

For small angles, $\sin\varphi = \varphi$ in radians, so:

$$G = \frac{4\,\pi}{\sin\phi \sin\theta} = \frac{4\,\pi}{\phi\,\theta\,(radians)} = \frac{4\,\pi}{\phi\,\theta}\left(\frac{360°}{2\,\pi}\frac{360°}{2\,\pi}\right) = \frac{41253}{\phi\,\theta\,(degrees)} \ or\ \frac{41253}{BW_\phi\,BW_\theta\,(degrees)} \qquad [5]$$

The second term in the equation above is identical to equation [3].

Converting to dB, $G_{max}(dB) = 10\ Log\left[\dfrac{41253}{BW_\phi\,BW_\theta}\right]$ with BW_ϕ and BW_θ in degrees $\qquad [6]$

For a very directional radar dish with a beamwidth of 1° and an average efficiency of 70%:
Ideally (in dB form): 10 log G =10 log 41253 = 46.2 dB.
With efficiency taken into account, G = 0.7(41253) = 28877, or in log form: 10 log G = 44.6 dB

Comparison between elliptical and rectangular areas for antenna pattern models:
By using the rectangular model there is a direct correlation between the development of gain in equation [5] and the ideal gain of equation [3]. The elliptical model has about one dB difference from the ideal calculation, but will yield the same real antenna gain when appropriate efficiencies are assumed.

The upper plot of Figure 4 shows the gain for an ideal antenna pattern using the elliptical model. The middle plot shows the gain for an ideal antenna using the rectangular model. The lower plot of Figure 4 shows the gain of a typical real antenna (rectangular model using an efficiency of 70% or elliptical model using an efficiency of 47%).

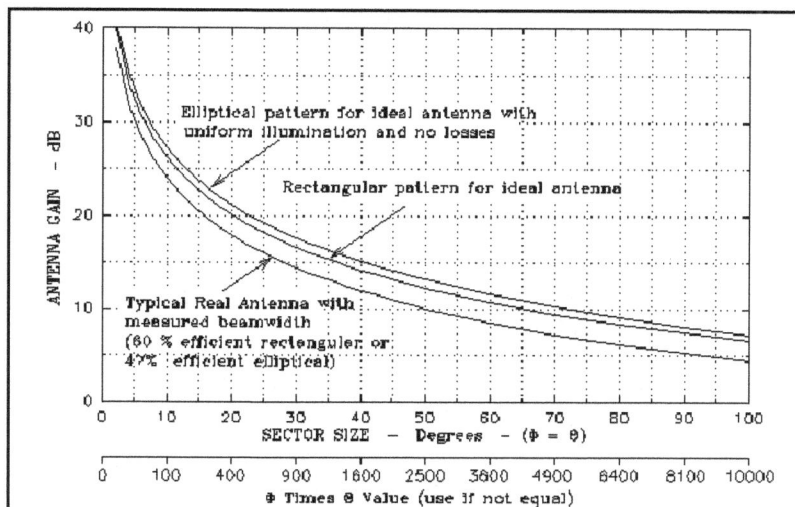

Figure 4. Antenna Sector Size vs, Gain

Gain as a function of λ:
When $\theta = 0$, each wave source in Figure 5 is in phase with one another and a maximum is produced in that direction.

Conversely, nulls to either side of the main lobe will occur when the waves radiating from the antenna cancel each other. The first null occurs when there is a phase difference of $\lambda/2$ in the wave fronts emanating from the aperture. To aid in visualizing what happens, consider each point in the antenna aperture, from A to C in Figure 5, as a point source of a spherical wave front. If viewed from infinity, the electromagnetic waves from each point interfere with each other, and when, for a particular direction, θ in Figure 5, each wave source has a corresponding point that is one-half wavelength out of phase, a null is produced in that direction due to destructive interference.

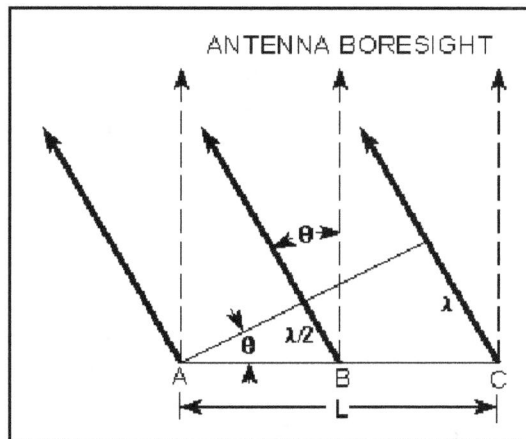

Figure 5. Directional Gain vs. Wavelength

In Figure 5, the wave emanating from point A is out of phase with the wave from point B by one-half of a wavelength. Hence, they cancel. Similarly, a point just to the right of point A cancels with a point just to the right of point B, and so on across the entire aperture. Therefore, the first null in the radiation pattern is given by:

$\text{Sin } \theta = \lambda/L$ and, in radians, $\theta = \lambda/L$ (for small angles)   [7]

3-1.4

As the angle off boresight is increased beyond the first null, the intensity of the radiation pattern rises then falls, until the second null is reached. This corresponds to a phase difference of two wavelengths between the left and right edges of the aperture. In this case, the argument proceeds as before, except now the aperture is divided into four segments (point A canceling with a point halfway between A and B, and so on).

The angle θ is the angle from the center (maximum) of the radiation pattern to the first null. The null-to-null beam width is 2θ. Generally, we are interested in the half-power (3 dB) beamwidth. It turns out that this beamwidth is approximately one-half of the null-to-null beamwidth, so that:

$$BW_{3\,dB} \approx (\tfrac{1}{2})(2\theta) = \lambda/L \qquad\qquad\qquad\qquad\qquad\qquad [8]$$

Therefore, beamwidth is a function of the antenna dimension "L" and the wavelength of the signal. It can be expressed as follows: Note: for circular antennas, L in the following equations = diameter

$$Bw_{\varphi(az)} = \lambda/L_{Az\ eff} \text{ and } BW_{\theta(el)} = \lambda/L_{El\ eff} \qquad\qquad\qquad\qquad [9]$$

Substituting the two variations of equation [9] into equation [3] and since $L_{Az\ eff}$ times $L_{El\ eff}$ = A_e (effective capture area of the antenna), we have:

$$G \approx \frac{4\pi}{BW_\phi\, BW_\theta\,(radians)} = \frac{4\pi\, L_{az}\, L_{el}}{\lambda^2} = \frac{4\pi\, A_e}{\lambda^2} \qquad\qquad\qquad [10]$$

Note: Equation is approximate since aperture efficiency isn't included as is done later in equation [12].

The efficiency (discussed later) will reduce the gain by a factor of 30-50%, i.e. real gain = .5 to .7 times theoretical gain.

Unity Gain Antenna.
If a square antenna is visualized and G=1, $A_e = \lambda^2 / 4\pi$. When a dimension is greater than 0.28 λ (~¼λ) it is known as an electrically large antenna, and the antenna will have a gain greater than one (positive gain when expressed in dB). Conversely, when the dimension is less than 0.28 λ (~¼λ)(an electrically small antenna), the gain will be less than one (negative gain when expressed in dB). Therefore, a unity gain antenna can be approximated by an aperture that is ¼λ by ¼λ.

Beamwidth as a Function of Aperture Length
It can be seen from Figure 5, that the wider the antenna aperture (L), the narrower the beamwidth will be for the same λ. Therefore, if you have a rectangular shaped horn antenna, the radiation pattern from the wider side will be narrower than the radiation pattern from the narrow side.

APERTURE EFFICIENCY, η

The Antenna Efficiency, η, is a factor which includes all reductions from the maximum gain. η can be expressed as a percentage, or in dB. Several types of "loss" must be accounted for in the efficiency, η:

    (1)   Illumination efficiency which is the ratio of the directivity of the antenna to the directivity of a uniformly illuminated antenna of the same aperture size,

    (2)   Phase error loss or loss due to the fact that the aperture is not a uniform phase surface,

    (3)   Spillover loss (Reflector Antennas) which reflects the energy spilling beyond the edge of the reflector into the back lobes of the antenna,

    (4)   Mismatch (VSWR) loss, derived from the reflection at the feed port due to impedance mismatch (especially important for low frequency antennas), and

    (5)   RF losses between the antenna and the antenna feed port or measurement point.

The aperture efficiency, η_a, is also known as the illumination factor, and includes items (1) and (2) above; it does not result in any loss of power radiated but affects the gain and pattern. It is nominally 0.6-0.8 for a planer array and 0.13 to 0.8 with a nominal value of 0.5 for a parabolic antenna, however η can vary significantly. Other antennas include the spiral (.002-.5), the horn (.002-.8), the double ridge horn (.005-.93), and the conical log spiral (.0017-1.0).

Items (3), (4), and (5) above represent RF or power losses which can be measured. The efficiency varies and generally gets lower with wider bandwidths. Also note that the gain equation is optimized for small angles - see derivation of wavelength portion of equation [7]. This explains why efficiency also gets lower for wider beamwidth antennas.

EFFECTIVE CAPTURE AREA
Effective capture area (A_e) is the product of the physical aperture area (A) and the aperture efficiency (η) or:

$$A_e = \eta\, A = \frac{\lambda^2 G}{4\pi} \hspace{3cm} \textbf{[11]}$$

GAIN AS A FUNCTION OF APERTURE EFFICIENCY
The Gain of an antenna with losses is given by:

$$G = \frac{4\pi\eta A}{\lambda^2} \quad \begin{aligned} &\eta = Aperture\ Efficiency \\ Where\ &A = Physical\ aperture\ area \\ &\lambda = wavelength \end{aligned} \hspace{2cm} \textbf{[12]}$$

Note that the gain is proportional to the aperture area and inversely proportional to the square of the wavelength. For example, if the frequency is doubled, (half the wavelength), the aperture could be decreased four times to maintain the same gain.

BEAM FACTOR
Antenna size and beamwidth are also related by the beam factor defined by:
Beam Factor = (D/λ)·(Beamwidth)   where D = antenna dimension in wavelengths.

The beam factor is approximately invariant with antenna size, but does vary with type of antenna aperture illumination or taper. The beam factor typically varies from 50-70°.

APERTURE ILLUMINATION (TAPER)
The aperture illumination or illumination taper is the variation in amplitude across the aperture. This variation can have several effects on the antenna performance:
    (1)   reduction in gain,
    (2)   reduced (lower) sidelobes in most cases, and
    (3)   increased antenna beamwidth and beam factor.

Tapered illumination occurs naturally in reflector antennas due to the feed radiation pattern and the variation in distance from the feed to different portions of the reflector. Phase can also vary across the aperture which also affects the gain, efficiency, and beamwidth.

CIRCULAR ANTENNA GAIN
Solving equation [12] in dB, for a circular antenna with area $\pi D^2/4$, we have:
10 Log G = 20 Log (D/λ) + 10 Log (η) + 9.94 dB ; where D = diameter   **[13]**

This data is depicted in the nomograph of Figure 6. For example, a six foot diameter antenna operating at 9 GHz would have approximately 44.7 dB of gain as shown by the dashed line drawn on Figure 6. This gain is for an antenna 100% efficient, and would be 41.7 dB for a typical parabolic antenna (50% efficient).

FREQUENCY (Gigahertz)

WAVELENGTH (Centimeters)

NOTE: The Gain below assumes $\eta = 1.0$
If $\eta = 0.79$, subtract 1 dB
If $\eta = 0.5$, subtract 3 dB (typical parabolic antenna)
If $\eta = 0.25$, subtract 6 dB

$$\text{GAIN (dB)} = 20 \text{ Log } D/\lambda + 10 \text{Log } \eta + 9.94 \text{ dB}$$

REFLECTOR DIAMETER (Meters)

REFLECTOR DIAMETER (Feet)

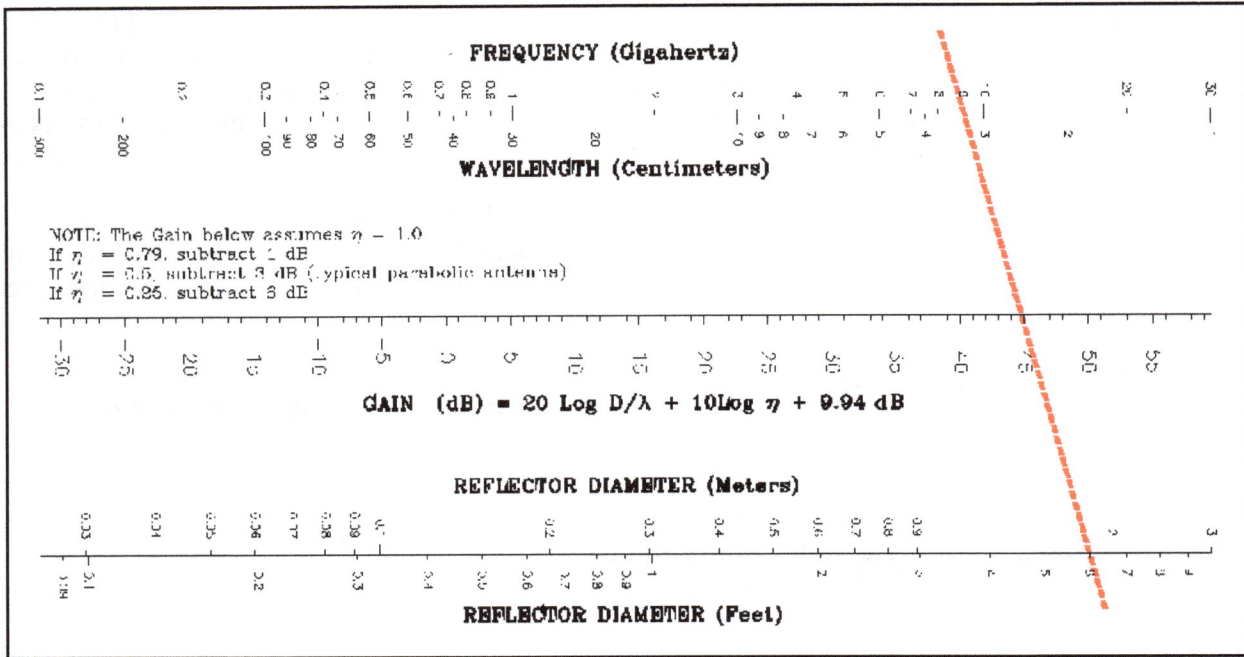

Figure 6. Antenna Gain Nomograph

An example of a typical antenna (with losses) showing the variation of gain with frequency is depicted in Figure 7, and the variation of gain with antenna diameter in Figure 8. The circle on the curves in Figure 7 and 8 correspond to the Figure 6 example and yields 42 dB of gain for the 6 ft dish at 9 GHz.

Example Problem: If the two antennas in the drawing are "welded" together, how much power will be measured at point A?   (Line loss $L_1 = L_2 = 0.5$, and $10 \log L_1$ or $L_2 = 3$ dB)

Multiple choice:
A. 16 dBm   b. 28 dBm   c. 4 dBm   d. 10 dBm   e. < 4 dBm

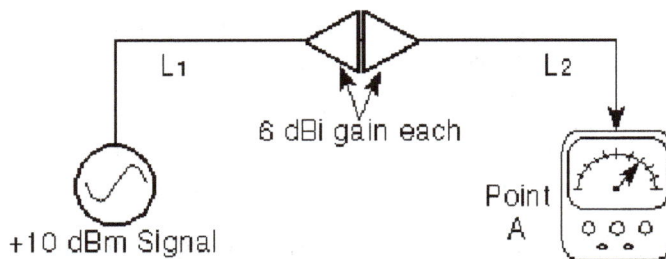

Answer:

L₁   6 dBi gain each   L₂

+10 dBm Signal   Point A

The antennas do not act as they normally would since the antennas are operating in the near field. They act as inefficient coupling devices resulting in some loss of signal. In addition, since there are no active components, you cannot end up with more power than you started with. The correct answer is "e. < 4 dBm."

10 dBm - 3 dB - small loss -3 dB = 4 dBm - small loss

If the antennas were separated by 5 ft and were in the far field, the antenna gain could be used with space loss formulas to calculate (at 5 GHz): 10 dBm - 3 dB + 6 dB - 50 dB (space loss) + 6 dB -3 dB = -34 dBm (a much smaller signal).

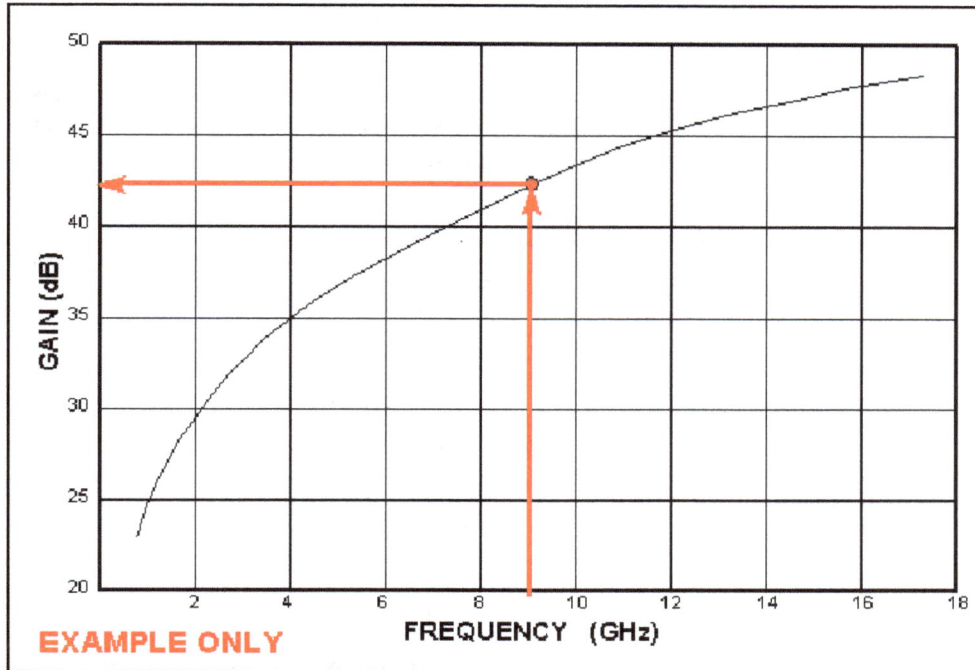

Figure 7. Gain of a Typical 6 Foot Dish Antenna (With Losses)

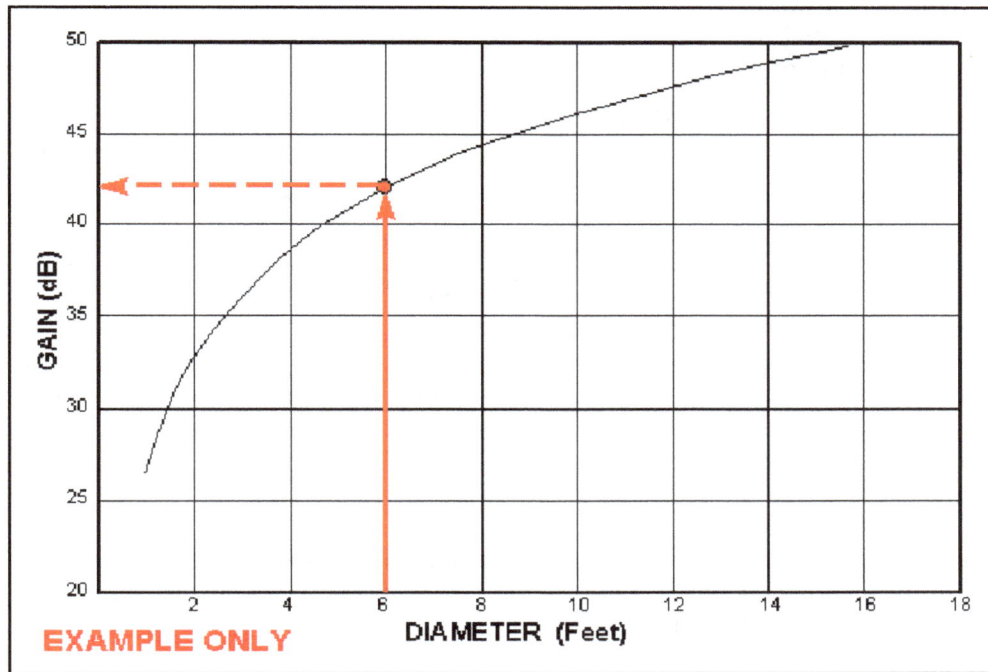

Figure 8. Gain of a Typical Dish at 9 GHz (With Losses)

POLARIZATION

Table 1 shows the **theoretical** ratio of power transmitted between antennas of different polarization. These ratios are seldom fully achieved due to effects such as reflection, refraction, and other wave interactions, so some practical ratios are also included.

Table 1. Polarization Loss for Various Antenna Combinations

Transmit Antenna Polarization	Receive Antenna Polarization	Ratio of Power Received to Maximum Power					
		Theoretical		Practical Horn		Practical Spiral	
		Ratio in dB	as Ratio	Ratio in dB	as Ratio	Ratio in dB	as Ratio
Vertical	Vertical	0 dB	1	*	*	N/A	N/A
Vertical	Slant (45° or 135°)	-3 dB	½	*	*	N/A	N/A
Vertical	Horizontal	- ∞ dB	0	-20 dB	1/100	N/A	N/A
Vertical	Circular (right-hand or left-hand)	-3 dB	½	*	*	*	*
Horizontal	Horizontal	0 dB	1	*	*	N/A	N/A
Horizontal	Slant (45° or 135°)	-3 dB	½	*	*	N/A	N/A
Horizontal	Circular (right-hand or left-hand)	-3 dB	½	*	*	*	*
Circular (right-hand)	Circular (right-hand)	0 dB	1	*	*	*	*
Circular (right-hand)	Circular (left-hand)	- ∞ dB	0	-20 dB	1/100	-10 dB	1/10
Circular (right or left)	Slant (45° or 135°)	-3 dB	½	*	*	*	*

* Approximately the same as theoretical
Note: Switching transmit and receive antenna polarization will give the same results.

The polarization of an electromagnetic wave is defined as the orientation of the electric field vector. Recall that the electric field vector is perpendicular to both the direction of travel and the magnetic field vector. The polarization is described by the geometric figure traced by the electric field

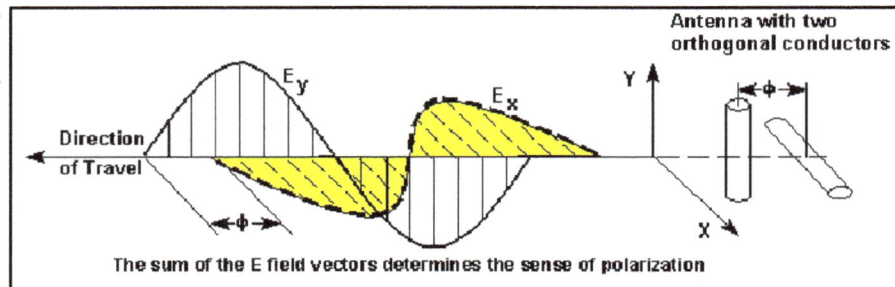

Figure1. Polarization Coordinates

vector upon a stationary plane perpendicular to the direction of propagation, as the wave travels through that plane. An electromagnetic wave is frequently composed of (or can be broken down into) two orthogonal components as shown in Figure 1. This may be due to the arrangement of power input leads to various points on a flat antenna, or due to an interaction of active elements in an array, or many other reasons.

The geometric figure traced by the sum of the electric field vectors over time is, in general, an ellipse as shown in Figure 2. Under certain conditions the ellipse may collapse into a straight line, in which case the polarization is called linear.

In the other extreme, when the two components are of equal magnitude and 90° out of phase, the ellipse will become circular as shown in Figure 3. Thus linear and circular polarization are the two special cases of elliptical polarization. Linear polarization may be further classified as being vertical, horizontal, or slant.

Figure 2 depicts plots of the E field vector while varying the relative amplitude and phase angle of its component parts.

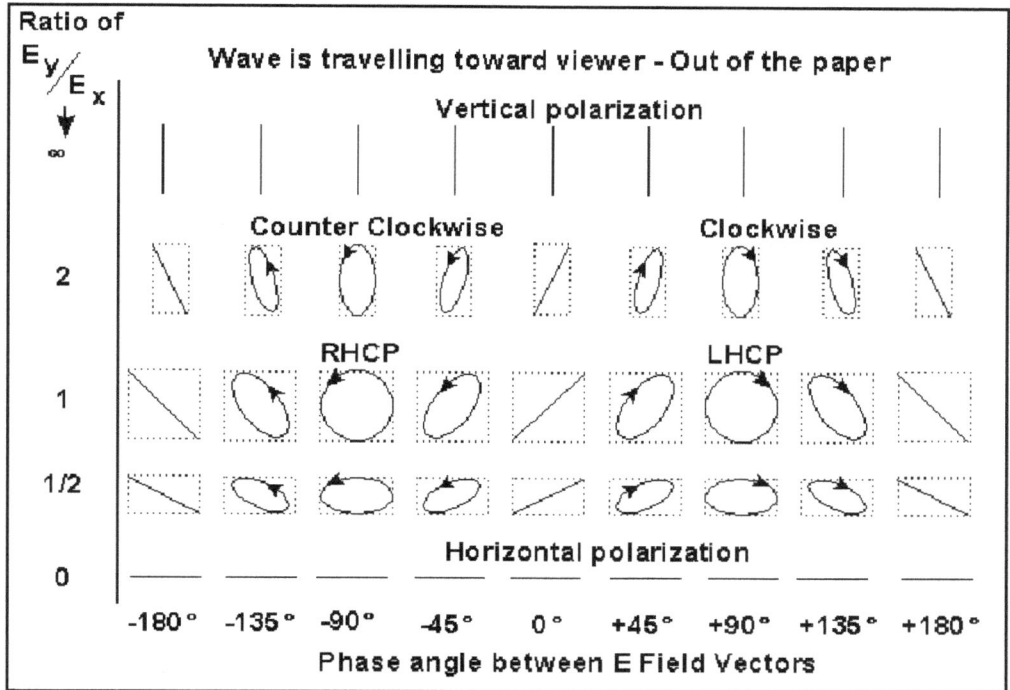

Figure 2. Polarization as a Function of E_y / E_x Ratio and Phase Angle
Adopted from J.D. Kraus, "Antennas", 2nd ed., Figure 2-37

For a linearly polarized antenna, the radiation pattern is taken both for a co-polarized and cross polarized response. The polarization quality is expressed by the ratio of these two responses. The ratio between the responses must typically be great (30 dB or greater) for an application such as cross-polarized jamming. For general applications, the ratio indicates system power loss due to polarization mismatch. For circularly polarized antennas, radiation patterns are usually taken with a rotating linearly polarized reference antenna. The reference antenna rotates many times while taking measurements around the azimuth of the antenna that is being tested. The resulting antenna pattern is the linear polarized gain with a cyclic ripple. The peak-to-peak value is the axial ratio, and represents the polarization quality for a circular polarized antenna. The typical RWR antenna has a maximum 3 dB axial ratio within 45° of boresight.

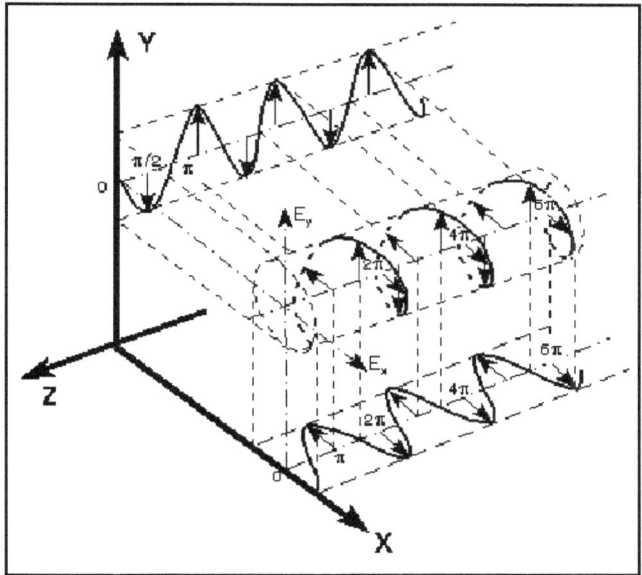

Figure 3. Circular Polarization – E Field

For any antenna with an aperture area, as the aperture is rotated, the viewed dimension along the axis remains constant, while the other viewed dimension decreases to zero at 90° rotation. The axial ratio of an antenna will get worse as the antenna is rotated off boresight because the field contribution from the axial component will remain fairly constant and the other orthogonal component will decrease with rotation.

The sense of antenna polarization is defined from a viewer positioned behind an antenna looking in the direction of propagation. The polarization is specified as a transmitting, not receiving antenna regardless of intended use.

We frequently use "hand rules" to describe the sense of polarization. The sense is defined by which hand would be used in order to point that thumb in the direction of propagation and point the fingers of the same hand in the direction of rotation of the E field vector. For example, referring to Figure 4, if your thumb is pointed in the direction of propagation and the rotation is counterclockwise looking in the direction of travel, then you have left hand circular polarization.

Optics people view an aperture from the front and therefore use the opposite reference.

The polarization of a linearly polarized horn antenna can be directly determined by the orientation of the feed probe, which is in the direction of the E-field.

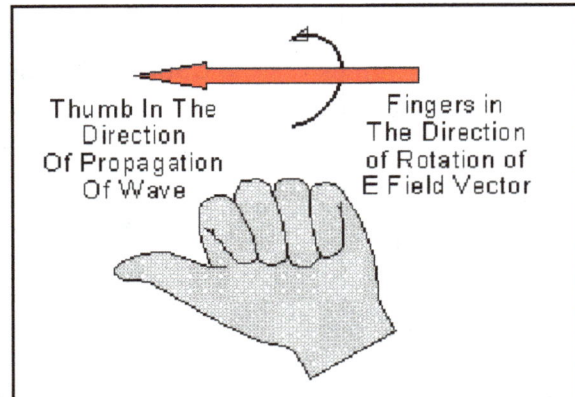

Figure 4. Left Hand Polarization

In general, a flat surface or sphere will reflect a linearly polarized wave with the same polarization as received. A horizontally polarized wave may get extended range because of water and land surface reflections, but signal cancellation will probably result in "holes" in coverage. Reflections will reverse the sense of circular polarization.

> If the desired antenna is used for receiving a direct transmission as shown in Figure 5 below, the same polarization sense (specified if transmitting) is required for maximum signal reception in this situation. Buy two right-hand or two left-hand circularly polarized antennas for this case. When you procure antennas, remember that the polarization is specified as if transmitting, regardless of intended use.

Wave propagation between two identical antennas is analogous to being able to thread a nut from one bolt to an identical opposite facing bolt.

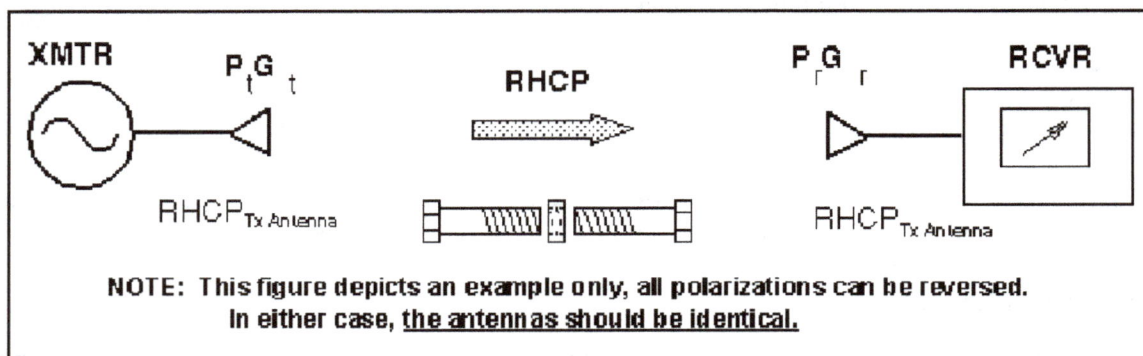

NOTE: This figure depicts an example only, all polarizations can be reversed. In either case, _the antennas should be identical._

Figure 5. Same Circular Polarization

If the desired antenna is used for a receiving a wave with a <u>single</u> or odd number of reflections, such as a bistatic radar where separate antennas are used for transmit and receive as shown in Figure 6, then opposite circularly polarized antennas would be used for maximum signal reception. In this case buy antennas of opposite polarization sense (one left hand and one right hand).

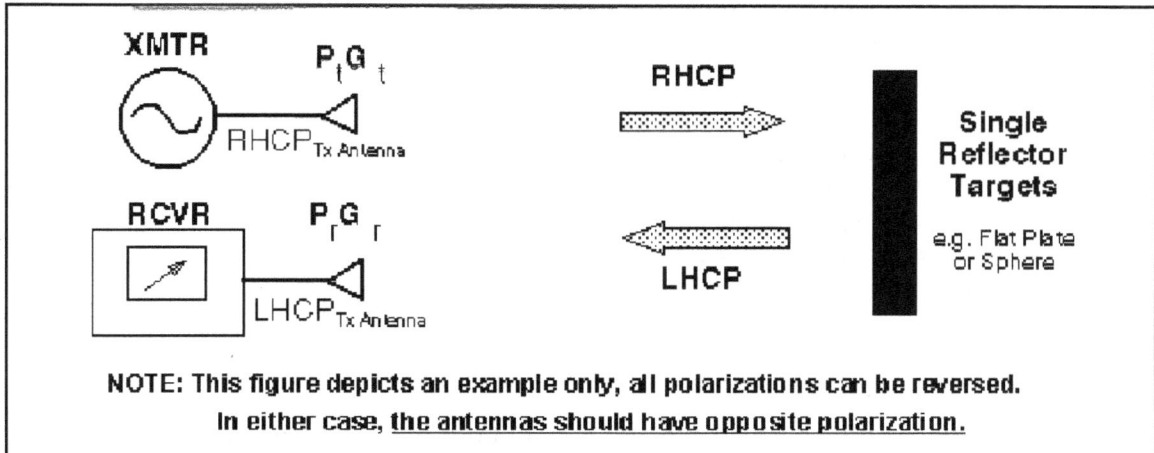

XMTR $P_t G_t$

$RHCP_{Tx\ Antenna}$

RCVR $P_r G_r$

$LHCP_{Tx\ Antenna}$

RHCP

LHCP

Single Reflector Targets

e.g. Flat Plate or Sphere

NOTE: This figure depicts an example only, all polarizations can be reversed.

In either case, <u>the antennas should have opposite polarization.</u>

Figure 6. Opposite Circular Polarization

In a corner reflector, waves reflect twice before returning to the receiver as shown in Figure 7, consequently they return with the same sense as they were transmitted. In this case (or any even number of reflections) buy antennas of the same polarization sense.

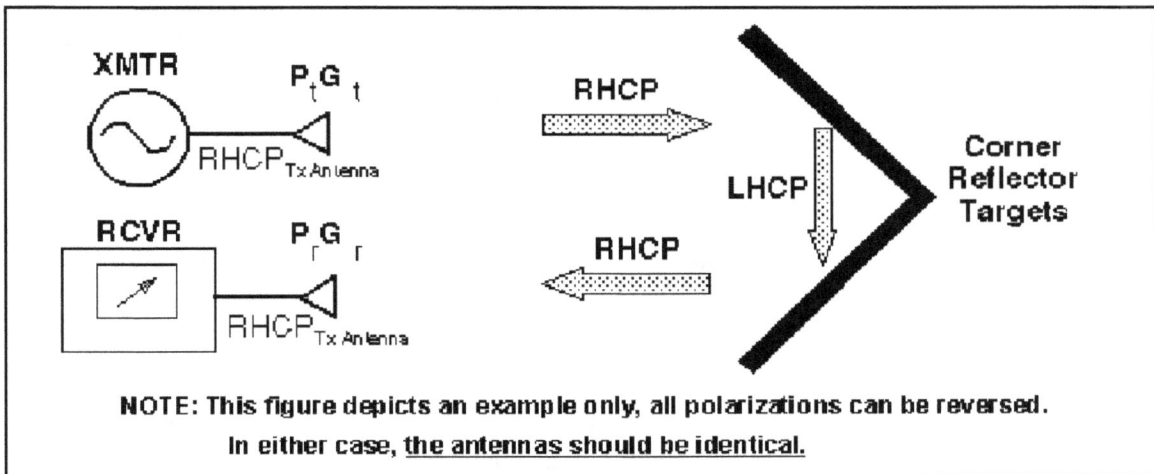

XMTR $P_t G_t$

$RHCP_{Tx\ Antenna}$

RCVR $P_r G_r$

$RHCP_{Tx\ Antenna}$

RHCP

RHCP

LHCP

Corner Reflector Targets

NOTE: This figure depicts an example only, all polarizations can be reversed.

In either case, <u>the antennas should be identical.</u>

Figure 7. Circular Polarization With Corner Reflector

An aircraft acts as both a corner reflector and a "normal" reflector so the return has mixed polarization. Most airborne radars use the same antenna for transmitting and receiving in order to receive the corner reflections and help exclude receipt of reflections from rain (single polarization reversal), however in doing so there is about a 5-9 dB loss from the ideal receiver case. It should be noted that the return from raindrops is attenuated by approximately 20 dB.

RADIATION PATTERNS

The radiation pattern is a graphical depiction of the relative field strength transmitted from or received by the antenna. Antenna radiation patterns are taken at one frequency, one polarization, and one plane cut. The patterns are usually presented in polar or rectilinear form with a dB strength scale. Patterns are normalized to the maximum graph value, 0 dB, and a directivity is given for the antenna. This means that if the side lobe level from the radiation pattern were down -13 dB, and the directivity of the antenna was 4 dB, then the sidelobe gain would be -9 dB.

Figures 1 to 14 on the pages following depict various antenna types and their associated characteristics. The patterns depicted are those which most closely match the purpose for which the given shape was intended. In other words, the radiation pattern can change dramatically depending upon frequency, and the wavelength to antenna characteristic length ratio. See section 3-4. Antennas are designed for a particular frequency. Usually the characteristic length is a multiple of $\lambda/2$ minus 2-15% depending on specific antenna characteristics.

The gain is assumed to mean directional gain of the antenna compared to an isotropic radiator transmitting to or receiving from all directions.

The half-power (-3 dB) beamwidth is a measure of the directivity of the antenna.

Polarization, which is the direction of the electric (not magnetic) field of an antenna is another important antenna characteristic. This may be a consideration for optimizing reception or jamming.

The bandwidth is a measure of how much the frequency can be varied while still obtaining an acceptable VSWR (2:1 or less) and minimizing losses in unwanted directions. See Glossary, Section 10.

A 2:1 VSWR corresponds to a 9.5dB (or 10%) return loss - see Section 6-2.

Two methods for computing antenna bandwidth are used:

Narrowband by %, $B = \left(\dfrac{F_U - F_L}{F_C} \right)(100)$, where F_C = Center frequency

Broadband by ratio, $B = \dfrac{F_U}{F_L}$

Bandwidth	
%	Ratio
5	1.05 : 1
10	1.11 : 1
20	1.22 : 1
30	1.35 : 1
40	1.50 : 1
50	1.67 : 1
60	1.85 : 1
67	2 : 1
100	3 : 1
120	4 : 1
133	5 : 1
150	7 : 1
160	9 : 1
163	10 : 1

An antenna is considered broadband if $F_U / F_L > 2$. The table at the right shows the equivalency of the two, however the shaded values are not normally used because of the aforementioned difference in broadband/narrowband.

Should there be ever a need to express bandwidth of an antenna in one or the other alternative formats, a conversion between the two narrowband and broadband bandwidth quantities can be easily calculated using the following relationships:

Calculate broadband ratio B_{bb} given narrowband B%,

$$B_{bb} = (200 + B\%)/(200 - B\%)$$ **[1]**

Calculate narrowband B% given broadband ratio B_{bb},

$$B\% = 200 * (Bbb - 1)/(Bbb + 1)$$ **[2]**

For an object that experiences a plane wave, the resonant mode is achieved when the dimension of the object is $n\lambda/2$, where n is an integer. Therefore, one can treat the apertures shown in the figure below as half wave length dipole antennas for receiving and reflecting signals. More details are contained in section 8-4.

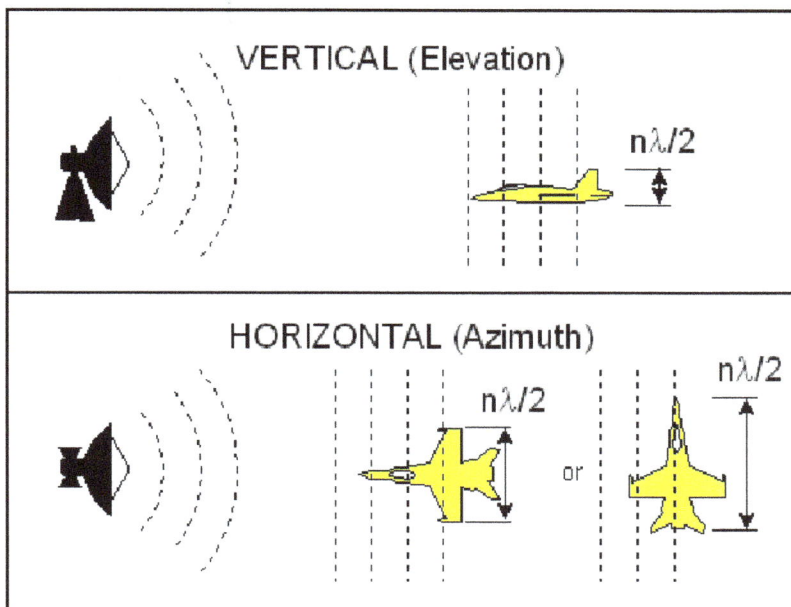

The following lists antenna types by page number. The referenced page shows frequency limits, polarizations, etc.

Type	Page	Type	Page
4 arm conical spiral	3-3.6	log periodic	3-3.8
alford loop	3-3.4	loop, circular	3-3.4
aperture synthesis	3-3.8	loop, alfred	3-3.4
array	3-3.8	loop, square	3-3.4
axial mode helix	3-3.5	luneberg lens	3-3.9
biconical w/polarizer	3-3.6	microstrip patch	3-3.9
biconical	3-3.6	monopole	3-3.3
cavity backed circuit fed slot	3-3.9	normal mode helix	3-3.5
cavity backed spiral	3-3.5	parabolic	3-3.7
circular loop	3-3.4	patch	3-3.9
conical spiral	3-3.5	reflector	3-3.9
corner reflector	3-3.9	rhombic	3-3.3
dipole array, linear	3-3.8	sinuous, dual polarized	3-3.6
dipole	3-3.3	slot, guide fed	3-3.9
discone	3-3.4	slot, cavity backed	3-3.9
dual polarized sinuous	3-3.6	spiral, 4 arm conical	3-3.6
guide fed slot	3-3.9	spiral, conical	3-3.5
helix, normal mode	3-3.5	spiral, cavity backed	3-3.5
helix, axial mode	3-3.5	square loop	3-3.4
horn	3-3.7	vee	3-3.3
linear dipole array	3-3.8	yagi	3-3.8

Antenna Type	Radiation Pattern	Characteristics
MONOPOLE 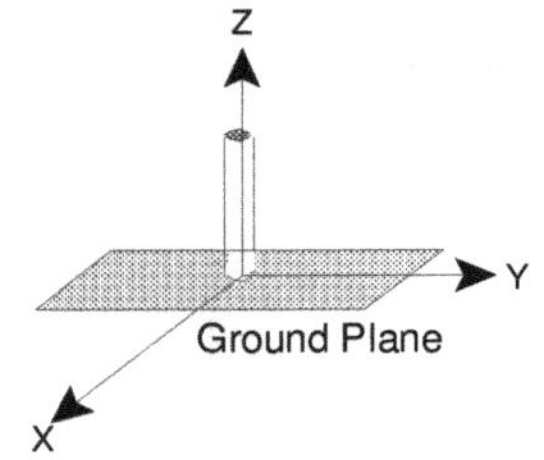 Z Y Ground Plane X	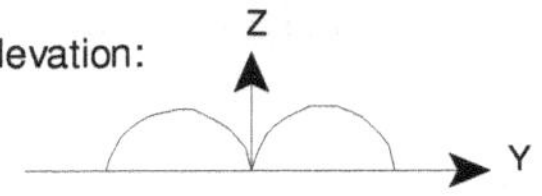 Elevation: Z Y Azimuth: Y X 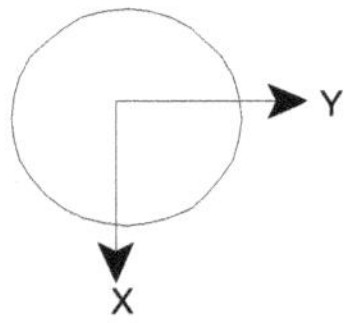	**Polarization: Linear** Vertical as shown **Typical Half-Power Beamwidth** 45 deg x 360 deg **Typical Gain:** 2-6 dB at best **Bandwidth:** 10% or 1.1:1 **Frequency Limit** **Lower:** None **Upper:** None **Remarks:** Polarization changes to horizontal if rotated to horizontal
λ/2 DIPOLE 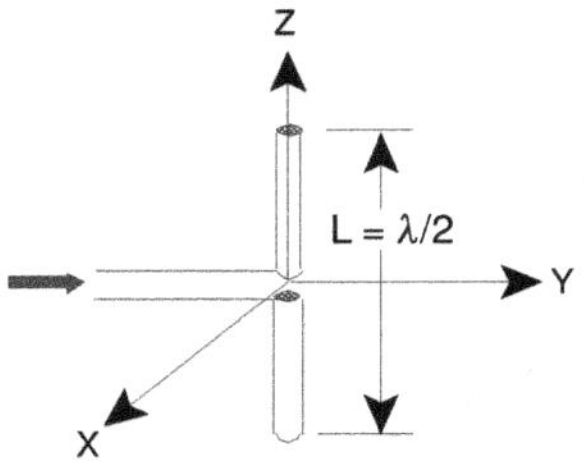 Z L = λ/2 Y X	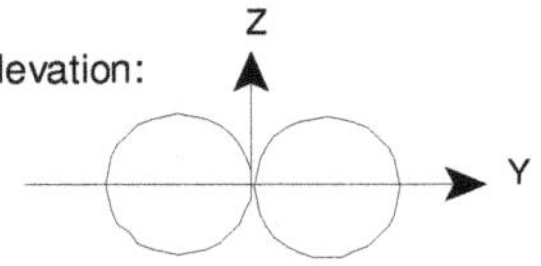 Elevation: Z Y Azimuth: Y X 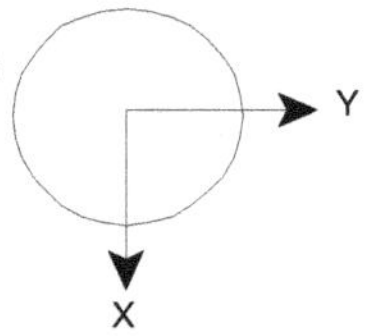	**Polarization:** Linear Vertical as shown **Typical Half-Power Beamwidth** 80 deg x 360 deg **Typical Gain:** 2 dB **Bandwidth:** 10% or 1.1:1 **Frequency Limit** **Lower:** None **Upper:** 8 GHz (practical limit) **Remarks:** Pattern and lobing changes significantly with L/f. Used as a gain reference < 2 GHz.

Figure 1

Antenna Type	Radiation Pattern	Characteristics
VEE 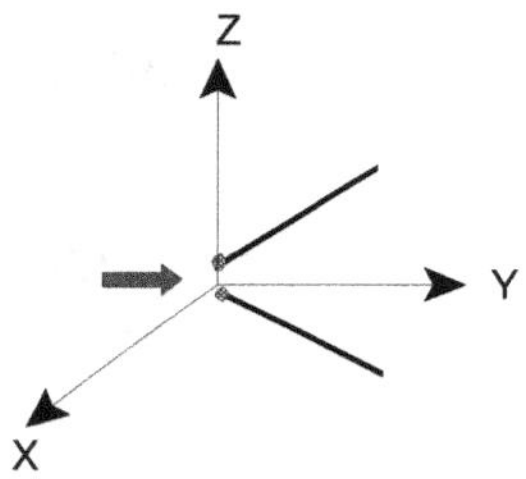 Z Y X	Elevation & Azimuth: Y 	**Polarization:** Linear Vertical as shown **Typical Half-Power Beamwidth** 60 deg x 60 deg **Typical Gain:** 2 to 7 dB **Bandwidth:** "Broadband" **Frequency Limit** **Lower:** 3 MHz **Upper:** 500 MHz (practical limits) **Remarks:** 24KHz versions are known to exist. Terminations may be used to reduce backlobes.
RHOMBIC 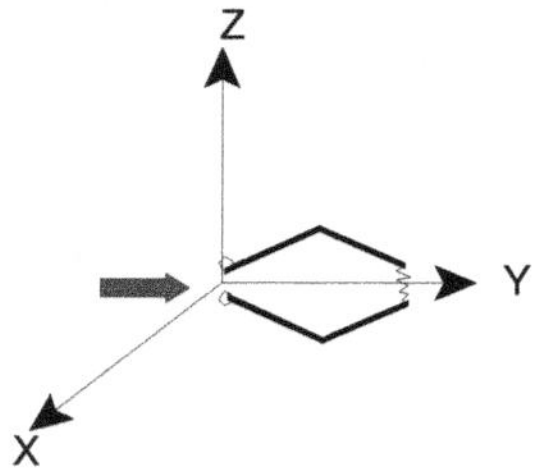 Z Y X	Elevation & Azimuth: Y 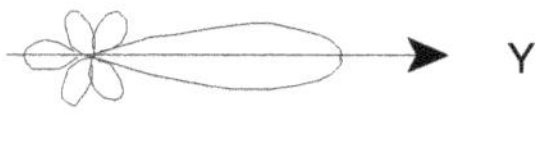	**Polarization:** Linear Vertical as shown **Typical Half-Power Beamwidth** 60 deg x 60 deg **Typical Gain:** 3 dB **Bandwidth:** "Broadband" **Frequency Limit** **Lower:** 3 MHz **Upper:** 500 MHz **Remarks:** Termination resistance used to reduce backlobes.

Figure 2

<table>
<tr><th>Antenna Type</th><th>Radiation Pattern</th><th>Characteristics</th></tr>
<tr><td>CIRCULAR LOOP
(Small)
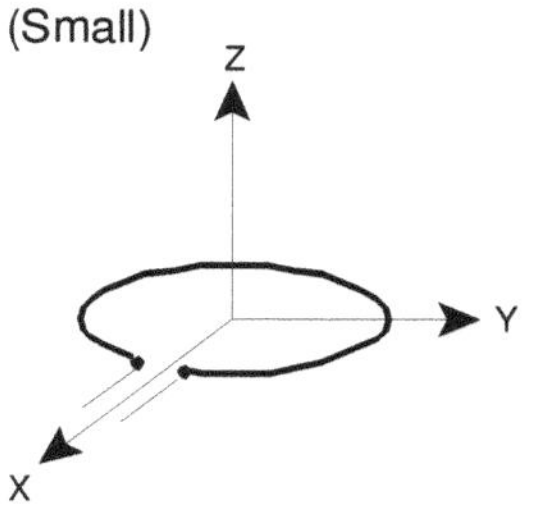Z
Y
X</td><td>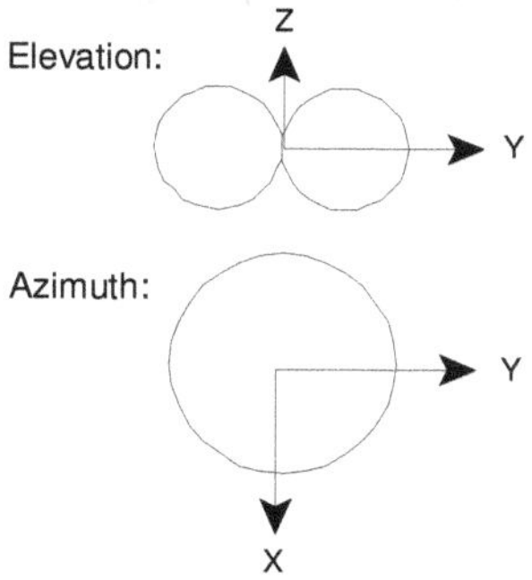Elevation:
Z
Y
Azimuth:
Y
X</td><td>Polarization: Linear
Horizontal as shown

Typical Half-Power Beamwidth:
80 deg x 360 deg

Typical Gain: -2 to 2 dB

Bandwidth: 10% or 1.1:1

Frequency Limit:
Lower: 50 MHz
Upper: 1 GHz</td></tr>
<tr><td>SQUARE LOOP
(Small)
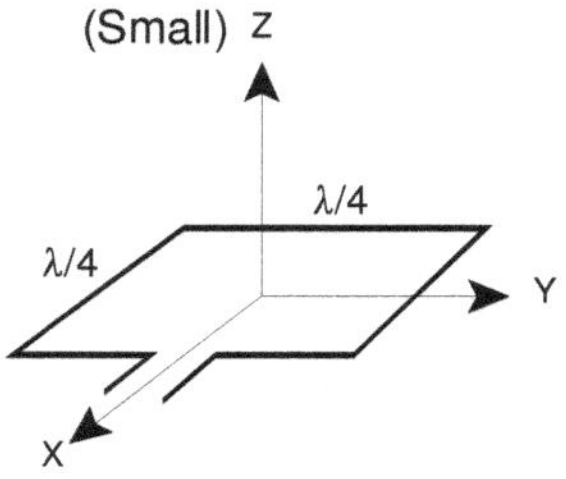Z
λ/4
λ/4
Y
X</td><td>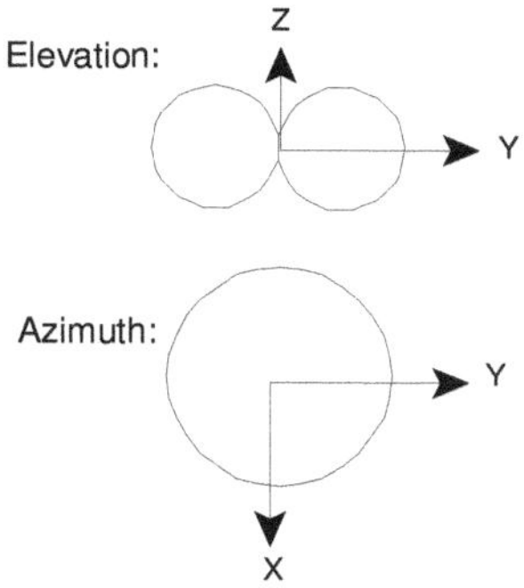Elevation:
Z
Y
Azimuth:
Y
X</td><td>Polarization: Linear
Horizontal as shown

Typical Half-Power Beamwidth:
100 deg x 360 deg

Typical Gain: 1-3 dB

Bandwidth: 10% or 1.1:1

Frequency Limit:
Lower: 50 MHz
Upper: 1 GHz</td></tr>
</table>

Figure 3

<table>
<tr><th>Antenna Type</th><th>Radiation Pattern</th><th>Characteristics</th></tr>
<tr><td>DISCONE
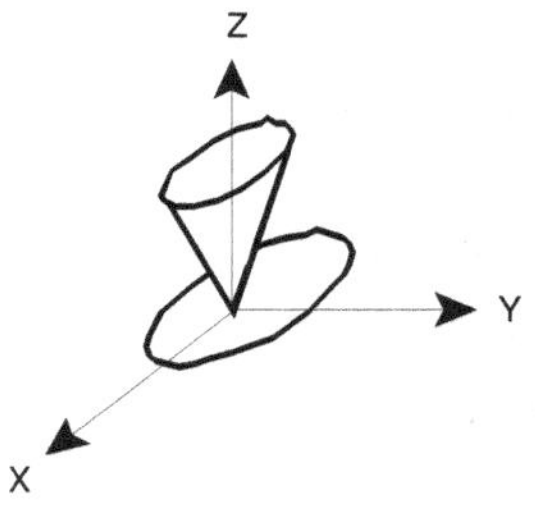Z
Y
X</td><td>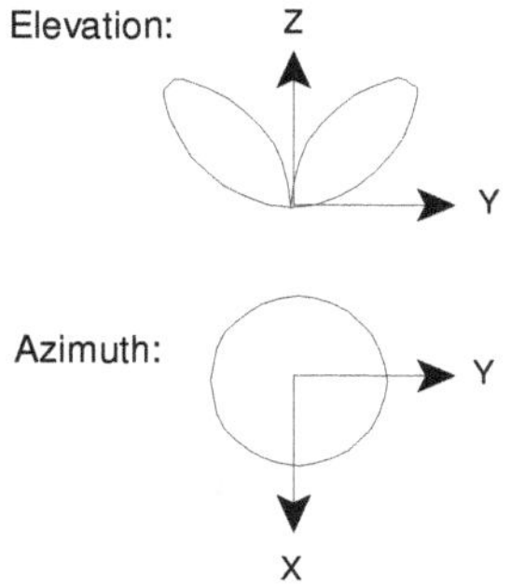Elevation:
Z
Y
Azimuth:
Y
X</td><td>Polarization: Linear
Vertical as shown

Typical Half-Power Beamwidth:
20-80 deg x 360 deg

Typical Gain: 0-4 dB

Bandwidth: 100% or 3:1

Frequency Limit:
Lower: 30 MHz
Upper: 3 GHz</td></tr>
<tr><td>ALFORD LOOP
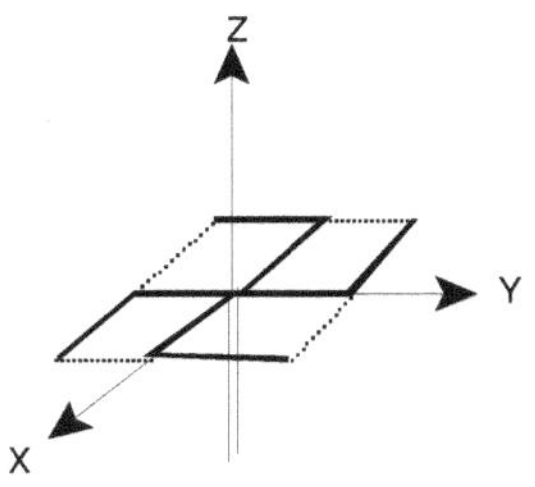Z
Y
X</td><td>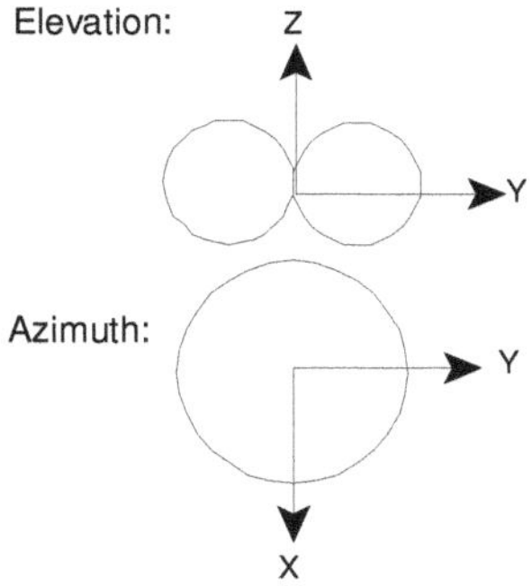Elevation:
Z
Y
Azimuth:
Y
X</td><td>Polarization: Linear
Horizontal as shown

Typical Half-Power Beamwidth:
80 deg x 360 deg

Typical Gain: -1 dB

Bandwidth: 67% or 2:1

Frequency Limit:
Lower: 100 MHz
Upper: 12 GHz</td></tr>
</table>

Figure 4

Antenna Type	Radiation Pattern	Characteristics
AXIAL MODE HELIX 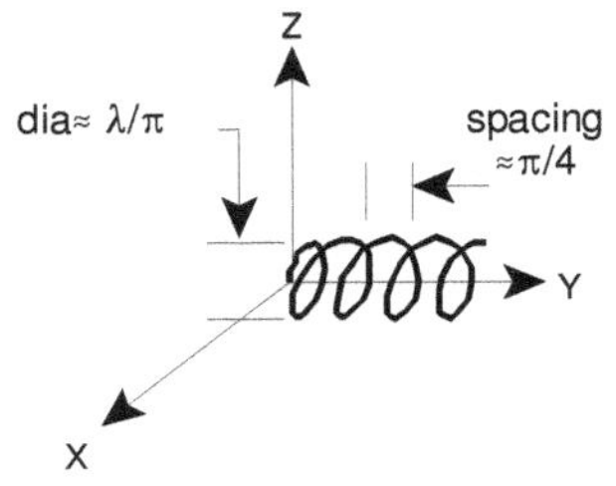 dia≈ λ/π spacing ≈π/4	Elevation & Azimuth 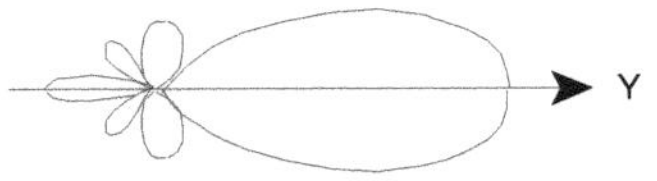	**Polarization:** Circular Left hand as shown **Typical Half-Power Beamwidth:** 50 deg x 50 deg **Typical Gain:** 10 dB **Bandwidth:** 52% or 1.7:1 **Frequency Limit** **Lower:** 100 MHz **Upper:** 3 GHz **Remarks:** Number of loops >3
NORMAL MODE HELIX 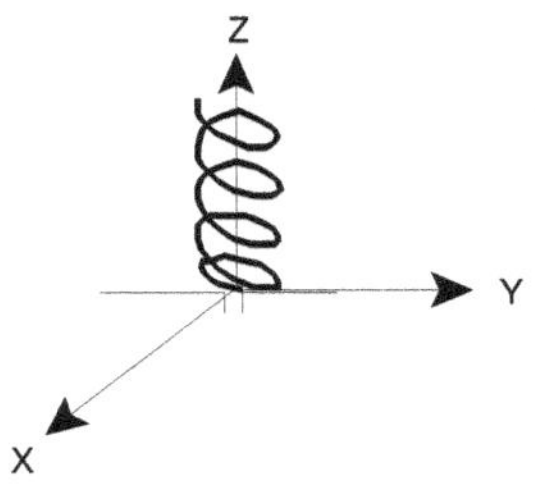	Elevation: Azimuth: 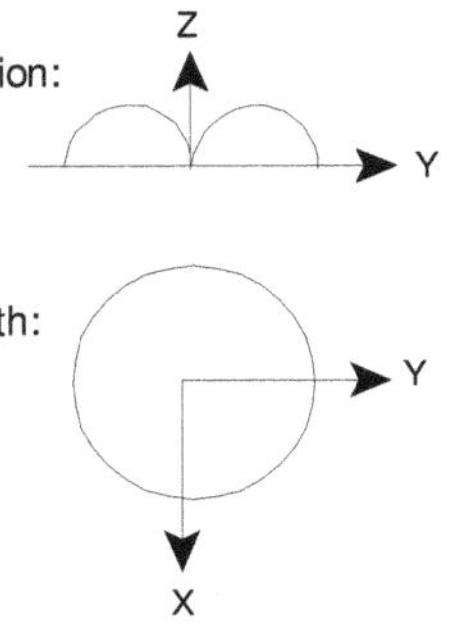	**Polarization:** Circular - with an ideal pitch to diameter ratio. **Typical Half-Power Beamwidth:** 60 deg x 360 deg **Typical Gain:** 0 dB **Bandwidth:** 5% or 1.05:1 **Frequency Limit** **Lower:** 100 MHz **Upper:** 3 GHz

Figure 5

Antenna Type	Radiation Pattern	Characteristics
CAVITY BACKED SPIRAL (Flat Helix) 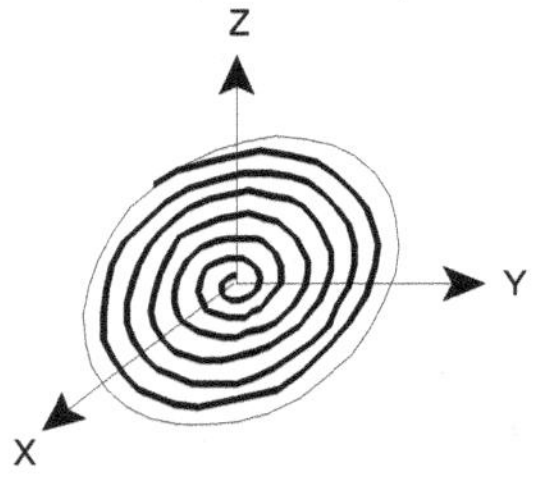	Elevation & Azimuth 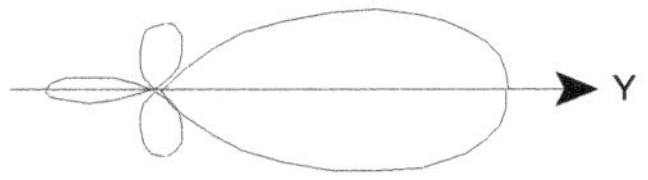	**Polarization:** Circular Left hand as shown **Typical Half-Power Beamwidth:** 60 deg x 90 deg **Typical Gain:** 2-4 dB **Bandwidth:** 160% or 9:1 **Frequency Limit:** **Lower:** 500 MHz **Upper:** 18 GHz
CONICAL SPIRAL 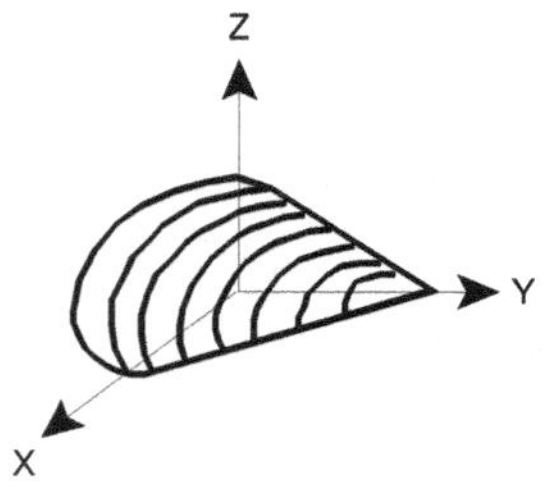	Elevation & Azimuth 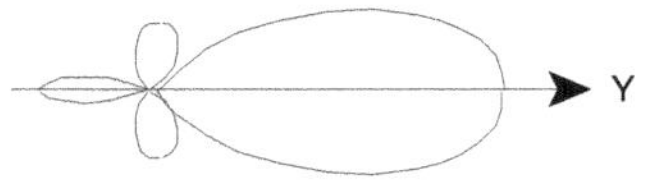	**Polarization:** Circular Left hand as shown **Typical Half-Power Beamwidth:** 60 deg x 60 deg **Typical Gain:** 5-8 dB **Bandwidth:** 120% or 4:1 **Frequency Limit:** **Lower:** 50 MHz **Upper:** 18 GHz

Figure 6

Antenna Type	Radiation Pattern	Characteristics
4 ARM CONICAL SPIRAL 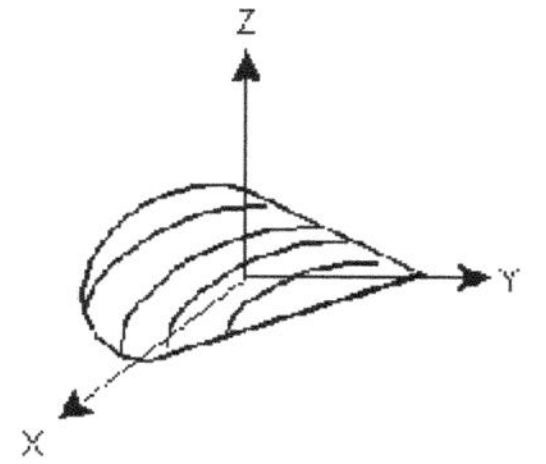	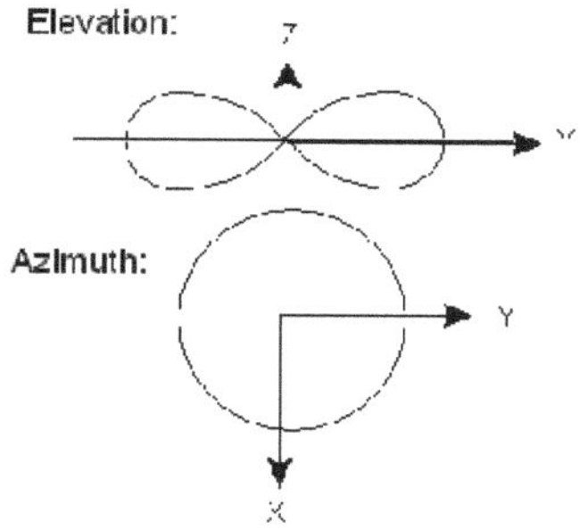 Elevation: Azimuth:	**Polarization:** Circular Left hand as shown **Typical Half-Power Beamwidth:** 50 deg x 360 deg **Typical Gain:** 0 dB **Bandwidth:** 120% or 4:1 **Frequency Limit:** **Lower:** 500 MHz **Upper:** 18 GHz
DUAL POLARIZED SINUOUS 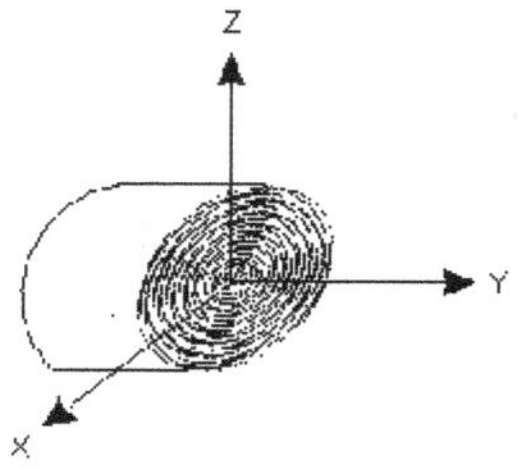	Elevation & Azimuth 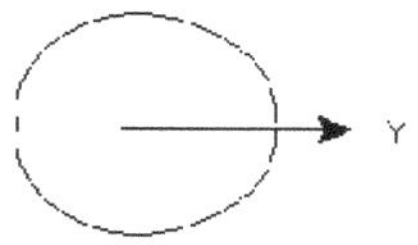	**Polarization:** Dual vertical or horizontal or dual Circular right hand or left hand with hybrid **Typical Half-Power Beamwidth:** 75 deg x 75 deg **Typical Gain:** 2 dB **Bandwidth:** 163% or 10:1 **Frequency Limit:** **Lower:** 500 MHz **Upper:** 18 GHz

Figure 7

Antenna Type	Radiation Pattern	Characteristics
BICONICAL 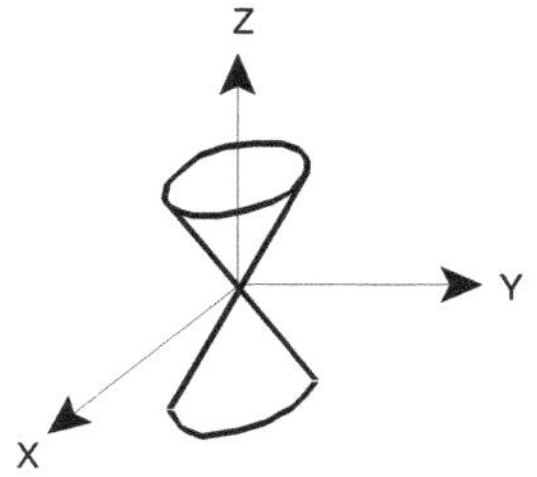	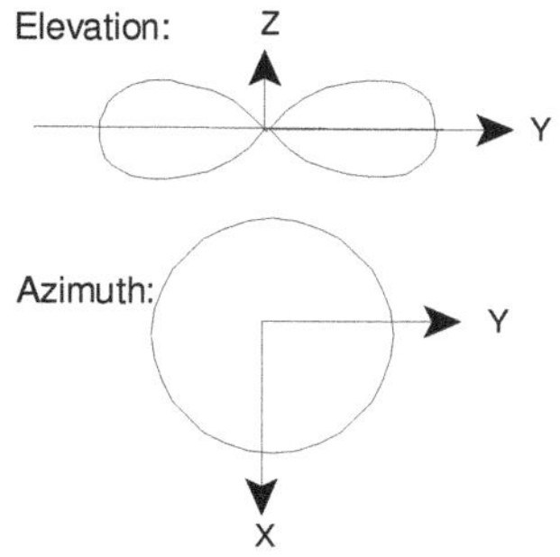 Elevation: Azimuth:	**Polarization:** Linear, Vertical as shown **Typical Half-Power Beamwidth:** 20-100 deg x 360 deg **Typical Gain:** 0-4 dB **Bandwidth:** 120% or 4:1 **Frequency Limit:** **Lower:** 500 MHz **Upper:** 40 GHz
BICONICAL W/POLARIZER 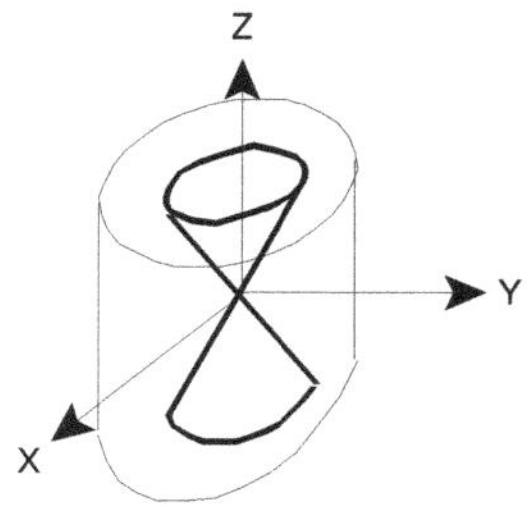	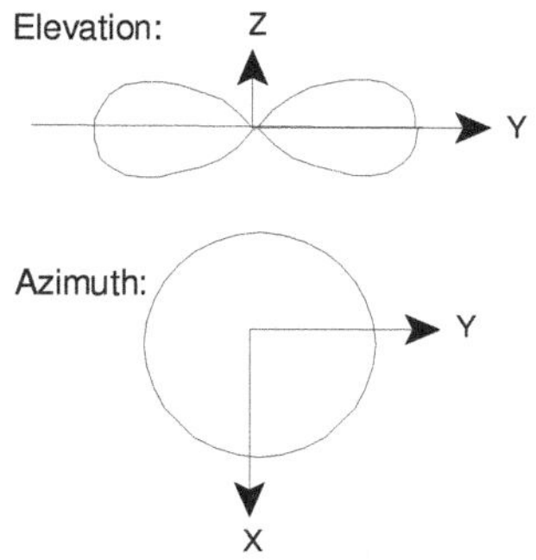 Elevation: Azimuth:	**Polarization:** Circular, Direction depends on polarization **Typical Half-Power Beamwidth:** 20-100 deg x 360 deg **Typical Gain:** -3 to 1 dB **Bandwidth:** 100% or 3:1 **Frequency Limit:** **Lower:** 2 GHz **Upper:** 18 GHz

Figure 8

Antenna Type	Radiation Pattern	Characteristics
HORN 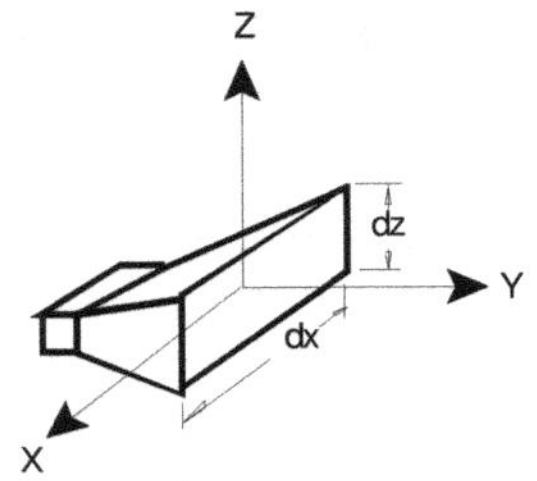	Elevation: 3 dB beamwidth = 56 $\lambda°$/dz Azimuth: 3 dB beamwidth = 70 $\lambda°$/dx	**Polarization:** Linear **Typical Half-Power Beamwidth:** 40 deg x 40 deg **Typical Gain:** 5 to 20 dB **Bandwidth:** If ridged: 120% or 4:1 If not ridged: 67% or 2:1 **Frequency Limit:** **Lower:** 50 MHz **Upper:** 40 GHz
HORN W / POLARIZER 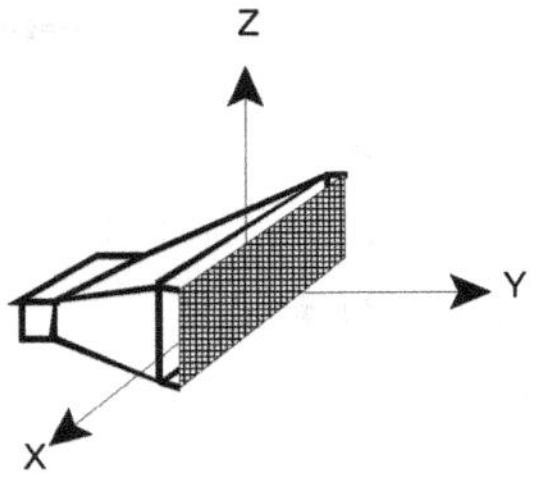	Elevation: Azimuth:	**Polarization:** Circular, Depends on polarizer **Typical Half-Power Beamwidth:** 40 deg x 40 deg **Typical Gain:** 5 to 10 dB **Bandwidth:** 60% or 2:1 **Frequency Limit:** **Lower:** 2 GHz **Upper:** 18 GHz

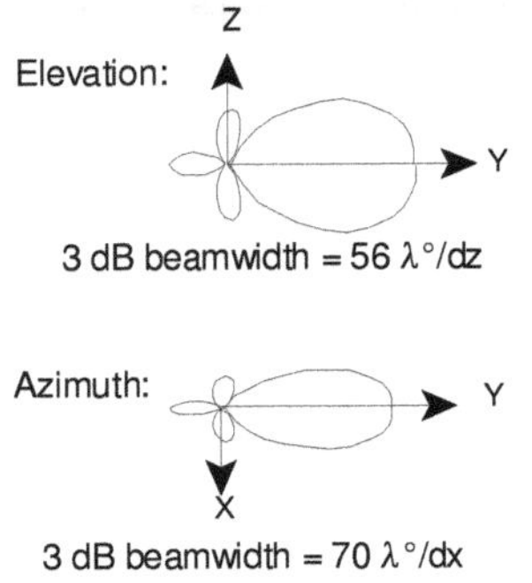

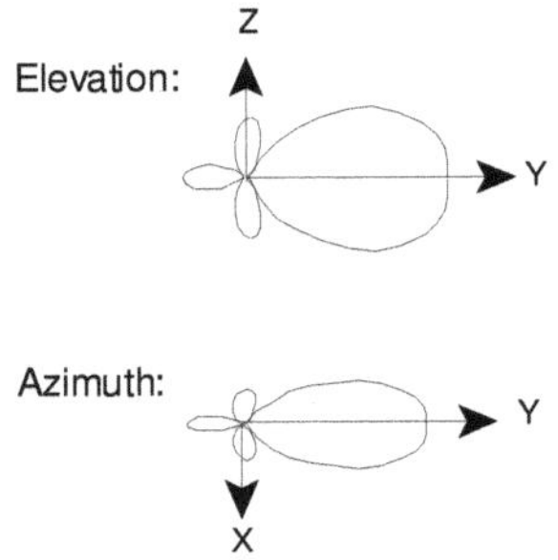

Figure 9

Antenna Type	Radiation Pattern	Characteristics
PARABOLIC (Prime) 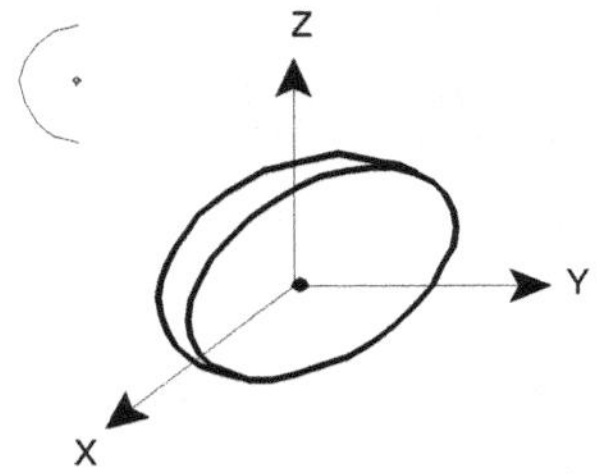	Elevation & Azimuth 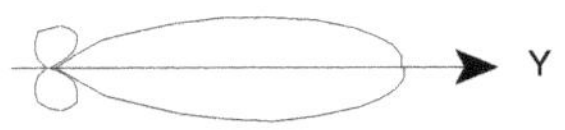	**Polarization:** Takes polarization of feed **Typical Half-Power Beamwidth:** 1 to 10 deg **Typical Gain:** 20 to 30 dB **Bandwidth:** 33% or 1.4:1 limited mostly by feed **Frequency Limit:** **Lower:** 400 MHz **Upper:** 13+ GHz
PARABOLIC 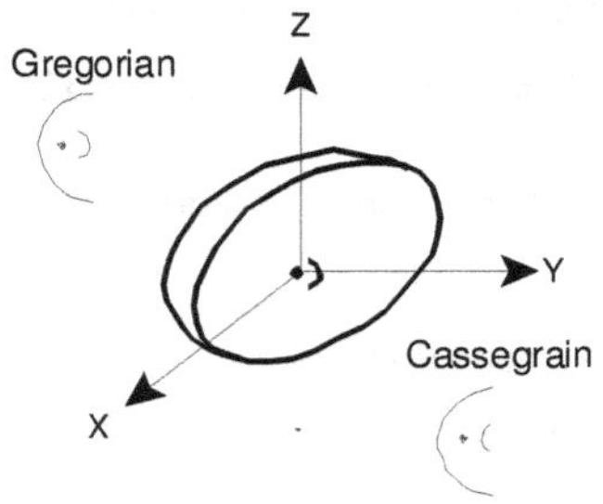 Gregorian Cassegrain	Elevation & Azimuth 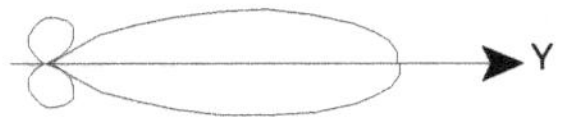	**Polarization:** Takes polarization of feed **Typical Half-Power Beamwidth:** 1 to 10 deg **Typical Gain:** 20 to 30 dB **Bandwidth:** 33% or 1.4:1 **Frequency Limit:** **Lower:** 400 MHz **Upper:** 13+ GHz

Figure 10

Antenna Type	Radiation Pattern	Characteristics
YAGI 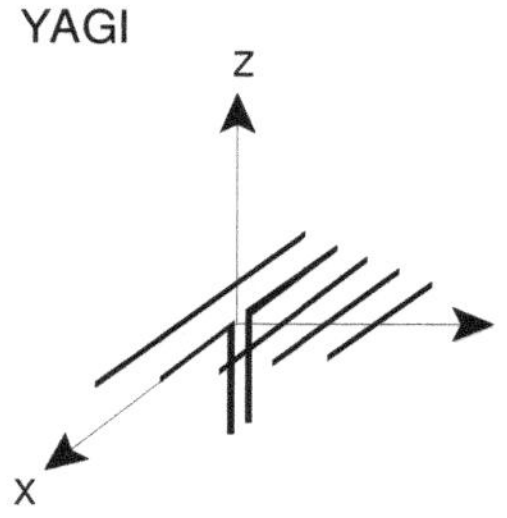	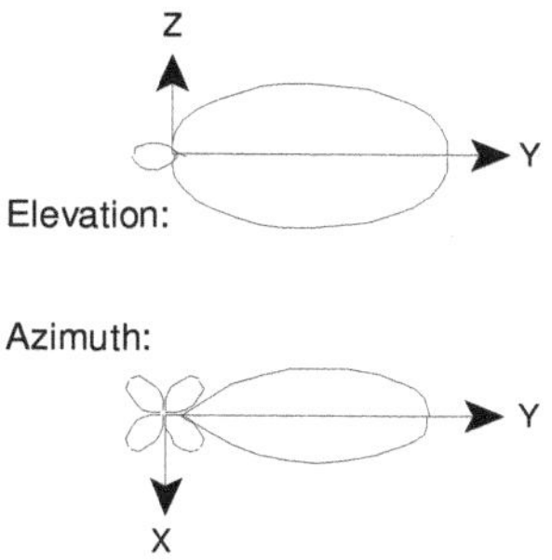 Elevation: Azimuth:	**Polarization:** Linear Horizontal as shown **Typical Half-Power Beamwidth** 50 deg X 50 deg **Typical Gain:** 5 to 15 dB **Bandwidth:** 5% or 1.05:1 **Frequency Limit:** **Lower:** 50 MHz **Upper:** 2 GHz
LOG PERIODIC 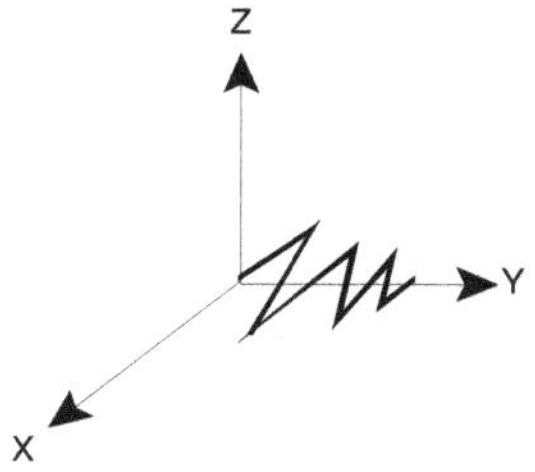	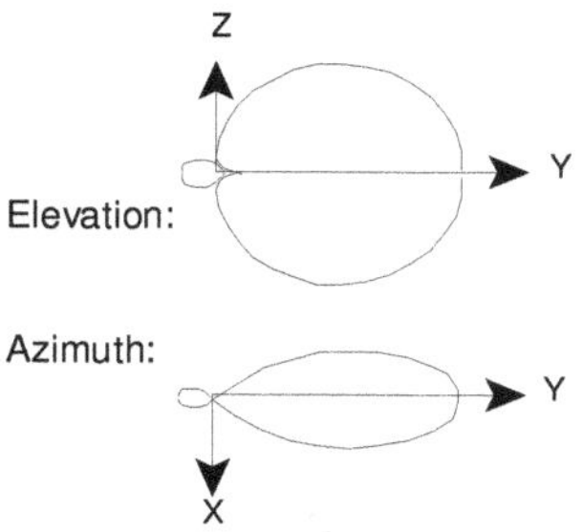 Elevation: Azimuth:	**Polarization:** Linear **Typical Half-Power Beamwidth:** 60 deg x 80 deg **Typical Gain:** 6 to 8 dB **Bandwidth:** 163% or 10:1 **Frequency Limit:** **Lower:** 3 MHz **Upper:** 18 GHz **Remarks:** This array may be formed with many shapes including dipoles or toothed arrays.

Figure 11

Antenna Type	Radiation Pattern	Characteristics
LINEAR DIPOLE ARRAY (Corporate Feed) 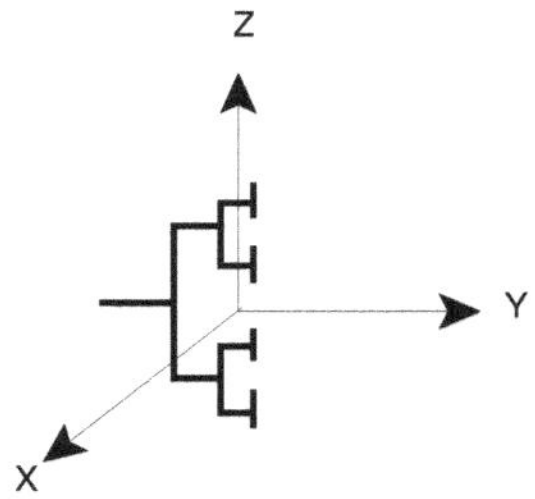	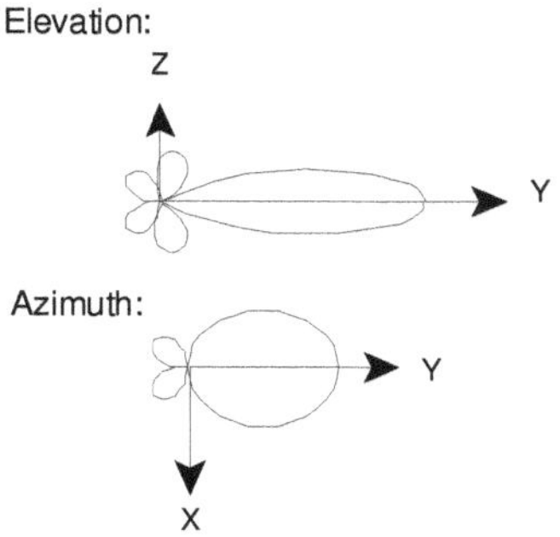 Elevation: Azimuth:	**Polarization:** Element dependent Vertical as shown **Typical Half-Power Beamwidth:** Related to gain **Typical Gain:** Dependent on number of elements **Bandwidth:** Narrow **Frequency Limit:** **Lower:** 10 MHz **Upper:** 10 GHz
APERTURE SYNTHESIS 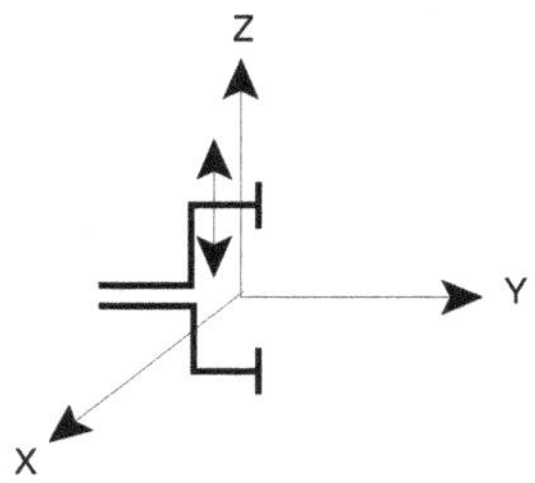	**Elevation &** **Azimuth** 	All characteristics dependent on elements **Remarks:** Excellent side-looking, ground mapping where the aircraft is a moving linear element.

Figure 12

Antenna Type	Radiation Pattern	Characteristics
CAVITY BACKED CIRCUIT FED SLOT (and Microstrip Patch 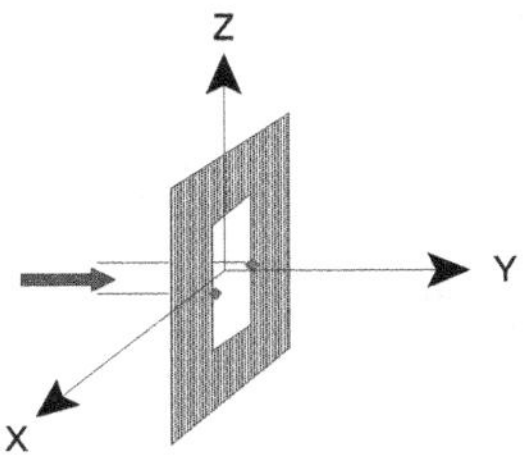) Elevation & Azimuth	Elevation & Azimuth 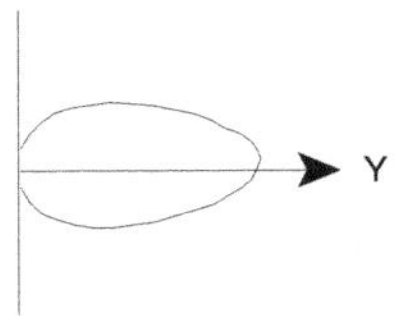	**Polarization:** Linear, vertical as shown **Typical Half-Power Beamwidth:** 80 deg x 80 deg **Typical Gain:** 6 dB **Bandwidth:** Narrow **Frequency Limit:** **Lower:** 50 MHz **Upper:** 18 GHz **Remarks:** The feed line is sometimes separated from the radiator by a dialetric & uses capacititive coupling. Large conformal phased arrays can be made this way.
GUIDE FED SLOT 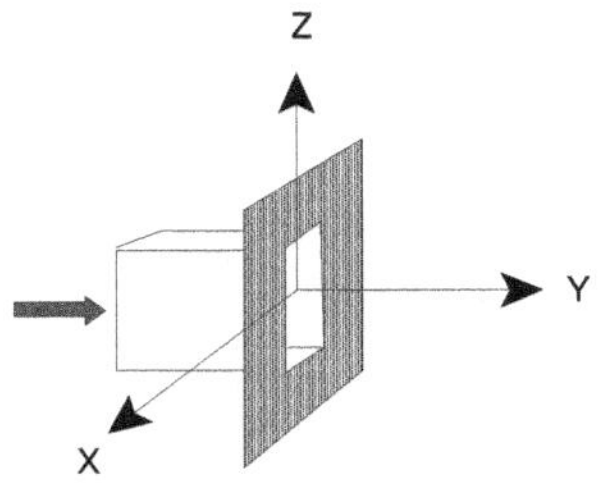	Elevation: Azimuth: 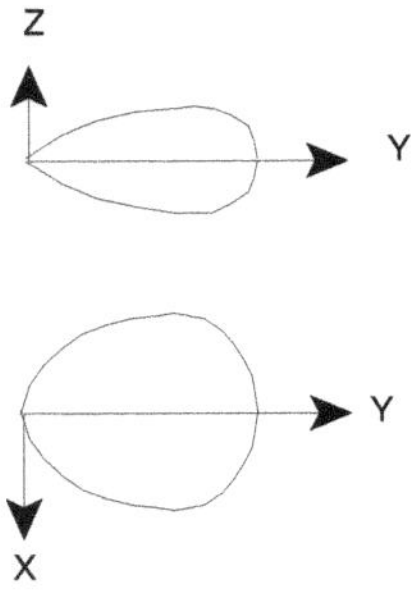	**Polarization:** Linear, **Typical Half-Power Beamwidth** Elevation: 45-50° Azimuth: 80° **Typical Gain:** 0 dB **Bandwidth:** Narrow **Frequency Limit:** **Lower:** 2 GHz **Upper:** 40 GHz **Remarks:** Open RF Waveguide

Figure 13

Antenna Type	Radiation Pattern	Characteristics
CORNER REFLECTOR 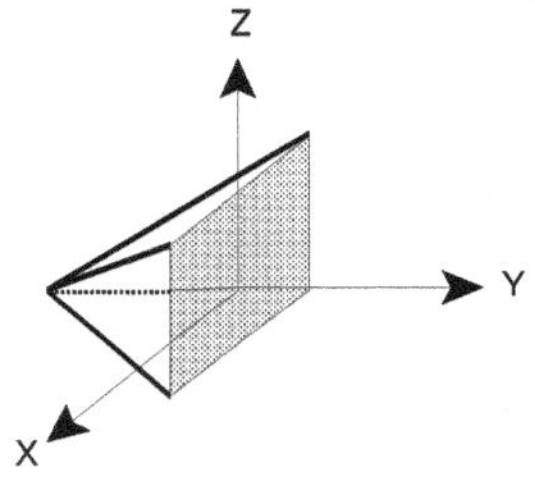	Elevation: (Z-Y) Azimuth: (X-Y) Dependent upon feed emitter	**Polarization:** Feed dependent **Typical Half-Power Beamwidth** 40 deg x variable **Typical Gain:** 10 dB above feed **Bandwidth:** Narrow **Frequency Limit** **Lower:** 1 GHz **Upper:** 40 GHz **Remarks:** Typically fed with a dipole or colinear array.
LUNEBURG LENS 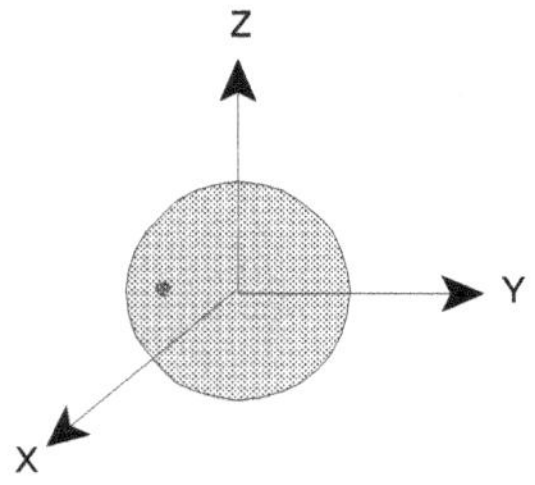	Elevation & Azimuth 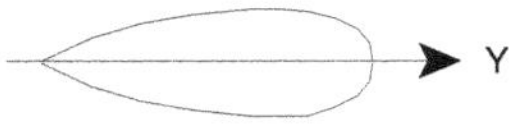	**Polarization:** Feed dependent **Typical Half-Power Beamwidth:** System dependent **Typical Gain:** System dependent **Bandwidth:** Narrow **Frequency Limit** **Lower:** 1 GHz **Upper:** 40 GHz **Remarks:** Variable index dielectric sphere.

Figure 14

FREQUENCY / PHASE EFFECTS OF ANTENNAS

The radiation patterns of the antennas presented in the previous section are for antenna geometries most commonly used. The antenna should be viewed as a matching network that takes the power from a transmission line (50 ohm, for example), and matches it to the free space "impedance" of 377 ohms. The most critical parameter is the change of VSWR with frequency. The pattern usually does not vary much from acceptable to the start of unacceptable VSWRs (> 2:1). For a given physical antenna geometric size, the actual radiation pattern varies with frequency.

The antenna pattern depicted in Figure 1 is for the dipole pictured in Section 3-3. The maximum gain is normalized to the outside of the polar plot and the major divisions correspond to 10 dB change. In this example, the dipole length (in wavelengths) is varied, but the same result can be obtained by changing frequency with a fixed dipole length. From the figure, it can be seen that side lobes start to form at 1.25λ and the side lobe actually has more gain than the main beam at 1.5λ. Since the radiation pattern changes with frequency, the gain also changes.

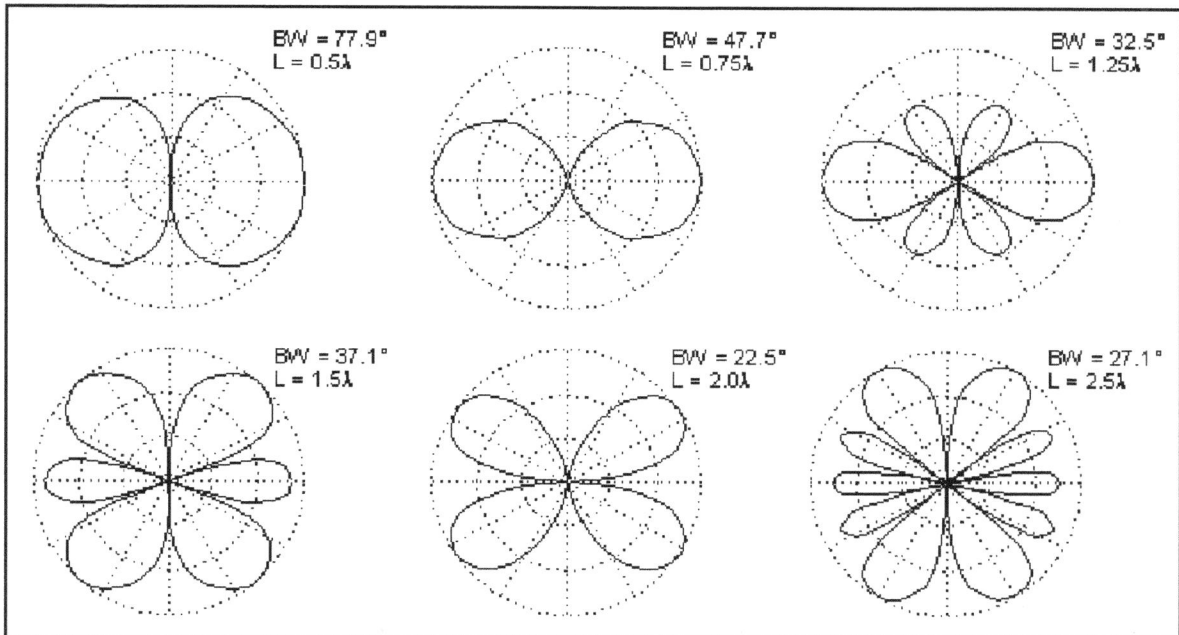

Figure 1. Frequency Effects

Figure 2 depicts phase/array effects, which are yet another method for obtaining varied radiation patterns. In the figure, parallel dipoles are viewed from the end. It can be seen that varying the phase of the two transmissions can cause the direction of the radiation pattern to change. This is the concept behind phased array antennas. Instead of having a system mechanically sweeping the direction of the antenna through space, the phase of radiating components is varied electronically, producing a moving pattern with no moving parts. It can also be seen that increasing the number of elements further increases the directivity of the array. In an array, the pattern does vary considerably with frequency due to element spacing (measured in wavelengths) and the frequency sensitivity of the phase shifting networks.

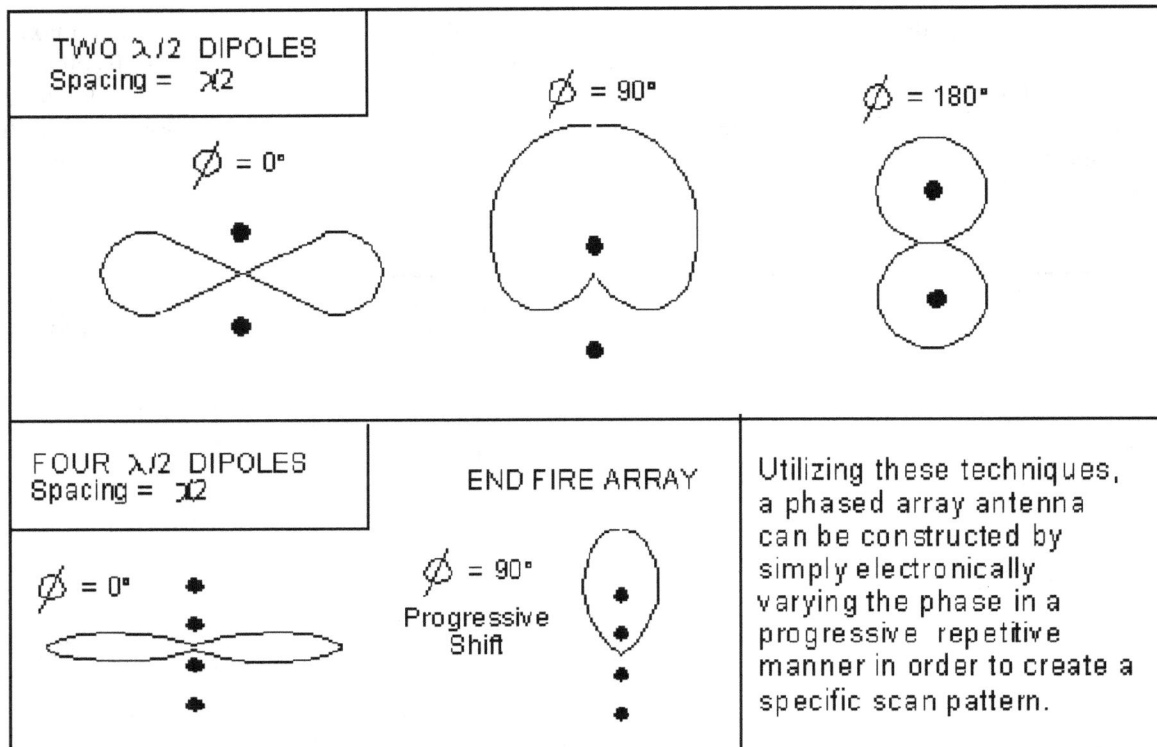

Figure 2. Phase / Array Effects*

* Note: Assuming Figure 2 depicts x-y plane antenna pattern cross section, to achieve the indicated array patterns using dipole antennas, the dipole antenna elements must be aligned with the z-axis.

Two antennas that warrant special consideration are the phased array and the Rotman bootlace type lens. Both of these antennas find wide application in EW, RADAR, and Communications. The phased array will be described first.

LINEAR PHASED ARRAY

The linear phased array with equal spaced elements is easiest to analyze and forms the basis for most array designs. Figure 3 schematically illustrates a corporate feed linear array with element spacing d.

It is the simplest and is still widely used. By controlling the phase and amplitude of excitation to each element, as depicted, we can control the direction and shape of the beam radiated by the array. The phase excitation, $\varphi(n)$, controls the beam pointing angle, θ_o, in a phased array. To produce a broadside beam, $\theta_o=0$, requires phase excitation, $\varphi(n)=0$. Other scan angles require an excitation, $\varphi(n) = nkd \sin(\theta_o)$, for the nth element where k is the wave number $(2\pi/\lambda)$. In this

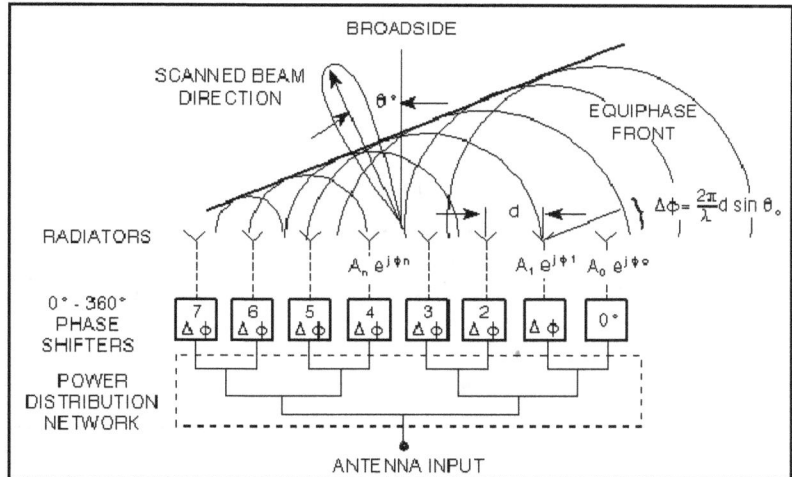

Figure 3. Corporate Fed Phased Array

manner a linear phased array can radiate a beam in any scan direction, θ_o, provided the element pattern has sufficient beamwidth. The amplitude excitation, A_n, can be used to control beam shape and sidelobe levels. Often the amplitude excitation is tapered in a manner similar to that used for aperture antennas to reduce the sidelobe levels. One of the problems that can arise with a phased array is insufficient bandwidth, since the phase shift usually is not obtained through the introduction of additional path length. However, it should be noted that at broadside the corporate feed does have equal path length and would have good bandwidth for this scan angle.

The linear array described above would yield a narrow fan beam in the plane normal to the plane containing the array and scan direction, with the narrow beamwidth in the plane of the array. To obtain a pencil beam it would be necessary to array several of these linear arrays in such a manner resulting in a planar array of radiating elements. A problem associated with all electronic scanning is beam distortion with scan angle. Figure 4 illustrates this phenomenon. It results in spread of the beam shape with a concomitant reduction in gain. This effect is known as "scan loss". For an ideal array element, scan loss is equal to the reduction in aperture size in the scan direction which varies as cos θ.,where θ is the scan angle measured from the planar array normal.

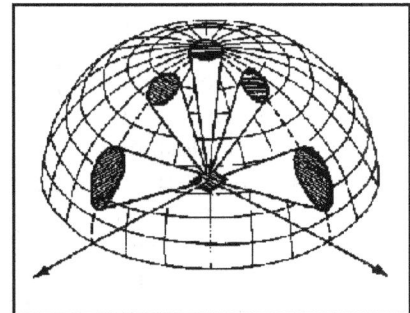

Figure 4. Beam Distortion

When elements are spaced greater than $\lambda/2$ apart, grating lobes are possible when scanning. As the beam is scanned further from broadside, a point is reached at which a second symmetrical main lobe is developed at the negative scan angle from broadside. This condition is not wanted because antenna gain is immediately reduced by 3 dB due to the second lobe. Grating lobes are a significant problem in EW applications because the broad frequency bandwidth requirements mean that at the high end of the frequency band, the elements may be spaced greater than $\lambda/2$. Therefore in order to avoid grating lobes over a large frequency bandwidth, element spacing must be no greater than $\lambda/2$ at the highest frequency of operation.

There are many other factors to consider with a phased array such as coning, where the beam curves at large scan angles, and mutual coupling between elements that affect match and excitation. Excessive mutual coupling will invariably result in blind scan angles where radiation is greatly attenuated. These issues will not be covered in detail here.

Of interest is the gain of the array which is given by:

$$Array\ Gain = G_e(\theta) \bullet \sum_{n=1}^{N} A(n)\, e^{j\,\phi(n)}\, e^{j\,n\,k\,d\,\sin\theta}$$

Where each element is as described in Section 3-4.

$G_e(\theta)$ is the element gain which in this case has been taken the same for all elements. Note that if we set A(n)=1, and ϕ(n)=0, then at broadside where sin(θ) = 0, the gain would be (N G_e). This represents the maximum gain of the array, which typically will not exceed nπ, and is a familiar figure. . It should be noted that in practical array design, the element pattern charastics are greatly influenced by mutual coupling and that the characteristic of elements at the edge of the array can deviate significantly from those near the center.

ROTMAN BOOTLACE LENS

Another method of feeding an array of elements is to use a lens such as the Rotman (rhymes with rotten) Bootlace type shown in Figure 5. The lens consists of a parallel plate region (nowadays microstrip or stripline construction) and cables of specified length connecting the array of elements to the parallel plate region. The geometry of the lens and the cable lengths are designed so that all ray paths traced from a beam port on the right side to its associated

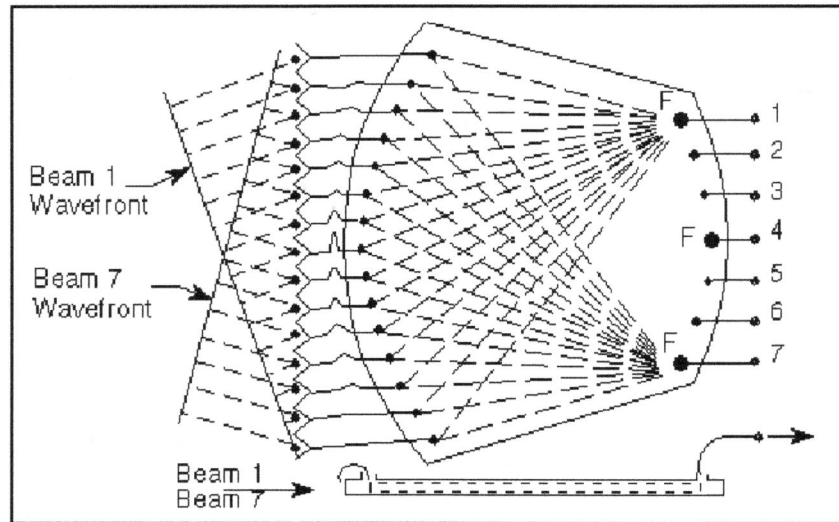

Figure 5. Rotman Bootlace Lens

wavefront on the left array port side, are equal. This tailoring of the design is accomplished at three focus points (beam ports 1, 4, and 7 in Figure 5). Departure from perfect focus at intermediate beam ports is negligible in most designs.

The Rotman lens provides both true time delay phase shift and amplitude taper in one lens component. The true time delay is one of the distinct advantages of the lens over the phase shifted array since that makes it independent of frequency. To understand how the taper is obtained requires knowledge of the parallel plate region. For a stripline design the unit would consist of a large flat plate-like center conductor sandwiched between two ground planes, and having a shape much like that of the plan view outline shown in Figure 5 with individual tapered launchers (connectors) attached to each beam port and array port. If the antenna is in the receive mode, the energy intercepted on the array port side can be controlled by the angle subtended by the tapered sections of the connector (launcher) much like a larger antenna would intercept a larger portion of energy from free space.

Unlike the phased array with its fine beam steering, the Rotman lens provides only a distinct set of beams. Fine steering is obtained by combining beams either equally or unequally to form intermediate beams. As can be seen in Figure 6, this results in a broader beam with less gain but lower side lobes than the primary beams.

High transmit power can be obtained using a Rotman lens by placing a low power amplifier between each lens output port and its antenna. In this case a separate Rotman lens would have to be used for receiving.

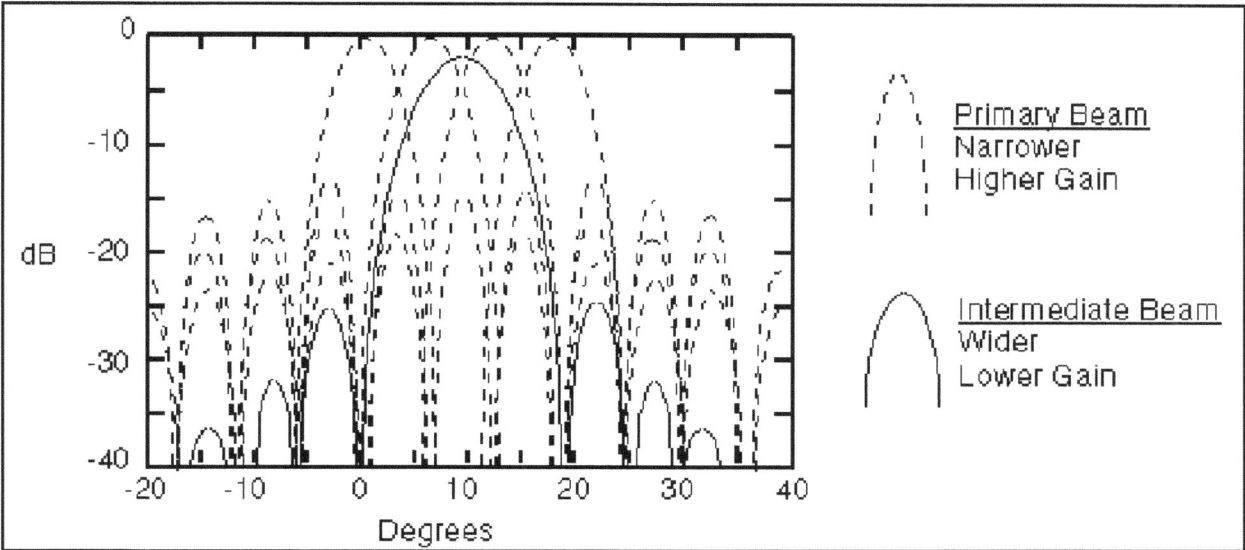

Figure 6. Primary and Intermediate Beam Formation in Lens Arrays

This Page Blank

ANTENNA NEAR FIELD

As noted in the sections on RF propagation and the radar equation, electromagnetic radiation expands spherically (Figure 1) and the power density at a long range (R) from the transmitting antenna is:

$$P_D = \frac{P_t G_t}{4\pi R^2} \qquad [1]$$

When the range is large, the spherical surface of uniform power density appears flat to a receiving antenna which is very small compared to the surface of the sphere. This is why the far field wave front is considered planar and the rays approximately parallel. Also, it is apparent that at some shorter range, the spherical surface no longer appears flat, even to a very small receiving antenna.

The planer, parallel ray approximation is valid for distances greater than the distance where the phase error is 1/16 of a wavelength or 22.5 degrees. This distance is given by

$$R_{ff} = \frac{2D^2}{\lambda} \quad \text{where } \lambda \text{ is the wavelength and D is the largest dimension of the transmit antenna.} \qquad [2]$$

Antenna measurements made at distances greater than R_{ff} generally result in negligible pattern error. Distances less than R_{ff} is termed the near-field.

If the same size antenna is used for multiple frequencies, R_{ff} will increase with increasing frequency. However, if various size antennas are used for different frequencies and each antenna is designed with D as a function of λ ($\lambda/2$ to 100λ), then R_{ff} will vary from $c/2f$ to $20000c/f$. In this case R_{ff} will decrease with increasing frequency. For example: a 10λ antenna at 3 GHZ has a D of 100 cm and corresponding R_{ff} of 20 m, while a 10λ antenna at 30 GHz has a D of 10 cm and corresponding R_{ff} of 2 m.

While the above analogy provides an image of the difference between the near and far fields, the relationship must be defined as a characteristic of the transmitting antenna.

Actual antennas, of course, are not ideal point source radiators but have physical dimensions. If the transmitting antenna placed at the origin of Figure 1 occupies distance D along the Z-axis and is boresighted along the Y-axis ($\varphi = 90$), then the geometry of point P on the sphere is represented in two dimensions by Figure 2. For convenience, the antenna is represented by a series of point sources in an array.

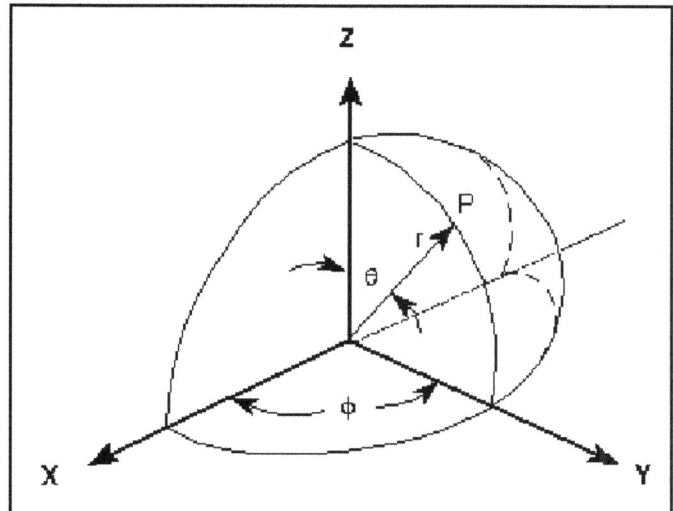

Figure 1. Spherical Radiation to Point "P" from an Ideal Point Source

When point P is close to the antenna, as in Figure 2, then the difference in distance of the two rays r and R taken respectively from the center of the antenna and the outer edge of the antenna varies as point P changes.

Derivation of equation [2] is given as follows:

From Figure 2, the following applies:

$$r^2 = z^2 + y^2 \qquad \text{[3]}$$

$$z = r \cos \theta \qquad \text{[4]}$$

$$y = r \sin \theta \qquad \text{and} \qquad \text{[5]}$$

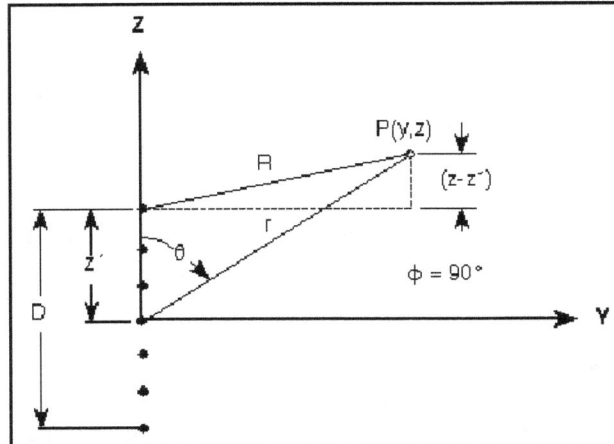

Figure 2. Near Field Geometry of Point "P" for a Non-Ideal Radiator With Dimension D.

$$R = \sqrt{y^2 + (z - z')^2} = \sqrt{y^2 + z^2 - 2z'z + (z')^2} \qquad \text{[6]}$$

Substituting [3] and [4] into [6] $\qquad R = \sqrt{r^2 + [-2(r \cos \theta)z' + (z')^2]} \qquad$ [7]
which puts point P into spherical coordinates.

Equation [7] can be expanded by the binomial theorem which for the first three terms, reduces to:

$$R = r - z' \cos \theta + \frac{(z')^2 \sin^2 \theta}{2r} + \dots \qquad \text{[8]}$$

In the parallel ray approximation for far field calculations (Figure 3) the third term of [8] is neglected.

The distance where the far field begins (R_{ff}) (or where the near field ends) is the value of r when the error in R due to neglecting the third term of equation [8], equals 1/16 of a wavelength.

R_{ff} is usually calculated on boresight, so $\theta = 90°$ and the second term of equation [8] equals zero (Cos 90° = 0), therefore from Figure 3, where D is the antenna dimension, R_{ff} is found by equating the third term of [8] to 1/16 wavelength.

$$\frac{(z')^2 \sin^2 \theta}{2 R_{ff}} = \frac{\lambda}{16}$$

Sin θ = Sin 90 = 1 and z' = D/2 $\qquad$ so: $\qquad \dfrac{\left(\dfrac{D}{2}\right)^2}{2 R_{ff}} = \dfrac{\lambda}{16}$

$$R_{ff} = \frac{16(D/2)^2}{2\lambda} = \frac{2 D^2}{\lambda} \qquad \text{[9]}$$

Equation [9] is the standard calculation of far field given in all references.

Besides [9] some general rules of thumb for far field conditions are:
$$r \gg D \quad \text{or} \quad r \gg \lambda$$

If the sphere and point P are a very great distance from the antenna, then the rays are very nearly parallel and this difference is small as in Figure 3.

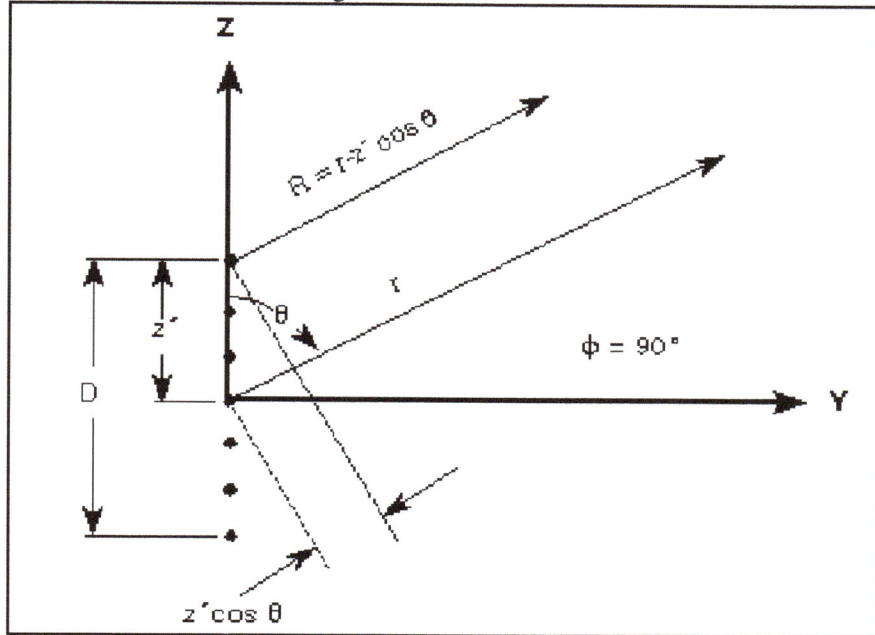

Figure 3. Far Field Parallel Ray Approximation for Calculations

For reference purposes, a simplified alternative method to derive minimum far field distance approximation for an antenna or an antenna array with aperture size "d" is presented in Figure 4 without the need to resort to the spherical coordinate system and the Binomial Theorem. This approach illustrates a simple application of the Pythagorean Theorem. The P_{FF} symbol represents an arbitrary point in the far field at antenna boresight. Refer to relevant quantities shown in Figure 4.

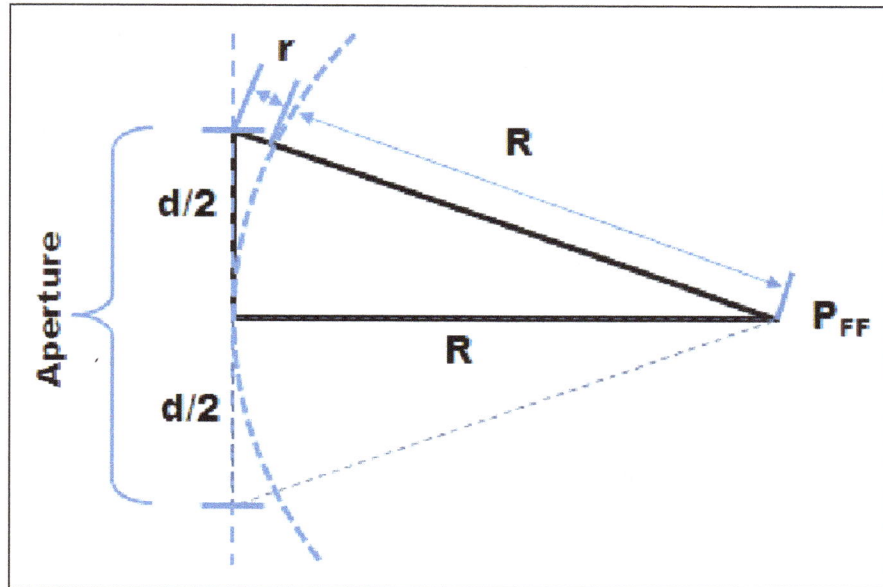

Figure 4. Alternative geometry for Far Field Estimation

From Figure 4, the distance "r" is the maximum propagation path difference across a given antenna aperture endpoint locations. Expressing this propagation path difference in terms of a wavelength provides means to derive an equation for generalized far field range relationship. Note a slight change of notation between Figures 3 and 4. A derivation of results already shown in Equations [2] & [9] is given as follows. From Figure 4, the following applies:

$$(R + r)^2 = (d/2)^2 + R^2 \qquad\qquad [10]$$

Collecting terms and cancelling,

$$2Rr + r2 = d2/4 \qquad\qquad [11]$$

Simplifying with a noted valid assumption,

$$2Rr \approx d2/4, \ \text{ since } r << R \qquad\qquad [12]$$

Solving for "r" (see Figure 4)

$$r \approx d2/(8R) \qquad\qquad [13]$$

For far field condition assume $r \leq \lambda/16$. Substituting "r" in terms of lambda,

$$\lambda/16 \geq r \approx d2/(8R) \ \rightarrow \ RFF \geq 2d2/\lambda \qquad\qquad [14]$$

As shown, Equation [14] represents the same result as Equation [9] derived earlier.

The power density within the near field varies as a function of the type of aperture illumination and is less than would be calculated by equation [1]. Thus, in the antenna near field there is stored energy. (The complex radiation field equations have imaginary terms indicating reactive power.) Figure 5 shows normalized power density for three different illuminations.

Curve A is for reference only and shows how power density would vary if it were calculated using equation [1].

Curve B shows power density variations on axis for an antenna aperture with a cosine amplitude distribution. This is typical of a horn antenna in the H-plane.

Curve C shows power density variations on axis for a uniformly illuminated antenna aperture or for a line source. This is typical of a horn antenna in the E-plane.

Curve D shows power density variations on axis for an antenna aperture with a tapered illumination. Generally the edge illumination is approximately -10 dB from the center illumination and is typical of a parabolic dish antenna.

Point E - For radiation safety purposes, a general rule of thumb for tapered illumination is that the maximum safe level of 10 mW/cm^2 (~200 V/m) is reached in the near field if the level at R_{ff} reaches 0.242 mW/cm^2 as can be verified by computing the power density at point E in Figure 5. (10 mW/cm^2 at point E extrapolates to 0.242 mW/cm^2 [16 dB lower] at R=R_{ff} , or Y axis value =1). Figure 1 in Section 3-6 depicts more precise values for radiation hazard exposure.

Point F - Far Field Point. At distances closer to the source than this point (near field), the power density from any given antenna is less than that predicted using Curve A. At farther distances, (far field) power densities from all types of antennas are the same.

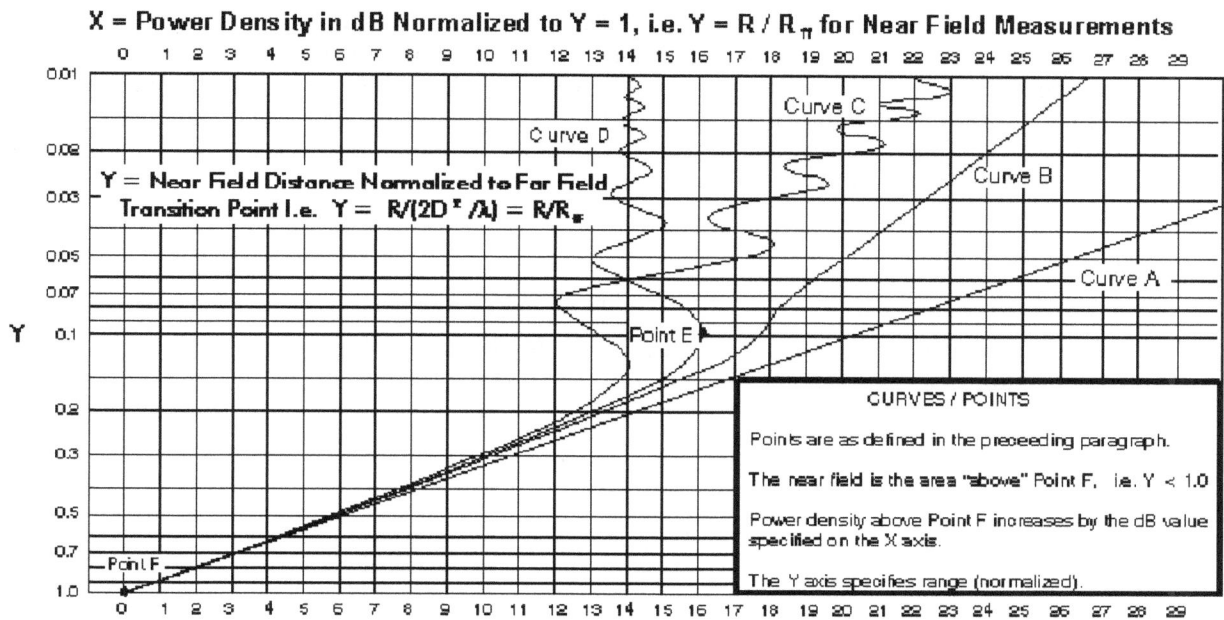

X = Power Density in dB Normalized to Y = 1, i.e. Y = R / R$_{n}$ for Near Field Measurements

Figure 5. Antenna Near-Field On-Axis Power Density (Normalized)
For Various Aperture Illuminations

FOR FAR FIELD MEASUREMENTS:

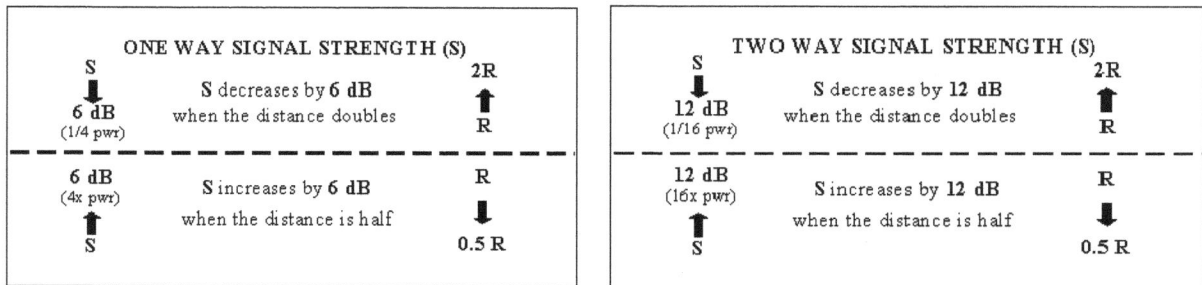

When free space measurements are performed at a known distance from a source, it is often necessary to know if the measurements are being performed in the far field. As can be seen from Curve A on Figure 5, if the distance is halved (going from 1.0 to 0.5 on the Y axis), the power density will increase by 6 dB (going from 0 to 6 dB on the X axis). Each reduction in range by ½ results in further 6 dB increases. As previously mentioned, Curve A is drawn for reference only in the near field region, since at distances less than R$_{ff}$ the power density increases less than 6 dB when the range is halved. In the far field, all curves converge and Equation [1] applies.

When a measurement is made in free space, a good check to ensure that is was performed in the far field is to repeat the measurement at twice the distance. The power should decrease by exactly 6 dB. A common error is to use 3 dB (the half power point) for comparison. Conversely, the power measurement can be repeated at half the distance, in which case you would look for a 6 dB increase, however the assumed extrapolation conclusion is not as certain, because the first measurement could have been made in the far field, and the second could have been made in the near field.

Care must be exercised in using the $2d^2/\lambda$ far field measurement criterion. For antennas with moderate sidelobe levels (>-25 dB) pattern errors are negligible and the error in directivity is less than 0.1 dB and this measurement distance suffices in most cases. However low sidelobe antennas require longer measurement distances. For example, maintaining a 1 dB or less sidelobe error for a linear array with a -40 dB sidelobe level requires a measurement distance of $6d^2/\lambda$ [1]. This corresponds to criterion for "r" in Equation [14] to be changed from $r \leq \lambda/16$ to $r \leq \lambda/48$ or 7.5 deg.

[1] R.C. Hansen; "Measurement Distance Effects on Low Sidelobe Patterns"; IEEE Transactions on Antennas and Propagation, **VOL.** AP-32, NO. *6,* JUNE **1984**

RADIATION HAZARDS

Radiation Hazard (RADHAZ) describes the hazards of electromagnetic radiation to fuels, electronic hardware, ordnance, and personnel. In the military these hazards are segregated as follows:
1) Hazards of Electromagnetic Radiation to Personnel (HERP)
2) Hazards of Electromagnetic Radiation to Ordnance (HERO)
3) Hazards of Electromagnetic Radiation to Fuel (HERF)

The current industrial specifications for RADHAZ are contained in ANSI/IEEE C95.1-1992 which was used as a reference to create the combined Navy regulation NAVSEA OP3565 / NAVAIR 16-1-529. Volume I contains HERP and HERF limits - its current version is REV 5. Volume II (REV 6) covers HERO. These limits are shown in Figure 1 although all values have been converted to <u>average</u> power density.

OP 3565 specifies HERO RADHAZ levels at frequencies below 1 GHz in <u>peak</u> value of electric field strength (V/m), while levels above 200 MHz are specified in <u>average</u> power density (mW/cm^2) - note the overlapping frequencies. Since Figure 1 depicts power density as the limits, you must convert the average values to peak field strength for use at lower frequencies. Also many applications of EMC work such as MIL-STD-461 use limits based on the electric (E) field strength in volts/meter. Remember that P=E^2/R, and from Section 4-2, we note that R=377Ω for free space. It can also be shown that the

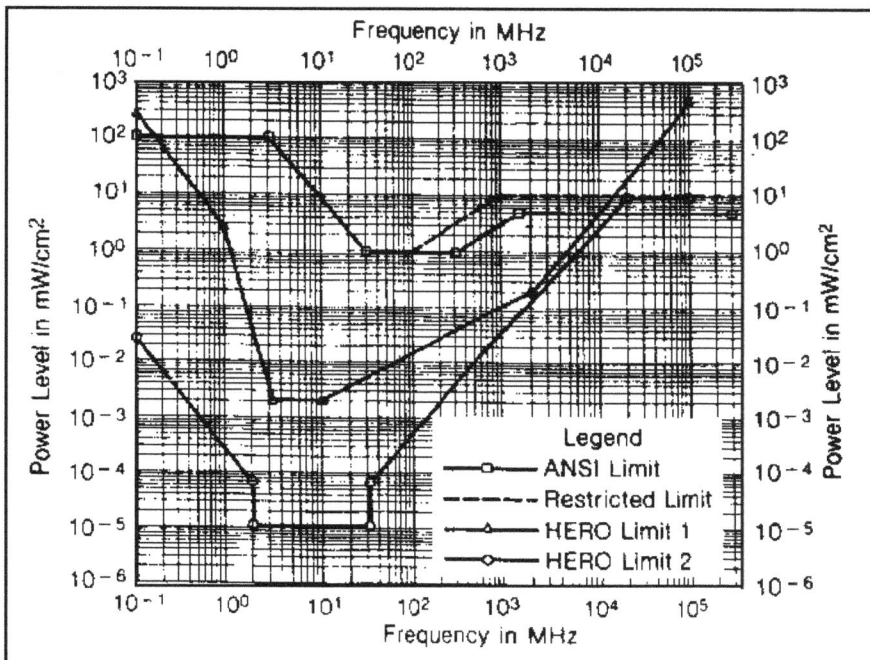

Figure 1. Radiation Hazards to Personnel and Ordnance

magnetic field strength (H field in Amps/meter) = I/m where I=E/R. Don't forget that RMS = 0.707 Peak. With the units of P$_D$ in mW/cm^2, E in V/m, and H in A/m, then P$_D$ (mW/cm^2) = E^2 / 3770 = 37.7 H^2. It should thus be noted that a 100 times increase in power (mW/cm^2) is only a 10 times increase in V/m.

The potential dangers to ordnance and fuels are obvious because there could be an explosive "chain reaction" by exploding; consequently, these limits are generally lower than personnel limits. There are three **HERO** categories. The HERO limit 2 is for HERO "unsafe" or "unreliable" explosive devices with exposed wires arranged in optimum (most susceptible) receiving orientation. This usually occurs during the assembly/disassembly of ordnance, but also applies to new/untested ordnance until proven "safe" or "susceptible." The HERO limit 1 is for HERO susceptible ordnance fully assembled undergoing normal handling and loading operations. HERO safe ordnance requires <u>no</u> RF radiation precautions. A list of which specific ordnance (by NALC) falls into each category can be found in OP 3565 along with specific frequency restrictions for each piece of ordnance. For example, all missiles of one variety are susceptible (HERO 1 limits), while another missile has both susceptible and safe variants (with no RADHAZ limits). Other ordnance may be HERO unsafe (HERO 2 limits).

The danger of **HERP** occurs because the body absorbs radiation and significant internal heating may occur without the individuals knowledge because the body does not have internal sensation of heat, and tissue damage may occur before the excess heat can be dissipated. As shown in Figure 1, the current "restricted" limit is for individuals more than 55" tall because they have more body mass. In other words, all people may be exposed to the lower limit, but only persons taller than 55" may be exposed to the higher limit of 10 mW/cm^2.

NAVSEA OP 3565 will be updated in the future to be compatible with DoD INST 6055.11 dated Feb 21, 1995 which supersedes it. The personnel radiation levels in Figures 2 and 3 were taken from the new release of DoD INST 6055.11.

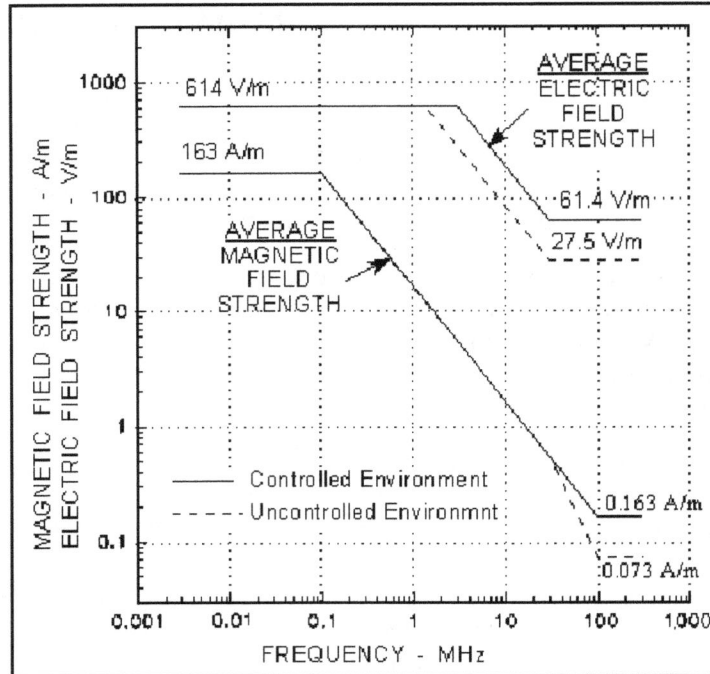

Figure 2. Lower Frequency HERP from DoD INST 6055.11

Unlike the existing "restricted limit" of NAVSEA OP 3565 discussed above, in the revised DoD instruction for personnel radiation hazards, a different approach to exposure was taken.

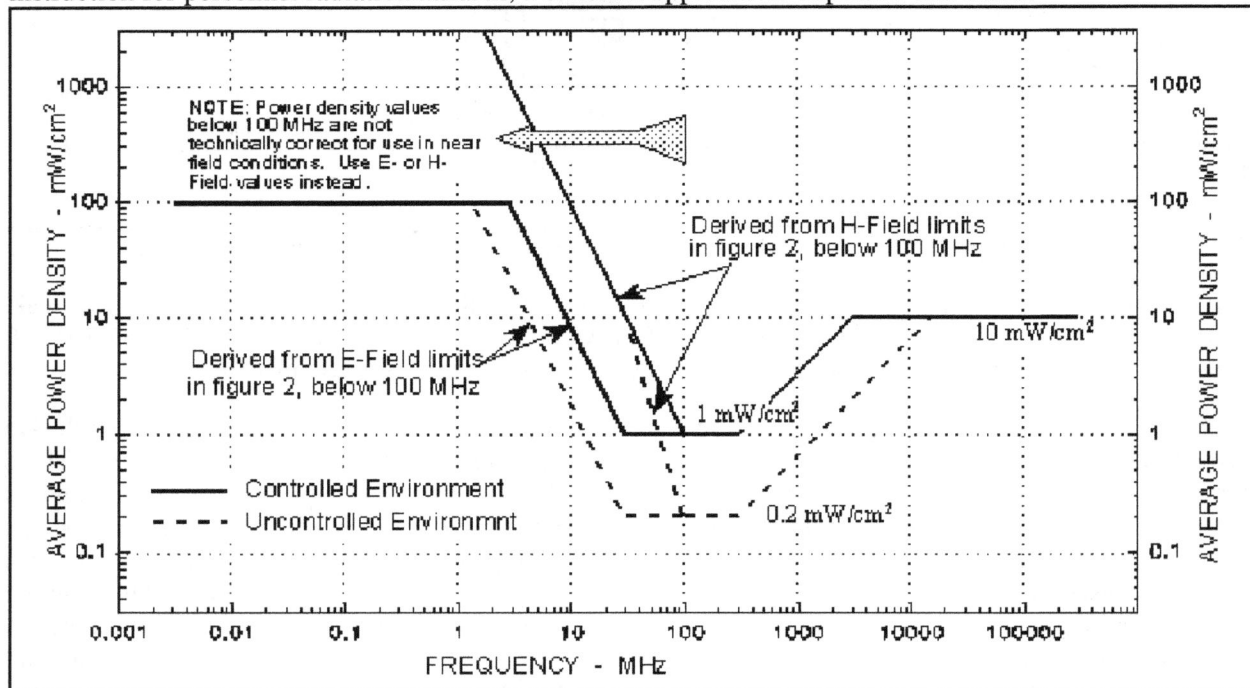

Figure 3. Radiation Hazards to Personnel from DoD INST 6055.11

Two maximum hazard limits are defined;

1) <u>Controlled Environments</u> - where personnel are aware of the potential danger of RF exposure concurrently with employment, or exposure which may occur due to incidental transient passage through an area, and;

2) <u>Uncontrolled Environments</u> - A lower maximum level where there is no expectation that higher levels should be encountered, such as living quarters.

These Personnel Exposure Limits (PELs) are based on a safety factor of ten times the Specific Absorption Rate (SAR) which might cause bodily harm. The term PEL is equivalent to the terms "Maximum Permissible Exposure (MPE)" and "Radio Frequency Protection Guides (RFPG)" in other publications.

There are several exceptions to the max limits in Figs 2 and 3 (in some cases higher levels are permitted):

- High Power Microwave (HPM) system exposure in a controlled environment, which has a single pulse or multiple pulses lasting less than 10 seconds, has a higher peak E-Field limit of 200 kV/m.
- EMP Simulation Systems in a controlled environment for personnel who are exposed to broad-band (0.1 MHz to 300 GHz) RF are limited to a higher peak E-Field of 100 kV/m.
- The given limits are also increased for pulsed RF fields. In this case the peak power density per pulse for pulse durations < 100 msec and no more than 5 pulses in the period is increased to: PELPulse = PEL x TAVG / 5 x Pulse Width, and the peak E-field is increased to 100 kV/m. If there are more than 5 pulses or they are greater then 100 msec, a time averaged PD should not exceed that shown in Figure 3.
- A rotating or scanning beam likewise reduces the hazard, so although an on-axis hazard might exist, there may be none with a moving beam. The power density may be approximated with:
  PDscan = PDfixed (2 x Beam Width / scan angle)
- Many other special limitations also apply, such as higher limits for partial body exposure, so if in doubt, read the DoD Inst 6055.11 in detail. Field measurements may be measured in accordance with IEEE C95.3-1991.

The PELs listed in Figures 2 and 3 were selected for an average RF exposure time at various frequencies. In a controlled environment, this averaging time was selected as 6 minutes for 0.003 to 15,000 MHz. If the exposure time is less than 6 minutes, then the level may be increased accordingly. Similar time weighted averages apply to uncontrolled environments, but it varies enough with frequency such that DoD INST 6055.11 should be consulted.

NAVSEA OP 3565 contains a list of Navy avionics which transmit RF as well as radars along with their respective hazard patterns. Special training is required for individuals who work in areas which emit RF levels which exceed the uncontrolled levels. Warning signs are also required in areas which exceed either the controlled or uncontrolled limits.

Although E-Field, H-Field, and power density can be mathematically converted in a far-field plane wave environment, the relations provided earlier do not apply in the near field, consequently the E- or H-field strength must be measured independently below 100 MHz. It should be noted that the specifications in NAVSEA OP 3565 for lower frequency HERO limits are listed as <u>peak</u> E-field values, whereas lower RF limits in DoD INST 6055.11 on HERP are in <u>average</u> (RMS) E-field values. Upper frequency restrictions are based on <u>average</u> (RMS) values of power density in both regulations except for certain circumstances.

HERF precautions are of more general concern to fuel truck operators. However, some general guidelines include:

- Do not energize a transmitter (radar/comm) on an aircraft or motor vehicle being fueled or on an adjacent aircraft or vehicle.
- Do not make or break any electrical, ground wire, or tie down connector while fueling.
- Radars able to illuminate fueling areas with a peak power density of 5 W/cm2 should be shut off.
- For shore stations, antennas radiating 250 watts or less should be installed at least 50 ft from fueling areas (at sea 500 watts is the relaxed requirement).
- For antennas which radiate more than 250 watts, the power density at 50 ft from the fueling operation should not be greater than the equivalent power density of a 250 watt transmitter at 50 ft.

ACTIVE ELECTRONICALLY SCANNED ARRAYS (AESA)

Scanning phased arrays employing electronically controlled phase shifters (e.g. PiN diode, ferrite) have been used in high power radar applications since the early 1960s. These are planar arrays and generally use a corporate type feed structure to array rows of elements in one dimension of the and use an analogous feed to array the rows in the orthogonal dimension of the array, resulting in a single feed point. A high power RF transmitter, usually employing some sort of liquid cooling, is used to excite the array. This architecture is cumbersome and difficult to package and the high power transmitter is a single point of failure for the system.

The maturation of solid state transmit/receive modules (T/R) using Gallium Arsenide (GaAs) technology or more recently Gallium Nitride (GaN) technology has made AESAs a practical reality. The T/R modules consist of a low noise amplifier (LNA) for the receive function and a solid state high power amplifier (SSPA) for realization of the transmit function. The T/R module may also contain the necessary phase control elements for beam scanning. A generic 8-element AESA for one dimension is shown in Figure 1 below.

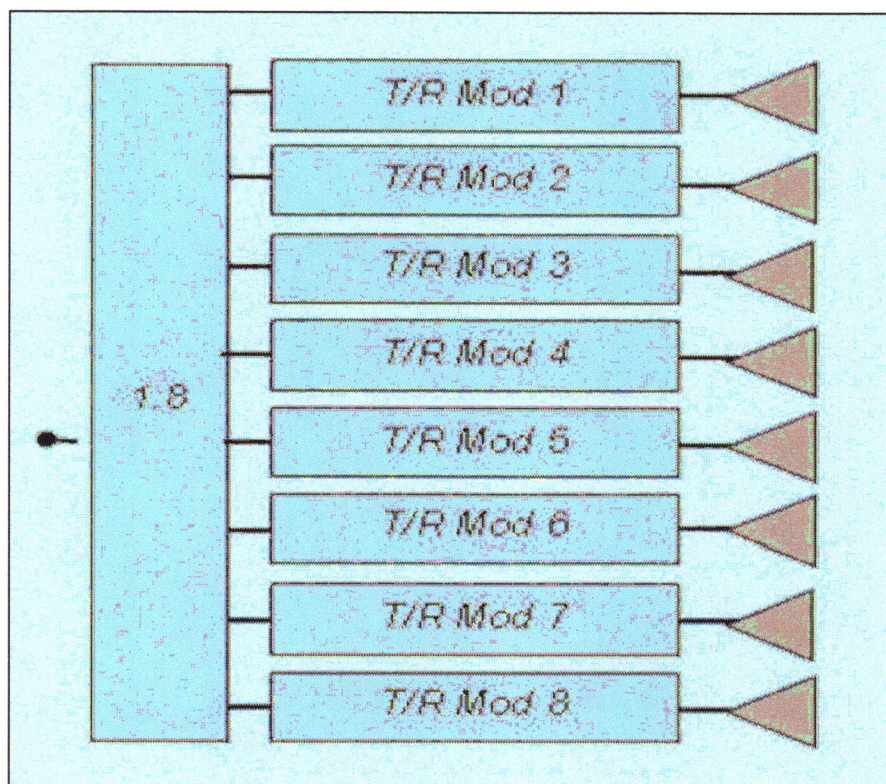

Figure 1: Eight Element AESA.

Electronically scanned phased array designs with decade bandwidths have been reported in the open literature for many years. However those reported are only lab versions and none have been implemented in a fielded system. In addition, the high average power requirement, particularly for EW systems, calls for the use of GaN SSPA technology and this, in conjunction with the dense array packaging requirement, poses severe heat dissipation/cooling issues.

AESAs for EW applications impose design considerations vastly different from an AESA for radar. In contradistinction with high **peak** power radar applications, e.g. AN/APG79, the high **average** power large duty factors (sometimes CW) requirements for EW impose severe design requirements and constraints for the wide bandwidth AESA. Incorporation of an AESA in airborne environment, with its concomitant restrictive volume, further exacerbates design constraints.

The heat dissipation problem is of major concern in modern high power AESAs. Due to the behavior of microwave transistor amplifiers, the power added efficiency (PAE) of a TR module transmitter is typically a relatively small fraction of the total prime power consumption. As a result, an AESA will dissipate a lot of heat which must be extracted. The reliability of GaAs and GaN MMIC chips improves if the RF amplifier system operates at reduced temperatures. Traditional air cooling used in most established avionic hardware is ill suited to the high packaging density of an AESA. As a result modern AESAs are predominantly liquid cooled. A typical liquid cooling system will use pumps to drive the coolant through channels in the cooling plenum of the array, and then route it to a heat exchanger. In comparison, with a conventional air cooled fighter radar, the AESA will be more reliable but will require more electrical power and more cooling infrastructure, and typically can produce much higher average transmit power.

RADAR EQUATIONS

FIELD INTENSITY and POWER DENSITY

Sometimes it is necessary to know the actual field intensity or power density at a given distance from a transmitter instead of the signal strength received by an antenna. Field intensity or power density calculations are necessary when estimating electromagnetic interference (EMI) effects, when determining potential radiation hazards (personnel safety), or in determining or verifying specifications.

Field intensity (field strength) is a general term that usually means the magnitude of the electric field vector, commonly expressed in volts per meter. At frequencies above 100 MHZ, and particularly above one GHz, power density (P_D) terminology is more often used than field strength.
Power density and field intensity are related by equation [1]:

$$P_D = \frac{E^2}{Z_0} = \frac{E^2}{120\pi} = \frac{E^2}{377} \qquad [1]$$

where P_D is in W/m^2, E is the RMS value of the field in volts/meter and 377 ohms is the characteristic impedance of free space. When the units of P_D are in mW/cm^2, then P_D (mW/cm^2) = E^2/3770.

Conversions between field strength and power density when the impedance is 377 ohms, can be obtained from Table 1. It should be noted that to convert dBm/m^2 to dBµV/m add 115.76 dB. Sample calculations for both field intensity and power density in the far field of a transmitting antenna are in Section 4-2 and Section 4-8. Refer to chapter 3 on antennas for the definitions of near field and far field.

Note that the "/" term before m, m^2, and cm^2 in Table 1 mean "per", i.e. dBm per m^2, not to be confused with the division sign which is valid for the Table 1 equation P=E^2/Z$_0$. Remember that in order to obtain dBm from dBm/m^2 given a certain area, you must add the logarithm of the area, not multiply. The values in the table are rounded to the nearest dBW, dBm, etc. per m^2 so the results are less precise than a typical handheld calculator and may be up to ½ dB off.

VOLTAGE MEASUREMENTS

Coaxial cabling typically has input impedances of 50, 75, and 93Ω, (±2) with 50Ω being the most common. Other types of cabling include the following: TV cable is 75Ω (coaxial) or 300Ω (twin-lead), audio public address (PA) is 600Ω, audio speakers are 3.2 (4), 8, or 16Ω.

In the 50Ω case, power and voltage are related by:

$$P = \frac{E^2}{Z_0} = \frac{E^2}{50} = 50\,I^2 \qquad [2]$$

Conversions between measured power, voltage, and current where the typical impedance is 50 ohms can be obtained from Table 2. The dBµA current values are given because frequently a current probe is used during laboratory tests to determine the powerline input current to the system .

MATCHING CABLING IMPEDANCE

In performing measurements, we must take into account an impedance mismatch between measurement devices (typically 50 ohms) and free space (377 ohms).

Table 1. Conversion Table - Field Intensity and Power Density
$$P_D = E^2/Z_0 \ (\text{ Related by free space impedance = 377 ohms })$$

E (Volts/m)	$20 \log 10^6$ (E) (dBμV/m)	P_D (watts/m^2)	$10 \log P_D$ (dBW/m^2)	Watts/cm^2	dBW/cm^2	mW/cm^2	dBm/cm^2	dBm/m^2
7,000	197	130,000	+51	13	+11	13,000	+41	+81
5,000	194	66,300	+48	6.6	+8	6,630	+38	+78
3,000	190	23,900	+44	2.4	+4	2,390	+34	+74
4,000	186	10,600	+40	1.1	0	1,060	+30	+70
1,000	180	2,650	+34	.27	-6	265	+24	+64
700	177	1,300	+31	.13	-9	130	+21	+61
500	174	663	+28	.066	-12	66	+18	+58
300	170	239	+24	.024	-16	24	+14	+54
200	166	106	+20	.011	-20	11	+10	+50
100	160	27	+14	.0027	-26	2.7	+4	+44
70	157	13	+11	1.3×10^{-3}	-29	1.3	+1	+41
50	154	6.6	+8	6.6×10^{-4}	-32	.66	-2	+38
30	150	2.4	+4	2.4×10^{-4}	-36	.24	-6	+34
20	146	1.1	+0	1.1×10^{-4}	-40	.11	-10	+30
10	140	.27	-6	2.7×10^{-5}	-46	.027	-16	+24
7	137	.13	-9	1.3×10^{-5}	-49	.013	-19	+21
5	134	.066	-12	6.6×10^{-6}	-52	66×10^{-4}	-22	+18
3	130	.024	-16	2.4×10^{-6}	-56	24×10^{-4}	-26	+14
2	126	.011	-20	1.1×10^{-6}	-60	11×10^{-4}	-30	+10
1	120	.0027	-26	2.7×10^{-7}	-66	2.7×10^{-4}	-36	+4
0.7	117	1.3×10^{-3}	-29	1.3×10^{-7}	-69	1.3×10^{-4}	-39	+1
0.5	114	6.6×10^{-4}	-32	6.6×10^{-8}	-72	66×10^{-4}	-42	-2
0.3	110	2.4×10^{-4}	-36	2.4×10^{-8}	-76	24×10^{-4}	-46	-6
0.2	106	1.1×10^{-4}	-40	1.1×10^{-8}	-80	11×10^{-4}	-50	-10
0.1	100	2.7×10^{-5}	-46	2.7×10^{-9}	-86	2.7×10^{-6}	-56	-16
70×10^{-3}	97	1.3×10^{-5}	-49	1.3×10^{-9}	-89	1.3×10^{-6}	-59	-19
50×10^{-3}	94	6.6×10^{-6}	-52	6.6×10^{-10}	-92	66×10^{-8}	-62	-22
30×10^{-3}	90	2.4×10^{-6}	-56	2.4×10^{-10}	-96	24×10^{-8}	-66	-26
20×10^{-3}	86	1.1×10^{-6}	-60	1.1×10^{-10}	-100	11×10^{-8}	-70	-30
10×10^{-3}	80	2.7×10^{-7}	-66	2.7×10^{-11}	-106	2.7×10^{-8}	-76	-36
7×10^{-3}	77	1.3×10^{-7}	-69	1.3×10^{-11}	-109	1.3×10^{-8}	-79	-39
5×10^{-3}	74	6.6×10^{-8}	-72	6.6×10^{-12}	-112	66×10^{-10}	-82	-42
3×10^{-3}	70	2.4×10^{-8}	-76	2.4×10^{-12}	-116	24×10^{-10}	-86	-46
2×10^{-3}	66	1.1×10^{-8}	-80	1.1×10^{-12}	-120	11×10^{-10}	-90	-50
1×10^{-3}	60	2.7×10^{-9}	-86	2.7×10^{-13}	-126	2.7×10^{-10}	-96	-56
7×10^{-4}	57	1.3×10^{-9}	-89	1.3×10^{-13}	-129	1.3×10^{-10}	-99	-59
5×10^{-4}	54	6.6×10^{-10}	-92	6.6×10^{-14}	-132	66×10^{-12}	-102	-62
3×10^{-4}	50	2.4×10^{-10}	-96	2.4×10^{-14}	-136	24×10^{-12}	-106	-66
2×10^{-4}	46	1.1×10^{-10}	-100	1.1×10^{-14}	-140	11×10^{-12}	-110	-70
1×10^{-4}	40	2.7×10^{-11}	-106	2.7×10^{-15}	-146	2.7×10^{-12}	-116	-76
7×10^{-5}	37	1.3×10^{-11}	-109	1.3×10^{-15}	-149	1.3×10^{-12}	-119	-79
5×10^{-5}	34	6.6×10^{-12}	-112	6.6×10^{-16}	-152	66×10^{-14}	-122	-82
3×10^{-5}	30	2.4×10^{-12}	-116	2.4×10^{-16}	-156	24×10^{-14}	-126	-86
2×10^{-5}	26	1.1×10^{-12}	-120	1.1×10^{-16}	-160	11×10^{-14}	-130	-90
1×10^{-5}	20	2.7×10^{-13}	-126	2.7×10^{-17}	-166	2.7×10^{-14}	-136	-96
7×10^{-6}	17	1.3×10^{-13}	-129	1.3×10^{-17}	-169	1.3×10^{-14}	-139	-99
5×10^{-6}	14	6.6×10^{-14}	-132	6.6×10^{-18}	-172	66×10^{-16}	-142	-102
3×10^{-6}	10	2.4×10^{-14}	-136	2.4×10^{-18}	-176	24×10^{-16}	-146	-106
2×10^{-6}	6	1.1×10^{-14}	-140	1.1×10^{-18}	-180	11×10^{-16}	-150	-110
1×10^{-6}	0	2.7×10^{-15}	-146	2.7×10^{-19}	-186	2.7×10^{-16}	-156	-116

NOTE: Numbers in table rounded off

FIELD STRENGTH APPROACH

To account for the impedance difference, the antenna factor (AF) is defined as: AF=E/V, where E is field intensity which can be expressed in terms taking 377 ohms into account and V is measured voltage which can be expressed in terms taking 50 ohms into account. Details are provided in Section 4-12.

POWER DENSITY APPROACH

To account for the impedance difference , the antenna's effective capture area term, A_e relates free space power density P_D with received power, P_r , i.e. $P_r = P_D A_e$. A_e is a function of frequency and antenna gain and is related to AF as shown in Section 4-12.

SAMPLE CALCULATIONS

Section 4-2 provides sample calculations using power density and power terms from Table 1 and Table 2, whereas Section 4-12 uses these terms plus field intensity and voltage terms from Table 1 and Table 2. Refer the examples in Section 4-12 for usage of the conversions while converting free space values of power density to actual measurements with a spectrum analyzer attached by coaxial cable to a receiving antenna.

Conversion Between Field Intensity (Table 1) and Power Received (Table 2).

Power received (watts or milliwatts) can be expressed in terms of field intensity (volts/meter or µv/meter) using equation [3]:

$$Power\ received\ (\ P_r\ ) = \frac{E^2}{480\ \pi^2}\ \frac{c^2}{f^2}\ G \qquad\qquad \textbf{[3]}$$

or in log form: $\quad 10 \log P_r = 20 \log E + 10 \log G - 20 \log f + 10 \log (c^2/480\pi^2)$ **[4]**

Then $\qquad\qquad 10 \log\ P_r = 20 \log E_1 + 10 \log G - 20 \log f_1 + K_4$ **[5]**

$$Where\ \ K_4 = 10 \log \left[\frac{c^2}{480\ \pi^2} \bullet \left(\begin{array}{c} conversions \\ as\ required \end{array} \frac{(Watts\ to\ mW)}{(volts\ to\ \mu v)^2\ (Hz\ to\ MHz\ or\ GHz)^2} \right) \right]$$

The derivation of equation [3] follows:

Values of K₄ (dB)

$P_D = E^2/120\pi \quad$ Eq [1], Section 4-1, terms (v^2/Ω)

$A_e = \lambda^2 G/4\pi \quad$ Eq [8], Section 3-1, terms (m^2)

$P_r = P_D A_e \quad$ Eq [2], Section 4-3,
terms (W/m^2)(m^2)

P_r	E_1	f_1 (Hz)	f_1 (MHz)	f_1 (GHz)
Watts (dBW)	volts/meter	132.8	12.8	-47.2
	µv/meter	12.8	-107.2	-167.2
mW (dBm)	volts/meter	162.8	42.8	-17.2
	µv/meter	42.8	-77.2	-137.7

$\therefore P_r = (\ E^2/120\pi\ )(\ \lambda^2 G/4\pi)\quad$ terms ($v^2/m^2\Omega$)(m^2)

$\lambda = c\ /f\quad$ Section 2-3, terms (m/sec)(sec)

$\therefore P_r = (\ E^2/480\pi^2\ )(\ c^2 G/f^2)$ which is equation [3]

terms ($v^2/m^2\Omega$)(m^2/sec^2)(sec^2) or v^2/Ω = watts

Table 2. Conversion Table - Volts to Watts and dBμA
($P_x = V_x^2/Z$ - Related by line impedance of 50 Ω)

Volts	dBV	dBμV	Watts	dBW	dBm	dBμA
700	56.0	176.0	9800	39.9	69.9	142.9
500	53.9	173.9	5000	37.0	67.0	140.0
300	49.5	169.5	1800	32.5	62.5	135.5
200	46.0	166.0	800	29.0	59.0	132.0
100	40.0	160.0	200	23.0	53.0	126.0
70	36.9	156.9	98	19.9	49.9	122.9
50	34.0	154.0	50	17.0	47.0	120.0
30	29.5	149.5	18	12.5	42.5	115.5
20	26.0	146.0	8	9.0	39.0	112.0
10	20.0	140.0	2	3.0	33.0	106.0
7	16.9	136.9	0.8	0	29.9	102.9
5	14.0	134.0	0.5	-3.0	27.0	100.0
3	9.5	129.5	0.18	-7.4	22.5	95.6
2	6.0	126.0	0.08	-11.0	19.0	92.0
1	0	120.0	0.02	-17.0	13.0	86.0
0.7	-3.1	116.9	9.8×10^{-3}	-20.1	9.9	82.9
0.5	-6.0	114.0	5.0×10^{-3}	-23.0	7.0	80.0
0.3	-10.5	109.5	1.8×10^{-3}	-27.4	2.6	75.6
0.2	-14.0	106.0	8.0×10^{-4}	-31.0	-1.0	72.0
0.1	-20.0	100.0	2.0×10^{-4}	-37.0	-7.0	66.0
.07	-23.1	96.9	9.8×10^{-5}	-40.1	-10.1	62.9
.05	-26.0	94.0	5.0×10^{-5}	-43.0	-13.0	60.0
.03	-30.5	89.5	1.8×10^{-5}	-47.4	-17.7	55.6
.02	-34.0	86.0	8.0×10^{-6}	-51.0	-21.0	52.0
.01	-40.0	80.0	2.0×10^{-6}	-57.0	-27.0	46.0
7×10^{-3}	-43.1	76.9	9.8×10^{-7}	-60.1	-30.1	42.9
5×10^{-3}	-46.0	74.0	5.0×10^{-7}	-63.0	-33.0	40.0
3×10^{-3}	-50.5	69.5	1.8×10^{-7}	-67.4	-37.4	35.6
2×10^{-3}	-54.0	66.0	8.0×10^{-8}	-71.0	-41.0	32.0
1×10^{-3}	-60.0	60.0	2.0×10^{-8}	-77.0	-47.0	26.0
7×10^{-4}	-64.1	56.9	9.8×10^{-9}	-80.1	-50.1	22.9
5×10^{-4}	-66.0	54.0	5.0×10^{-9}	-83.0	-53.0	20.0
3×10^{-4}	-70.5	49.5	1.8×10^{-9}	-87.4	-57.4	15.6
2×10^{-4}	-74.0	46.0	8.0×10^{-10}	-91.0	-61.0	12.0
1×10^{-4}	-80.0	40.0	2.0×10^{-10}	-97.0	-67.0	6.0
7×10^{-5}	-84.1	36.9	9.8×10^{-11}	-100.1	-70.1	2.9
5×10^{-5}	-86.0	34.0	5.0×10^{-11}	-103.0	-73.0	0
3×10^{-5}	-90.5	29.5	1.8×10^{-11}	-107.4	-77.4	-4.4
2×10^{-5}	-94.0	26.0	8.0×10^{-12}	-111.0	-81.0	-8.0
1×10^{-5}	-100.0	20.0	2.0×10^{-12}	-117.0	-87.0	-14.0
7×10^{-6}	-104.1	16.9	9.8×10^{-13}	-120.1	-90.1	-17.1
5×10^{-6}	-106.0	14.0	5.0×10^{-13}	-123.0	-93.0	-20.0
3×10^{-6}	-110.5	9.5	1.8×10^{-13}	-127.4	-97.4	-24.4
2×10^{-6}	-114.0	6.0	8.0×10^{-14}	-131.0	-101.0	-28.0
1×10^{-6}	-120.0	0	2.0×10^{-14}	-137.0	-107.0	-34.0
7×10^{-7}	-124.1	-3.1	9.8×10^{-15}	-140.1	-110.1	-37.1
5×10^{-7}	-126.0	-6.0	5.0×10^{-15}	-143.0	-113.0	-40.0
3×10^{-7}	-130.5	-10.5	1.8×10^{-15}	-147.4	-117.4	-44.4
2×10^{-7}	-134.0	-14.0	8.0×10^{-16}	-151.0	-121.0	-48.0
1×10^{-7}	-140.0	-20.0	2.0×10^{-16}	-157.0	-127.0	-54.0

POWER DENSITY

Radio Frequency (RF) propagation is defined as the travel of electromagnetic waves through or along a medium. For RF propagation between approximately 100 MHz and 10 GHz, radio waves travel very much as they do in free space and travel in a direct line of sight. There is a very slight difference in the dielectric constants of space and air. The dielectric constant of space is one. The dielectric constant of air at sea level is 1.000536. In all but the highest precision calculations, the slight difference is neglected.

From chapter 3, Antennas, an isotropic radiator is a theoretical, lossless, omnidirectional (spherical) antenna. That is, it radiates uniformly in all directions. The power of a transmitter that is radiated from an isotropic antenna will have a uniform power density (power per unit area) in all directions. The power density at any distance from an isotropic antenna is simply the transmitter power divided by the surface area of a sphere $(4\pi R^2)$ at that distance. The surface area of the sphere increases by the square of the radius, therefore the power density, P_D, (watts/square meter) decreases by the square of the radius.

$$\begin{matrix} Power\ density\ from \\ an\ isotropic\ antenna \end{matrix} = P_D = \frac{P_t}{4\pi R^2} \qquad \begin{matrix} where: P_t = Transmitter\ Power \\ R = Range\ From\ Antenna\,(i.e.\ radius\ of\ sphere) \end{matrix} \qquad [1]$$

P_t is either peak or average power depending on how P_D is to be specified.

Radars use directional antennas to channel most of the radiated power in a particular direction. The Gain (G) of an antenna is the ratio of power radiated in the desired direction as compared to the power radiated from an isotropic antenna, or:

$$G = \frac{Maximum\ radiation\ intensity\ of\ actual\ antenna}{Radiation\ intensity\ of\ isotropic\ antenna\ with\ same\ power\ input}$$

The power density at a distant point from a radar with an antenna gain of G_t is the power density from an isotropic antenna multiplied by the radar antenna gain.

Power density from radar, $\qquad P_D = \frac{P_t G_t}{4\pi R^2}$ $\qquad\qquad\qquad\qquad$ [2]

P_t is either peak or average power depending on how P_D is to be specified.

Another commonly used term is effective isotropic radiated power (EIRP), where **EIRP = P_t G_t.** ERP is also used but EIRP is preferred because it specifically defines the type of reference antenna as isotropic.

A receiving antenna captures a portion of this power determined by it's effective capture Area (A_e). The received power available at the antenna terminals is the power density times the effective capture area (A_e) of the receiving antenna.

> e.g. If the power density at a specified range is one microwatt per square meter and the antenna's effective capture area is one square meter then the power captured by the antenna is one microwatt.

For a given receiver antenna size the capture area is constant no matter how far it is from the transmitter, as illustrated in Figure 1. Also notice from Figure 1 that the received signal power decreases by 1/4 (6 dB) as the distance doubles. This is due to the R^2 term in the denominator of equation [2].

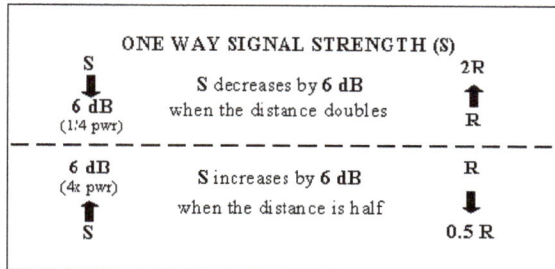

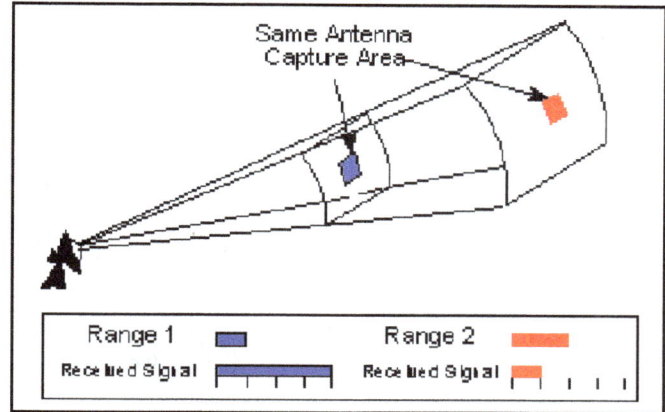

Figure 1. Power Density vs. Range

Sample Power Density Calculation - Far Field (Refer to Section 3-5 for the definition of near field and far field)

Calculate the power density at 100 feet for 100 watts transmitted through an antenna with a gain of 10.

Given: $P_t = 100$ watts $\quad G_t = 10$ (dimensionless ratio) $\quad R = 100$ ft

This equation produces power density in watts per square range unit.

$$P_D = \frac{P_t G_t}{4\pi R^2} = \frac{(100\ watts)(10)}{4\pi(100\ ft)^2} = 0.0080\ watts/\ ft^2$$

For safety (radiation hazard) and EMI calculations, power density is usually expressed in milliwatts per square cm. That's nothing more than converting the power and range to the proper units.

100 watts = 1 x 10^2 watts = 1 x 10^5 mW

100 feet = 30.4785 meters = 3047.85 cm.

$$P_D = \frac{P_t G_t}{4\pi R^2} = \frac{(10^5\ mW) \bullet (10)}{4\pi(3047.85cm)^2} = 0.0086\ mW/\ cm^2$$

However, antenna gain is almost always given in dB, not as a ratio. It's then often easier to express EIRP in dBm.

$$P_t(dBm) = 10\ Log\left[\frac{P_t\ watts}{1\ mW}\right] = 10\ Log\left[\frac{100}{.001}\right] = 50\ dBm$$

$$G_t(dB) = 10\ Log\left[\frac{G_t}{1}\right] = 10\ Log(10) = 10\ dB$$

EIRP (dBm) = P_t (dBm) + G_t (dB) = 50 + 10 = 60 dBm

To reduce calculations, the graph in Figure 2 can be used. It gives EIRP in dBm, range in feet and power density in mW/cm^2. Follow the scale A line for an EIRP of 60 dBm to the point where it intersects the 100 foot range scale. Read the power density directly from the A-scale x-axis as 0.0086 mW/cm^2 (confirming our earlier calculations).

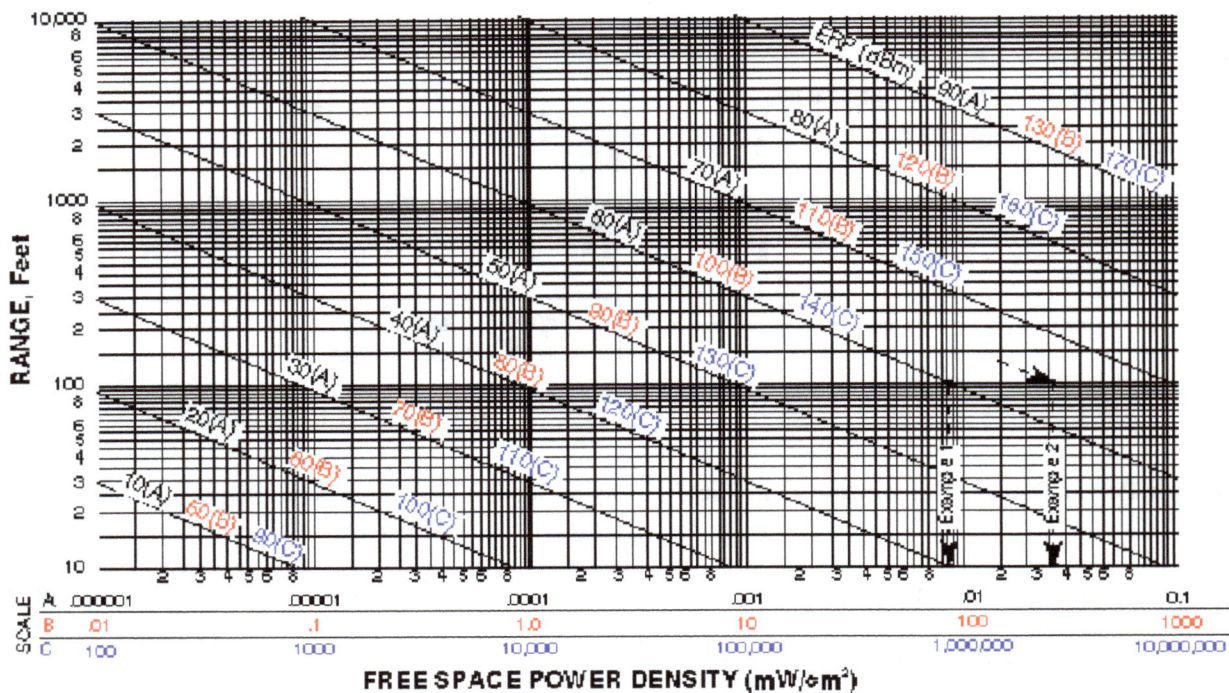

Figure 2. Power Density vs. Range and EIRP

Example 2

When antenna gain and power (or EIRP) are given in dB and dBm, it's necessary to convert back to ratios in order to perform the calculation given in equation [2]. Use the same values as in example 1 except for antenna gain.

Suppose the antenna gain is given as 15 dB: **G_t (dB) = 10 Log (G_t)**

$$Therefore: \quad G_t = [10]^{\left(\frac{G_t(dB)}{10}\right)} = [10]^{\left(\frac{15}{10}\right)} = 31.6228$$

$$P_D = \frac{P_t G_t}{4\pi R^2} = \frac{(10^5 \, mW)(31.6228)}{4\pi (3047.85)^2} = 0.0271 \, mW/cm^2$$

Follow the 65 dBm (extrapolated) EIRP line and verify this result on the A-scale X-axis.

Example 3 - Sample Real Life Problem

Assume we are trying to determine if a jammer will damage the circuitry of a missile carried onboard an aircraft and we cannot perform an actual measurement. Refer to the diagram at the right.

Given the following:

Jammer power: 500 W ($P_t = 500$)

Jammer line loss and antenna gain: 3 dB ($G_t = 2$)

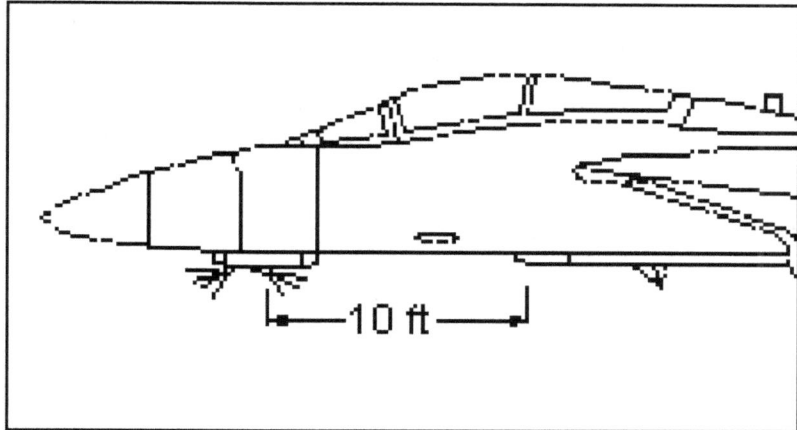

Missile antenna diameter: 10 in

Missile antenna gain: Unknown

Missile limiter protection (maximum antenna power input): 20 dBm (100mW) average and peak.

The power density at the missile antenna caused by the jammer is computed as follows:

$$P_D = \frac{P_t\,G_t}{4\pi\,R^2} = \frac{500W\,(2)}{4\pi\,[(10ft)(.3048m/ft)\,]^2} = 8.56W/m^2$$

The maximum input power actually received by the missile is either:
$P_r = P_D\,A_e$ (if effective antenna area is known) or
$P_r = P_D\,G_m\lambda^2/4\pi$ (if missile antenna gain is known)

To cover the case where the missile antenna gain is not known, first assume an aperture efficiency of 0.7 for the missile antenna (typical). Then:

$P_r = P_D\,A\,\eta = 8.56\ W/m^2\,(\pi)[\,(10/2\ in)(.0254\ m/in)\,]^2\,(0.7) = 0.3$ watts

Depending upon missile antenna efficiency, we can see that the power received will be about 3 times the maximum allowable and that either better limiter circuitry may be required in the missile or a new location is needed for the missile or jammer. Of course if the antenna efficiency is 0.23 or less, then the power will not damage the missile's receiver.

If the missile gain were known to be 25 dB, then a more accurate calculation could be performed. Using the given gain of the missile (25 dB= numeric gain of 316), and assuming operation at 10 GHz (λ = .03m)

$P_r = P_D\,G_m\,\lambda^2\,/\,4\pi = 8.56\ W/m^2\,(316)(.03)^2/\,4\pi = .19$ watts   (still double the allowable tolerance)

ONE-WAY RADAR EQUATION / RF PROPAGATION

The one-way (transmitter to receiver) radar equation is derived in this section. This equation is most commonly used in RWR, communications, or ESM type of applications. The following is a summary of the important equations explored in this section:

ONE-WAY RADAR EQUATION

Peak Power at Receiver Input, $P_r \, (or \, S) = P_D \, A_e = \dfrac{P_t \, G_t \, A_e}{4\pi \, R^2}$ and Antenna Gain, $G = \dfrac{4\pi \, A_e}{\lambda^2}$

or : Equivalent Area, $A_e = \dfrac{G \, \lambda^2}{4\pi}$

So the one-way radar equation is :

$S \, (or \, P_r) = \dfrac{P_t \, G_t \, G_r \, \lambda^2}{(4\pi R)^2} = P_t \, G_t \, G_r \left[\dfrac{c^2}{(4\pi f \, R)^2} \right]^*$ (Note : $\lambda = \dfrac{c}{f}$)

* keep λ, c, and R in the same units

On reducing to log form this becomes:
$10\log P_r = 10\log P_t + 10\log G_t + 10\log G_r - 20\log f R + 20\log (c/4\pi)$

or in simplified terms:
$10\log P_r = 10\log P_t + 10\log G_t + 10\log G_r - \alpha_1$ (in dB)

Where: α_1 = one-way free space loss = $20\log (f_1 R) + K_1$ (in dB) and:
$K_1 = 20\log [(4\pi/c)(\text{Conversion factors if units if not in m/sec, m, and Hz})]$

Note: To avoid having to include additional terms for these calculations, always combine any transmission line loss with antenna gain

Values of K_1 (in dB)		
Range (units)	f_1 in MHz $K_1 =$	f_1 in GHz $K_1 =$
NM	37.8	97.8
km	32.45	92.45
m	-27.55	32.45
yd	-28.33	31.67
ft	-37.87	22.13

Note: Losses due to antenna polarization and atmospheric absorption (Sections 3-2 & 5-1) are not included in any of these equations.

Recall from Section 4-2 that the power density at a distant point from a radar with an antenna gain of G_t is the power density from an isotropic antenna multiplied by the radar antenna gain.

Power density from radar, $P_D = \dfrac{P_t \, G_t}{4\pi \, R^2}$ [1]

If you could cover the entire spherical segment with your receiving antenna you would theoretically capture all of the transmitted energy. You can't do this because no antenna is large enough. (A two degree segment would be about a mile and three-quarters across at fifty miles from the transmitter.)

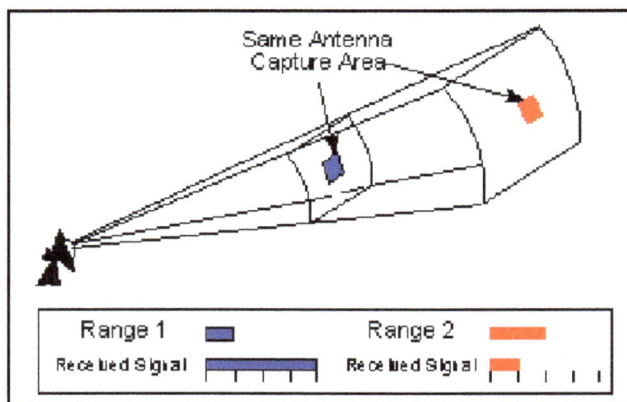

Figure 1. Power Density vs. Range

A receiving antenna captures a portion of this power determined by it's effective capture Area (A_e). The received power available at the antenna terminals is the power density times the effective capture area (A_e) of the receiving antenna.

For a given receiver antenna size the capture area is constant no matter how far it is from the transmitter, as illustrated in Figure 1. This concept is shown in the following equation:

Peak Power at Receiver input,

$$P_R \text{ (or S)} = P_D A_e = \frac{P_t G_t A_e}{4\pi R^2} \quad \text{which is known as the one-way (beacon) equation} \quad [2]$$

In order to maximize energy transfer between an antenna and transmitter or receiver, the antenna size should correlate with frequency. For reasonable antenna efficiency, the size of an antenna will be greater than $\lambda/4$. Control of beamwidth shape may become a problem when the size of the active element exceeds several wavelengths.

The relation between an antenna's effective capture area (A_e) or effective aperture and it's Gain (G) is:

$$\text{Antenna Gain, } G = \frac{4\pi A_e}{\lambda^2} \quad [3]$$

$$\text{or : Equivalent Area, } A_e = \frac{G \lambda^2}{4\pi} \quad [4]$$

Since the effective aperture is in units of length squared, from equation [3], it is seen that gain is proportional to the effective aperture normalized by the wavelength. This physically means that to maintain the same gain when doubling the frequency, the area is reduced by 1/4. This concept is illustrated in Figure 2.

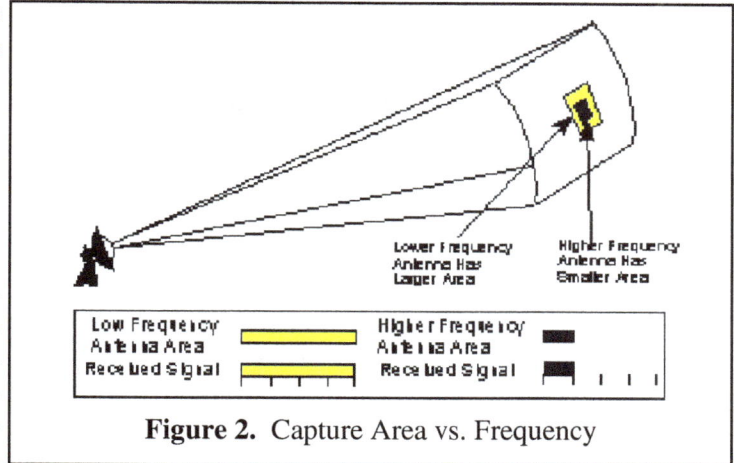

Figure 2. Capture Area vs. Frequency

If equation [4] is substituted into equation [2], the following relationship results:

$$\text{Peak Power at Receiver Input} = S \text{ (or } P_R) = \frac{P_t G_t G_r \lambda^2}{(4\pi)^2 R^2} = \frac{P_t G_t G_r \lambda^2}{(4\pi R)^2} \quad [5]$$

This is the signal calculated one-way from a transmitter to a receiver. For instance, a radar application might be to determine the signal received by a RWR, ESM, or an ELINT receiver. It is a general purpose equation and could be applied to almost any line-of-sight transmitter to receiver situation if the RF is higher than 100 MHZ.

The free space travel of radio waves can, of course, be blocked, reflected, or distorted by objects in their path such as buildings, flocks of birds, chaff, and the earth itself.

As illustrated in Figure 1, as the distance is doubled the received signal power decreases by 1/4 (6 dB). This is due to the R^2 term in equation [5].

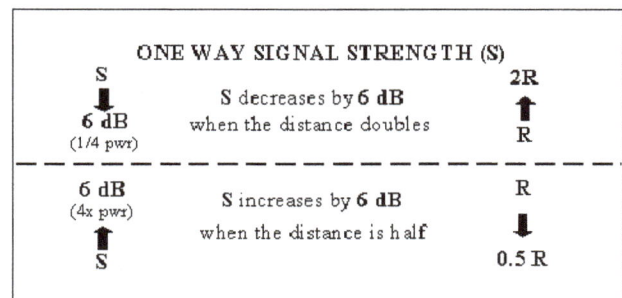

To illustrate this, blow up a round balloon and draw a square on the side of it. If you release air so that the diameter or radius is decreased by 1/2, the square shrinks to 1/4 the size. If you further blow up the balloon, so the diameter or radius is doubled, the square has quadrupled in area.

The one-way free space loss factor (α_1), (sometimes called the path loss factor) is given by the term $(4\pi R^2)(4\pi/\lambda^2)$ or $(4\pi R/\lambda)^2$. As shown in Figure 3, the loss is due to the ratio of two factors (1) the effective radiated area of the transmit antenna, which is the surface area of a sphere $(4\pi R^2)$ at that distance (R), and (2) the effective capture area (A_e) of the receive antenna which has a gain of one. If a receiving antenna could capture the whole surface area of the sphere, there would be no spreading loss, but a practical antenna will capture only a small part of the spherical radiation. Space loss is calculated using isotropic antennas for both transmit and receive, so α_1 is independent of the actual antenna. Using Gr = 1 in equation [11] in section 3-1, $A_e = \lambda^2/4\pi$. Since this term is in the denominator of α_1, the higher the frequency (lower λ) the more the space loss. Since G_t and G_r are part of the one-way radar equation, S (or P_r) is adjusted according to actual antennas as shown in the last portion of Figure 3. The value of the received signal (S) is:

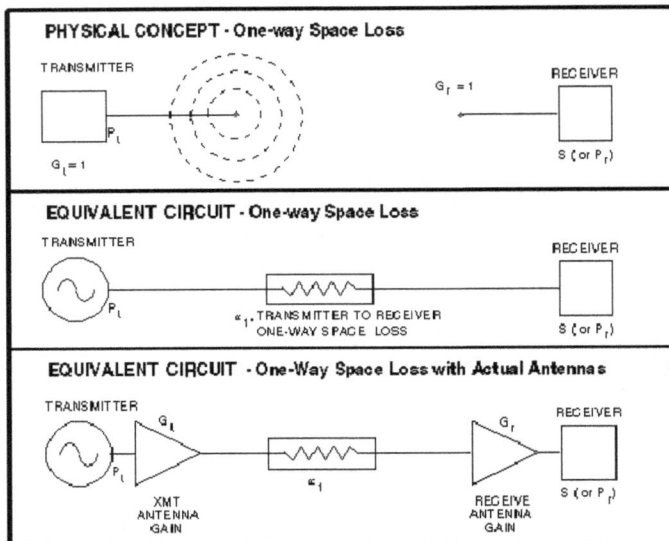

Figure 3. Concept of One-Way Space Loss

$$S \ (or \ P_R) = \frac{P_t G_t G_r \lambda^2}{(4\pi)^2 R^2} = P_t G_t G_r \left[\frac{\lambda^2}{(4\pi)^2} \right] \qquad [6]$$

To convert this equation to dB form, it is rewritten as:

$$10 \ Log \ (S \ or \ P_r) = 10 \ Log \ (P_t G_t G_r) + 20 \ Log \left[\frac{\lambda}{4\pi f R} \right] \qquad [7]$$

Since $\lambda = c / f$, equation [7] can be rewritten as:

$$10 \ Log \ (S \ or \ P_r) = 10 \ Log(P_t \, G_t \, G_r) - \alpha_1 \qquad [8]$$

Where the one-way free space loss, α_1, is defined as: $\alpha_1 = 20 \ Log \left[\frac{4\pi f R}{c} \right] *$ $\qquad$ [9]

The signal received equation in dB form is: $10 \log \ (P_r \ or \ S) = 10 \log P_t + 10 \log G_t + 10 \log G_r - \alpha_1$ $\qquad$ [10]

The one-way free space loss, α_1, can be given in terms of a variable and constant term as follows:

$$\alpha_1 = 20 \ Log \left[\frac{4\pi f \ R}{c} \right]^* = 20 \ Log \ f_1 R + K_1 \quad (in \ dB) \qquad [11]$$

The value of f_1 can be either in MHz or GHz as shown with commonly used units of R in the adjoining table.

$where \ K_1 = 20 \ Log \left[\frac{4\pi}{c} \bullet (Conversion \ units \ if \ not \ in \ m/ \sec, m, \ and \ Hz) \right]$

Note: To avoid having to include additional terms for these calculations, always combine any transmission line loss with antenna gain.

Values of K_1 (dB)		
Range (units)	f_1 in MHz $K_1 =$	f_1 in GHz $K_1 =$
NM	37.8	97.8
km	32.45	92.45
m	-27.55	32.45
yd	-28.33	31.67
ft	-37.87	22.13

A value for the one-way free space loss (α_1) can be obtained from:

(a) The One-way Free Space Loss graph (Figure 4). Added accuracy can be obtained using the Frequency Extrapolation graph (Figure 5)

(b) The space loss nomograph (Figure 6 or 7)

(c) The formula for α_1, equation [11].

FOR EXAMPLE:

Find the value of the one-way free space loss, α_1, for an RF of 7.5 GHz at 100 NM.

(a) From Figure 4, find 100 NM on the X-axis and estimate where 7.5 GHz is located between the 1 and 10 GHz lines (note dot). Read α_1 as 155 dB. An alternate way would be to read the α_1 at 1 GHz (138 dB) and add the frequency extrapolation value (17.5 dB for 7.5:1, dot on Figure 5) to obtain the same 155 dB value.

(b) From the nomogram (Figure 6), the value of α_1 can be read as 155 dB (Note the dashed line).

(c) From the equation 11, the precise value of α_1 is 155.3 dB.

Remember, α_1 is a free space value. If there is atmospheric attenuation because of absorption of RF due to certain molecules in the atmosphere or weather conditions etc., the atmospheric attenuation is in addition to the space loss (refer to Section 5-1).

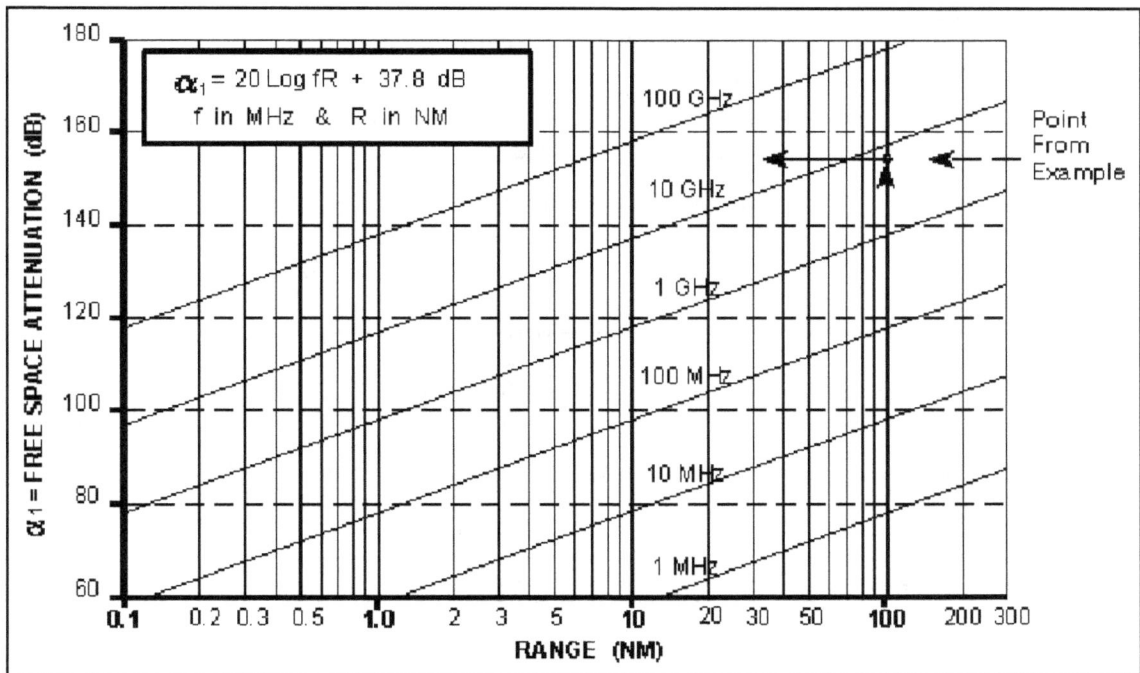

Figure 4. One-Way Free Space Loss

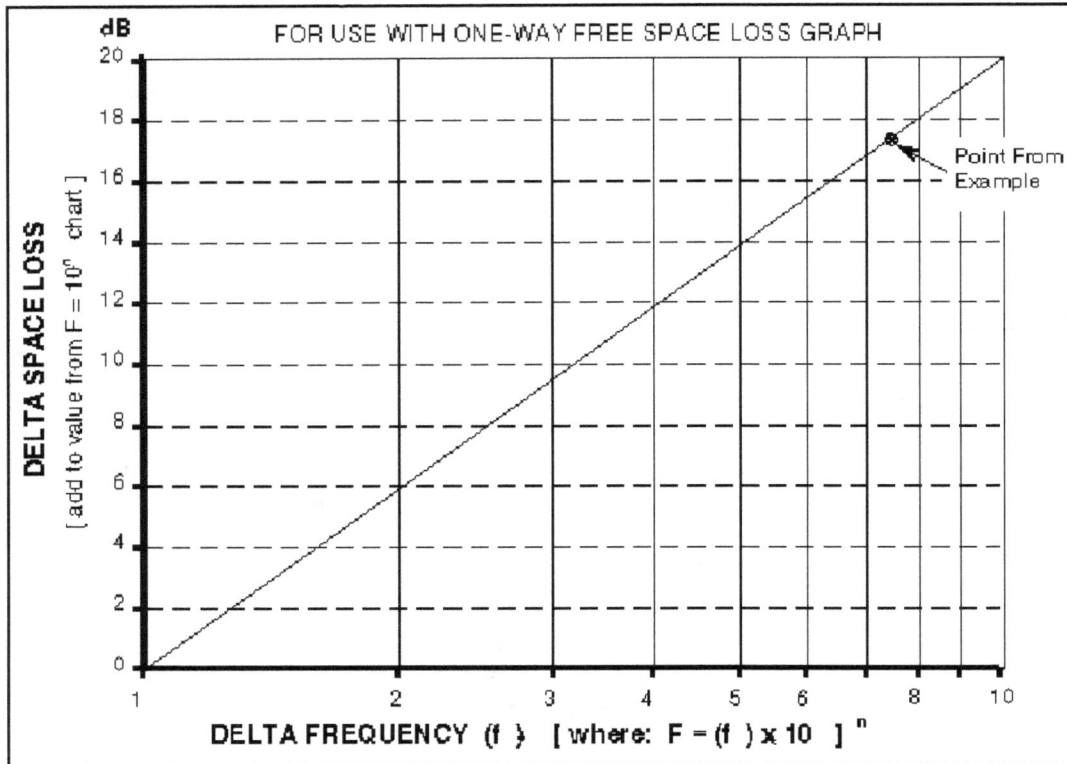

Figure 5. Frequency Extrapolation

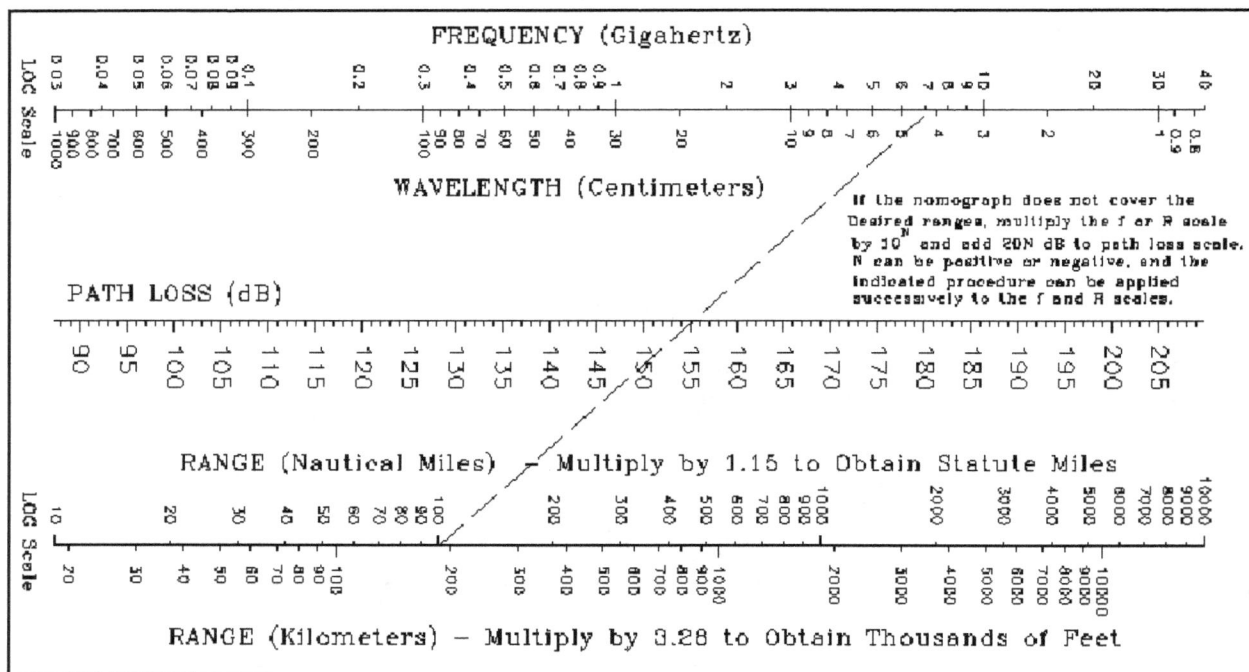

Figure 6. One-Way Space Loss Nomograph For Distances Greater Than 10 Nautical Miles

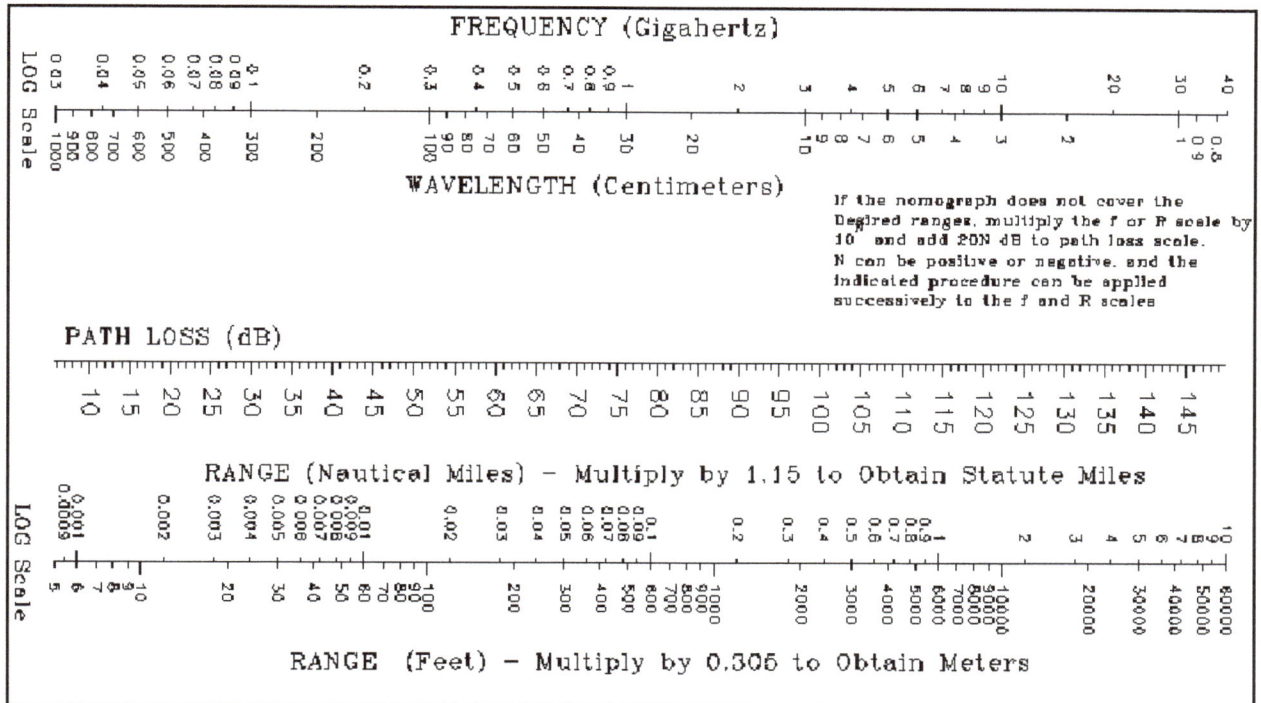

FREQUENCY (Gigahertz)
WAVELENGTH (Centimeters)

If the nomograph does not cover the
Desired ranges, multiply the f or R scale by
10ᴺ and add 20N dB to path loss scale.
N can be positive or negative, and the
indicated procedure can be applied
successively to the f and R scales

PATH LOSS (dB)

RANGE (Nautical Miles) – Multiply by 1.15 to Obtain Statute Miles

RANGE (Feet) – Multiply by 0.305 to Obtain Meters

Figure 7. One-Way Space Loss Nomograph For Distances Less Than 10 Nautical Miles

Figure 8 is the visualization of the losses occurring in one-way radar equation. Note: To avoid having to include additional terms, always combine any transmission line loss with antenna gain. Losses due to antenna polarization and atmospheric absorption also need to be included.

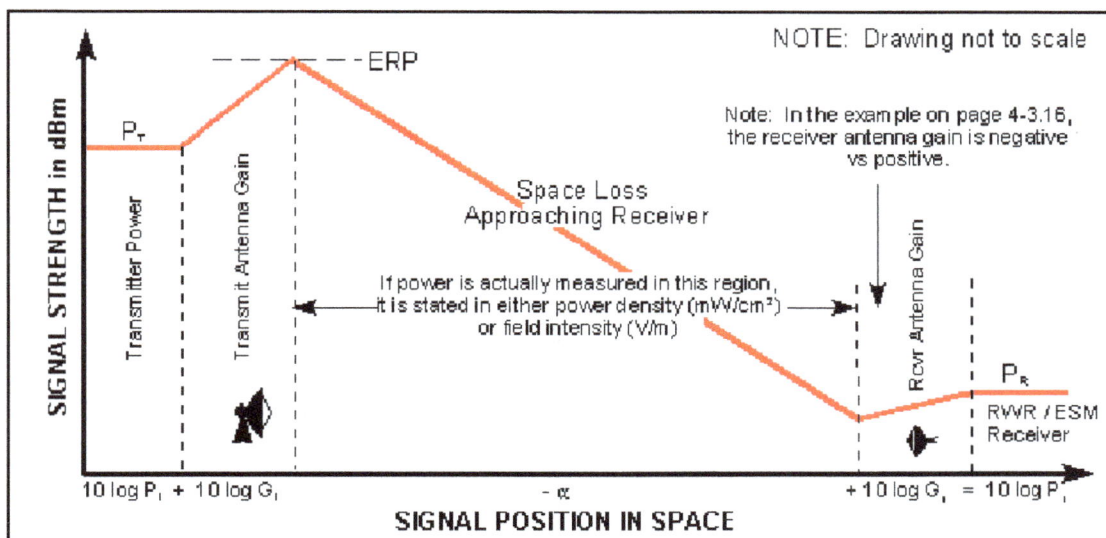

Figure 8. Visualization of One-Way Radar Equation

4-3.6

RWR/ESM RANGE EQUATION (One-Way)

The one-way radar (signal strength) equation [5] is rearranged to calculate the maximum range R_{max} of RWR/ESM receivers. It occurs when the received radar signal just equals S_{min} as follows:

$$R_{max} \cong \left[\frac{P_t G_t G_r \lambda^2}{(4\pi)^2 S_{min}} \right]^{\frac{1}{2}} \quad or \quad \left[\frac{P_t G_t G_r c^2}{(4\pi f)^2 S_{min}} \right]^{\frac{1}{2}} \quad or \quad \left[\frac{P_t G_t A_e}{4\pi S_{min}} \right]^{\frac{1}{2}} \tag{12}$$

In log form:
$$20\log R_{max} = 10\log P_t + 10\log G_t - 10\log S_{min} - 20\log f + 20\log(c/4\pi) \tag{13}$$

and since $K_1 = 20\log\{4\pi/c$ times conversion units if not in m/sec, m, and Hz$\}$
(Refer to section 4-3 for values of K_1).

$$10\log R_{max} = \tfrac{1}{2}[\ 10\log P_t + 10\log G_t - 10\log S_{min} - 20\log f - K_1]\quad \text{(keep } P_t \text{ and } S_{min} \text{ in same units)} \tag{14}$$

If you want to convert back from dB, then $R_{max} \cong \dfrac{10^{M\ dB}}{20}$, where M dB is the resulting number in the brackets of equation 14.

From Section 5-2, Receiver Sensitivity / Noise, S_{min} is related to the noise factor S:
$$S_{min} = (S/N)_{min} (NF)KT_oB \tag{15}$$
The one-way RWR/ESM range equation becomes:

$$R_{max} \cong \left[\frac{P_t G_t G_r \lambda^2}{(4\pi)^2 (S/N)_{min}(NF)KT_oB} \right]^{\frac{1}{2}} \quad or \quad \left[\frac{P_t G_t G_r c^2}{(4\pi f)^2 (S/N)_{min}(NF)KT_oB} \right]^{\frac{1}{2}} \quad or \quad \left[\frac{P_t G_t A_e}{4\pi (S/N)_{min}(NF)KT_oB} \right]^{\frac{1}{2}} \tag{16}$$

RWR/ESM RANGE INCREASE AS A RESULT OF A SENSITIVITY INCREASE
As shown in equation [12] $S_{min}^{-1} \propto R_{max}^2$ Therefore, $-10 \log S_{min} \propto 20 \log R_{max}$ and the table below results:

% Range Increase: Range + (% Range Increase) x Range = New Range
      i.e., for a 6 dB sensitivity increase, 500 miles +100% x 500 miles = 1,000 miles
Range Multiplier: Range x Range Multiplier = New Range
      i.e., for a 6 dB sensitivity increase 500 miles x 2 = 1,000 miles

dB Sensitivity Increase	% Range Increase	Range Multiplier	dB Sensitivity Increase	% Range Increase	Range Multiplier
+ 0.5	6	1.06	10	216	3.16
1.0	12	1.12	11	255	3.55
1.5	19	1.19	12	298	3.98
2	26	1.26	13	347	4.47
3	41	1.41	14	401	5.01
4	58	1.58	15	462	5.62
5	78	1.78	16	531	6.31
6	100	2.0	17	608	7.08
7	124	2.24	18	694	7.94
8	151	2.51	19	791	8.91
9	182	2.82	20	900	10.0

RWR/ESM RANGE DECREASE AS A RESULT OF A SENSITIVITY DECREASE

As shown in equation [12] $S_{min}^{-1} \propto R_{max}^2$ Therefore, $-10 \log S_{min} \propto 20 \log R_{max}$ and the table below results:

% Range Decrease: Range - (% Range decrease) x Range = New Range
　　　　i.e., for a 6 dB sensitivity decrease, 500 miles - 50% x 500 miles = 250 miles
Range Multiplier: Range x Range Multiplier = New Range
　　　　i.e., for a 6 dB sensitivity decrease 500 miles x .5 = 250 miles

dB Sensitivity Decrease	% Range Decrease	Range Multiplier	dB Sensitivity Decrease	% Range Decrease	Range Multiplier
- 0.5	6	0.94	-10	68	0.32
- 1.0	11	0.89	- 11	72	0.28
- 1.5	16	0.84	- 12	75	0.25
- 2	21	0.79	- 13	78	0.22
- 3	29	0.71	- 14	80	0.20
- 4	37	0.63	- 15	82	0.18
- 5	44	0.56	- 16	84	0.16
- 6	50	0.50	- 17	86	0.14
- 7	56	0.44	- 18	87	0.13
- 8	60	0.4	- 19	89	0.11
- 9	65	0.35	- 20	90	0.10

Example of One-Way Signal Strength: A 5 (or 7) GHz radar has a 70 dBm signal fed through a 5 dB loss transmission line to an antenna that has 45 dB gain. An aircraft that is flying 31 km from the radar has an aft EW antenna with -1 dB gain and a 5 dB line loss to the EW receiver (assume all antenna polarizations are the same).
Note: The respective transmission line losses will be combined with antenna gains, i.e.:
　　　　　　　-5 +45 = 40 dB, -5 - 1 = -6 dB, -10 + 5 = -5 dB.

　　　(1) What is the power level at the input of the EW receiver?

　　　Answer (1): P_r at the input to the EW receiver = Transmitter power - xmt cable loss + xmt antenna gain - space loss + rcvr antenna gain - rcvr cable loss.
Space loss (from section 4-3) @ 5 GHz = $20 \log f R + K_1 = 20 \log (5 \times 31) + 92.44 = 136.25$ dB.
Therefore:
　　$P_r = 70 + 40 - 136.25 - 6 = -32.25$ dBm @ 5 GHz ($P_r = -35.17$ dBm @ 7 GHz since $\alpha_1 = 139.17$ dB)

　　　(2) If the received signal is fed to a jammer with a gain of 60 dB, feeding a 10 dB loss transmission line which is connected to an antenna with 5 dB gain, what is the power level from the jammer at the input to the receiver of the 5 (or 7) GHz radar?

　　　Answer (2):　P_r at the input to the radar receiver = Power at the input to the EW receiver+ Jammer gain - jammer cable loss + jammer antenna gain - space loss + radar rcvr antenna gain - radar rcvr cable loss .
Therefore:
　　$P_r = -32.25 + 60 - 5 - 136.25 + 40 = -73.5$ dBm @ 5 GHz. ($P_r = -79.34$ dBm @ 7 GHz
　　　　since $\alpha_1 = 139.17$ dB and $P_t = -35.17$ dBm).

This problem continues in section 4-4, 4-7, and 4-10.

TWO-WAY RADAR EQUATION (MONOSTATIC)

In this section the radar equation is derived from the one-way equation (transmitter to receiver) which is then extended to the two-way radar equation. The following is a summary of the important equations to be derived here:

TWO-WAY RADAR EQUATION (MONOSTATIC)

Peak power at the radar receiver input is:

$$P_r = \frac{P_t G_t G_r \lambda^2 \sigma}{(4\pi)^3 R^4} = P_t G_t G_r \left[\frac{\sigma c^2}{(4\pi)^3 f^2 R^4} \right]^*$$

Note : $\lambda = c/f$ and $\sigma = RCS$

* keep λ or $c, \sigma,$ and R in the same units

On reducing the above equation to log form we have:

$10\log P_r = 10\log P_t + 10\log G_t + 10\log G_r + 10\log \sigma - 20\log f - 40\log R - 30\log 4\pi + 20\log c$

or in simplified terms: $10\log P_r = 10\log P_t + 10\log G_t + 10\log G_r + G_\sigma - 2\alpha_1$ (in dB)

Note: Losses due to antenna polarization and atmospheric absorption (Sections 3-2 and 5-1) are not included in these equations.

Target gain factor, $G_\sigma = 10\log \sigma + 20\log f_1 + K_2$ (in dB)				One-way free space loss, $\alpha_1 = 20\log (f_1 R) + K_1$ (in dB)				

K_2 Values (dB)	RCS (σ) (units)	f_1 in MHz $K_2 =$	f_1 in GHz $K_2 =$	K_1 Values (dB)	Range (units)	f_1 in MHz $K_1 =$	f_1 in GHz $K_1 =$
	m^2	-38.54	21.46		NM	37.8	97.8
	ft^2	-48.86	11.14		Km	32.45	92.45
					m	-27.55	32.45
					yd	-28.33	31.67
					ft	-37.87	22.13

Figure 1 illustrates the physical concept and equivalent circuit for a target being illuminated by a monostatic radar (transmitter and receiver co-located). Note the similarity of Figure 1 to Figure 3 in Section 4-3. Transmitted power, transmitting and receiving antenna gains, and the one-way free space loss are the same as those described in Section 4-3. The physical arrangement of the elements is different, of course, but otherwise the only difference is the addition of the equivalent gain of the target RCS factor.

From Section 4-3, One-Way Radar Equation / RF Propagation, the power in the receiver is:

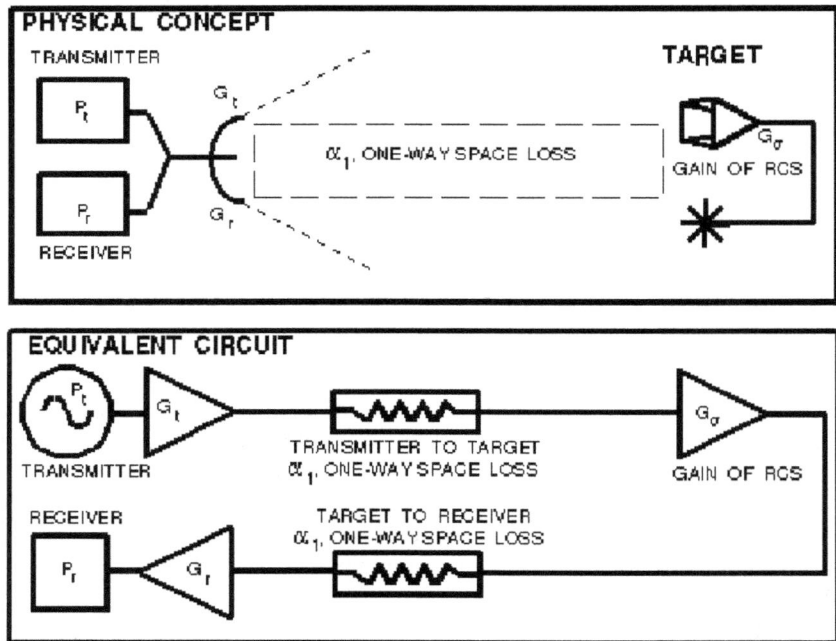

Figure 1. The Two-Way Monostatic Radar Equation Visualized

$$\frac{Received\ Signal}{at\ Target} = \frac{P_t\,G_t\,G_r\,\lambda^2}{(4\pi R\,)^2} \qquad [1]$$

From equation [3] in Section 4-3: $\qquad Antenna\ Gain,\ G = \dfrac{4\pi\,A_e}{\lambda^2} \qquad [2]$

Similar to a receiving antenna, a radar target also intercepts a portion of the power, but reflects (reradiates) it in the direction of the radar. The amount of power reflected toward the radar is determined by the Radar Cross Section (RCS) of the target. RCS is a characteristic of the target that represents its size as seen by the radar and has the dimensions of area (σ) as shown in Section 4-11. RCS area is not the same as physical area. But, for a radar target, the power reflected in the radar's direction is equivalent to re-radiation of the power captured by an antenna of area σ (the RCS). Therefore, the effective capture area (A_e) of the receiving antenna is replaced by the RCS (σ).

$$G_r = \frac{4\pi\sigma}{\lambda^2} \qquad [3] \qquad \text{so we now have:} \qquad \frac{Reflected\ Signal}{from\ target} = \frac{P_t\,G_t\,\lambda^2\,4\pi\sigma}{(4\pi R\,)^2\,\lambda^2} \qquad [4]$$

The equation for the power reflected in the radar's direction is the same as equation [1] except that P_t G_t, which was the original transmitted power, is replaced with the reflected signal power from the target, from equation [4]. This gives:

$$\frac{Reflected\ Signal\ Received\ Back}{at\ Input\ to\ Radar\ Receiver} = \frac{P_t\,G_t\,\lambda^2\,4\pi\sigma}{(4\pi R\,)^2\,\lambda^2} \ x \ \frac{G_r\,\lambda^2}{(4\pi R\,)^2} \qquad [5]$$

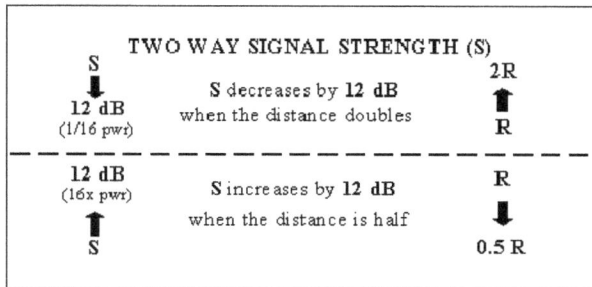

TWO WAY SIGNAL STRENGTH (S)

S

↓

12 dB
(1/16 pwr)

S decreases by 12 dB
when the distance doubles

2R
↑
R

– – – – – – – – – – – – – – – – – – –

12 dB
(16x pwr)
↑
S

S increases by 12 dB
when the distance is half

R
↓
0.5 R

If like terms are cancelled, the two-way radar equation results. The peak power at the radar receiver input is:

$$P_r = \frac{P_t\,G_t\,G_r\,\lambda^2\,\sigma}{(4\pi\,)^3\,R^4} = P_t\,G_t\,G_r\left[\frac{\sigma\,c^2}{(4\pi\,)^3\,f^2\,R^4}\right]^* \qquad [6]$$

* Note: $\lambda = c/f$ and σ = RCS. Keep λ or c, σ, and R in the same units.

On reducing equation [6] to log form we have:

$10\log P_r = 10\log P_t + 10\log G_t + 10\log G_r + 10\log \sigma - 20\log f - 40\log R - 30\log 4\pi + 20\log c \qquad [7]$

Target Gain Factor

If Equation [5] terms are rearranged instead of cancelled, a recognizable form results:

$$S\ (or\ P_r) = (\,P_t\,G_t\,G_r\,) \bullet \left[\frac{\lambda^2}{(4\pi R\,)^2}\right] \bullet \left[\frac{4\pi\sigma}{\lambda^2}\right] \bullet \left[\frac{\lambda^2}{(4\pi R\,)^2}\right] \qquad [8]$$

In log form:

$$10\log[S\ (or\ P_r)] = 10\log P_t + 10\log G_t + 10\log G_r + 20\log\left[\frac{\lambda}{4\pi R}\right] + 10\log\left[\frac{4\pi\sigma}{\lambda^2}\right] + 20\log\left[\frac{\lambda}{4\pi R}\right] \qquad [9]$$

The fourth and sixth terms can each be recognized as $-\alpha$, where α is the one-way free space loss factor defined in Section 4-3. The fifth term containing RCS (σ) is the only new factor, and it is the "Target Gain Factor".

In simplified terms the equation becomes:

$$10\log [S \text{ (or } P_r)] = 10\log P_t + 10\log G_t + 10\log G_r + G_\sigma - 2\alpha_1 \qquad (\text{in dB}) \qquad [10]$$

Where α_1 and G_σ are as follows:

From Section 4-3, equation [11], the space loss in dB is given by:

$$\alpha_1 = 20\log\left[\frac{4\pi f\, R}{c}\right]^* = 20\log f_1 R + K_1 \quad \text{where } K_1 = 20\log\left[\frac{4\pi}{c} \bullet (\textit{Conversion units if not in m / s ec, m, and Hz)}\right] [11]$$

* Keep c and R in the same units. The table of values for K_1 is again presented here for completeness. The constant, K_1, in the table includes a range and frequency unit conversion factor.

One-way free space loss, $\alpha_1 = 20\log(f_1 R) + K_1$ (in dB)			
K_1 Values (dB)	Range (units)	f_1 in MHz $K_1 =$	f_1 in GHz $K_1 =$
	NM	37.8	97.8
	Km	32.45	92.45
	m	-27.55	32.45
	yd	-28.33	31.67
	ft	-37.87	22.13

While it's understood that RCS is the antenna aperture area equivalent to an isotropically radiated target return signal, the target gain factor represents a gain, as shown in the equivalent circuit of Figure 1. The Target Gain Factor expressed in dB is G_σ as shown in equation [12].

$$G_\sigma = 10\log\left[\frac{4\pi\sigma}{\lambda^2}\right] = 10\log\left[\frac{4\pi\sigma f^2}{c^2}\right] = 10\log\sigma + 20\log f_1 + K_2 \quad (\textit{in dB}) \qquad [12]$$

$$\textit{where}: K_2 = 10\log\left[\frac{4\pi}{c^2} \bullet \left(\frac{\textit{Frequency and RCS}}{\textit{conversions as required}} \frac{(\textit{Hz to MHz or GHz })^2}{(\textit{meters to feet })^2}\right)\right]$$

The "Target Gain Factor" (G_σ) is a composite of RCS, frequency, and dimension conversion factors and is called by various names: "Gain of RCS", "Equivalent Gain of RCS", "Gain of Target Cross Section", and in dB form "Gain-sub-Sigma".

If frequency is given in MHz and RCS (σ) is in m^2, the formula for G_σ is:

$$G_\sigma = 10\log\sigma + 20\log f_1 + 10\log\left[4\pi \bullet \left(\frac{\textit{sec}}{3 \times 10^8 \, m}\right)^2 \bullet m^2 \bullet \left(\frac{1 \times 10^6}{\textit{sec}}\right)^2\right] \qquad [13]$$

or:

$$G_\sigma = 10\log\sigma + 20\log f_1 - 38.54 \quad (\textit{in dB}) \qquad [14]$$

For this example, the constant K_2 is -38.54 dB. K_2 values for various area and frequency and frequency units are summarized in the adjoining table.

Target gain factor, $G_\sigma = 10\log\sigma + 20\log f_1 + K_2$ (in dB)			
K_2 Values (dB)	RCS (σ) (units)	f_1 in MHz $K_2 =$	f_1 in GHz $K_2 =$
	m^2	-38.54	21.46
	ft^2	-48.86	11.14

In the two-way radar equation, the one-way free space loss factor (α_1) is used twice, once for the radar transmitter to target path and once for the target to radar receiver path. The radar illustrated in Figure 1 is monostatic so the two path losses are the same and the values of the two α_1's are the same.

If the transmission loss in Figure 1 from P_t to G_t equals the loss from G_r to P_r, and $G_r = G_t$, then equation [10] can be written as:

$$10\log [S \text{ or } P_r] = 10\log P_t + 20\log G_{tr} - 2\alpha_1 + G_\sigma \quad \text{(in dB)} \qquad [15]$$

The space loss factor (α_1) and the target gain factor (G_σ) include all the necessary unit conversions so that they can be used directly with the most common units. Because the factors are given in dB form, they are more convenient to use and allow calculation without a calculator when the factors are read from a chart or nomograph.

Most radars are monostatic. That is, the radar transmitting and receiving antennas are literally the same antenna. There are some radars that are considered "monostatic" but have separate transmitting and receiving antennas that are co-located. In that case, equation [10] could require two different antenna gain factors as originally derived:

$$10\log [S \text{ or } P_r] = 10\log P_t + 10\log G_t + 10\log G_r - 2\alpha_1 + G_\sigma \quad \text{(in dB)} \qquad [16]$$

Note: To avoid having to include additional terms for these calculations, always combine any transmission line loss with antenna gain.

Figure 2 is the visualization of the path losses occurring with the two-way radar equation. **Note:** to avoid having to include additional terms, always combine any transmission line loss with antenna gain. Losses due to antenna polarization and atmospheric absorption also need to be included.

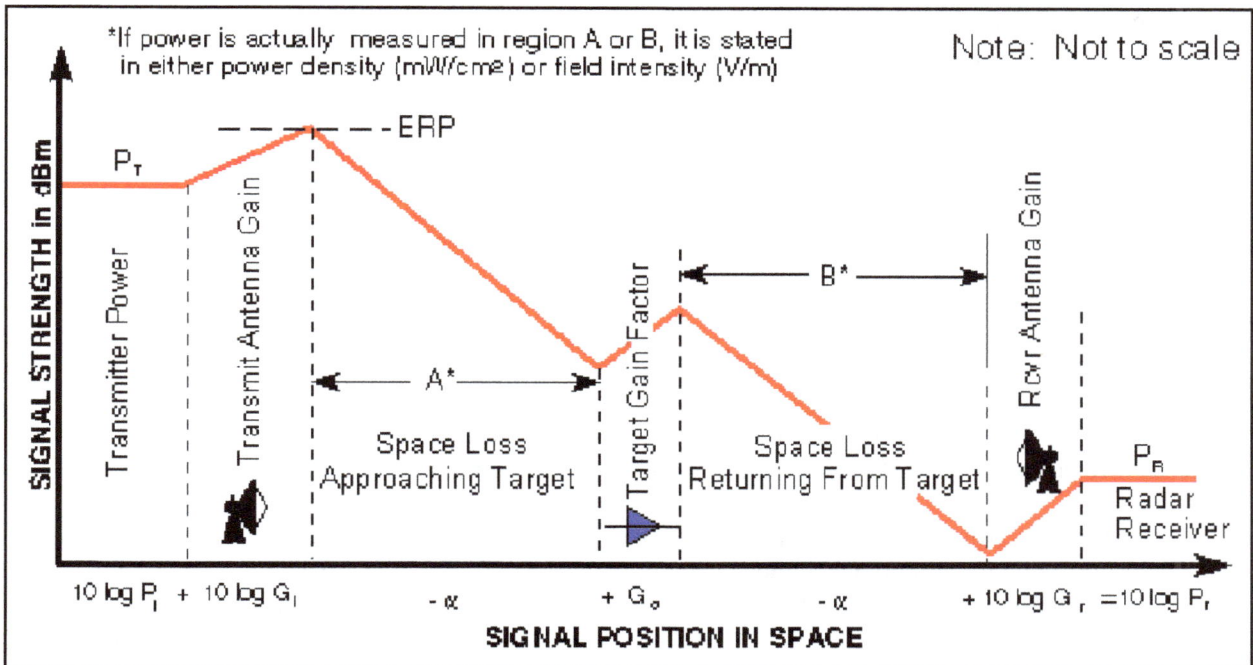

Figure 2. Visualization of Two-Way Radar Equation

RADAR RANGE EQUATION (Two-Way Equation)

The Radar Equation is often called the "Radar Range Equation". The Radar Range Equation is simply the Radar Equation rewritten to solve for maximum Range. The maximum radar range (R_{max}) is the distance beyond which the target can no longer be detected and correctly processed. It occurs when the received echo signal just equals S_{min}.

The Radar Range Equation is then:

$$R_{max} \cong \left[\frac{P_t G_t G_r \lambda^2 \sigma}{(4\pi)^3 S_{min}} \right]^{\frac{1}{4}} \quad or \quad \left[\frac{P_t G_t G_r c^2 \sigma]}{(4\pi)^3 f^2 S_{min}} \right]^{\frac{1}{4}} \quad or \quad \left[\frac{P_t G_t A_e \sigma}{(4\pi)^2 S_{min}} \right]^{\frac{1}{4}} \qquad [17]$$

The first equation, of the three above, is given in Log form by:

$$40\log R_{max} \cong 10\log P_t + 10\log G_t + 10\log G_r + 10\log \sigma - 10\log S_{min} - 20\log f - 30\log 4\pi + 20\log c \qquad [18]$$

As shown previously, Since $K_1 = 20\log$ [$(4\pi/c)$ times conversion units if not in m/sec, m, and Hz], we have:

$$10\log R_{max} \cong \tfrac{1}{4} [10\log P_t + 10\log G_t + 10\log G_r + 10\log \sigma - 10\log S_{min} - 20\log f_1 - K_1 - 10.99 \text{ dB}] \qquad [19]$$

If you want to convert back from dB, then

$$R_{max} \cong 10^{\frac{M \, dB}{40}}$$

Where M dB is the resulting number within the brackets of equation 19.

One-way free space loss, α_1 = 20log $(f_1 R)$ + K_1 (in dB)			
K_1 Values (dB)	Range (units)	f_1 in MHz K_1 =	f_1 in GHz K_1 =
	NM	37.8	97.8
	Km	32.45	92.45
	m	-27.55	32.45
	yd	-28.33	31.67
	ft	-37.87	22.13

From Section 5-2, Receiver Sensitivity / Noise, S_{min} is related to the noise factors by:

$$S_{min} = (S/N)_{min} (NF) k T_0 B \qquad [20]$$

The Radar Range Equation for a tracking radar (target continuously in the antenna beam) becomes:

$$R_{max} \cong \left[\frac{P_t G_t G_r \lambda^2 \sigma}{(4\pi)^3 (S/N)_{min}(NF)kT_0 B} \right]^{\frac{1}{4}} \quad or \quad \left[\frac{P_t G_t G_r c^2 \sigma}{(4\pi)^3 f^2 (S/N)_{min}(NF)kT_0 B} \right]^{\frac{1}{4}} \quad or \quad \left[\frac{P_t G_t A_e \sigma}{(4\pi)^2 (S/N)_{min}(NF)kT_0 B} \right]^{\frac{1}{4}} \qquad [21]$$

P_t in equations [17], [19], and [21] is the peak power of a CW or pulse signal. For pulse signals these equations assume the radar pulse is square. If not, there is less power since P_t is actually the average power within the pulse width of the radar signal. Equations [17] and [19] relate the maximum detection range to S_{min}, the minimum signal which can be detected and processed (the receiver sensitivity). The bandwidth (B) in equations [20] and [21] is directly related to S_{min}. B is approximately equal to 1/PW. Thus a wider pulse width means a narrower receiver bandwidth which lowers S_{min}, assuming no integration.

One cannot arbitrarily change the receiver bandwidth, since it has to match the transmitted signal. The "widest pulse width" occurs when the signal approaches a CW signal (see Section 2-11). A CW signal requires a very narrow bandwidth (approximately 100 Hz). Therefore, receiver noise is very low and good sensitivity results (see Section 5-2). If the radar pulse is narrow, the receiver filter bandwidth must be increased for a match (see Section 5-2), i.e. a 1 μs pulse requires a bandwidth of approximately 1 MHz. This increases receiver noise and decreases sensitivity.

If the radar transmitter can increase its PRF (decreasing PRI) and its receiver performs integration over time, an increase in PRF can permit the receiver to "pull" coherent signals out of the noise thus reducing S/N_{min} thereby increasing the detection range. Note that a PRF increase may limit the maximum range due to the creation of overlapping return echoes (see Section 2-10).

There are also other factors that limit the maximum practical detection range. With a scanning radar, there is loss if the receiver integration time exceeds the radar's time on target. Many radars would be range limited by line-of-sight/radar horizon (see Section 2-9) well before a typical target faded below S_{min}. Range can also be reduced by losses due to antenna polarization and atmospheric absorption (see Sections 3-2 and 5-1).

Two-Way Radar Equation (Example)

Assume that a 5 GHz radar has a 70 dBm (10 kilowatt) signal fed through a 5 dB loss transmission line to a transmit/receive antenna that has 45 dB gain. An aircraft that is flying 31 km from the radar has an RCS of 9 m^2. What is the signal level at the input to the radar receiver? (There is an additional loss due to any antenna polarization mismatch but that loss will not be addressed in this problem). This problem continues in Sections 4-3, 4-7, and 4-10.

Answer:
Starting with: $10\log S = 10\log P_t + 10\log G_t + 10\log G_r + G_\sigma - 2\alpha_1$ (in dB)

We know that: $\alpha_1 = 20\log f R + K_1 = 20\log (5 \times 31) + 92.44 = 136.25$ dB

and that: $G_\sigma = 10\log \sigma + 20\log f_1 + K_2 = 10\log 9 + 20\log 5 + 21.46 = 44.98$ dB (see Table 1)
(Note: The aircraft transmission line losses (-5 dB) will be combined with the antenna gain (45 dB) for both receive and transmit paths of the radar)

So, substituting in we have: $10\log S = 70 + 40 + 40 + 44.98 - 2(136.25) =$ **-77.52 dBm** @ 5 GHz

The answer changes to -80.44 dBm if the tracking radar operates at 7 GHz provided the antenna gains and the aircraft RCS are the same at both frequencies.

$\alpha_1 = 20\log (7 \times 31) + 92.44 = 139.17$ dB, $G_\sigma = 10\log 9 + 20\log 7 + 21.46 = 47.9$ dB (see Table 1)

$10\log S = 70 + 40 + 40 + 47.9 - 2(139.17) =$ **-80.44 dBm** @ 7 GHz

Table 1. Values of the Target Gain Factor (G_σ) in dB for Various Values of Frequency and RCS

Frequency (GHz)	RCS - Square meters						
	0.05	5	9	10	100	1,000	10,000
0.5 GHz	2.44	22.42	24.98	25.44	35.44	45.44	55.44
1 GHz	8.46	28.46	31.0	31.46	41.46	51.46	61.46
5 GHz	22.44	42.44	44.98	45.44	55.44	65.44	75.44
7 GHz	25.36	45.36	47.9	48.36	58.36	68.36	78.36
10 GHz	28.46	48.46	51.0	51.46	61.46	71.46	81.46
20 GHz	34.48	54.48	57.02	57.48	67.48	77.48	87.48
40 GHz	40.50	60.48	63.04	63.5	73.5	83.5	93.5

Note: Shaded values were used in the examples.

TWO-WAY RADAR RANGE INCREASE AS A RESULT OF A SENSITIVITY INCREASE

As shown in equation [17] $S_{min}^{-1} \propto R_{max}^{4}$ Therefore, $-10 \log S_{min} \propto 40 \log R_{max}$ and the table below results:

% Range Increase: Range + (% Range Increase) x Range = New Range
      i.e., for a 12 dB sensitivity increase, 500 miles +100% x 500 miles = 1,000 miles
Range Multiplier: Range x Range Multiplier = New Range
      i.e., for a 12 dB sensitivity increase 500 miles x 2 = 1,000 miles

Table 2. Effects of Sensitivity Increase

dB Sensitivity Increase	% Range Increase	Range Multiplier	dB Sensitivity Increase	% Range Increase	Range Multiplier
+ 0.5	3	1.03	10	78	1.78
1.0	6	1.06	11	88	1.88
1.5	9	1.09	12	100	2.00
2	12	1.12	13	111	2.11
3	19	1.19	14	124	2.24
4	26	1.26	15	137	2.37
5	33	1.33	16	151	2.51
6	41	1.41	17	166	2.66
7	50	1.50	18	182	2.82
8	58	1.58	19	198	2.98
9	68	1.68	20	216	3.16

TWO-WAY RADAR RANGE DECREASE AS A RESULT OF A SENSITIVITY DECREASE

As shown in equation [17] $S_{min}^{-1} \propto R_{max}^{4}$ Therefore, $-10 \log S_{min} \propto 40 \log R_{max}$ and the table below results:

% Range Decrease: Range - (% Range Decrease) x Range = New Range
      i.e., for a 12 dB sensitivity decrease, 500 miles - 50% x 500 miles = 250 miles

Range Multiplier: Range x Range Multiplier = New Range
      i.e., for a 12 dB sensitivity decrease 500 miles x 0.5 = 250 miles

Table 3. Effects of Sensitivity Decrease

dB Sensitivity Decrease	% Range Decrease	Range Multiplier	dB Sensitivity Decrease	% Range Decrease	Range Multiplier
- 0.5	3	0.97	-10	44	0.56
- 1.0	6	0.94	- 11	47	0.53
- 1.5	8	0.92	- 12	50	0.50
- 2	11	0.89	- 13	53	0.47
- 3	16	0.84	- 14	55	0.45
- 4	21	0.79	- 15	58	0.42
- 5	25	0.75	- 16	60	0.40
- 6	29	0.71	- 17	62	0.38
- 7	33	0.67	- 18	65	0.35
- 8	37	0.63	- 19	67	0.33
- 9	40	0.60	- 20	68	0.32

ALTERNATE TWO-WAY RADAR EQUATION

In this section the same radar equation factors are grouped differently to create different constants as is used by some authors.

TWO-WAY RADAR EQUATION (MONOSTATIC)

Peak power at the radar receiver input is:

$$P_r = \frac{P_t G_t G_r \lambda^2 \sigma}{(4\pi)^3 R^4} = \frac{P_t G_t G_{rsigma} c^2}{(4\pi)^3 f^2 R^4}{}^* \quad \left(\, Note: \lambda = \frac{c}{f} \; and \; \sigma \; is \; RCS \, \right) \qquad [1]$$

* Keep λ or c, σ, and R in the same units. On reducing the above equation to log form we have:

or: **10log P_r = 10log P_t + 10log G_t + 10log G_r - α_2** (in dB)

Where: $\alpha_2 = 20\log f_1 R^2$ - 10log σ + K_3 , and $K_3 = -10\log c^2/(4\pi)^3$

Note:
Losses due to antenna polarization and atmospheric absorption (Sections 3-2 and 5-1) are not included in these equations

K_3 Values:
(dB)

Range Units	f_1 in MHz σ in m^2	f_1 in GHz σ in m^2	f_1 in MHz σ in ft^2	f_1 in GHz σ in ft^2
NM	114.15	174.15	124.47	184.47
km	103.44	163.44	113.76	173.76
m	-16.56	43.44	-6.24	53.76
yd	-18.1	41.9	-7.78	52.22
ft	-37.2	22.8	-26.88	33.12

In the last section, we had the basic radar equation given as equation [6] and it is repeated as equation [1] in the table above.

In section 4-4, in order to maintain the concept and use of the one-way space loss coefficient, α_1 , we didn't cancel like terms which was done to form equation [6] there. Rather, we regrouped the factors of equation [5]. This resulted in two minus α_1 terms and we defined the remaining term as G_σ, which accounted for RCS (see equation [8] & [9]).

Some authors take a different approach, and instead develop an entirely new single factor α_2 , which is used instead of the combination of α_1 and G_σ.

If equation [1] is reduced to log form, (and noting that $f = c/\lambda$) it becomes:
10log P_r = 10log P_t + 10log G_t + 10log G_r - 20log (fR^2) + 10log σ + 10log ($c^2/(4\pi)^3$)   **[2]**

We now call the last three terms on the right minus α_2 and use it as a single term instead of the two terms α_1 and G_σ. The concept of dealing with one variable factor may be easier although we still need to know the range, frequency and radar cross section to evaluate α_2. Additionally, we can no longer use a nomograph like we did in computing α_1 and visualize a two-way space loss consisting of two times the one-way space loss, since there are now 3 variables vs two.

Equation [2] reduces to: **10log P_r = 10log P_t + 10log G_t + 10log G_r - α_2**   (in dB)   **[3]**

Where $\alpha_2 = 20\log (f_1 R^2)$ - 10log σ + K_3 and where f_1 is the MHz or GHz value of frequency

and $K_3 = -10\log (c^2/(4\pi)^3)$ + 20log (conversion for Hz to MHz or GHz)+ 40log (range unit conversions if not in meters) - 20log (RCS conversions for meters to feet)

The values of K_3 are given in the table above.
Comparing equation [3] to equation [10] in Section 4-4, it can be seen that $\alpha_2 = 2\alpha_1 - G_\sigma$.

This Page Blank

TWO-WAY RADAR EQUATION (BISTATIC)

The following table contains a summary of the equations developed in this section.

TWO-WAY RADAR EQUATION (BISTATIC)

Peak power at the radar receiver input is:

$$P_r = \frac{P_t\, G_t\, G_r\, \lambda^2\, \sigma}{(4\pi)^3\, R_{Tx}^2\, R_{Rx}^2} = P_t\, G_t\, G_r \left[\frac{\sigma\, c^2}{(4\pi)^3\, f^2\, R_{Tx}^2\, R_{Rx}^2} \right]^*$$

Note: $\lambda = c/f$ and $\sigma = RCS$

* keep λ or c, σ, and R in the same units

On reducing the above equation to log form we have:

$10\log P_r = 10\log P_t + 10\log G_t + 10\log G_r + 10\log \sigma - 20\log f + 20\log c - 30\log 4\pi - 20\log R_{Tx} - 20\log R_{Rx}$

or in simplified terms: $10\log P_r = 10\log P_t + 10\log G_t + 10\log G_r + G_\sigma - \alpha_{Tx} - \alpha_{Rx}$ (in dB)

Where α_{Tx} corresponds to transmitter to target loss and α_{Rx} corresponds to target to receiver loss.

Note: Losses due to antenna polarization and atmospheric absorption (Sections 3-2 and 5-1) are not included in these equations.

Target gain factor, $G_\sigma = 10\log \sigma + 20\log f_1 + K_2$ (in dB)			One-way free space loss, $\alpha_{Tx\ or\ Rx} = 20\log (f_1 R_{Tx\ or\ Rx}) + K_1$ (in dB)			
K_2 Values (dB)	f_1 in MHz	f_1 in GHz	**K_1 Values** (dB)	Range	f_1 in MHz	f_1 in GHz
RCS (σ) (units)	$K_2 =$	$K_2 =$		(units)	$K_1 =$	$K_1 =$
m^2	-38.54	21.46		NM	37.8	97.8
ft^2	-48.86	11.14		Km	32.45	92.45
				m	-27.55	32.45
				yd	-28.33	31.67
				ft	-37.87	22.13

BISTATIC RADAR

There are also true bistatic radars - radars where the transmitter and receiver are in different locations as is depicted in Figure 1. The most commonly encountered bistatic radar application is the semi-active missile. The transmitter is located on, or near, the launch platform (surface or airborne), and the receiver is in the missile which is somewhere between the launch platform and the target.

The transmitting and receiving antennas are not the same and are not in the same location. Because the target-to-radar range is different from the target-to-missile range, the target-to-radar and target-to-missile space losses are different.

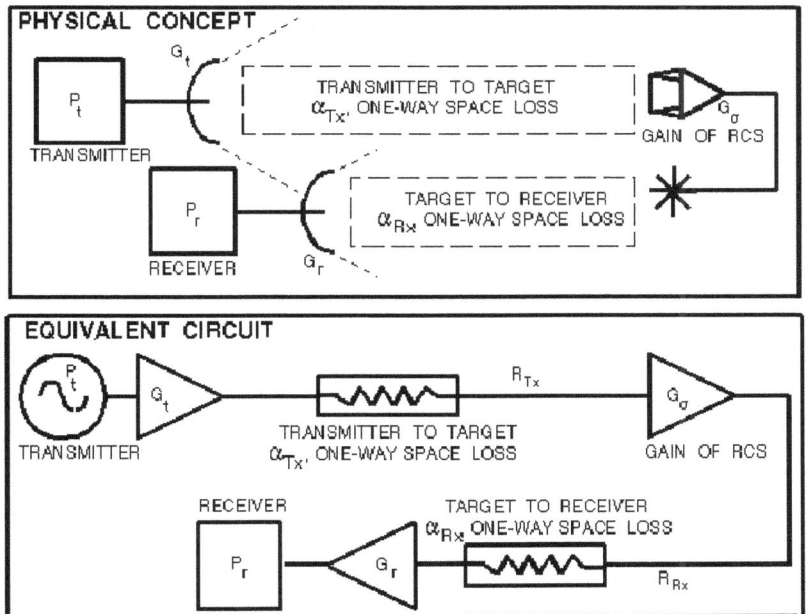

Figure 1. Bistatic Radar Visualized

The peak power at the radar receiver input is:

$$P_r = \frac{P_t\,G_t\,G_r\,\lambda^2\,\sigma}{(4\pi)^3\,R_{Tx}^2\,R_{Rx}^2} = P_t\,G_t\,G_r\left[\frac{\sigma\,c^2}{(4\pi)^3\,f^2\,R_{Tx}^2\,R_{Rx}^2}\right] \qquad (\ Note: \lambda = \frac{c}{f}\ and\ \sigma = RCS) \qquad [1]$$

Keep λ or c, σ, and R in the same units.

On reducing the above equation to log form we have:

$10\log P_r = 10\log P_t + 10\log G_t + 10\log G_r + 10\log \sigma - 20\log f + 20\log c - 30\log 4\pi - 20\log R_{Tx} - 20\log R_{Rx}$ **[2]**

or in simplified terms:

$$10\log P_r = 10\log P_t + 10\log G_t + 10\log G_r + G_\sigma - \alpha_{Tx} - \alpha_{Rx} \qquad (in\ dB) \qquad [3]$$

Where α_{Tx} corresponds to transmitter to target loss and α_{Rx} corresponds to target to receiver loss, or:

$\alpha_{Tx} = 20\log(f_1 T_{Tx}) + K_1$ (in dB)   and $\alpha_{Rx} = 20\log(f_1 T_{Rx}) + K_1$ (in dB)

with K_1 values provided on page 4-6.1 and with f_1 being the MHz or GHz value of frequency.

Therefore, the difference between monostatic and bistatic calculations is that two α's are calculated for two different ranges and different gains may be required for transmit and receive antennas.

To avoid having to include additional terms for these calculations, always combine any transmission line loss with antenna gain.

As shown in Figure 2, it should also be noted that the bistatic RCS received by the missile is not always the same as the monostatic RCS. In general, the target's RCS varies with angle. Therefore, the bistatic RCS and monostatic RCS will be equal for receive and transmit antennas at the same angle to the target (but only if all three are in a line, as RCS also varies with elevation angle).

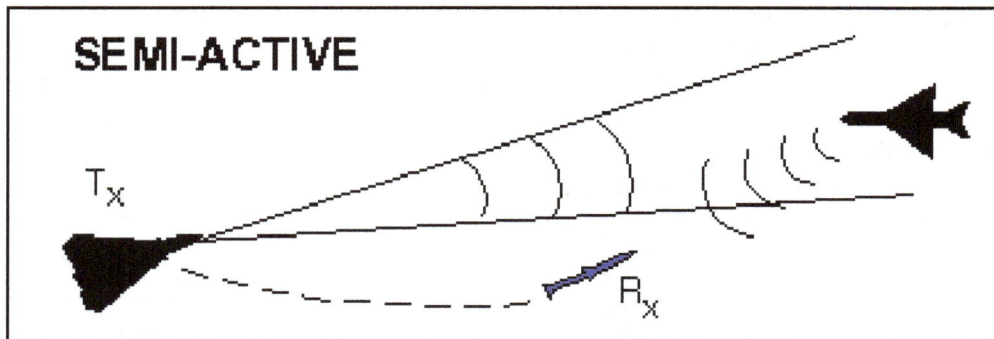

Figure 2. Bistatic RCS Varies

JAMMING TO SIGNAL (J/S) RATIO - CONSTANT POWER [SATURATED] JAMMING

The following table contains a summary of the equations developed in this section.

JAMMING TO SIGNAL (J/S) RATIO (MONOSTATIC)	* Keep R and σ in same units
$J/S = (P_j\, G_{ja}\, 4\pi\, R^2)\, /\, (P_t\, G_t\, \sigma)$ (ratio form)* or: $10\log J/S = 10\log P_j + 10\log G_{ja} - 10\log P_t - 10\log G_t - 10\log\sigma^* + 10.99\ dB + 20\log R^*$ Note (1): Neither f nor λ terms are part of these equations	Target gain factor, (in dB) $G_\sigma = 10\log\sigma + 20\log f_1 + K_2$ K_2 Values (dB):
If simplified radar equations developed in previous sections are used: $10\log J/S = 10\log P_j + 10\log G_{ja} - 10\log P_t - 10\log G_t - G_\sigma + \alpha_1$ (in dB) Note (2): the $20\log f_1$ term in $-G_\sigma$ cancels the $20\log f_1$ term in α_1	RCS (σ) f₁ in MHz f₁ in GHz (units) $K_2 =$ $K_2 =$ m2 -38.54 21.46 ft2 -48.86 11.14

JAMMING TO SIGNAL (J/S) RATIO (BISTATIC) R_{Tx} is the range from the radar transmitter to the target. See note (1).	One-way free space loss (dB)
$J/S = (P_j\, G_{ja}\, 4\pi\, R_{Tx}^2)\, /\, (P_t\, G_t\, \sigma)$ (ratio form) * or: $10\log J/S = 10\log P_j + 10\log G_{ja} - 10\log P_t - 10\log G_t - 10\log\sigma^* + 10.99\ dB + 20\log R_{Tx}^*$	α_1 or $\alpha_{Tx} =$ $20\log (f_1\, R) + K_1$ K_1 Values (dB):
If simplified radar equations developed in previous sections are used: see note (2). $10\log J/S = 10\log P_j + 10\log G_{ja} - 10\log P_t - 10\log G_t - G_\sigma + \alpha_{Tx}$ (in dB)	Range f₁ in MHz f₁ in GHz (units) $K_1 =$ $K_1 =$ NM 37.8 97.8 km 32.45 92.45 m -27.55 32.45 ft -37.87 22.13

This section derives the J/S ratio from the one-way range equation for J and the two-way range equation for S, and deals exclusively with active (transmitting) Electronic Attack (EA) devices or systems. Furthermore, the only purpose of EA is to prevent, delay, or confuse the radar processing of target information.

By official definition, EA can be either Jamming or Deception. This may be somewhat confusing because almost any type of active EA is commonly called "jamming", and the calculations of EA signal in the radar compared to the target signal in the radar commonly refer to the "jamming-to-signal" ratio ("J-to-S" ratio). Therefore this section uses the common jargon and the term "jammer" refers to any EA transmitter, and the term "jamming" refers to any EA transmission, whether Deception or Concealment.

Jamming: "Official" jamming should more aptly be called Concealment or Masking. Essentially, Concealment uses electronic transmissions to swamp the radar receiver and hide the targets. Concealment (Jamming) usually uses some form of noise as the transmitted signal. In this section, Concealment will be called "noise" or "noise jamming".

Deception: Deception might be better called Forgery. Deception uses electronic transmissions to forge false target signals that the radar receiver accepts and processes as real targets.

"J" designates the EA signal strength whether it originates from a noise jammer or from a deception system.

Basically, there are two different methods of employing active EA against hostile radars:

Self Protection EA
Support EA

For most practical purposes, Self Protection EEA is usually Deception and Support EA is usually noise jamming. As the name implies, Self Protection EA is EA that is used to protect the platform that it is on. Self Protection EA is often called "self screening jamming", "Defensive EA" or historically "Deception ECM". The top half of Figure 1 shows self-screening jamming.

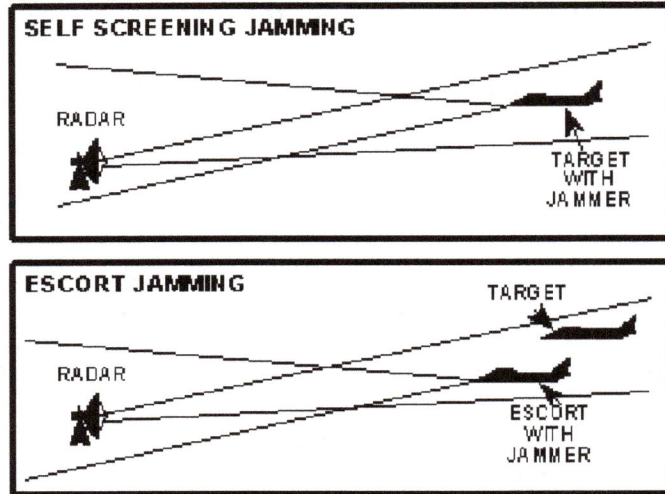

Figure 1. Self Protection and Escort Jamming

The bottom half of Figure 1 illustrates escort jamming which is a special case of support jamming. If the escort platform is sufficiently close to the target, the J-to-S calculations are the same as for self protection EA.

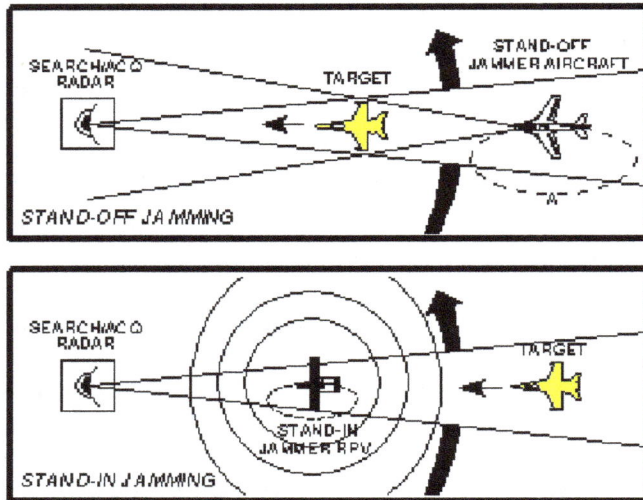

Figure 2. Support Jamming

Support EA is electronic transmissions radiated from one platform and is used to protect other platforms or fulfill other mission requirements, like distraction or conditioning. Figure 2 illustrates two cases of support jamming protecting a striker - stand-off jamming (SOJ) and stand-in jamming (SIJ). For SOJ the support jamming platform is maintaining an orbit at a long range from the radar - usually beyond weapons range. For SIJ, a remotely piloted vehicle is orbiting very close to the victim radar. Obviously, the jamming power required for the SOJ to screen a target is much greater than the jamming power required for the SIJ to screen the same target.

When factoring EA into the radar equation, the quantities of greatest interest are "J-to-S" and Burn- Through Range.

"J-to-S" is the ratio of the signal strength of the jammer signal (J) to the signal strength of the target return signal (S). It is expressed as "J/S" and, in this section, is always in dB. J usually (but not always) must exceed S by some amount to be effective, therefore the desired result of a J/S calculation in dB is a positive number. Burn-through Range is the radar to target range where the target return signal can first be detected through the jamming and is usually slightly farther than crossover range where J=S. It is usually the range where the J/S just equals the minimum effective J/S (See Section 4-8).

The significance of "J-to-S" is sometimes misunderstood. The effectiveness of EA **is not** a direct mathematical function of "J-to-S". The magnitude of the "J-to-S" required for effectiveness is a function of the particular EA technique and of the radar it is being used against.   Different EA techniques may very well require different "J-to-S" ratios against the same radar. When there is sufficient "J-to-S" for effectiveness, increasing it will rarely increase the effectiveness at a given range. Because modern radars can have sophisticated signal processing and/or EP capabilities, in certain radars too much "J-to-S" could cause the signal processor to ignore the jamming, or activate special anti-jamming modes. Increasing "J-to-S" (or the jammer power) does, however, allow the target aircraft to get much closer to the threat radar before burn-through occurs, which essentially means more power is better if it can be controlled when desired.

IMPORTANT NOTE: If the signal S is CW or PD and the Jamming J is amplitude modulated, then the J used in the formula has to be reduced from the peak value (due to sin x/x frequency distribution). The amount of reduction is dependent upon how much of the bandwidth is covered by the jamming signal. To get an exact value, integrals would have to be taken over the bandwidth. As a rule of thumb however:
  • If the frequency of modulation is less than the BW of the tracking radar reduce J/S by 10 Log (duty cycle).
  • If the frequency of modulation is greater than the BW of the tracking radar reduce J/S by 20 Log(duty cycle).

For example; if your jamming signal is square wave chopped (50% duty cycle) at a 100 Hz rate while jamming a 1 kHz bandwidth receiver, then the J/S is reduced by 3 dB from the maximum. If the duty cycle was 33%, then the reduction would be 4.8 dB. If the 50% and 33% duty cycle jamming signals were chopped at a 10 kHz (vice the 100 Hz) rate, the rule of thumb for jamming seen by the receiver would be down 6 dB and 9.6 dB, respectively, from the maximum since the 10 kHz chopping rate is greater than the 1 kHz receiver BW.

J/S for SELF PROTECTION EA vs. MONOSTATIC RADAR

Figure 3 is radar jamming visualized. The Physical concept of Figure 3 shows a monostatic radar that is the same as Figure 1, Section 4-4, and a jammer (transmitter) to radar (receiver) that is the same as Figure 3, Section 4-3. In other words, Figure 3 is simply the combination of the previous two visual concepts where there is only one receiver (the radar's).

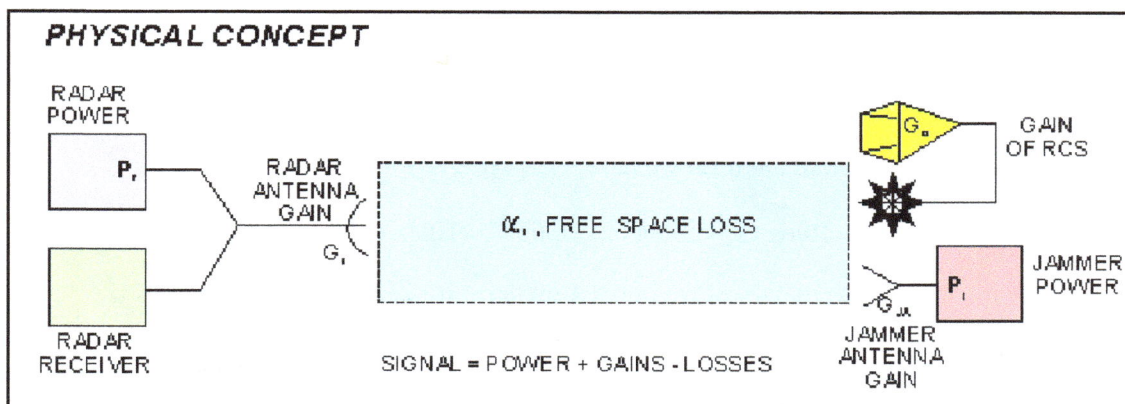

Figure 3. Radar Jamming Visualized

The equivalent circuit shown in Figure 4 applies to jamming monostatic radars with either self protect EA or support EA. For self protect (or escort) v.s. a monostatic radar, the jammer is on the target and the radar receive and transmit antennas are collocated so the three ranges and three space loss factors (α's) are the same.

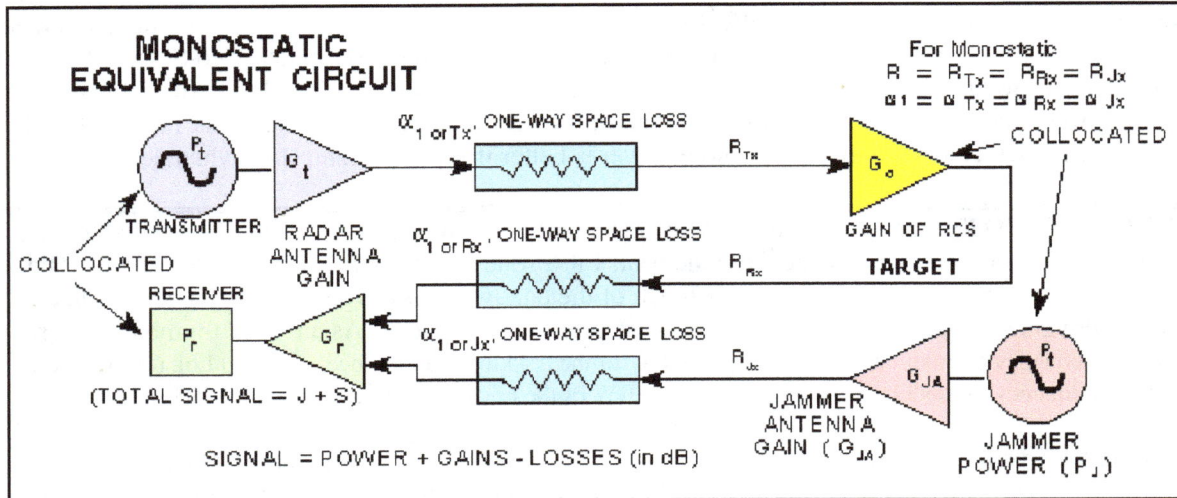

Figure 4. Monostatic Radar EA Equivalent Circuit

<u>J-S Ratio (Monostatic)</u> The ratio of the power received (P_{r1} or J) from the jamming signal transmitted from the target to the power received (P_{r2} or S) from the radar skin return from the target equals J/S.

From the one way range equation in Section 4-3:
$$P_r1 \; or \; J = \frac{P_j \, G_{ja} \, G_r \, \lambda^2}{(4\pi R)^2} \qquad\qquad [1]$$

From the two way range equation in Section 4.4:
$$P_r2 \; or \; S = \frac{P_t \, G_t \, G_r \, \lambda^2 \, \sigma}{(4\pi)^3 \, R^4} \qquad\qquad [2]$$

so
$$\frac{J}{S} = \frac{P_j \, G_{ja} \, G_r \, \lambda^2 (4\pi)^3 \, R^4}{P_t \, G_t \, G_r \, \lambda^2 \, \sigma (4\pi R)^2} = \frac{P_j \, G_{ja} \, 4\pi \, R^2}{P_t \, G_t \, \sigma}^* \qquad (ratio \; form) \qquad [3]$$

    * Keep R and σ in the same units.

On reducing the above equation to log form we have:

       10log J/S = 10log P_j + 10log G_{ja} - 10log P_t - 10log G_t - 10log σ + 10log 4π + 20log R   [4]

or   10log J/S = 10log P_j + 10log G_{ja} - 10log P_t - 10log G_t - 10log σ + 10.99 dB + 20log R   [5]

Note: Neither *f* nor λ terms are part of the final form of equation [3] and equation [5].

<u>J/S Calculations (Monostatic) Using a One Way Free Space Loss</u> - The simplified radar equations developed in previous sections can be used to express J/S.

From the one way range equation Section 4-3:

$$10\log (P_{r1} \text{ or } J) = 10\log P_j + 10\log G_{ja} + 10\log G_r - \alpha_1 \qquad \text{(in dB)} \qquad [6]$$

From the two way range equation in Section 4.4:

$$10\log (P_{r2} \text{ or } S) = 10\log P_t + 10\log G_t + 10\log G_r + G_\sigma - 2\alpha_1 \qquad \text{(in dB)} \qquad [7]$$

$$10\log (J/S) = 10\log P_j + 10\log G_{ja} - 10\log P_t - 10\log G_t - G_\sigma + \alpha_1 \quad \text{(in dB)} \qquad [8]$$

Note: To avoid having to include additional terms for these calculations, always combine any transmission line loss with antenna gain. The $20\log f_1$ term in $-G_\sigma$ cancels the $20\log f_1$ term in α_1.

Target gain factor, $G_\sigma = 10\log \sigma + 20\log f_1 + K_2$ (in dB)			One-way free space loss, $\alpha_1 = 20\log (f_1 R) + K_1$ (in dB)			
K_2 Values (dB)			K_1 Values (dB)	Range (units)	f_1 in MHz $K_1 =$	f_1 in GHz $K_1 =$
RCS (σ) (units)	f_1 in MHz $K_2 =$	f_1 in GHz $K_2 =$		NM	37.8	97.8
m^2	-38.54	21.46		km	32.45	92.45
ft^2	-48.86	11.14		m	-27.55	32.45
				yd	-28.33	31.67
				ft	-37.87	22.13

J/S for SELF PROTECTION EA vs. BISTATIC RADAR

The semi-active missile illustrated in Figure 5 is the typical bistatic radar which would require the target to have self protection EA to survive. In this case, the jammer is on the target and the target to missile receiver range is the same as the jammer to receiver range, but the radar to target range is different. Therefore, only two of the ranges and two of the α's (Figure 6.) are the same.

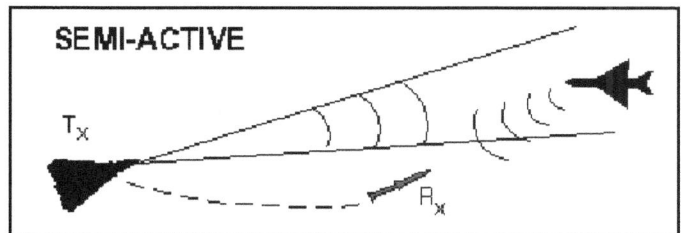

Figure 5. Bistatic Radar

In the following equations:
$\alpha_{Tx} = $ The one-way space loss from the radar transmitter to the target for range R_{Tx}
$\alpha_{Rx} = $ The one-way space loss from the target to the missile receiver for range R_{Rx}

Like the monostatic radar, the bistatic jamming and reflected target signals travel the same path from the target and enter the receiver (missile in this case) via the same antenna. In both monostatic and bistatic J/S equations this common range cancels, so both J/S equations are left with an R_{Tx}^2 or 20 log R_{Tx} term. Since in the monostatic case $R_{Tx} = R_{Rx}$ and $\alpha_{Tx} = \alpha_{Rx}$, only R or α_1 is used in the equations. Therefore, the bistatic J/S equations [11], [13], or [14] will work for monostatic J/S calculations, but the opposite is only true if bistatic R_{Tx} and α_{Tx} terms are used for R or α_1 terms in monostatic equations [3], [5], and [8].

The equivalent circuit shown in Figure 6 applies to jamming bistatic radar. For self protect (or escort) vs. a bistatic radar, the jammer is on the target and the radar receive and transmit antennas are at separate locations so only two of the three ranges and two of the three space loss factors (α's) are the same.

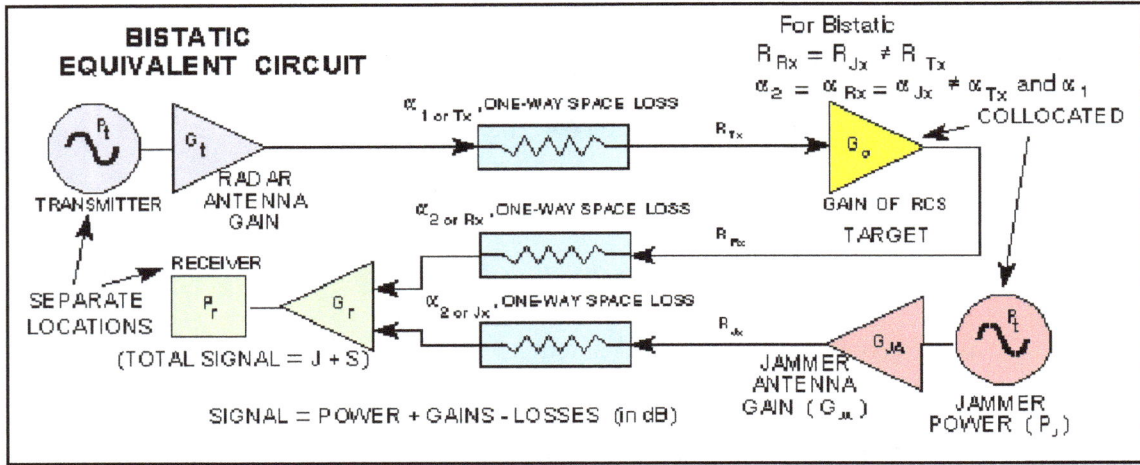

Figure 6. Bistatic Radar EA Equivalent Circuit

J-to-S Ratio (Bistatic) When the radar's transmit antenna is located remotely from the receiving antenna (Figure 6), the ratio of the power received (P_{r1} or J) from the jamming signal transmitted from the target to the power received (P_{r2} or S) from the radar skin return from the target equals J/S. For jammer effectiveness J normally has to be greater than S.

From the one way range equation in Section 4-3: $P_r 1 \; or \; J = \dfrac{P_j \, G_{ja} \, G_r \, \lambda^2}{(4\pi \, R_{Rx})^2}$ $(R_{Jx} = R_{Rx})$ **[9]**

From the two way range equation in Section 4.4: $P_r 2 \; or \; S = \dfrac{P_t \, G_t \, G_r \, \lambda^2 \, \sigma}{(4\pi)^3 \, R_{Tx}^2 \, R_{Rx}^2}$ **[10]**

so $\dfrac{J}{S} = \dfrac{P_j \, G_{ja} \, G_r \, \lambda^2 (4\pi)^3 \, R_{Tx}^2 \, R_{Rx}^2}{P_t \, G_t \, G_r \, \lambda^2 \, \sigma (4\pi \, R_{Rx})^2} = \dfrac{P_j \, G_{ja} \, 4\pi \, R_{Tx}^2}{P_t \, G_t \, \sigma}^{*}$ *(ratio form)*   **[11]**

* Keep R and σ in the same units.

On reducing the above equation to log form we have:

10log J/S = 10log P_j + 10log G_{ja} - 10log P_t - 10log G_t - 10log σ + 10log 4π + 20log R_{Tx} **[12]**

or   10log J/S = 10log P_j + 10log G_{ja} - 10log P_t - 10log G_t - 10log σ + 10.99 dB + 20log R_{Tx} **[13]**

Note: To avoid having to include additional terms for these calculations, always combine any transmission line loss with antenna gain. Neither *f* nor λ terms are part of the final form of equation [11] and equation [13].

<u>Bistatic J/S Calculations (Bistatic) Using a One Way Free Space Loss</u> - The simplified radar equations developed in previous sections can be used to express J/S.

From the one way range equation in Section 4-3:

$$10\log (P_{r1} \text{ or } J) = 10\log P_j + 10\log G_{ja} + 10\log G_r - \alpha_{Rx} \qquad \text{(all factors dB)} \qquad [14]$$

From the two way range equation in Section 4-4:

$$10\log (P_{r2} \text{ or } S) = 10\log P_t + 10\log G_t + 10\log G_r + G_\sigma - \alpha_{Tx} - \alpha_{Rx} \qquad \text{(all factors dB)} \qquad [15]$$

$$10\log (J/S) = 10\log P_j + 10\log G_{ja} - 10\log P_t - 10\log G_t - G_\sigma + \alpha_{Tx} \qquad \text{(all factors dB)} \qquad [16]$$

Note: To avoid having to include additional terms for these calculations, always combine any transmission line loss with antenna gain. The $20\log f_1$ term in $-G_\sigma$ cancels the $20\log f_1$ term in α_1.

Target gain factor, $G_\sigma = 10\log \sigma + 20\log f_1 + K_2$ (in dB)			One-way free space loss $\alpha_{Tx \text{ or } Rx} = 20\log f_1 R_{Tx \text{ or } Rx} + K_1$ (in dB)			
K_2 Values (dB)	RCS (σ) (units)	f_1 in MHz $K_2 =$ f_1 in GHz $K_2 =$	K_1 Values (dB)	Range (units)	f_1 in MHz $K_1 =$	f_1 in GHz $K_1 =$
	m^2	-38.54 21.46		NM	37.8	97.8
	ft^2	-48.86 11.14		km	32.45	92.45
				m	-27.55	32.45
				yd	-28.33	31.67
				ft	-37.87	22.13

<u>Saturated J/S (Monostatic) Example (Constant Power Jamming)</u>

Assume that a 5 GHz radar has a 70 dBm signal fed through a 5 dB loss transmission line to an antenna that has 45 dB gain. An aircraft is flying 31 km from the radar. The aft EW antenna has -1 dB gain and a 5 dB line loss to the EW receiver (there is an additional loss due to any antenna polarization mismatch but that loss will not be addressed in this problem). The aircraft has a jammer that provides 30 dBm saturated output if the received signal is above -35 dBm. The jammer feeds a 10 dB loss transmission line which is connected to an antenna with a 5 dB gain. If the RCS of the aircraft is 9 m^2, what is the J/S level received by the tracking radar?

Answer: The received signal at the jammer is the same as the example in Section 4-3, i.e. answer (1) = -32.3 dBm @ 5 GHz. Since the received signal is above -35 dBm, the jammer will operate in the saturated mode, and equation [5] can be used. (See Section 4-10 for an example of a jammer in the linear region.)

$$10\log J/S = 10\log P_j + 10\log G_{ja} - 10\log P_t - 10\log G_t - 10\log \sigma + 10.99 \text{ dB} + 20\log R$$

Note: the respective transmission line losses will be combined with antenna gains,
i.e. -5 + 45 = 40 dB & -10 +5 = -5 dB.

$$10\log J/S = 30 - 5 - 70 - 40 - 9.54 + 10.99 + 89.8 = 6.25 \text{ dB @ 5 GHz*}$$

* The answer is still 6.25 dB if the tracking radar operates at 7 GHz provided the antenna gains and the aircraft RCS are the same at both frequencies.

In this example, there is inadequate jamming power at each frequency if the J/S needs to be 10 dB or greater to be effective. One solution would be to replace the jammer with one that has a greater power output. If the antenna of the aircraft and the radar are not the proper polarization, additional power will also be required (see Section 3-2).

This Page Blank

BURN-THROUGH / CROSSOVER RANGE

The burn-through equations are derived in this section. These equations are most commonly used in jammer type of applications. The following is a summary of the important equations explored in this section:

J/S CROSSOVER RANGE (MONOSTATIC) (J = S)	* Keep P_t & P_j in same units Keep R and σ in same units
$R_{J=S} = [(P_t\ G_t\ \sigma) / (P_j\ G_{ja}\ 4\pi)]^{1/2}$ (dB Ratio) or $20 \log R_{J=S} = 10\log P_t + 10\log G_t + 10\log \sigma - 10\log P_j - 10\log G_{ja} - 10.99\ dB$	K_1 Values (dB): Range f_1 in MHz in GHz (units) $\underline{K_1 =}$ $\underline{K_1 =}$ m -27.55 32.45 ft -37.87 22.13
If simplified radar equations already converted to dB are used: $20 \log R_{J=S} = 10\log P_t + 10\log G_t + G_\sigma - 10\log P_j - 10\log G_{ja} - K_1 - 20\log f_1\ (\ dB)$	
BURN-THROUGH RANGE (MONOSTATIC) The radar to target range where the target return signal (S) can first be detected through the EA (J). $R_{BT} = [(P_t\ G_t\ \sigma\ J_{min\ eff}) / (P_j\ G_{ja}\ 4\pi\ S)]^{1/2}$ (dB Ratio) or $20\log R_{BT} = 10\log P_t + 10\log G_t + 10\log\sigma - 10\log P_j - 10\log G_{ja} + 10\log(J_{min\ eff}/S) - 10.99$	Target gain factor (dB) $G_\sigma = 10\log \sigma + 20\log f_1 + K_2$ K_2 Values (dB): RCS (σ)f_1 in MHz in GHz (units) $\underline{K_2 =}$ $\underline{K_2 =}$ m^2 -38.54 21.46 ft^2 -48.86 11.14
If simplified radar equations already converted to dB are used: $20\log R_{BT} = 10\log P_t + 10\log G_t + G_\sigma - 10\log P_j - 10\log G_{ja} - K_1 + 10\log(J_{min\ eff}/S) - 20\log f_1$(in dB)* f_1 is MHz or GHz value of frequency	
BURN-THROUGH RANGE (BISTATIC) R_{Tx} is the range from the radar transmitter to the target and is different from R_{Rx} which is the range from the target to the receiver. Use Monostatic equations and substitute R_{Tx} for R	

CROSSOVER RANGE and BURN-THROUGH RANGE

To present the values of J and S, (or J/S) over a minimum to maximum radar to target range of interest, equation [1], section 4-7. which has a slope of 20 log for J vs. range and equation [2], section 4-7, which has a slope of 40 log for S vs. range are plotted. When plotted on semi-log graph paper, J and S (or J/S) vs. range are straight lines as illustrated in Figure 1.

Figure 1 is a sample graph - it cannot be used for data.

The crossing of the J and S lines (known as crossover) gives the range where J = S (about 1.29 NM), and shows that shorter ranges will produce target signals greater than the jamming signal.

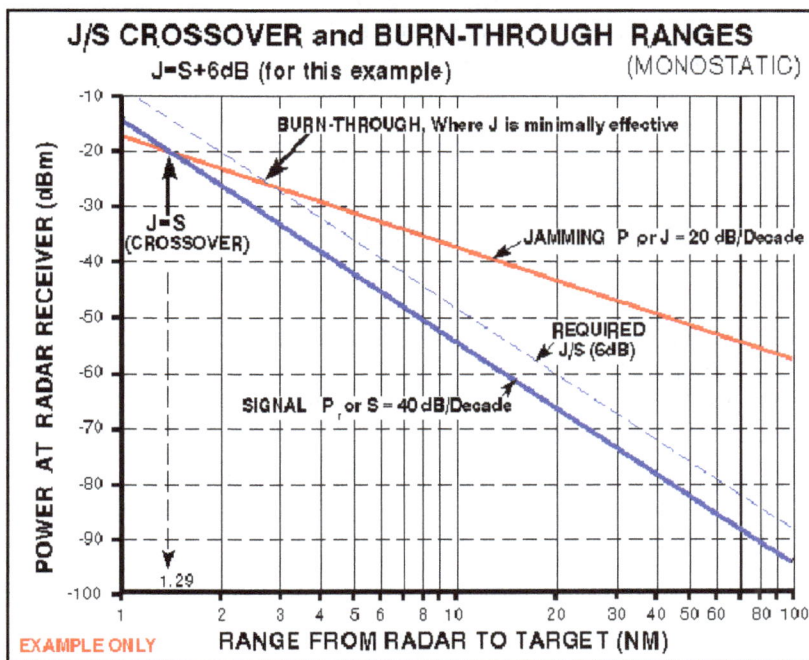

Figure 1. Sample J and S Graph

The point where the radar power overcomes the jamming signal is known as burn-through. The crossover point where J = S could be the burn-through range, but it usually isn't because normally J/S > 0 dB to be effective due to the task of differentiating the signal from the jamming noise floor (see receiver sensitivity section). For this example, the J/S required for the EA to be effective is given as 6 dB, as shown by the dotted line. This required J/S line crosses the jamming line at about 2.8 NM which, in this example, is the burn-through range.

In this particular example, we have: P_t = 80 dBm G_t = 42 dB
P_j = 50 dBm G_{ja} = 6 dB
σ = 18 m^2 f = 5.9 GHz (not necessary for all calculations)

A radar can be designed with higher than necessary power for earlier burn-through on jamming targets. Naturally that would also have the added advantage of earlier detection of non-jamming targets as well.

> Note: To avoid having to include additional terms for the following calculations, always combine any transmission line loss with antenna gain.

CROSSOVER AND BURN-THROUGH RANGE EQUATIONS (MONOSTATIC) - To calculate the crossover range or burn-through range the J/S equation must be solved for range. From equation [3], section 4-7:

$$\frac{J}{S} = \frac{P_j\, G_{ja}\, 4\pi\, R^2}{P_t\, G_t\, \sigma} \qquad \text{(ratio form)} \qquad \text{Solving for R:} \quad R = \sqrt{\frac{P_t\, G_t\, \sigma\, J}{P_j\, G_{ja}\, 4\pi\, S}} \qquad [1]$$

BURN-THROUGH RANGE (MONOSTATIC) - Burn-through Range (Monostatic) is the radar to target range where the target return signal (S) can first be detected through the jamming (J). It is usually the range when the J/S just equals the minimum effective J/S.

$$R_{BT} = \sqrt{\frac{P_t\, G_t\, \sigma\, J_{min\,eff}}{P_j\, G_{ja}\, 4\pi\, S}} \qquad \text{(burn-through range)} \qquad [2]$$

or in dB form, (using 10log 4π = 10.99 dB):

20log R_{BT} = 10log P_t + 10log G_t + 10log σ - 10log P_j - 10log G_{ja} + 10log ($J_{min\,eff}$/S) - 10.99 dB   [3]

RANGE WHEN J/S CROSSOVER OCCURS (MONOSTATIC) - The crossover of the jammer's 20 dB/decade power line and the skin return signal's 40 dB/decade power line of Figure 1 occurs for the case where J = S in dB or J/S=1 in ratio. Substituting into equation [1] yields:

$$R_{(J=S)} = \sqrt{\frac{P_t\, G_t\, \sigma}{P_j\, G_{ja}\, 4\pi}} \qquad \text{(Crossover range)} \qquad [4]$$

or in dB form:

20log $R_{J=S}$ = 10log P_t + 10log G_t + 10log σ - 10log P_j - 10log G_{ja} - 10.99 dB   [5]

Note: keep R and σ in same units in all equations.

CROSSOVER AND BURN-THROUGH EQUATIONS (MONOSTATIC)
USING α - ONE WAY FREE SPACE LOSS

The other crossover burn-through range formulas can be confusing because a frequency term is subtracted (equations [6], [7] and [8]), but both ranges are independent of frequency. This subtraction is necessary because when J/S is calculated directly as previously shown, λ^2 or $(c/f)^2$ terms canceled, whereas in the simplified radar equations, a frequency term is part of the G_σ term and has to be cancelled if one solves for R. From equation [8], section 4-7:

$$10\log J/S = 10\log P_j + 10\log G_{ja} - 10\log P_t - 10\log G_t - G_\sigma + \alpha_1 \quad \text{(factors in dB)}$$

or rearranging: $\quad \alpha_1 = 10\log P_t + 10\log G_t + G_\sigma - 10\log P_j - 10\log G_{ja} + 10\log (J/S)$

from section 4-4: $\qquad \alpha_1 = 20\log f_1 R_1 + K_1 \qquad\qquad$ or $\qquad\qquad 20\log R_1 = \alpha_1 - K_1 - 20\log f_1$

then substituting for α_1:

$$20\log R_1 = 10\log P_t + 10\log G_t + G_\sigma - 10\log P_j - 10\log G_{ja} - K_1 + 10\log (J/S) - 20\log f_1 \quad \text{(in dB)} \qquad \textbf{[6]}$$

<u>EQUATION FOR BURN-THROUGH RANGE (MONOSTATIC)</u> - Burn-through occurs at the range when the J/S just equals the minimum effective J/S. G_σ and K_1 are as defined on page 4-8.1.

$$20\log R_{BT} = 10\log P_t + 10\log G_t + G_\sigma - 10\log P_j - 10\log G_{ja} - K_1 + 10\log (J_{min\,eff}/S) - 20\log f_1 \quad \text{(in dB)} \qquad \textbf{[7]}$$

<u>EQUATION FOR THE RANGE WHEN J/S CROSSOVER OCCURS (MONOSTATIC)</u>

The J/S crossover range occurs for the case where J = S , substituting into equation [6] yields:

$$20\log R_{J=S} = 10\log P_t + 10\log G_t + G_\sigma - 10\log P_j - 10\log G_{ja} - K_1 - 20\log f_1 \quad \text{(factors in dB)} \qquad \textbf{[8]}$$

BURN-THROUGH RANGE (BISTATIC)

Bistatic J/S crossover range is the radar-to-target range when the power received (S) from the radar skin return from the target equals the power received (J) from the jamming signal transmitted from the target. As shown in Figure 6, section 4-7, the receive antenna that is receiving the same level of J and S is remotely located from the radar's transmit antenna. Bistatic equations [11], [13], and [14] in section 4-7 show that J/S is only a function of radar to target range, therefore J/S is not a function of wherever the missile is in its flight path provided the missile is in the antenna beam of the target's jammer. The missile is closing on the target at a very much higher rate than the target is closing on the radar, so the radar to target range will change less during the missile flight.

It should be noted that for a very long range air-to-air missile shot, the radar to target range could typically decrease to 35% of the initial firing range during the missile time-of-flight, i.e. A missile shot at a target 36 NM away, may be only 12 NM away from the firing aircraft at missile impact.

Figure 2 shows both the jamming radiated from the target and the power reflected from the target as a function of radar-to-target range. In this particular example, the RCS is assumed to be smaller, 15 m^2 vice 18m^2 in the monostatic case, since the missile will be approaching the target from a different angle. This will not, however, always be the case.

In this plot, the power reflected is:

$$P_{ref} = \frac{P_t\, G_t\, 4\,\pi\,\sigma}{(4\pi R)^2}$$

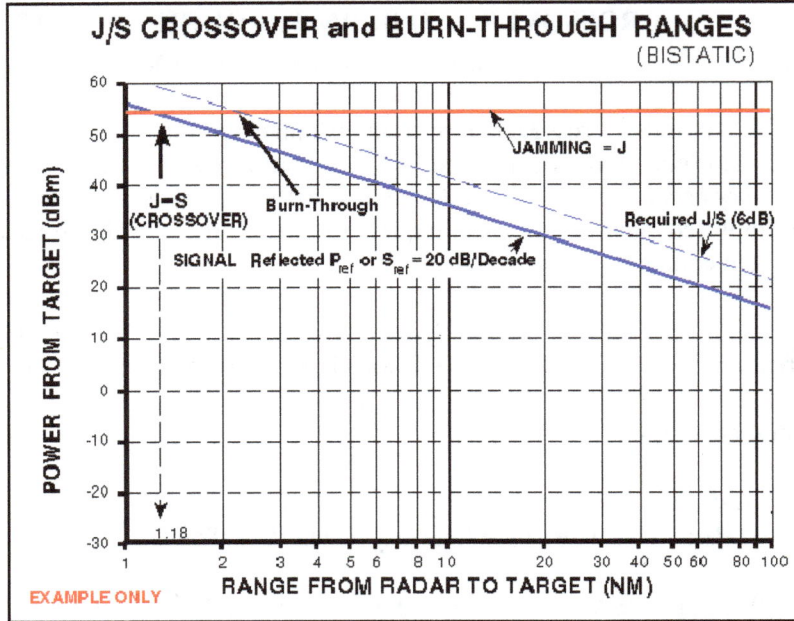

Figure 2. Bistatic Crossover and Burnthrough

Substituting the values given previously in the example on page 4-8.1, we find that the crossover point is at 1.18 NM (due to the assumed reduction in RCS).

CROSSOVER AND BURN-THROUGH RANGE EQUATIONS (BISTATIC)

To calculate the radar transmitter-to-target range where J/S crossover or burn-through occurs, the J/S equation must be solved for range. From equation [11] in section 4-7:

$$\frac{J}{S} = \frac{P_j\, G_{ja}\, 4\pi\, R_{Tx}^2}{P_t\, G_t\, \sigma} \qquad (ratio\ form)$$

Solving for R_{Tx}:
$$R_{Tx} = \sqrt{\frac{P_t\, G_t\, \sigma\, J}{P_j\, G_{ja}\, 4\pi\, S}} \qquad\qquad [9]$$

Note:
Bistatic equation [10] is identical to monostatic equation [1] except R_{Tx} must be substituted for R and a bistatic RCS (σ) will have to be used since RCS varies with aspect angle. The common explanations will not be repeated in this section.

BURN-THROUGH RANGE (BISTATIC) - Burn-through Range (Bistatic) occurs when J/S just equals the minimum effective J/S. From equation [9]:

$$R_{Tx(BT)} = \sqrt{\frac{P_t\, G_t\, \sigma\, J_{min\,eff}}{P_j\, G_{ja}\, 4\pi\, S}} \qquad (ratio\ form) \qquad [10]$$

or in dB form:

20log $R_{Tx(BT)}$ = 10log P_t + 10log G_t + 10log σ - 10log P_j - 10log G_{ja} + 10log ($J_{min\,eff}$/S) - 10.99 dB   [11]

If using the simplified radar equations (factors in dB):

20log $R_{Tx(BT)}$ = 10log P_t + 10log G_t + G_σ - 10log P_j - 10log G_{ja} - K_1 + 10log ($J_{min\,eff}$/S) - 20log f_1 [12]

   Where G_σ and K_1 are defined on page 4-8.1

RANGE WHEN J/S CROSSOVER OCCURS (BISTATIC)

The crossover occurs when J = S in dB or J/S = 1 in ratio.

$$R_{Tx(J=S)} = \sqrt{\frac{P_t\, G_t\, \sigma}{P_j\, G_{ja}\, 4\pi}} \quad (ratio) \qquad\qquad [13]$$

or in log form:

$$20\log R_{Tx(J=S)} = 10\log P_t + 10\log G_t + 10\log \sigma - 10\log P_j - 10\log G_{ja} - 10.99 \text{ dB} \qquad [14]$$

If simplified equations are used (with G_σ and K_1 as defined on page 4-8.1) we have:

$$20\log R_{Tx(J=S)} = 10\log P_t + 10\log G_t + G_\sigma - 10\log P_j - 10\log G_{ja} - K_1 - 20\log f_1 \qquad (factors\ in\ dB) \qquad [15]$$

Note: keep R and σ in same units in all equations.

DETAILS OF SEMI-ACTIVE MISSILE J/S

Unless you are running a large scale computer simulation that includes maneuvering, antenna patterns, RCS, etc., you will seldom calculate the variation in J/S that occurs during a semi-active missile's flight. Missiles don't fly straight lines at constant velocity. Targets don't either - they maneuver. If the launch platform is an aircraft, it maneuvers too. A missile will accelerate to some maximum velocity above the velocity of the launch platform and then decelerate.

The calculation of the precise variation of J/S during a missile flight for it to be effective requires determination of all the appropriate velocity vectors and ranges at the time of launch, and the accelerations and changes in relative positions during the fly out. In other words, it's too much work for too little return. The following are simplified examples for four types of intercepts.

	J/S (dB)	ΔJ/S (dB)
At Launch:	29	n/a
Intercept Type	At 2 sec. to Intercept:	
AAM Head-on:	23	-6
SAM Incoming Target:	25	-4
AAM Tail Chase:	29	0
SAM Outbound Target:	35	+6

In these examples, all velocities are constant, and are all along the same straight line. The missile velocity is 800 knots greater than the launch platform velocity which is assumed to be 400 kts. The missile launch occurs at 50 NM.

For the AAM tail chase, the range from the radar to the target remains constant and so does the J/S. **In these examples** the maximum variation from launch J/S is ± 6 dB. That represents the difference in the radar to target range closing at very high speed (AAM head on) and the radar to target range opening at moderate speed (SAM outbound target). The values shown above are examples, not rules of thumb, every intercept will be different.

Even for the simplified linear examples shown, graphs of the J and S will be curves - not straight lines. Graphs could be plotted showing J and S vs. radar to target range, or J and S vs. missile to target range, or even J/S vs. time of flight. If the J/S at launch is just barely the minimum required for effectiveness, and increasing it is difficult, then a detailed graph may be warranted, but in most cases this isn't necessary.

This Page Blank

SUPPORT JAMMING

The following table contains a summary of equations developed in this section:

MAIN LOBE JAMMING TO SIGNAL (J/S) RATIO (For SOJ/SIJ)	Target gain factor,

$$J/S = (P_j \, G_{ja} \, 4\pi \, R_{Tx}^4) / (P_t \, G_t \, \sigma \, [BW_J/BW_R] \, R_{Jx}^2) \quad \text{(ratio form)}*$$

$$10\log J/S = 10\log P_j - 10\log[BW_J/BW_R] + 10\log G_{ja} - 10\log P_t - 10\log G_t - 10\log \sigma + \downarrow$$
$$10.99 \text{ dB} + 40\log R_{Tx} - 20\log R_{Jx} * \qquad \leftarrow \leftarrow$$

or if simplified radar equations are used:

$$10\log J/S = 10\log P_j - BF + 10\log G_{ja} - \alpha_{jx} - 10\log P_t - 10\log G_t - G_\sigma + 2\alpha_1 \quad \text{(in dB)}*$$

Target gain factor,
$G_\sigma = 10\text{Log}\sigma + 20\text{Log} f_1 + K_2$ (in dB)
K_2 Values (dB):

RCS (σ) (units)	f_1 in MHz K_2 =	f_1 in GHz K_2 =
m²	-38.54	21.46
ft²	-48.86	11.14

SIDE LOBE JAMMING TO SIGNAL (J/S) RATIO (For SOJ/SIJ)

$$J/S = (P_j \, G_{ja} \, G_{r(SL)} \, 4\pi \, R_{Tx}^4) / (P_t \, G_t \, G_{r(ML)} \, \sigma \, [BW_J/BW_R] \, R_{Jx}^2) \quad \text{(ratio form)}*$$

$$10\log J/S = 10\log P_j - BF + 10\log G_{ja} + 10\log G_{r(SL)} - 10\log P_t - 10\log G_t - 10\log G_{r(ML)} + \rightarrow \downarrow$$
$$10.99 \text{ dB} - 10\log \sigma + 40\log R_{Tx} - 20\log R_{Jx} * \qquad \leftarrow \leftarrow$$

or if simplified radar equations are used (in dB)*:

$$10\log J/S = 10\log P_j - BF + 10\log G_{ja} + 10\log G_{r(SL)} - \alpha_{jx} - 10\log P_t - 10\log G_t - 10\log G_{r(ML)} - G_\sigma + 2\alpha_1$$

One-way free space loss,
α_1 or $\alpha_{Tx} = 20\text{Log}(f_1 R) + K_1$ (in dB)
K_1 Values (dB):

Range (units)	f_1 in MHz K_1 =	f_1 in GHz K_1 =
NM	37.8	97.8
Km	32.45	92.45
m	-27.55	32.45
yd	-28.33	31.67
ft	-37.87	22.13

R_{Jx}	Range from the support jammer transmitter to the radar receiver
R_{Tx}	Range between the radar and the target
BF	10 Log of the ratio of BW_J of the noise jammer to BW_R of the radar receiver
$G_{r(SL)}$	Side lobe antenna gain
$G_{r(ML)}$	Main lobe antenna gain
α_{JX}	One way free space loss between SOJ transmitter and radar receiver
α_1	One way space loss between the radar and the target

* Keep R and σ in same units

Support jamming adds a few geometric complexities. A SOJ platform usually uses high gain, directional antennas. Therefore, the jamming antenna must not only be pointed at the victim radar, but there must be alignment of radar, targets, and SOJ platform for the jamming to be most effective. Two cases will be described, main lobe-jamming and side-lobe jamming.

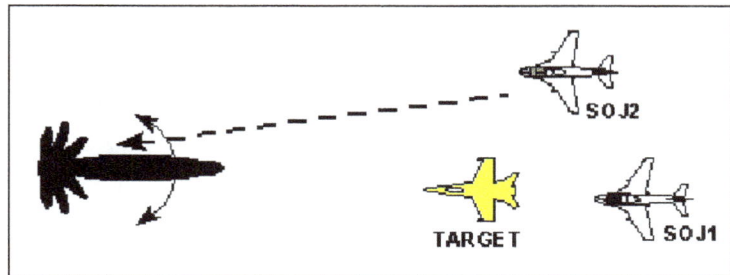

Figure 1. Radar Antenna Pattern

Support jamming is usually applied against search and acquisition radars which continuously scan horizontally through a volume of space. The scan could cover a sector or a full 360°. The horizontal antenna pattern of the radar will exhibit a main lobe and side lobes as illustrated in Figure 1. The target is detected when the main lobe sweeps across it. For main lobe jamming, the SOJ platform and the target(s) must be aligned with the radar's main lobe as it sweeps the target(s).

For side lobe jamming, the SOJ platform may be aligned with one or more of the radar's side lobes when the main lobe sweeps the target. The gain of a radar's side lobes are many tens of dB less (usually more than 30 dB less) than the gain of the main lobe, so calculations of side lobe jamming must use the gain of the side lobe for the radar receive antenna gain, not the gain of the main lobe. Also, because many modern radars employ some form of side lobe blanking or side lobe cancellation, some knowledge of the victim radar is required to predict the effectiveness of side lobe jamming.

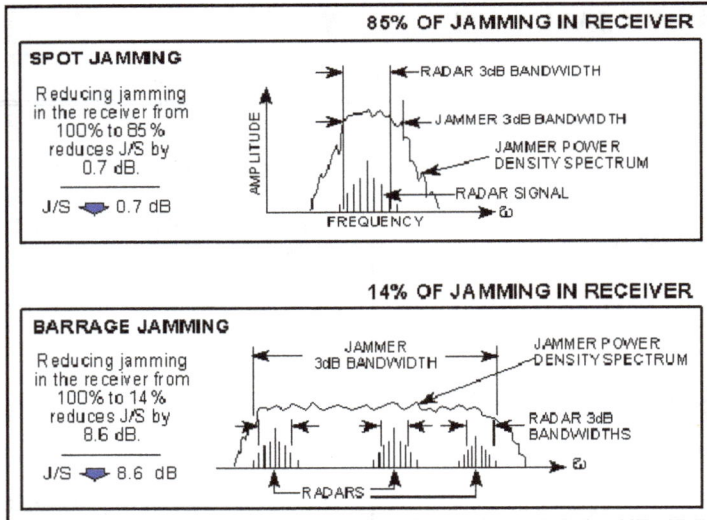

Figure 2. Noise Jamming

All radar receivers are frequency selective. That is, they are filters that allow only a narrow range of frequencies into the receiver circuitry. Deceptive EA, by definition, creates forgeries of the real signal and, ideally, are as well matched to the radar receiver as the real signal. On the other hand, noise jamming probably **will not match** the radar receiver bandwidth characteristics. Noise jamming is either spot jamming or barrage jamming. As illustrated in Figure 2, spot jamming is simply narrowing the bandwidth of the noise jammer so that as much of the jammer power as possible is in the radar receiver bandwidth. Barrage jamming is using a wide noise bandwidth to cover several radars with one jammer or to compensate for any uncertainty in the radar frequency. In both cases some of the noise power is "wasted" because it is not in the radar receiver filter.

In the past, noise jammers were often described as having so many "watts per MHz". This is nothing more than the power of the noise jammer divided by the noise bandwidth. That is, a 500 watt noise jammer transmitting a noise bandwidth of 200 MHz has 2.5 watts/MHz. Older noise jammers often had noise bandwidths that were difficult, or impossible, to adjust accurately. These noise jammers usually used manual tuning to set the center frequency of the noise to the radar frequency. Modern noise jammers can set on the radar frequency quite accurately and the noise bandwidth is selectable, so the noise bandwidth is more a matter of choice than it used to be, and it is possible that all of the noise is placed in the victim radar's receiver.

If, in the example above, the 500 watt noise jammer were used against a radar that had a 3 MHz receiver bandwidth, the noise jammer power applicable to that radar would be:

$$3\ MHz\ x\ 2.5\ watts/MHz = 7.5\ watts \_ 38.75\ dBm \qquad [1]$$

The calculation must be done as shown in equation [1] - multiply the watts/MHz by the radar bandwidth first and then convert to dBm. You can't convert to dBm/MHz and then multiply. (See derivation of dB in Section 2-4)

An alternate method for dB calculations is to use the bandwidth reduction factor (BF).

The BF is: $\qquad BF_{dB} = 10\ \text{Log} \left[\dfrac{BW_J}{BW_R} \right] \qquad\qquad\qquad [2]$

where: BW_J is the bandwidth of the noise jammer, and BW_R is the bandwidth of the radar receiver.

The power of the jammer in the jamming equation (P_J) can be obtained by either method. If equation [1] is used then P_J is simply 38.75 dBm. If equation [2] is used then the jamming equation is written using (P_J - BF). All the following discussion uses the second method. Which ever method is used, it is required that $BW_J \geq BW_R$. If $BW_J < BW_R$, then all the available power is in the radar receiver and equation [1] does not apply and the BF = 0.

> Note: To avoid having to include additional terms for the following calculations, always combine any transmission line loss with antenna gain.

MAIN LOBE STAND-OFF / STAND-IN JAMMING

The equivalent circuit shown in Figure 3 applies to main lobe jamming by a stand-off support aircraft or a stand-in RPV. Since the jammer is not on the target aircraft, only two of the three ranges and two of the three space loss factors (α's) are the same. Figure 3 differs from the J/S monostatic equivalent circuit shown in Figure 4 in Section 4-7 in that the space loss from the jammer to the radar receiver is different.

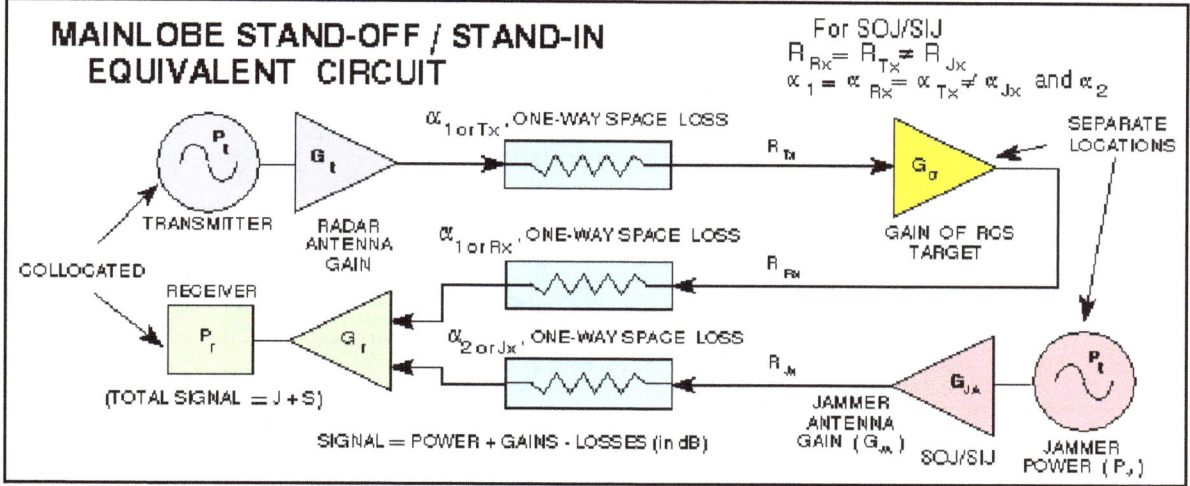

Figure 3. Main Lobe Stand-Off / Stand-In EA Equivalent Circuit

The equations are the same for both SOJ and SIJ. From the one way range equation in Section 4-3, and with inclusion of BF losses:

$$P_r 1 \; or \; J = \frac{P_j \, G_{ja} \, G_r \, \lambda^2 \, BW_R}{(4\pi \, R_{Jx})^2 \, BW_J} \tag{3}$$

From the two way range equation in Section 4.4:

$$P_r 2 \; or \; S = \frac{P_t \, G_t \, G_r \, \lambda^2 \, \sigma}{(4\pi)^3 \, R_{Tx}^4} \tag{4}$$

so

$$\frac{J}{S} = \frac{P_j \, G_{ja} \, G_r \, \lambda^2 (4\pi)^3 \, R_{Tx}^4 \, BW_R}{P_t \, G_t \, G_r \, \lambda^2 \, \sigma (4\pi \, R_{Jx})^2 \, BW_J} = \frac{P_j \, G_{ja} \, 4\pi \, R_{Tx}^4 \, BW_R}{P_t \, G_t \, \sigma \, R_{Jx}^2 \, BW_J} \quad (ratio \; form) \tag{5}$$

Note: Keep R and σ in the same units. Converting to dB and using 10 log 4π = 10.99 dB:

10log J/S = 10log P_j -10log [BW_j/BW_R] +10log G_{ja} -10log P_t -10log G_t - 10log σ + 10.99 dB +40log R_{Tx} -20log R_{Jx} **[6]**

If the simplified radar equation is used, the free space loss from the SOJ/SIJ to the radar receiver is α_{Jx}, then equation [7] is the same as monostatic equation [6] in Section 4-7 except α_{Jx} replaces α, and the bandwidth reduction factor [BF] losses are included:

10log J = 10log P_j - BF + 10log G_{ja} + 10log G_r - α_{Jx} (factors in dB)   **[7]**

Since the free space loss from the radar to the target and return is the same both ways, $\alpha_{Tx} = \alpha_{Rx} = \alpha_1$, equation [8] is the same as monostatic equation [7] in Section 4-7.

10log S = 10log P_t + 10log G_t + 10log G_r + G_σ - $2\alpha_1$ (factors in dB)   **[8]**

and   10log J/S = 10log P_j - BF + 10log G_{ja} - α_{Jx} - 10log P_t - 10log G_t - G_σ + $2\alpha_1$ (factors in dB)   **[9]**

Notice that unlike equation [8] in Section 4-7, there are two different α's in [9] because the signal paths are different.

SIDE LOBE STAND-OFF / STAND-IN JAMMING

The equivalent circuit shown in Figure 4. It differs from Figure 3, (main lobe SOJ/SIJ) in that the radar receiver antenna gain is different for the radar signal return and the jamming.

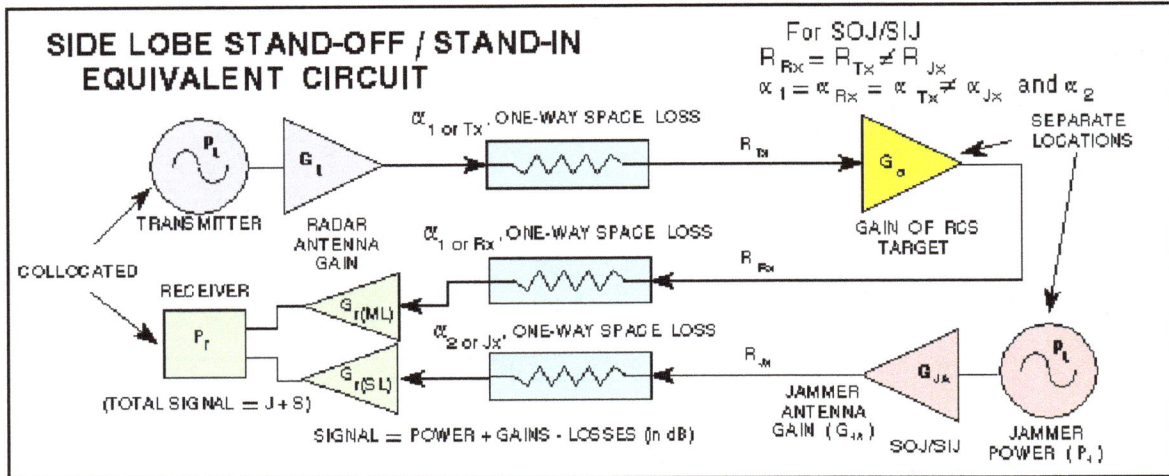

Figure 4. Side Lobe Stand-Off / Stand-In EA Equivalent Circuit

To calculate side lobe jamming, the gain of the radar antenna's side lobes must be known or estimated. The gain of each side lobe will be different than the gain of the other side lobes. If the antenna is symmetrical, the first side lobe is the one on either side of the main lobe, the second side lobe is the next one on either side of the first side lobe, and so on. The side lobe gain is G_{SLn}, where the 'n' subscript denotes side lobe number: 1, 2, ..., n.

The signal is the same as main lobe equations [4] and [8], except $G_r = G_{r(ML)}$

$$P_r \, 2 \ or \ S = \frac{P_t \, G_t \, G_{r(ML)} \, \lambda^2 \, \sigma}{(4\pi)^3 \, R_{Tx}^4} \quad \textit{(ratio form)} \tag{10}$$

If simplified radar equations are used:

$10\log S = 10\log P_t + 10\log G_t + 10\log G_{r(ML)} + G_\sigma - 2\alpha_1$ (factors in dB)

The jamming equation is the same as main lobe equations [3] and [7] except $G_r = G_{r(SL)}$:

$$J = \frac{P_j \, G_{ja} \, G_{r(SL)} \, \lambda^2 \, BW_R}{(4\pi \, R_{Jx})^2 \, BW_J} \tag{11}$$

$10\log J = 10\log P_j - BF + 10\log G_{ja} + 10\log G_{r(SL)} - \alpha_{Jx}$ (factors in dB)   [12]

so $$\frac{J}{S} = \frac{P_j \, G_{ja} \, G_{r(SL)} \, 4\pi \, R_{Tx}^4 \, BW_R}{P_t \, G_t \, G_{r(ML)} \, \sigma \, R_{Jx}^2 \, BW_J} \quad \textit{(ratio form)} \tag{13}$$

Note: keep R and σ in same units. Converting to dB and using $10\log 4\pi = 10.99$ dB:

$10\log J/S = 10\log P_j - BF + 10\log G_{ja} + 10\log G_{r(SL)} - 10\log P_t - 10\log G_t - 10\log G_{r(ML)} - 10\log\sigma + 10.99 \ dB + 40\log R_{Tx} - 20\log R_{Jx}$
(factors in dB)   [14]

If simplified radar equations are used:

$10\log J/S = 10\log P_j - BF + 10\log G_{ja} + 10\log G_{r(SL)} - \alpha_{Jx} - 10\log P_t - 10\log G_t - 10\log G_{r(ML)} - G_\sigma + 2\alpha_1$ [15]
(in dB)

JAMMING TO SIGNAL (J/S) RATIO - CONSTANT GAIN [LINEAR] JAMMING

JAMMING TO SIGNAL (J/S) RATIO (MONOSTATIC)

$$\frac{J}{S} = \frac{G_{ja(Rx)}\, G_j\, G_{ja(Tx)}\, \lambda^2}{4\pi\sigma} = \frac{G_{ja(Rx)}\, G_j\, G_{ja(Tx)}\, c^2}{4\pi\sigma\, f^2} \quad \text{(ratio form)}$$

$G_{ja(Rx)}$ = The Gain of the jammer receive antenna
G_j = The gain of the jammer
$G_{ja(Tx)}$ = The Gain of the jammer transmit antenna

or:

$10\log J/S = 10\log G_{ja(Rx)} + 10\log G_j + 10\log G_{ja(Tx)} - 10\log (4\pi\sigma/\lambda^2)$

or if simplified radar equations developed in previous sections are used:

$10\log J/S = 10\log G_{ja(Rx)} + 10\log G_j + 10\log G_{ja(Tx)} - G_\sigma \quad$ (dB)

* Keep λ and σ in same units.   Note: $\lambda = c/f$

Target gain factor,
$G_\sigma = 10\log \sigma + 20\log f_1 + K_2$ (dB)

K_2 Values (dB):

RCS (σ) (units)	f_1 in MHz K_2 =	f_1 in GHz K_2 =
m²	-38.54	21.46
ft²	-48.86	11.14

JAMMING TO SIGNAL (J/S) RATIO (BISTATIC)

Same as the monostatic case except G_σ will be different since RCS (σ) varies with aspect angle.

Since the jammer on the target is amplifying the received radar signal before transmitting it back to the radar, both J and S experience the two way range loss. Figure 1 shows that the range for both the signal and constant gain jamming have a slope that is 40 dB per decade. Once the jammer output reaches maximum power, that power is constant and the jamming slope changes to 20 dB per decade since it is only a function of one way space loss and the J/S equations for constant power (saturated) jamming must be used.

Normally the constant gain (linear) region of a repeater jammer occurs only at large

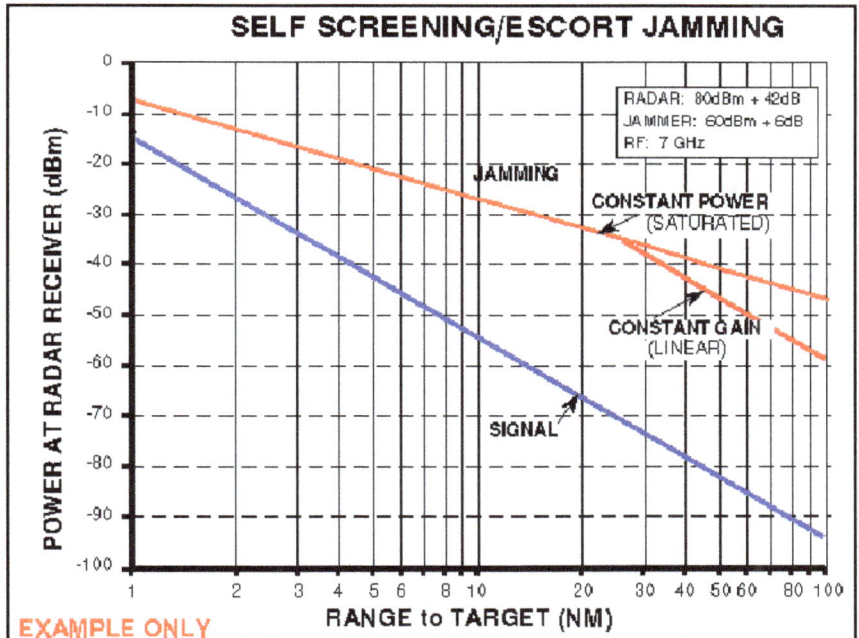

Figure 1. Sample Constant Gain / Constant Power Graph

distances from the radar and the constant power (saturated) region is reached rapidly as the target approaches the radar. When a constant gain jammer is involved it may be necessary to plot jamming twice - once using J from the constant power (saturated) equation [1] in Section 4-7 and once using the constant gain (linear) equation [4], as in the example shown in Figure 1.

CONSTANT GAIN SELF PROTECTION EA

Most jammers have a constant power output - that is, they always transmit the maximum available power of the transmitter (excepting desired EA modulation). Some jammers also have a constant gain (linear) region. Usually these are coherent repeaters that can amplify a low level radar signal to a power that is below the level that results in maximum available (saturated) power output. At some radar to target range, the input signal is sufficiently high that the full jammer gain would exceed the maximum available power and the jammer ceases to be constant gain and becomes constant power.

To calculate the power output of a constant gain jammer where:
S_{Rj} = The Radar signal at the jammer input (receive antenna terminals)
$G_{ja(Rx)}$ = The Gain of the jammer receive antenna
G_j = The gain of the jammer
α_{Tx} = The one-way free space loss from the radar to the target
P_{jCG} = The jammer constant gain power output
P_j = The maximum jammer power output
L_R = The jammer receiving line loss; combine with antenna gain $G_{ja(Rx)}$

From equation [10], Section 4-3, calculate the radar power received by the jammer.
$$10\log S_{Rj} = 10\log P_t + 10\log G_t - \alpha_{Tx} + 10\log G_{ja(Rx)} \qquad \text{(factors in dB)} \qquad [1]$$

The jammer constant gain power output is: $\qquad 10\log P_{jCG} = 10\log S_{Rj} + 10\log G_{ja} \qquad [2]$
and, by definition: $\qquad\qquad\qquad\qquad\qquad P_{jCG} \le P_j \qquad\qquad\qquad\qquad\qquad [3]$

MONOSTATIC

The equivalent circuit shown in Figure 2 is different from the constant power equivalent circuit in Figure 4 in Section 4-7. With constant gain, the jamming signal experiences the gain of the jammer and its antennas plus the same space loss as the radar signal.

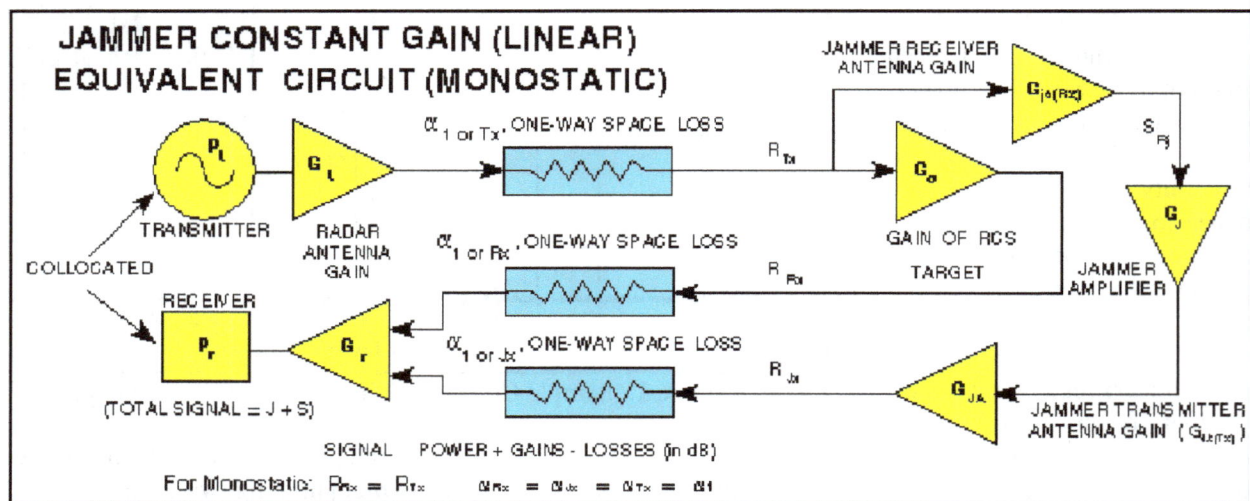

Figure 2. Jammer Constant Gain EA Equivalent Circuit (Monostaic)

To calculate J, the one way range equation from Section 4-3 is used twice:

$$J = \frac{P_t \, G_t \, G_{ja(Rx)} \, \lambda^2}{(4\pi R)^2} \, \frac{G_j \, G_{ja(Tx)} \, G_r \, \lambda^2}{(4\pi R)^2} \qquad\qquad [4]$$

From the two way range equation in Section 4-4:
$$S = \frac{P_t \, G_t \, G_r \, \lambda^2 \, \sigma}{(4\pi)^3 \, R^4} \qquad\qquad [5]$$

Terms cancel when combined: $\quad \dfrac{J}{S} = \dfrac{G_{ja(Rx)} \, G_j \, G_{ja(Tx)} \, \lambda^2}{4\pi\sigma} \quad$ *Keep λ and σ in same units* **[6]**

Or in dB form: $\quad 10\log J/S = 10\log G_{ja(Rx)} + 10\log G_j + 10\log G_{ja(Tx)} - 10\log (4\pi\sigma/\lambda^2) \qquad$ **[7]**

Since the last term can be recognized as minus G_σ from equation [10] in Section 4-4, where the target gain factor, $G_\sigma = 10\log (4\pi\sigma/\lambda^2) = 10\log (4\pi\sigma f^2/c^2)$, it follows that:

$$10\log J/S = 10\log G_{ja(Rx)} + 10\log G_j + 10\log G_{ja(Tx)} - G_\sigma \quad \text{(factors in dB)} \qquad [8]$$

Target gain factor, $G_\sigma = 10\log \sigma + 20\log f_1 + K_2$ (in dB)		
K$_2$ Values		
(dB) RCS (σ)	f_1 in MHz	f_1 in GHz
(units)	K$_2$ =	K$_2$ =
m^2	-38.54	21.46
ft^2	-48.86	11.14

BISTATIC

The bistatic equivalent circuit shown in Figure 3 is different from the monostatic equivalent circuit shown in Figure 2 in that the receiver is separately located from the transmitter, $R_{Tx} \neq R_{Rx}$ or R_{Jx} and G_σ will be different since the RCS (σ) varies with aspect angle.

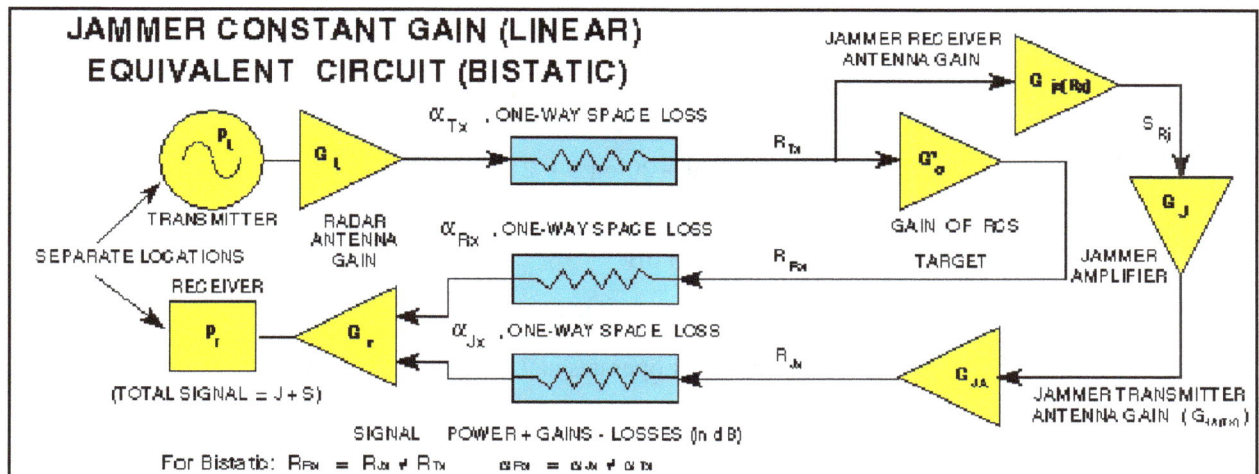

Figure 3. Jammer Constant Gain EA Equivalent Circuit (Bistatic)

To calculate J, the one way range equation from Section 4-3 is used twice:

$$J = \frac{P_t\, G_t\, G_{ja(Rx)}\, \lambda^2}{(4\pi\, R_{Tx}\,)^2}\, \frac{G_j\, G_{ja(Tx)}\, G_r\, \lambda^2}{(4\pi\, R_{Rx}\,)^2} \qquad (R_{Jx} = R_{Rx}) \tag{9}$$

From the two way range equation in Section 4-4: $S = \dfrac{P_t\, G_t\, G_r\, \lambda^2\, \sigma'}{(4\pi\,)^3\, R_{Tx}^2\, R_{Rx}^2}$ (σ' is bistatic RCS) **[10]**

Terms cancel when combined: $\dfrac{J}{S} = \dfrac{G_{ja(Rx)}\, G_j\, G_{ja(Tx)}\, \lambda^2}{4\pi\sigma'}$ *Keep λ and σ in same units* **[11]**

Or in dB form: $10\log J/S = 10\log G_{ja(Rx)} + 10\log G_j + 10\log G_{ja(Tx)} - 10\log (4\pi\sigma'/\lambda^2)$ **[12]**

Since the last term can be recognized as minus G_σ from equation [10] in Section 4-4, where the target gain factor, $G_\sigma = 10\log (4\pi\sigma'/\lambda^2) = 10\log (4\pi\sigma' f^2/c^2)$, it follows that:

$$10\log = 10\log G_{ja(Rx)} + 10\log G_j + 10\log G_{ja(Tx)} - G_\sigma' \qquad \text{(factors in dB)} \tag{13}$$

Target gain factor, $G_\sigma = 10\log \sigma + 20\log f_1 + K_2$ (in dB)		
K_2 Values		
(dB) RCS (σ)	f_1 in MHz	f_1 in GHz
(units)	$K_2 =$	$K_2 =$
m^2	-38.54	21.46
ft^2	-48.86	11.14

Linear J/S (Monostatic) Example (Linear Power Jamming)

Assume that a 5 GHz radar has a 70 dBm signal fed through a 5 dB loss transmission line to an antenna that has 45 dB gain. An aircraft that is flying 31 km from the radar has an aft EW antenna with -1 dB gain and a 5 dB line loss to the EW receiver (there is an additional loss due to any antenna polarization mismatch but that loss will not be addressed in this problem). The received signal is fed to a jammer with a gain of 60 dB, feeding a 10 dB loss transmission line which is connected to an antenna with 5 dB gain.

If the RCS of the aircraft is 9 m^2, what is the J/S level received at the input to the receiver of the tracking radar?

Answer:

$10\log J/S = 10\log G_{ja(Rx)} + 10\log G_j + 10\log G_{ja(Tx)} - G_\sigma$

$G_\sigma = 10\log \sigma + 20\log f_1 + K_2 = 10\log 9 + 20\log 5 + 21.46 = 44.98$ dB

Note: The respective transmission line losses will be combined with antenna gains,
i.e. -1 -5 = -6 dB and -10 + 5 = -5 dB

$10\log J/S = -6 + 60 - 5 - 44.98 = 4.02$ dB @ 5 GHz

The answer changes to 1.1 dB if the tracking radar operates at 7 GHz provided the antenna gains and aircraft RCS are the same at both 5 and 7 GHz.

$G_\sigma = 10\log 9 + 20\log 7 + 21.46 = 47.9$ dB

$10\log J/S = -6 + 60 - 5 - 47.9 = 1.1$ dB @ 7 GHz

Separate J (-73.5 dBm @ 5 GHz and -79.34 dBm @ 7 GHz) and S (-77.52 dBm @ 5 GHz and -80.44 dBm @ 7 GHz) calculations for this problem are provided in Sections 4-3 and 4-4, respectively. A saturated gain version of this problem is provided in Section 4-7.

RADAR CROSS SECTION (RCS)

Radar cross section is the measure of a target's ability to reflect radar signals in the direction of the radar receiver, i.e. it is a measure of the ratio of backscatter power per steradian (unit solid angle) in the direction of the radar (from the target) to the power density that is intercepted by the target.

The RCS of a target can be viewed as a comparison of the strength of the reflected signal from a target to the reflected signal from a perfectly smooth sphere of cross sectional area of 1 m² as shown in Figure 1 .

The conceptual definition of RCS includes the fact that not all of the radiated energy falls on the target. A target's RCS (σ) is most easily visualized as the product of three factors:

σ = Projected cross section x **Reflectivity** x **Directivity** .

RCS(σ) is used in Section 4-4 for an equation representing power reradiated from the target.

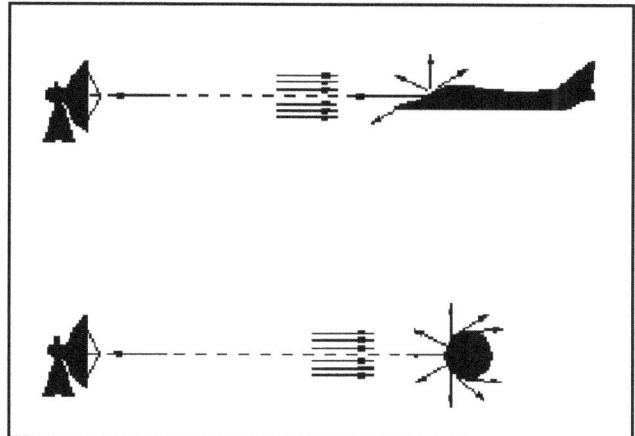

Figure 1. Concept of Radar Cross Section

Reflectivity: The percent of intercepted power reradiated (scattered) by the target.

Directivity: The ratio of the power scattered back in the radar's direction to the power that would have been backscattered had the scattering been uniform in all directions (i.e. isotropically).

Figures 2 and 3 show that RCS does not equal geometric area. For a sphere, the RCS, **σ = πr²**, where r is the radius of the sphere.

The RCS of a sphere is independent of frequency if operating at sufficiently high frequencies where λ<<Range, and λ<< radius (r). Experimentally, radar return reflected from a target is compared to the radar return reflected from a sphere which has a frontal or projected area of one square meter (i.e. diameter of about 44 in). Using the spherical shape aids in field or laboratory measurements since orientation or positioning of the sphere will not affect radar reflection intensity measurements as a flat plate would. If calibrated, other sources (cylinder, flat plate, or corner reflector, etc.) could be used for comparative measurements.

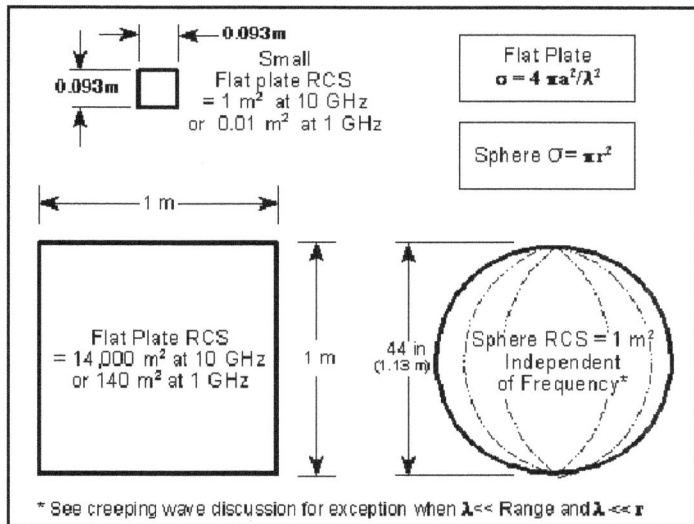

Figure 2. RCS vs Physical Geometry

To reduce drag during tests, towed spheres of 6", 14" or 22" diameter may be used instead of the larger 44" sphere, and the reference size is 0.018, 0.099 or 0.245 m² respectively instead of 1 m². When smaller sized spheres are used for tests you may be operating at or near where λ~radius. If the results are then scaled to a 1 m² reference, there may be some perturbations due to creeping waves. See the discussion at the end of this section for further details.

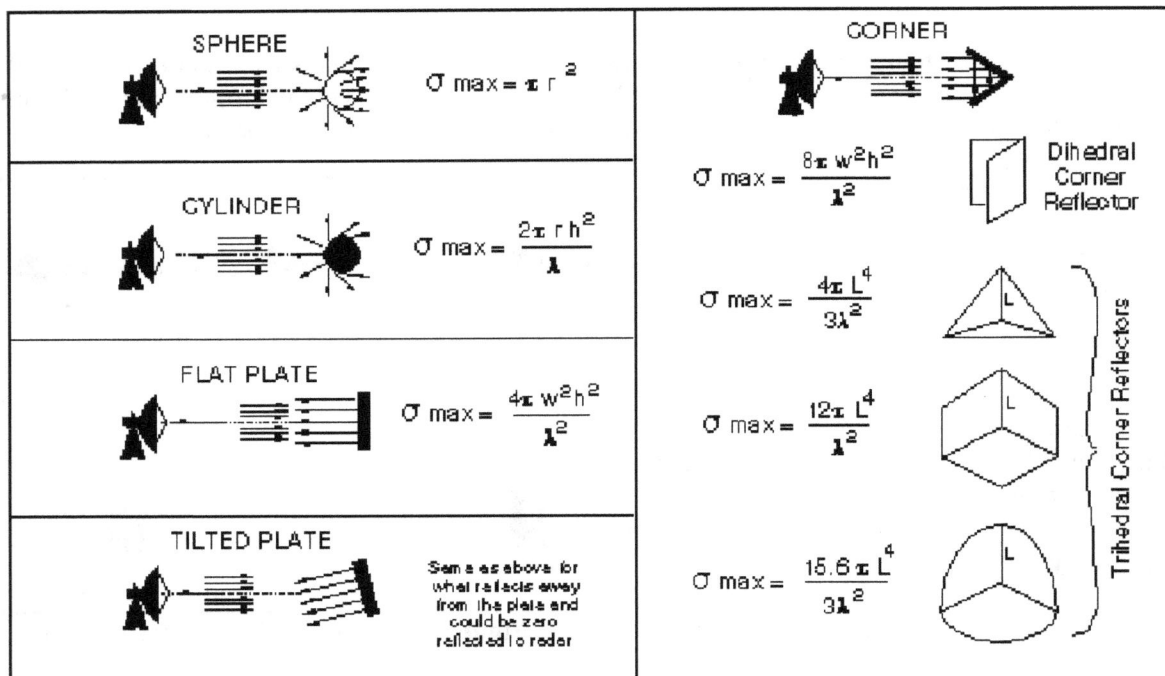

Figure 3. Backscatter From Shapes

In Figure 4, RCS patterns are shown as objects are rotated about their vertical axes (the arrows indicate the direction of the radar reflections).

The sphere is essentially the same in all directions.

The flat plate has almost no RCS except when aligned directly toward the radar.

The corner reflector has an RCS almost as high as the flat plate but over a wider angle, i.e., over

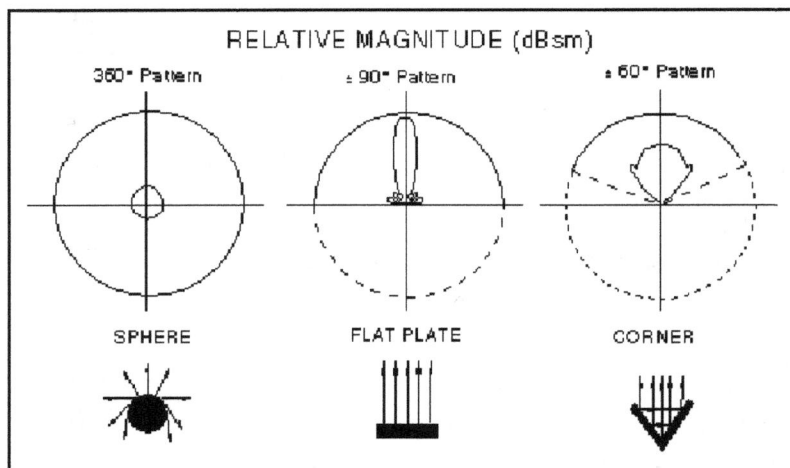

Figure 4. RCS Patterns

±60°. The return from a corner reflector is analogous to that of a flat plate always being perpendicular to your collocated transmitter and receiver.

Targets such as ships and aircraft often have many effective corners. Corners are sometimes used as calibration targets or as decoys, i.e. corner reflectors.

An aircraft target is very complex. It has a great many reflecting elements and shapes. The RCS of real aircraft must be measured. It varies significantly depending upon the direction of the illuminating radar.

Figure 5 shows a typical RCS plot of a jet aircraft. The plot is an azimuth cut made at zero degrees elevation (on the aircraft horizon). Within the normal radar range of 3-18 GHz, the radar return of an aircraft in a given direction will vary by a few dB as frequency and polarization vary (the RCS may change by a factor of 2-5). It does not vary as much as the flat plate.

As shown in Figure 5, the RCS is highest at the aircraft beam due to the large physical area observed by the radar and perpendicular aspect (increasing reflectivity). The next highest RCS area is the nose/tail area, largely because of reflections off the engines or propellers. Most self-protection jammers cover a field of view of +/- 60 degrees about the aircraft nose and tail, thus the high RCS on the beam does not have coverage.

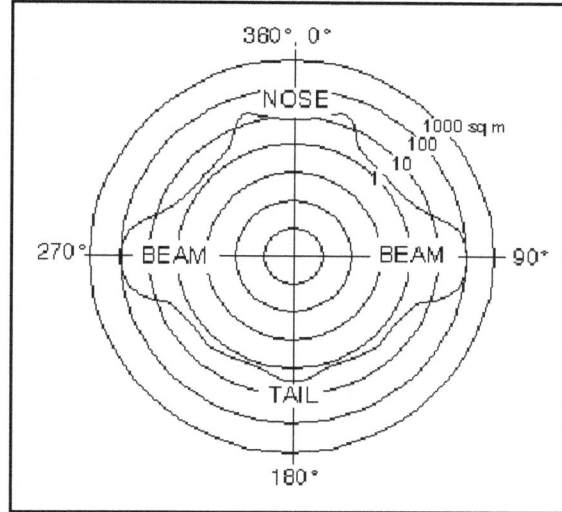

Figure 5. Typical Aircraft RCS

Beam coverage is frequently not provided due to inadequate power available to cover all aircraft quadrants, and the side of an aircraft is theoretically exposed to a threat 30% of the time over the average of all scenarios.

Typical radar cross sections are as follows: Missile 0.5 sq m; Tactical Jet 5 to 100 sq m; Bomber 10 to 1000 sq m; and ships 3,000 to 1,000,000 sq m. RCS can also be expressed in decibels referenced to a square meter (dBsm) which equals 10 log (RCS in m^2).

Again, Figure 5 shows that these values can vary dramatically. The strongest return depicted in the example is 100 m^2 in the beam, and the weakest is slightly more than 1 m^2 in the 135°/225° positions. These RCS values can be very misleading because other factors may affect the results. For example, phase differences, polarization, surface imperfections, and material type all greatly affect the results. In the above typical bomber example, the measured RCS may be much greater than 1000 square meters in certain circumstances (90°, 270°).

SIGNIFICANCE OF THE REDUCTION OF RCS

If each of the range or power equations that have an RCS (σ) term is evaluated for the significance of decreasing RCS, Figure 6 results. Therefore, an RCS reduction can increase aircraft survivability. The equations used in Figure 6 are as follows:

Range (radar detection):
From the 2-way range equation in Section 4-4:
$$P_r = \frac{P_t\, G_t\, G_r\, \lambda^2\, \sigma}{(4\pi)^3\, R^4} \quad \text{Therefore, } R^4 \propto \sigma \text{ or } \sigma^{1/4} \propto R$$

Range (radar burn-through):
The crossover equation in Section 4-8 has:
$$R_{BT}^2 = \frac{P_t\, G_t\, \sigma}{P_j\, G_j\, 4\pi} \quad \text{Therefore, } R_{BT}^2 \propto \sigma \text{ or } \sigma^{1/2} \propto R_{BT}$$

Power (jammer): Equating the received signal return (P$_r$) in the two way range equation to the received jammer signal (P$_r$) in the one way range equation, the following relationship results:

$$P_r = \underset{\underset{S}{\uparrow}}{\frac{P_t\, G_t\, G_r\, \lambda^2\, \sigma}{(4\pi)^3\, R^4}} = \underset{\underset{J}{\uparrow}}{\frac{P_j\, G_j\, G_r\, \lambda^2}{(4\pi R)^2}}$$

Therefore, P$_j \propto \sigma$ or $\sigma \propto$ P$_j$
Note: jammer transmission line loss is combined with the jammer antenna gain to obtain G$_t$.

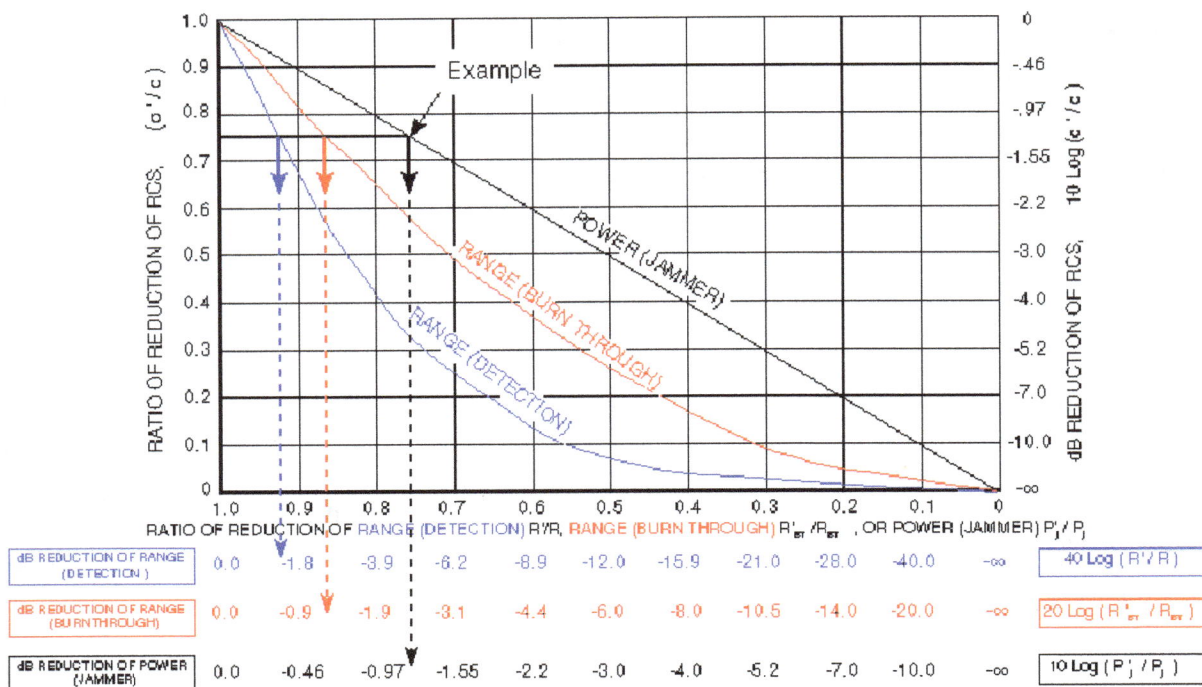

Figure 6. Reduction of RCS Affects Radar Detection, Burn-through, and Jammer Power

Example of Effects of RCS Reduction - As shown in Figure 6, if the RCS of an aircraft is reduced to 0.75 (75%) of its original value, then (1) the jammer power required to achieve the same effectiveness would be 0.75 (75%) of the original value (or -1.25 dB). Likewise, (2) If Jammer power is held constant, then burn-through range is 0.87 (87%) of its original value (-1.25 dB), and (3) the detection range of the radar for the smaller RCS target (jamming not considered) is 0.93 (93%) of its original value (-1.25 dB).

OPTICAL / MIE / RAYLEIGH REGIONS

Figure 7 shows the different regions applicable for computing the RCS of a sphere. The optical region ("far field" counterpart) rules apply when $2\pi r/\lambda > 10$. In this region, the RCS of a sphere is independent of frequency. Here, the RCS of a sphere, $\sigma = \pi r^2$. The RCS equation breaks down primarily due to creeping waves in the area where $\lambda \sim 2\pi r$. This area is known as the Mie or resonance region. If we were using a 6" diameter sphere, this frequency would be 0.6 GHz. (Any frequency ten times higher, or above 6 GHz, would give expected results). The largest positive perturbation (point A) occurs at <u>exactly</u> 0.6 GHz where the RCS would be 4 times higher than the RCS computed using the optical region formula. Just slightly above 0.6 GHz a minimum occurs (point B) and the actual RCS would be 0.26 times the value calculated by using the optical region formula. If we used a one meter diameter sphere, the perturbations would occur at 95 MHz, so any frequency above 950 MHz (~1 GHz) would give predicted results.

CREEPING WAVES

The initial RCS assumptions presume that we are operating in the optical region (λ<<Range and λ<<radius). There is a region where specular reflected (mirrored) waves combine with back scattered creeping waves both constructively and destructively as shown in Figure 8. Creeping waves are tangential to a smooth surface and follow the "shadow" region of the body. They occur when the circumference of the sphere $\sim \lambda$ and typically add about 1 m^2 to the RCS at certain frequencies.

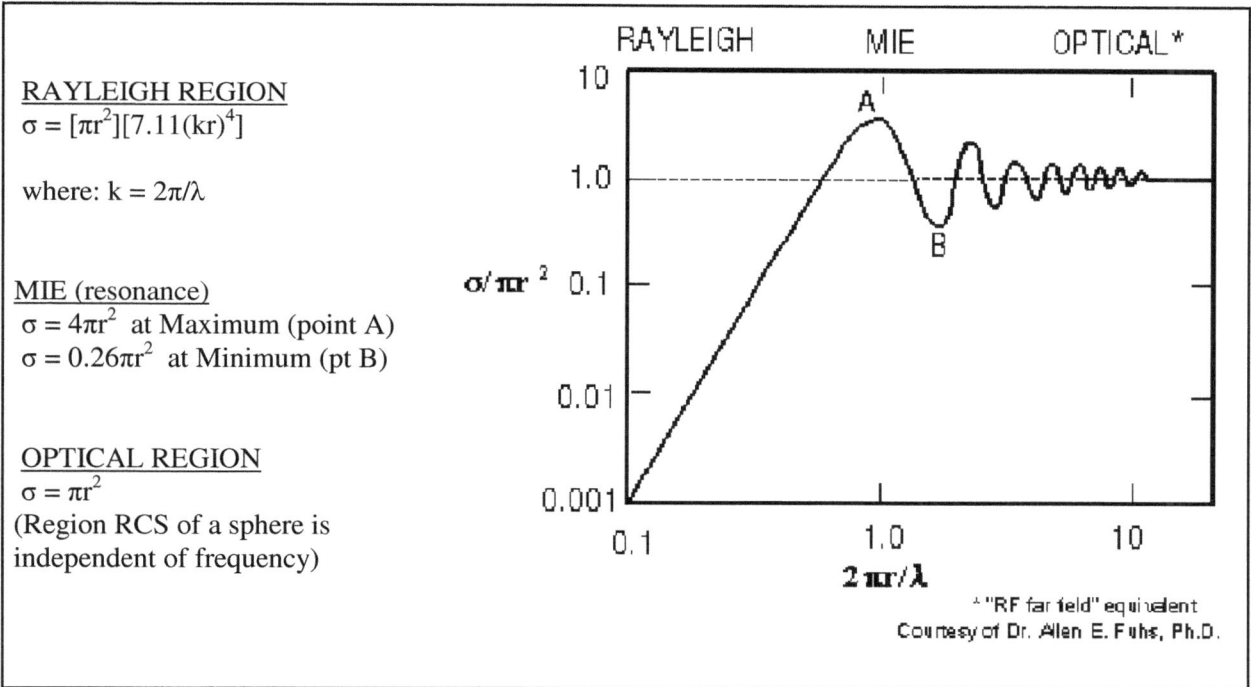

Figure 7. Radar Cross Section of a Sphere

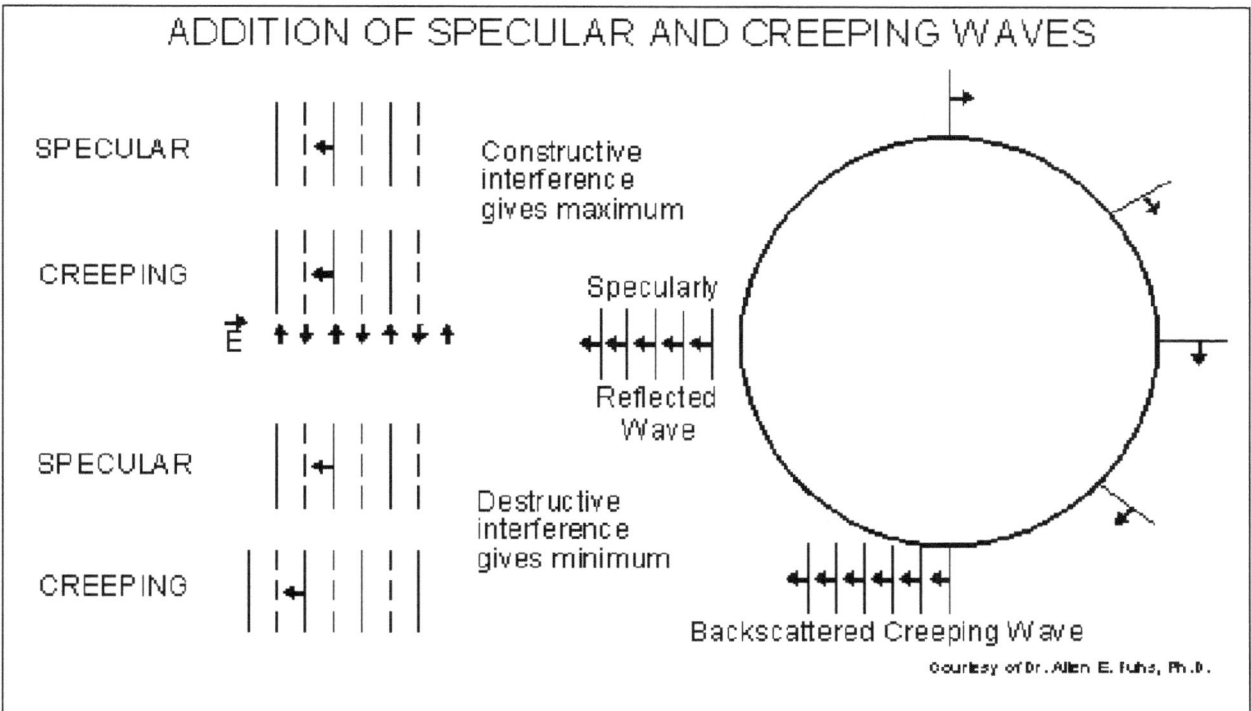

RAYLEIGH REGION
$$\sigma = [\pi r^2][7.11(kr)^4]$$

where: $k = 2\pi/\lambda$

MIE (resonance)
$\sigma = 4\pi r^2$ at Maximum (point A)
$\sigma = 0.26\pi r^2$ at Minimum (pt B)

OPTICAL REGION
$\sigma = \pi r^2$
(Region RCS of a sphere is independent of frequency)

RAYLEIGH   MIE   OPTICAL*

$\sigma/\pi r^2$

$2\pi r/\lambda$

* "RF far field" equivalent
Courtesy of Dr. Allen E. Fuhs, Ph.D.

Figure 7. Radar Cross Section of a Sphere

ADDITION OF SPECULAR AND CREEPING WAVES

SPECULAR

CREEPING

Constructive interference gives maximum

$\vec{E}$

SPECULAR

CREEPING

Destructive interference gives minimum

Specularly Reflected Wave

Backscattered Creeping Wave

Courtesy of Dr. Allen E. Fuhs, Ph.D.

Figure 8. Addition of Specular and Creeping Waves

EMISSION CONTROL (EMCON)

When EMCON is imposed, RF emissions must not exceed -110 dBm/meter^2 at one nautical mile. It is best if systems meet EMCON when in either the Standby or Receive mode versus just the Standby mode (or OFF). If one assumes antenna gain equals line loss, then emissions measured at the port of a system must not exceed -34 dBm (i.e. the stated requirement at one nautical mile is converted to a measurement at the antenna of a point source - see Figure 1). If antenna gain is greater than line loss (i.e. gain 6 dB, line loss 3 dB), then the -34 dBm value would be lowered by the difference and would be -37 dBm for the example. The opposite would be true if antenna gain is less.

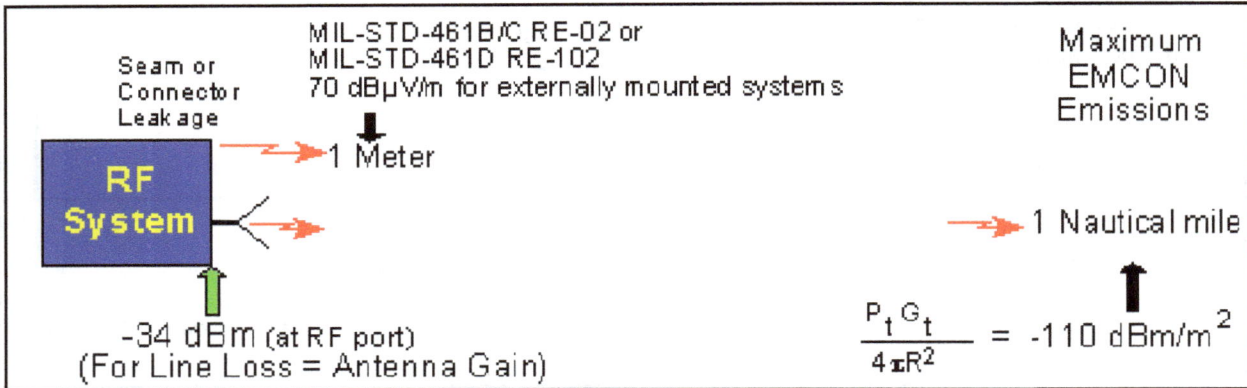

Figure 1. EMCON Field Intensity / Power Density Measurements

To compute the strength of emissions at the antenna port in Figure 1, we use the power density equation (see Section 4-2)

$$P_D = \frac{P_t G_t}{4\pi R^2} \qquad [1] \qquad\qquad \text{or rearranging} \qquad\qquad P_t G_t = P_D (4\pi R^2) \qquad [2]$$

Given that P_D = -110 dBm/m^2 = $(10)^{-11}$ mW/m^2, and R = 1 NM = 1852 meters.

$$P_t G_t = P_D (4\pi R^2) = (10^{-11} \text{mW/m}^2)(4\pi)(1852\text{m})^2 = 4.31(10)^{-4} \text{ mW} = -33.65 \approx -34 \text{ dBm}$$
at the RF system antenna as given.

or, the equation can be rewritten in Log form and each term multiplied by 10:
$$10\log P_t + 10\log G_t = 10\log P_D + 10\log (4\pi R^2) \qquad [3]$$

Since the m^2 terms on the right side of equation [3] cancel, then:
$\quad$ 10log P_t + 10log G_t = -110 dBm + 76.35 dB = -33.65 dBm $\approx$ -34 dBm as given in Figure 1.

If MIL-STD-461B/C RE02 (or MIL-STD-461D RE-102) measurements (see Figure 2) are made on seam/connector leakage of a system, emissions below 70 dBμV/meter which are measured at <u>one meter</u> will meet the EMCON requirement. Note that the airframe provides attenuation so portions of systems mounted inside an aircraft that measure 90 dBμV/meter will still meet EMCON if the airframe provides 20 dB of shielding (note that the requirement at one nm is converted to what would be measured at one meter from a point source)

The narrowband emission limit shown in Figure 2 for RE02/RE102 primarily reflect special concern for local oscillator leakage during EMCON as opposed to switching transients which would apply more to the broadband limit.

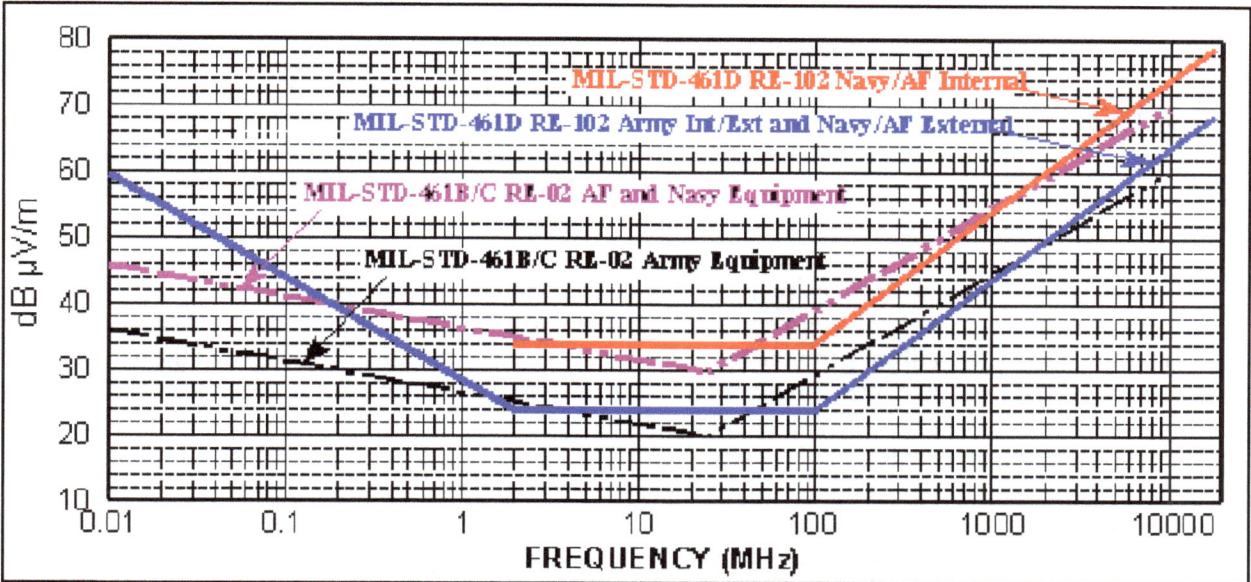

Figure 3. MIL-STD-461 Narrowband Radiated Emissions Limits

Note that in MIL-STD-461D, the narrowband radiated emissions limits were retitled RE-102 from the previous RE-02 and the upper frequency limit was raised from 10 GHz to 18 GHz. The majority of this section will continue to reference RE02 since most systems in use today were built to MIL-STD-461B/C.

For the other calculation involving leakage (to obtain 70 dBμV/m) we again start with: $P_D = \dfrac{P_t G_t}{4\pi R^2}$

and use the previous fact that: $10\log(P_t G_t) = -33.6$ dBm $= 4.37 \times 10^{-4}$ mW (see Section 2-4).

The measurement is at one meter so $R^2 = 1\ m^2$

we have: $\dfrac{4.37 \times 10^{-4}}{4\pi}\ mW/m^2 = .348 \times 10^{-4}\ mW/m^2 = -44.6\ dBm/m^2 = P_D$ @ 1 meter

Using the field intensity and power density relations (see Section 4-1)

$E = \sqrt{P_D Z} = \sqrt{3.48 \times 10^{-8} \bullet 377\Omega} = 36.2 \times 10^{-4}\ V/m$

Changing to microvolts ($1V = 10^6\ \mu V$) and converting to logs we have:

$20\log(E) = 20\log(10^6 \times 36.2 \times 10^{-4}) = 20\log(.362 \times 10^4) = 71.18$ dBμV/m $\approx \underline{70\ \text{dBμV/m}}$ as given in Figure 1.

Some words of Caution

A common error is to <u>only</u> use the one-way free space loss coefficient α_1 directly from Figure 6, Section 4-3 to calculate what the output power would be to achieve the EMCON limits at 1 NM. This is incorrect since the last term on the right of equation [3] ($10 \, \text{Log}(4\pi R^2)$) is simply the Log of the surface area of a sphere - **it is NOT** the one-way free space loss factor α_1. You cannot interchange power (watts or dBW) with power density (watts/m^2 or dBW/m^2).

The equation uses power density (P_D), **NOT** received power (P_r). It is independent of RF and therefore varies only with range. If the source is a transmitter and/or antenna, then the power-gain product (or EIRP) is easily measured and it's readily apparent if $10 \log (P_t \, G_t)$ is less than -34 dBm. If the output of the measurement system is connected to a power meter in place of the system transmission line and antenna, the -34 dBm value must be adjusted. The measurement on the power meter (dBm) minus line loss (dB) plus antenna gain (dB) must not be higher than -34 dBm.

However, many sources of radiation are through leakage, or are otherwise inaccessible to direct measurement and P_D must be measured with an antenna and a receiver. The measurements must be made at some RF(s), and received signal strength is a function of the antenna used therefore measurements must be scaled with an appropriate correction factor to obtain correct power density.

RE-02 Measurements

When RE-02 measurements are made, several different antennas are chosen dependent upon the frequency range under consideration. The voltage measured at the output terminals of an antenna is not the actual field intensity due to actual antenna gain, aperture characteristics, and loading effects. To account for this difference, the antenna factor is defined as:

$$AF = E/V \qquad\qquad\qquad \textbf{[4]}$$

where   E = Unknown electric field to be determined in V/m (or μV/m)
   V = Voltage measured at the output terminals of the measuring antenna

For an antenna loaded by a 50 Ω line (receiver), the theoretical antenna factor is developed as follows:

$$P_D \, A_e = P_r = V^2/R = V_r^2/50 \ \text{ or } \ V_r = \sqrt{50 \, P_D A_e}$$

From Section 4-3 we see that $A_e = G_r \lambda^2/4\pi$, and from Section 4-1, $E^2 = 377 \, P_D$ therefore we have:

$$AF = \frac{E}{V} = \frac{\sqrt{377 \, P_D}}{\sqrt{50 \, P_D (\lambda^2 \, G_r / 4\pi)}} = \frac{9.73}{\lambda \sqrt{G_r}} \qquad\qquad \textbf{[5]}$$

Reducing this to decibel form we have:

$$20 \log AF = 20 \log E - 20 \log V = 20 \log \left[\frac{9.73}{\lambda \sqrt{G_{SUBr}}} \right] \quad \textit{with } \lambda \textit{ in meters and Gain numeric ratio (not dB)} \qquad \textbf{[6]}$$

This equation is plotted in Figure 3.

Since all of the equations in this section were developed using far field antenna theory, use only the indicated region.

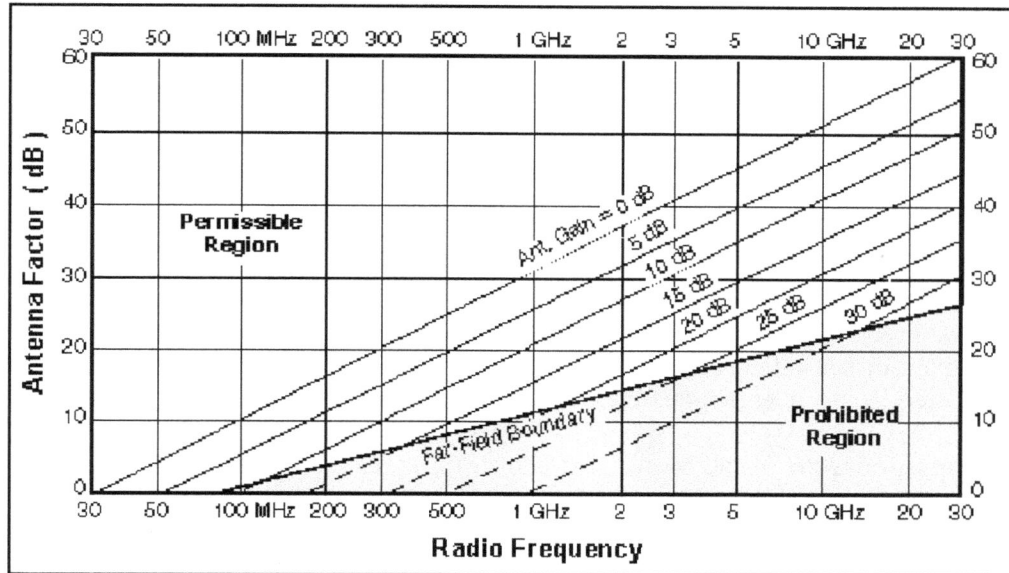

Figure 3. Antenna Factor vs Frequency for Indicated Antenna Gain

In practice the electric field is measured by attaching a field intensity meter or spectrum analyzer with a narrow bandpass preselector filter to the measuring antenna, recording the actual reading in volts and applying the antenna factor.

$$20\log E = 20\log V + 20\log AF \qquad\qquad [7]$$

Each of the antennas used for EMI measurements normally has a calibration sheet for both gain and antenna factor over the frequency range that the antenna is expected to be used. Typical values are presented in Table 1.

Table 1. Typical Antenna Factor Values

Frequency Range	Antenna(s) used	Antenna Factor	Gain(dB)
14 kHz - 30 MHz	41" rod	22-58 dB	0 - 2
20 MHz - 200 MHz	Dipole or Biconical	0-18 dB	0 - 11
200 MHz - 1 GHz	Conical Log Spiral	17-26 dB	0 - 15
1 GHz - 10 GHz	Conical Log Spiral or Ridged Horn	21-48 dB	0 - 28
1 GHz - 18 GHz	Double Ridged Horn	21-47 dB	0 - 32
18 GHz - 40 GHz	Parabolic Dish	20-25 dB	27 - 35

The antenna factor can also be developed in terms of the receiving antenna's effective area. This can be shown as follows:

$$AF = \frac{E}{V} = \frac{\sqrt{377\,P_D}}{\sqrt{50\,P_D\,A_e}} = \frac{2.75}{\sqrt{A_e}} \qquad \text{[8]}$$

Or in log form:

$$20\log AF = 20\log E - 20\log V = 20\log\left[\frac{2.75}{\sqrt{A_e}}\right] \qquad \text{[9]}$$

While this relation holds for any antenna, many antennas (spiral, dipole, conical etc.) which do not have a true "frontal capture area" do not have a linear or logarithmic relation between area and gain and in that respect the parabolic dish is unique in that the antenna factor does not vary with frequency, only with effective capture area. Consequently a larger effective area results in a smaller antenna factor.

A calibrated antenna would be the first choice for making measurements, followed by use of a parabolic dish or "standard gain" horn. A standard gain horn is one which was designed such that it closely follows the rules of thumb regarding area/gain and has a constant antenna factor. If a calibrated antenna, parabolic dish, or "standard horn" is not available, a good procedure is to utilize a flat spiral antenna (such as the AN/ALR-67 high band antennas). These antennas typically have an average gain of 0 dB (typically -4 to +4 dB), consequently the antenna factor would not vary a lot and any error would be small.

EXAMPLE:

Suppose that we want to make a very general estimation regarding the ability of a system to meet EMCON requirements. We choose to use a spiral antenna for measurements and take one of our samples at 4 GHz. Since we know the gain of the spiral is relatively flat at 4 GHz and has a gain value of approximately one (0 dB) in that frequency range. The antenna is connected to a spectrum analyzer by 25 feet of RG9 cable. We want to take our measurements at 2 meters from the system so our setup is shown below:

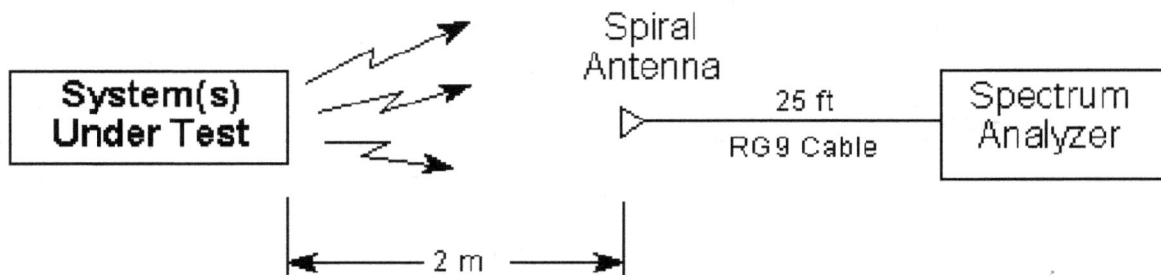

Our RG9 cable has an input impedance of 50Ω, and a loss of 5 dB (from Figure 5, Section 6-1).

First, let's assume that we measure -85 dBm at the spectrum analyzer and we want to translate this into the equivalent strength at 1 NM. Our power received by the antenna is: P_r = -85 dBm + 5 dB line loss = -80 dBm

also $P_D = P_r/A_e$ and $A_e = G\lambda^2/4\pi = (G/4\pi)\bullet(c/f)^2 = (1/4\pi)\bullet(3x10^8/4x10^9)^2 = 4.47x10^{-4}$ m^2

in log form: 10 Log P_D = 10 Log P_r - 10 Log A_e = -80 dBm + 33.5 = -46.5 dBm/m^2
                at our 2 meter measuring point

To convert this to a value at 1 NM, we use
        $P_t G_t = P_{D@1\ nm} 4\pi R_1^2 = P_{D@2\ m} 4\pi R_2^2$ and we solve for $P_{D@1\ nm}$

in log form after cancelling the 4π terms:

        10 Log $P_{D@1\ nm}$ = 10 Log $P_{D@2\ m}$ + 10 Log $(R_{2m}/R_{1nm})^2$ = -46.5 dBm/m^2 - 59.3 dB = -105.8 dBm/m^2
which is more power than the maximum value of -110 dBm/m^2 specified.

        If we are making repetitive measurement as we might do when screening an aircraft on the flight line with numerous systems installed, or when we want to improve (reduce) the leakage on a single system by changing antennas, lines, connectors, or EMI gaskets or shielding, this mathematical approach would be unnecessarily time consuming since it would have to be repeated after each measurement. A better approach would be to convert the -110 dBm/m^2 value at 1 NM to the maximum you can have at the measuring instrument (in this case a spectrum analyzer), then you could make multiple measurements and know immediately how your system(s) are doing. It should be noted that -90 to -100 dBm is about the minimum signal level that can be detected by a spectrum analyzer, so you couldn't take measurements much further away unless you used an antenna with a much higher gain.

In order not to exceed EMCON, the power density must not exceed -110 dBm/m^2 at 1 NM, which is 10^{-11} mW/m^2.

        $P_t G_t = P_{D@1\ nm} 4\pi R_1^2 = P_{D@2\ m} 4\pi R_2^2$

we solve for $P_{D@2\ m} = 10^{-11}(1852m)^2/(2m)^2 = 8.57$ x 10^{-6} mW/m^2 = -50.7 dBm/m^2

        We'll be using a spectrum analyzer, so we want to compute what the maximum power or voltage may be.

<u>Method 1 - Using the Power Density Approach</u>
        Using logs/dB and the values of $P_{D@2\ m}$ and A_e determined previously:
        10 Log P_r = 10 Log P_D + 10 Log A_e = -50.7 - 33.5 = -84.2 dBm

taking line loss into account we have: -84.2 - 5 dB = <u>- 89.2 dBm</u> as the maximum measurement reading.

If we wanted to calculate it in volts, and take into account our line impedance we would have the following:

        $P_r = P_D A_e = V^2/R = V^2/50\Omega$ also $A_e = G\lambda^2/4\pi$ so solving for V we have:

$$V = \sqrt{P_D\left[\frac{G_r\lambda^2}{4\pi}\right]R} = \sqrt{P_D\left[\frac{G_r}{4\pi}\left(\frac{c}{f}\right)^2\right]R} = \sqrt{8.57x10^{-9}\left[\frac{1}{4\pi}\left(\frac{3x10^8}{4x10^9}\right)^2\right]50\Omega} = 1.38x10^{-5}\ volts\ (before\ line\ loss)$$

since our line loss is 5 dB, we have -5dB = 20 Log V_2/V_1 . Solving for V_2 we get <u>7.79x10^{-6} volts</u> or <u>-89 dBm</u> as a maximum at our measurement device input. We can see immediately that our value of -85 dBm that we measured on the previous page would not meet specifications, and neither would any signal with more power than -89 dBm.

Method 2 - Using the Antenna Factor Approach

Starting with the same value of power density that we obtained above (8.57×10^{-9} W/m^2), we find the field intensity from Table 1, Section 4-1 to be approximately 65 dBμv/m. Also from Figure 3 in this section, AF = 43 dB @ 4 GHz.
(by calculating with equation [6], the exact value is 42.3 dB)

From equation [6]:
20log V = 20log E - 20log AF
20log V = 65 - 43 = 22 dBμv/m.

Since dBμv/m = 20 log (V)(10^6) = 20 log V + 20 log 10^6 = 20 log V + 120 , we see that to get an answer in dBv we must subtract 120 from the dBμv/m value so: V_{dB} = 22 - 120 = -98dBv. We then subtract our line loss (-5dB) and we have:

V = -98 - 5 = -103 dBv = 17 dBμv = 7.1x10^{-6} volts

using the fact that P = V^2/R and for the input line R = 50Ω, P = 1×10^{-12} W = -120 dBW = -90 dBm

Although this method is just as accurate as that obtained using method 1, the values obtained in Table 1, Section 4-1, and Figure 3 must be interpolated, and may not result in values which are as precise as the appropriate formulas would produce.

Sample Problem: What is the approximate transmit power from a receiver?

A.	1 nanowatt (nW)	F.	100 μW	K.	10 W
B.	10 nW	G.	1 milliwatt (mW)	L.	100 W
C.	100 nW	H.	10 mW	M.	1 kilowatt (kW)
D.	1 microwatt (μW)	I.	100 mW	N.	10 kW
E.	10 μW	J.	1 watt (W)	O.	100 kW

The question may seem inappropriate since a receiver is supposedly a passive device which only receives a signal. If the receiver was a crystal video receiver as shown in Section 5-3, it wouldn't transmit power unless a built-in-test (BIT) signal was injected after the antenna to periodically check the integrity of the microwave path and components. The potential exists for the BIT signal to leak across switches and couple back through the input path and be transmitted by the receiver's antennas.

If the receiver uses a local oscillator (LO) and a mixer to translate the signal to an intermediate frequency (IF) for processing (such as a superhet shown in Section 5-3), there is the potential for the CW LO signal to couple back through the signal input path and be transmitted by the receiver's antenna. Normally a mixer has 20 dB of rejection for the reverse direction. In addition, the LO may be further attenuated by receiver front end filters.

In both cases, the use of isolators described in Section 6-7 could be used to further attenuate any signals going in the reverse direction, i.e. back to the antenna. A good receiver design should ensure that any RF leakage radiated by the receiver will not exceed the EMCON level.

In answer to the initial question, "transmit" leakage power should be less than -34 dBm (0.4 μW) to meet EMCON. Therefore, the real answer may be "A", "B", or "C" if EMCON is met and could be "D" through possibly "G" if EMCON is not met.

EW JAMMING TECHNIQUES

INTRODUCTION

Electronic jamming is a form of Electronic Attack where jammers radiate interfering signals toward an enemy's radar, blocking the receiver with highly concentrated energy signals. The two main technique styles are noise techniques and repeater techniques. The three types of noise jamming are spot, sweep, and barrage. Repeater techniques can be further subdivided into categories as shown in Figure 1.

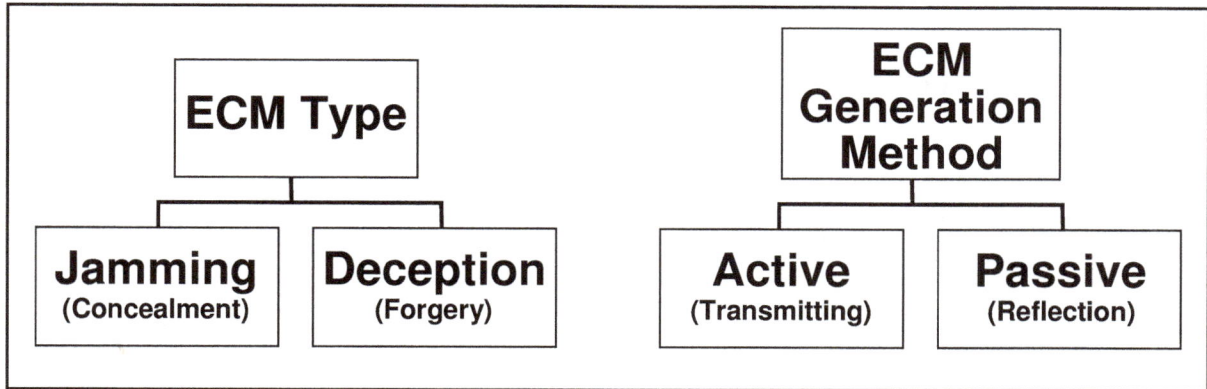

Figure 1. EA Repeater Technique Divisions

ASYNCHRONOUS SWEPT WAVE MODULATION (ASWM)

ASWM is synonymous with A-SWM. Asynchronous indicates that the waveform is free running - also see SSWM. A swept wave modulation is essentially a swept amplitude modulation (SAM). It is a waveform that is swept between two frequencies that are usually chosen to bracket a radar's passive angle scanning rate. The modulation amplitude can be either down modulated or On-Off Keyed (OOK). The down modulated shape can be square wave, rectangular wave, linear (e.g. a sine wave), or a combination. The OOK modulated shape can be square wave or rectangular wave.

BARRAGE JAMMING

The jamming of multiple frequencies at once by a single jammer. The advantage is that multiple frequencies can be jammed simultaneously; however, the jamming effect can be limited because this requires the jammer to spread its full power between these frequencies. So the more frequencies being jammed, the less effectively each is jammed.

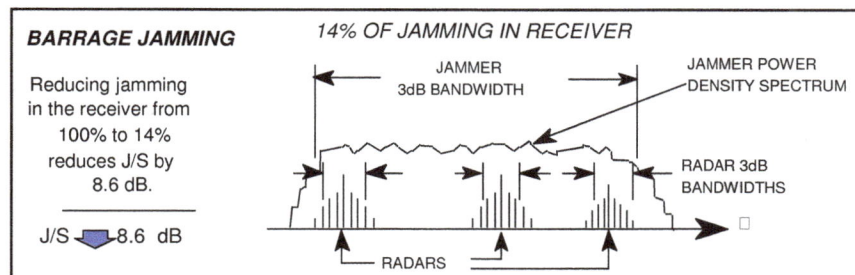

Figure 2. Barrage Jamming

CROSS POLARIZATION (X-POL)

(1) A self-screening or support EA technique that causes angle errors in tracking radars and sensing errors in jamming suppression EP systems of surveillance radars by radiating a signal that is orthogonally polarized to the principal polarization of the victim radar. (2) A technique used against monopulse and other passive lobe tracking radars. Requires a strong jam-to-signal ratio or the skin echo will show up in the pattern nulls.

HOME ON JAM (HOJ)

A means whereby a missile guidance receiver utilizes the self-screening target jamming signal to develop angular steering information so that the missile can home on that target.

IMAGE JAMMING

Jamming at the image frequency of the radar receiver. Barrage jamming is made most effective by generating energy at both the normal operating and image frequency of the radar. Image jamming inverts the phase of the response and is thereby useful as an angle deception technique. Not effective if the radar uses image rejection.

INVERSE CON SCAN (ICS)

One method of confusing a radar operator or fire control radar system is to provide erroneous target bearings. This is accomplished by first sensing the radar antenna or antenna dipole scan rate and then modulating repeater amplifier gain so that the weapons system will fire at some bearing other than the true target bearing.

J/S REQUIREMENT

The jamming to signal ratio for effective coverage of the true target. Usually on the order of zero dB (J/S=1).

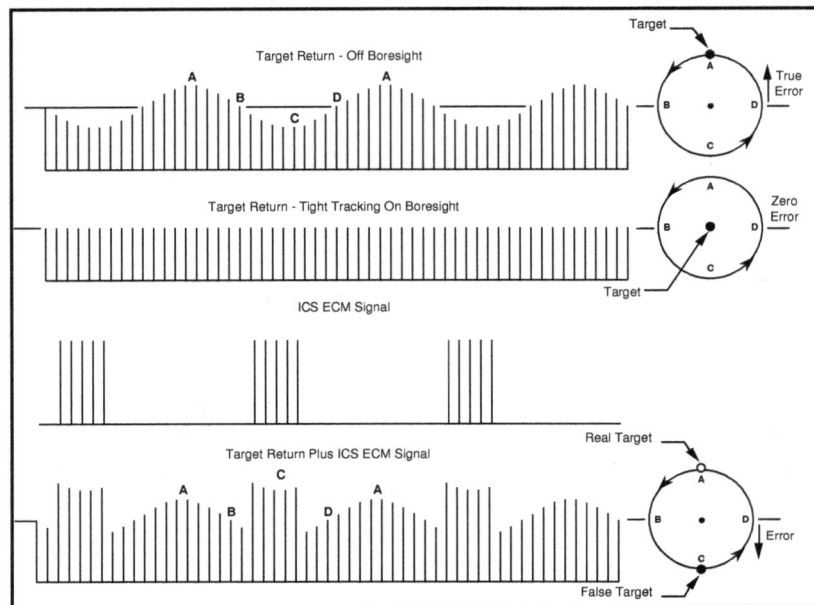

Figure 3. Inverse Con Scan

LINEAR MODULATION

A modulation technique in which the output varies in a straight line (linear) manner with the input (modulating) waveform. This is different from a discontinuous (On-Off) modulation. Typically the modulating waveform is a ramp or triangular wave but could also be a sine wave modulating input.

LOBE ON RECEIVE ONLY (LORO)

Predicting the future location of a target (to track) requires that the radar look at areas where the target is not located. When this scanning is accomplished with the radiated beam, large angle targets such as chaff clouds can create complete radar "white-out". RWR indications can be obtained before actual target lock-on. These problems can be overcome by scanning only the receiving antennas and using a separate transmitting antenna pointed only at the target.

In a LORO system, a transmitting antenna emits a few "exploratory" pulses along a direction obtained from an acquisition radar. These exploratory pulses are the acquisition mode of the TTR. That is, in its acquisition mode the small beamed TTR must scan the large location segment provided by the acquisition radar. In radars equipped with Fast Time Constant, the return pulse is applied to a differentiator of extremely short time constant. When the pulse is received, it is "cut-off" on the leading edge and only that portion is fed to the computer. This allows the radar to effectively track on the leading edge of the target. FTC does not improve the range resolution but it can prevent any countermeasures aft of the target which are in the same resolution cell as the target (such as chaff) from interfering with the radar receiver.

The receiving antennas scan their sector for the target return due to these exploratory pulses; as the power centroid is located, the center of the receiving pattern is brought onto the target. The transmitting antenna, which is slaved to the receiving antenna, is then pointing directly at the desired target and only that target is radiated during tracking. This approach allows a very small radiated beam, but the resolution cell of the system is still that of the receiving antenna.

NOISE JAMMING

The transmission of noise-like signals in the target system's radar receiver bandpass. At low power levels, noise jamming has the characteristics of receiver noise and can be mistaken by the radar operator as a problem with the radar. The object of noise jamming is to introduce a disturbing signal into the hostile electronic equipment so that the actual signal is obscured by the interference. The victim of this disturbance might be a radar receiver, a communications network or a data link.

See also Barrage Jamming and Spot Jamming

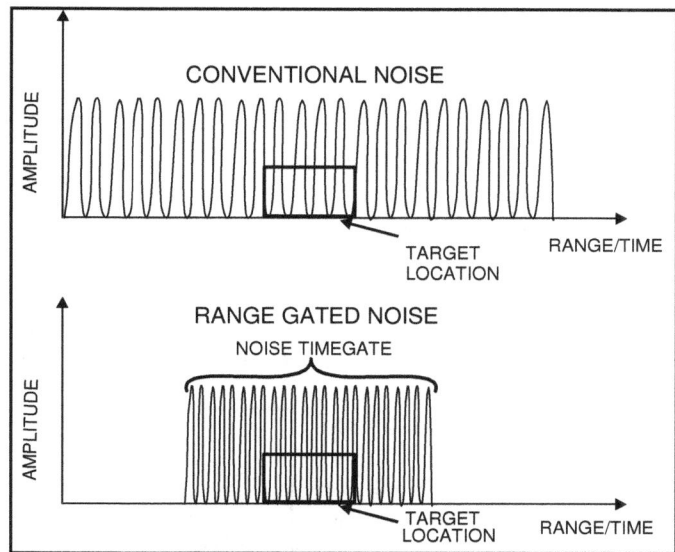

Figure 4. Noise Jamming

ON-OFF MODULATION

On-Off Modulation is any modulation which switches rapidly between two states. This definition includes pulse radar operation.

On-Off Keying (OOK) is the envelope modulation of a jamming signal with a rectangular wave. The modulation rate and duty cycle are adjusted commensurate with the victim radar's processing time constants. These can be related to AGC time constants, logic time-outs, data sampling cycles. or any other data processing response times.

An important distinction must be made between the terms On-Off Keying and Blinking. While both terms involve envelope modulations that turn a jamming signal on and off, the term OOK is used as the envelope modulation of a single jamming signal, and the term blinking is used as the tactical application of OOK involving two or more cooperative jamming platforms.

PSEUDO RANDOM NOISE (PRN)

A controlled, noise-like, pulse pattern repeated in synchronism with the victim radar pulse repetition frequency. Synonymous with quasi-noise jamming.

RANDOM DUAL LINE (RDL)

RDL is a coherent repeater technique that is essentially the same as velocity noise (VN), narrow band noise (NBN), and pseudo random noise (PRN) except that no false Doppler frequencies are stepped directly over the target return. The objective is to prevent the momentary additions of the EA signal and target signal that might highlight the presence of a coherent return.

RANDOM RANGE PROGRAM (RANRAP)

Description TBD

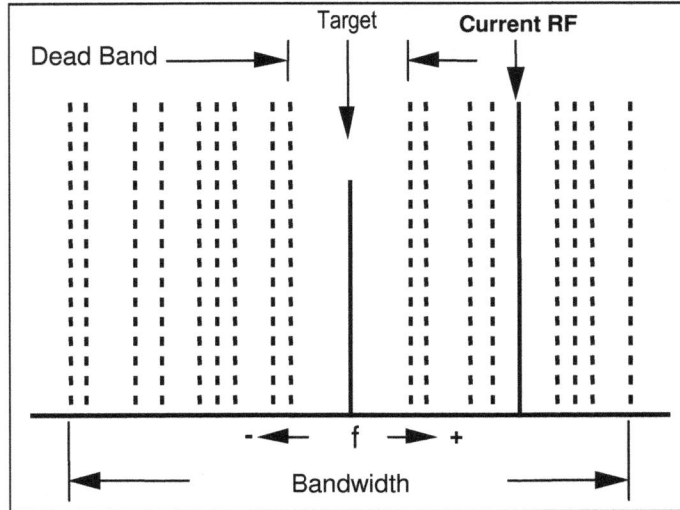

Figure 5. Random Dual Line

RANGE GATE PULL OFF (RGPO)

Once a tracking radar has detected a target, it will place *range gates* to either side of it. Range gates essentially blank out all signals which originate from ranges outside a narrow window, substantially increasing the signal-to-noise ratio and protecting the radar against unsynchronised jamming pulses. The radar 'concentrates' on a short range interval which encloses the target's location, and it no longer looks out for other targets. This state is known as 'lock-on'. But range gates can be 'stolen', and it is the objective of the Range Gate Pull-Off (RGPO) technique to *break lock* and escape from out of the window.

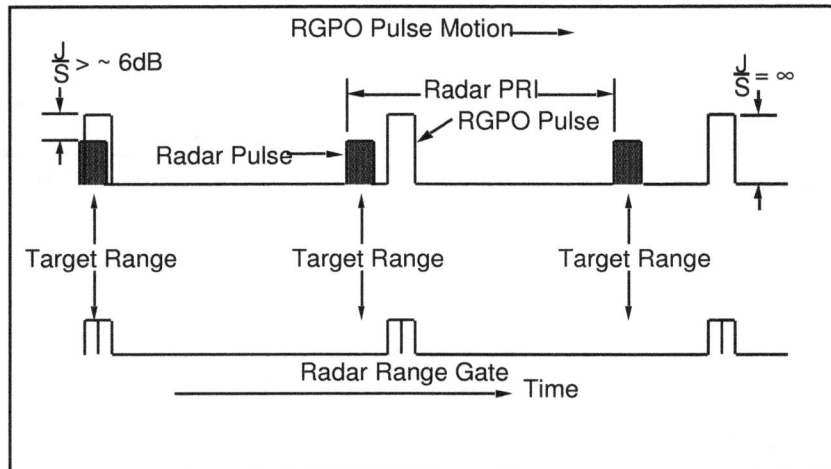

Figure 6. Range Gate Pull Off

RGPO works as follows:

Upon detection (or assumption) that a tracking radar has locked on, the on-board jammer is switched on and starts to work in a couple of phases:

1. First, a sample of the illuminating pulse signal is taken and the radar's <u>pulse repetition frequency</u> (PRF) is determined. This sample is amplified and retransmitted simultaneously when further pulses are received. The aircraft actually highlights itself on the radar screen. The jamming power is steadily increased, and this continues until the replica is much stronger than the echo from the aircraft skin return. At this time, the sensitivity of the tracking radar's receiver is usually reduced in order to avoid overload. This causes the skin echo vanishes below the noise floor.

2. *Another* replica is transmitted after each of the 'dummy' skin echoes. The power of the second replica is increased while the dummy is made weaker.

3. Next, the tracker has locked on to the *delayed replica*, whereas the skin return has decreased into the noise. With respect to each of the radar's pulses, the replica is now being delayed by small, but increasing amounts of time. The range gates, of course, follow the dummy target which appears to be receding. This continues until the range gates have been moved away from the target's real position. The result is that the radar is tracking a *phantom target* and the skin return is being blanked out by the range gates.

4. Finally, the jammer is switched off and leaves the radar with just *nothing* but noise inside the window between its range gates. Break-lock was successfully achieved and the "tracking" radar needs to switch back into a search or acquisition mode and loses time. The whole cycle will start again if the target is still within range and is reacquired.

As described above, RGPO creates only false targets which appear at greater ranges than the real target because the deceptive signal is transmitted *after* the skin echo. However, if the victim radar's PRF is constant then the time of incidence of the next radar pulse can be calculated and jamming pulses can be placed such that false targets at closer ranges are also produced.

SCINTILLATION (SCINT)

Scintillation is not an EA technique by itself, it is an implementation of an EA technique. Scintillation is simply superimposing a small, pseudo random amplitude modulation on the EA signal to make it appear more realistic to a manual operator.

SPOT JAMMING

Occurs when a jammer focuses all of its power on a single frequency. While this would severely degrade the ability to track on the jammed frequency, a frequency agile radar would hardly be affected because the jammer can only jam one frequency. While multiple jammers could

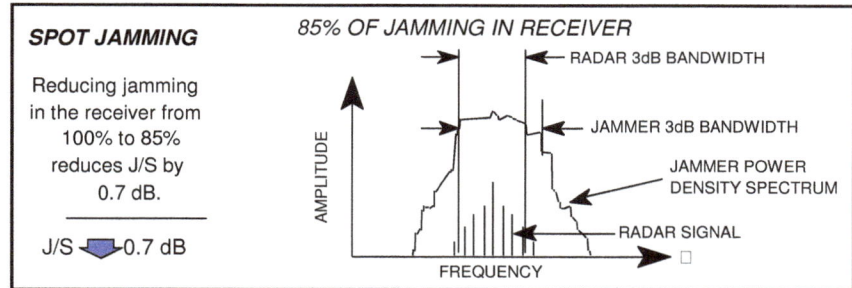

Figure 7. Spot Jamming

possibly jam a range of frequencies, this would consume a great deal of resources to have any effect on a frequency-agile radar, and would probably still be ineffective

SWEPT JAMMING

This happens when a jammer's full power is shifted from one frequency to another. While this has the advantage of being able to jam multiple frequencies in quick succession, it does not affect them all at the same time, and thus limits the effectiveness of this type of jamming. Although, depending on the error checking in the receiver(s) this can render a wide range of receivers effectively useless.

SWEPT AMPLITUDE MODULATION (SAM)

The OOK frequency is linearly varied in a sawtooth fashion between preset frequency limits while the duty factor is held constant.

SWEPT WAVE MODULATION (SWM)

A swept wave modulation (SWM) is essentially a swept amplitude modulation (SAM). It is a waveform that is swept between two frequencies that are usually chosen to bracket a radar's passive angle scanning rate. SWM can be either Synchronous Swept Wave Modulation (S-SWM) or Asynchronous Swept Wave Modulation (A-SWM). The modulation amplitude can be either down modulated or On-Off Keyed (OOK). The down modulating shape can be square wave, rectangular wave, linear (e.g. a sine wave), or a combination. The OOK modulating shape can be square wave or rectangular wave.

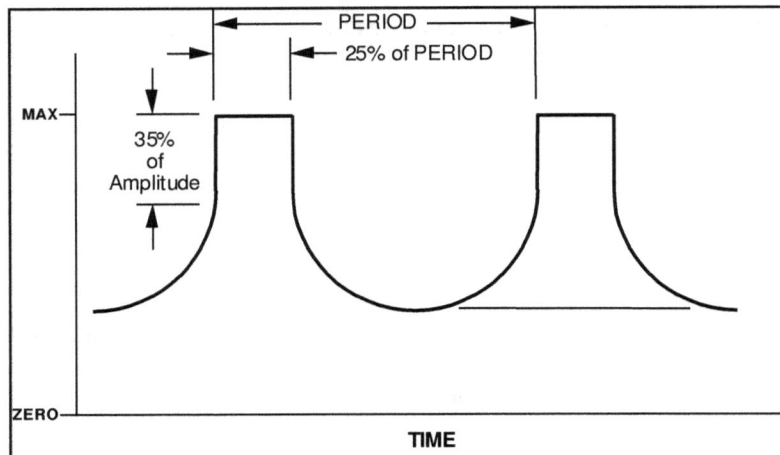

Figure 8. Typical Linear SWM Modulation

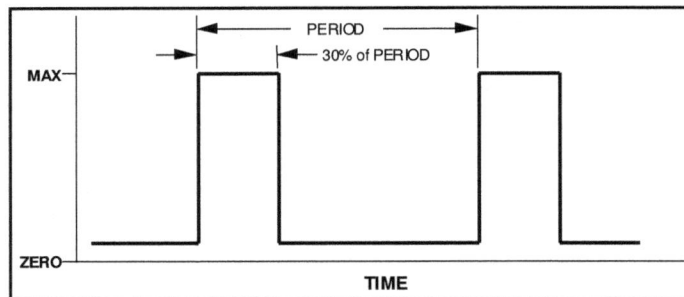

Figure 9. Rectangular SWM

SYNCHRONOUS SWEPT WAVE MODULATION (SSWM)

SSWM is synonymous with S-SWM. S-SWM and A-SWM are essentially the same except that asynchronous means that the waveform is free running and synchronous means that when a radar scan or TWS beam can be detected, the modulation waveform is synchronized to the detected beam. For programming purposes A-SWM sets sweep limits and rate by frequency (Hz), while S-SWM sets them by period (mSec).

Active Con-Scan radars will not have a detectable modulation if the target is being tightly tracked in the center of the beam. Therefore, a SWM can "jog" the tracking sufficiently to detect the modulation and allowing subsequent synchronization of the SWM waveform.

TRACK WHILE SCAN JAMMING

The technique of shifting or "walking" EA pulses off target. Many angle jamming techniques are effective.

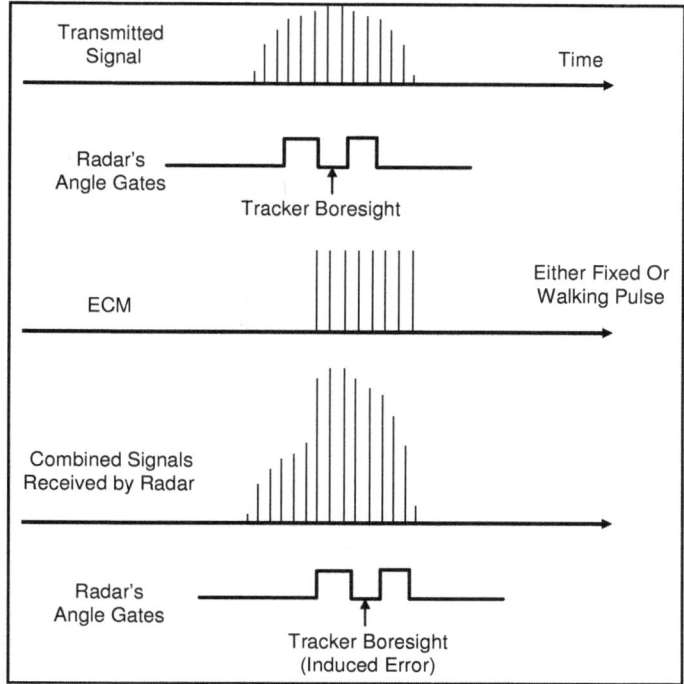

Figure 10. TWS Jamming

VELOCITY FALSE TARGETS (VFT)

VFT is a pseudo-random false Doppler target concealment technique. It is designed for use against radars that acquire Doppler targets with a bank of contiguous narrow band filters. A false Doppler target is programmed to remain in a Doppler filter long enough for the radar processing to declare it a valid target return, but not long enough for the radar processing to establish tracking.

The false Doppler target is then switched to the next pseudo-randomly selected frequency and repeated. It is intended to overload the radar processing and/or the operator's ability to identify an actual target. In the illustration, the VFTs jump around in the indicated numerical order.

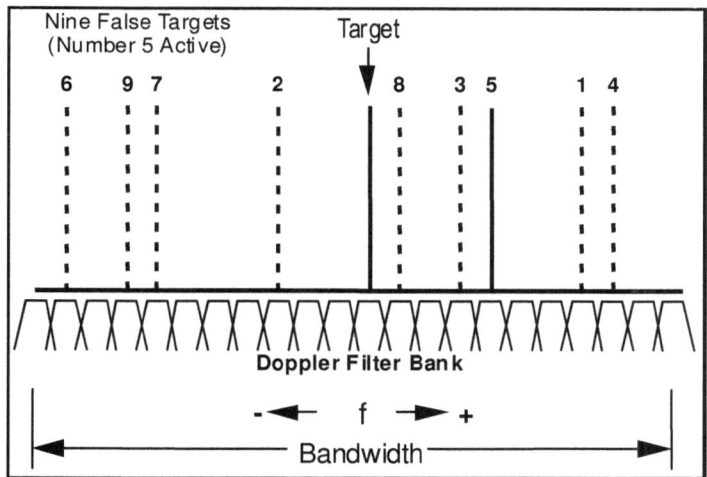

Figure 11. Velocity False Targets

VELOCITY GATE PULL OFF

This is a method of capturing the velocity gate of a Doppler radar and moving it away from the skin echo. Similar to the RGPO, but used against CW or Doppler velocity tracking radar systems. The CW or pulse doppler frequency, which is amplified and retransmitted, is shifted in frequency (velocity) to provide an apparent rate change or Doppler shift.

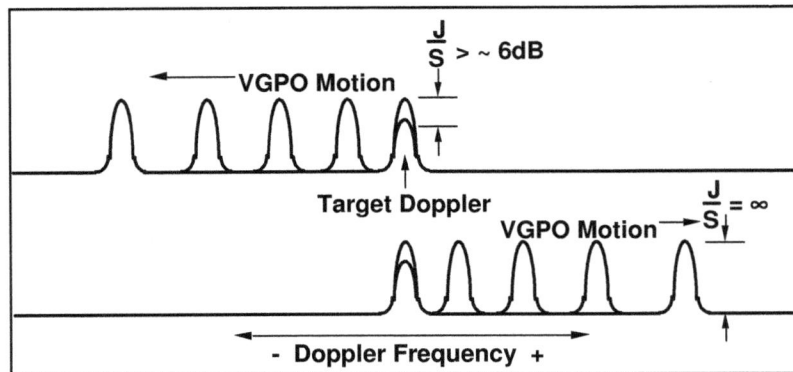

Figure 12. Velocity Gate Pull Off

VELOCITY NOISE (VN)

VN is a coherent repeater technique. The objective is to create noise centered on a coherent radar's RF, with a noise bandwidth that is close to, or less than, the radar's bandwidth and conceal the target, or destroy target signal coherency. VN is generated by pseudo randomly stepping a frequency over the victim radar's bandwidth. The dashed RF lines represent possible frequencies and the solid line represents the frequency currently active.

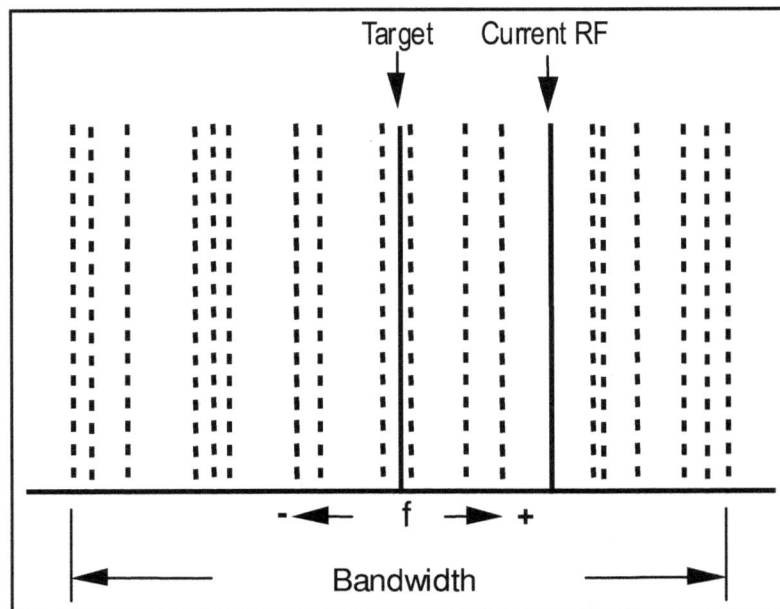

Figure 13. Velocity Noise

RADAR AND RECEIVER CHARACTERISTICS & TEST

RF ATMOSPHERIC ABSORPTION / DUCTING

Signal losses are associated with each stage of signal processing in both the transmitting and receiving portions of the system. The transmitting losses include power transmission efficiency, waveguide and antenna losses, and duplexer losses. In the receiver, losses include antenna, waveguide, RF amplifier, mixer, and IF amplifier.

In addition to these losses, energy traveling through the atmosphere suffers from atmospheric attenuation caused primarily by absorption by the gasses. For lower frequencies (below 10 GHz), the attenuation is reasonably predictable. For high frequencies in the millimeter wave range, the attenuation not only increases, but becomes more dependent upon peculiar absorbing characteristics of H_2O, O_2, and the like.

Figure 1 shows the areas of peak absorption in the millimeter wave spectrum. Figure 2 shows how the intensity of precipitation can affect atmospheric attenuation.

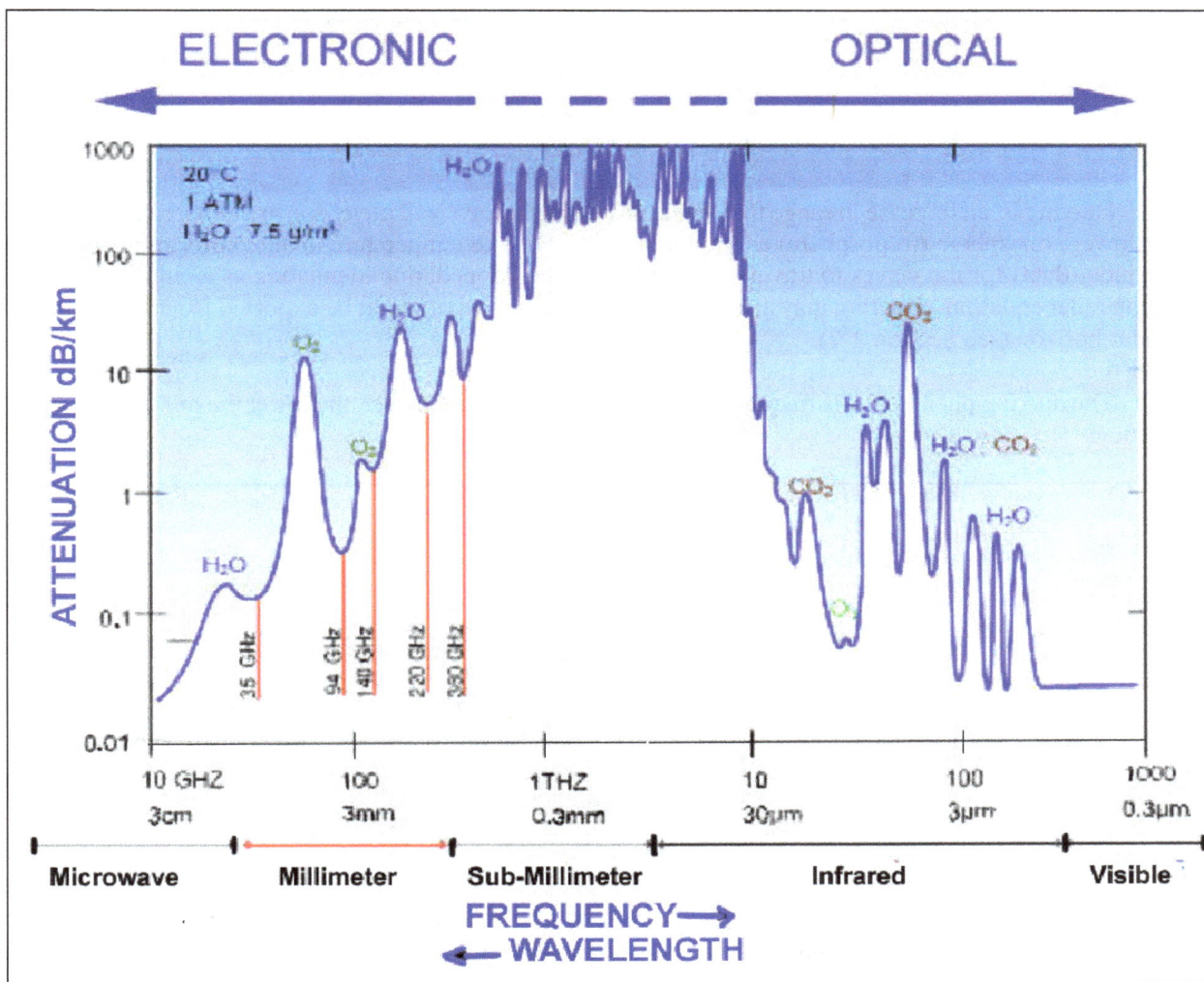

Figure 1. Atmospheric Absorption of Millimeter Waves

ATMOSPHERIC ATTENUATION

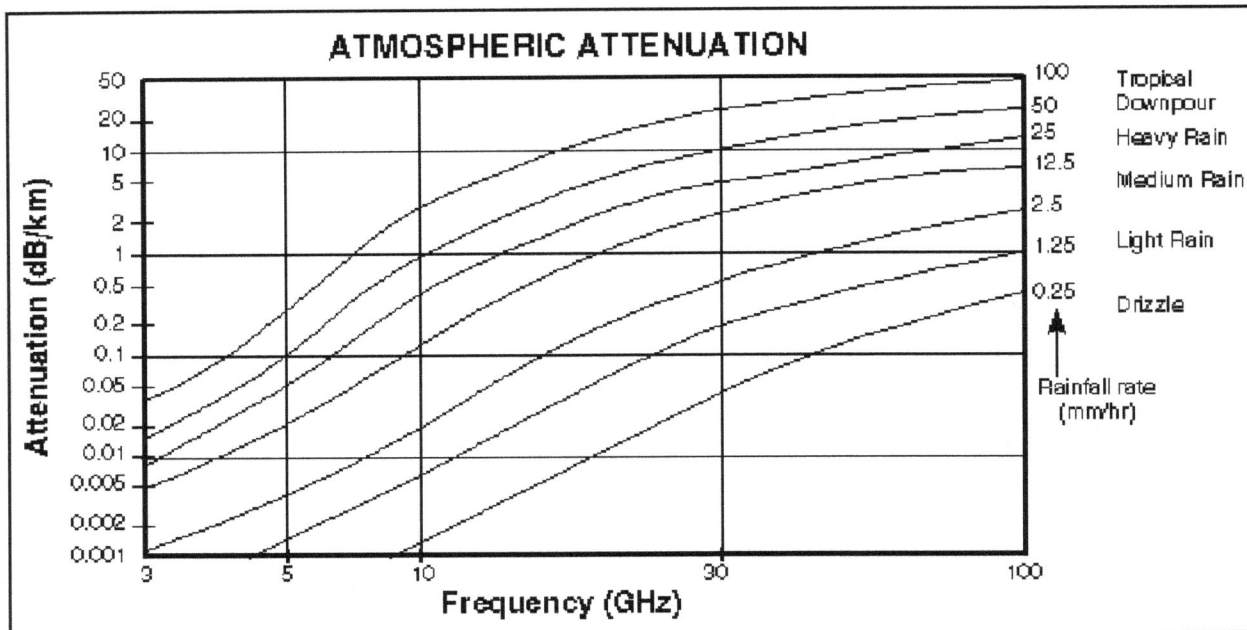

Figure 2. Atmospheric Attenuation

Ducting is an increase in range that an electromagnetic wave will travel due to a temperature inversion of the lower atmosphere (troposphere) as shown in Figure 3. The temperature inversion forms a channel or waveguide (duct) for the waves to travel in, and they can be trapped, not attenuating as would be expected from the radar equation. Ducting may also extend range beyond what might be expected from limitations of the radar horizon (see Section 2-9).

The ducting phenomena is frequency sensitive. The thicker the duct, the lower the minimum trapped frequency.

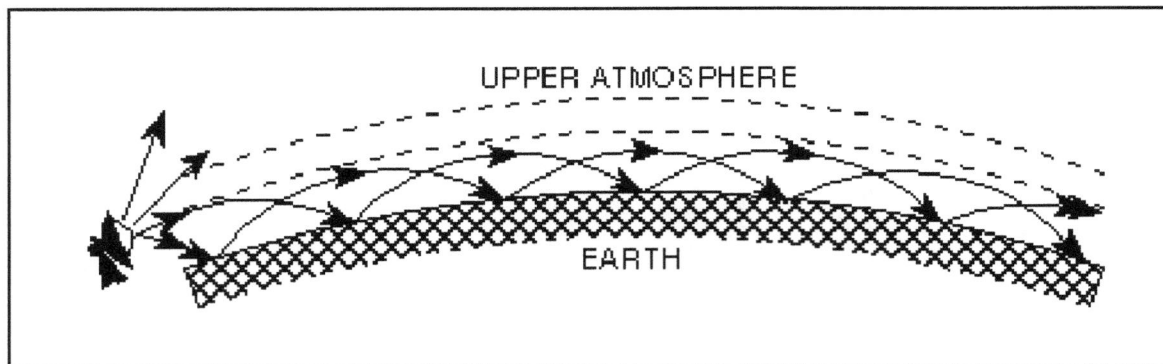

Figure 3. Ducting

A similar occurrence takes place with ionospheric refraction, however the greatest increase in range occurs in the lower frequencies. This is familiar to amateur radio operators who are able to contact counterparts "around the world".

RECEIVER SENSITIVITY / NOISE

RECEIVER SENSITIVITY

Sensitivity in a receiver is normally taken as the minimum input signal (S_{min}) required to produce a specified output signal having a specified signal-to-noise (S/N) ratio and is defined as the minimum signal-to-noise ratio times the mean noise power, see equation [1]. For a signal impinging on the antenna (system level) sensitivity is known as minimum operational sensitivity (MOS), see equation [2]. Since MOS includes antenna gain, it may be expressed in dBLi (dB referenced to a linear isotropic antenna). When specifying the sensitivity of receivers intended to intercept and process pulse signals, the minimum pulse width at which the specified sensitivity applies must also be stated. See the discussion of post-detection bandwidth (B_V) in Section 5-2 for significance of minimum pulsewidth in the receiver design.

$$\mathbf{S_{min} = (S/N)_{min}kT_oB(NF)} \quad \text{receiver sensitivity ("black box" performance parameter)} \quad [1]$$

or $\quad \mathbf{MOS = (S/N)_{min}kT_oB(NF)/G} \quad$ system sensitivity i.e. the receiver is connected to an antenna $\quad$ [2]
(transmission line loss included with antenna gain)

where:
S/N_{min}	=	Minimum signal-to-noise ratio needed to <u>process</u> (vice just detect) a signal
NF	=	Noise figure/factor
k	=	Boltzmann's Constant = 1.38×10^{-23} Joule/°K
T_o	=	Absolute temperature of the receiver input (°Kelvin) = 290°K
B	=	Receiver Bandwidth (Hz)
G	=	Antenna/system gain

We have a lower MOS if temperature, bandwidth, NF, or S/N_{min} decreases, or if antenna gain increases. For radar, missile, and EW receivers, sensitivity is usually stated in dBm. For communications and commercial broadcasting receivers, sensitivity is usually stated in micro-volts or dBμv. See Section 4-1.

There is no standard definition of sensitivity level. The term minimum operational sensitivity (MOS) can be used in place of S_{min} at the system level where aircraft installation characteristics are included. The "black box" term minimum detectable signal (MDS) is often used for S_{min} but can cause confusion because a receiver may be able to detect a signal, but not properly process it. MDS can also be confused with minimum discernable signal, which is frequently used when a human operator is used to interpret the reception results. A human interpretation is also required with minimum visible signal (MVS) and tangential sensitivity (discussed later). To avoid confusion, the terms S_{min} for "black box" minimum sensitivity and MOS for system minimum sensitivity are used in this section. All receivers are designed for a certain sensitivity level based on requirements. One would not design a receiver with more sensitivity than required because it limits the receiver bandwidth and will require the receiver to process signals it is not interested in. In general, while processing signals, the higher the power level at which the sensitivity is set, the fewer the number of false alarms which will be processed. Simultaneously, the probability of detection of a "good" (low-noise) signal will be decreased.

Sensitivity can be defined in two opposite ways, so discussions can frequently be confusing. It can be the ratio of response to input or input to response. In using the first method (most common in receiver discussions and used herein), it will be a negative number (in dBm), with the more negative being "better" sensitivity, e.g. -60 dBm is "better" than -50 dBm sensitivity. If the second method is used, the result will be a positive number, with higher being "better." Therefore the terms low sensitivity or high sensitivity can be very confusing. The terms S_{min} and MOS avoid confusion.

SIGNAL-TO-NOISE (S/N) RATIO

The Signal-to-Noise Ratio (S/N) (a.k.a. SNR) in a receiver is the signal power in the receiver divided by the mean noise power of the receiver. All receivers require the signal to exceed the noise by some amount. Usually if the signal power is less than or just equals the noise power it is not detectable. For a signal to be detected, the signal energy plus the noise energy must exceed some threshold value. Therefore, just because N is in the denominator doesn't mean it can be increased to lower the MOS. S/N is a <u>required</u> minimum ratio, if N is increased, then S must also be increased to maintain that threshold. The threshold value is chosen high enough above the mean noise level so that the probability of random noise peaks exceeding the threshold, and causing false alarms, is acceptably low.

Figure 1 depicts the concept of required S/N. It can be seen that the signal at time A exceeds the S/N ratio and indicates a false alarm or target. The signal at time B is just at the threshold, and the signal at time C is clearly below it. In the sample, if the temperature is taken as room temperature ($T_o = 290°K$), the noise power input is -114 dBm for a one MHz bandwidth. Normally S/N_{min} may be set higher than S/N shown in Figure 1 to meet false alarm specifications.

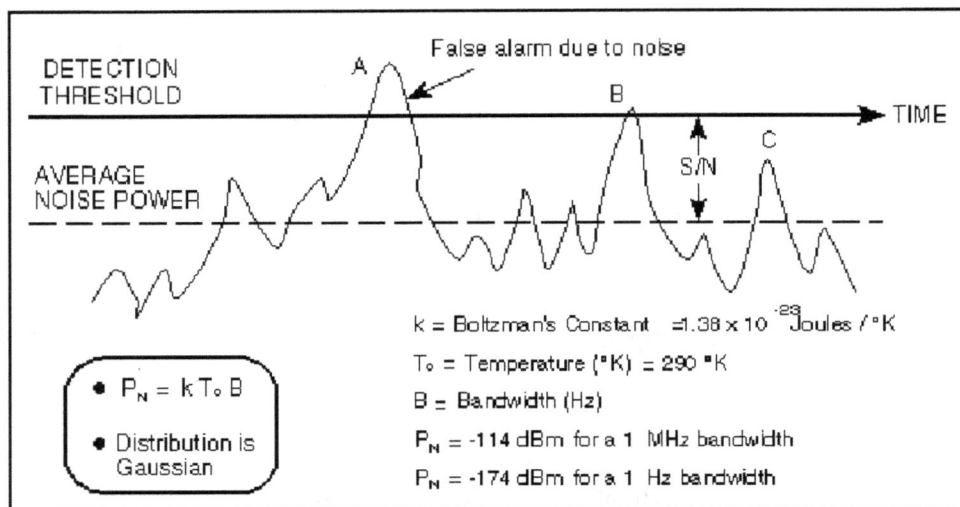

DETECTION THRESHOLD

AVERAGE NOISE POWER

False alarm due to noise

A

B

C

S/N

TIME

$P_N = kT_o B$

Distribution is Gaussian

k = Boltzman's Constant $= 1.38 \times 10^{-23}$ Joules / °K

T_o = Temperature (°K) = 290 °K

B = Bandwidth (Hz)

P_N = -114 dBm for a 1 MHz bandwidth

P_N = -174 dBm for a 1 Hz bandwidth

Figure 1. Receiver Noise Power at Room Tempeerature

The acceptable minimum Signal-to-Noise ratio (or think of it as Signal **above** Noise) for a receiver depends on the intended use of the receiver. For instance, a receiver that had to detect a single radar pulse would probably need a higher minimum S/N than a receiver that could integrate a large number of radar pulses (increasing the total signal energy) for detection with the same probability of false alarms. Receivers with human operators using a video display may function satisfactorily with low minimum S/N because a skilled operator can be very proficient at picking signals out of a noise background. As shown in Table 1, the setting of an acceptable minimum S/N is highly dependant on the required characteristics of the receiver and of the signal.

Table 1. Typical Minimum S/N Required				
Skilled Operator	Auto-Detection	Auto-detection with Amplitude, TOA, and Frequency Measurements	AOA Phase Interferometer	AOA Amplitude Comparison
3 to 8 dB	10 to 14 dB	14 to 18 dB	14 to 18 dB	16 to 24 dB

A complete discussion of the subject would require a lengthy dissertation of the probability and statistics of signal detection, which is beyond the scope of this handbook, however a simplified introduction follows. Let's assume that we have a receiver that we want a certain probability of detecting a single pulse with a specified false alarm probability. We can use Figure 2 to determine the required signal-to-noise ratio.

S/N EXAMPLE

If we are given that the desired probability of detecting a single pulse (P_d) is 98%, and we want the false alarm rate (P_n) to be no more than 10^{-3}, then we can see that S/N must be 12 dB (see Figure 2).

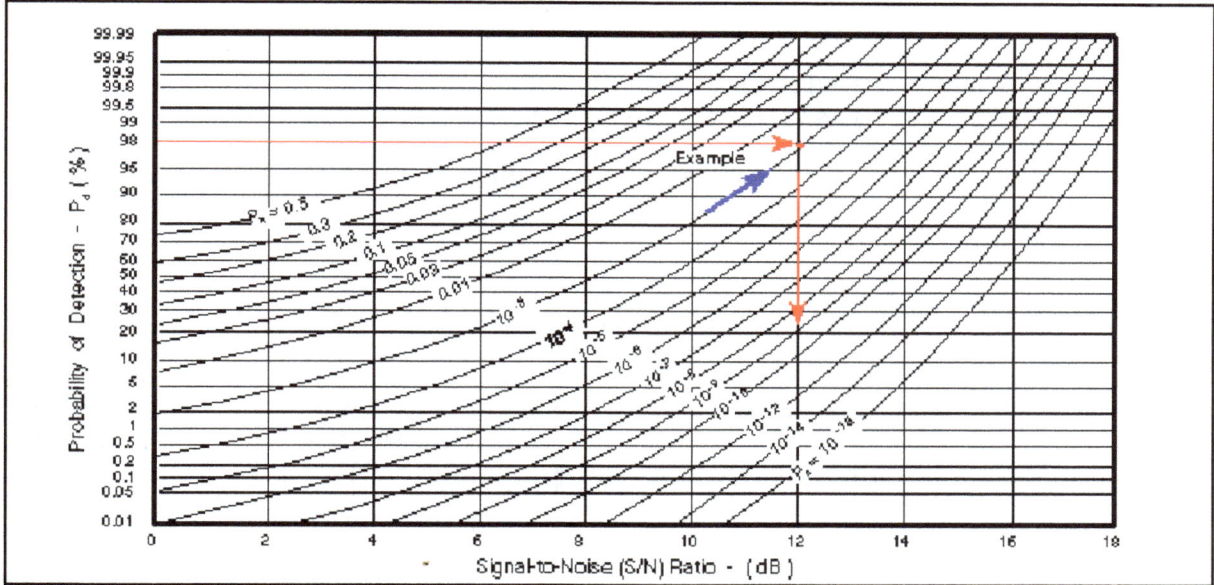

Figure 2. Nomograph of Signal-to-Noise (S/N) Ratio as a Function of Probability of Detection (P_d) and Probability of False Alarm Rate (P_n)

MAXIMUM DETECTION RANGE (ONE-WAY)

From Section 4-3, the one way signal strength from a transmitter to a receiver is: $S\ (or\ P_R) = \dfrac{P_t\, G_t\, G_r\, \lambda^2}{(4\pi\ )^2\, R^2}$

For calculations involving receiver sensitivity the "S" can be replaced by S_{min}. Since $S_{min} = (S/N)_{min}\, kT_oB(NF)$, given by equation [1], the one-way radar equation can be solved for any of the other variables in terms of receiver parameters. In communication, radar, and electronic warfare applications, you might need to solve for the maximum range (R_{max}) where a given radar warning receiver could detect a radiated signal with known parameters. We would then combine and rearrange the two equations mentioned to solve for the following one-way equation:

$$R_{max} \cong \sqrt{\frac{P_t\, G_t\, G_r\, \lambda^2}{(4\pi\ )^2 (S/N)_{min}\, k\, T_o\, B\,(NF)}} \quad or \quad \sqrt{\frac{P_t\, G_t\, G_r\, c^2}{(4\pi f\ )^2 (S/N)_{min}\, k\, T_o\, B\,(NF)}} \quad or \quad \sqrt{\frac{P_t\, G_t\, A_e}{4\pi (S/N)_{min}\, k\, T_o\, B\,(NF)}} \quad [3]$$

We could use standard room temperature of 290° K as T_o, but NF would have to be determined as shown later.

In this calculation for receiver R_{max} determination, P_t, G_t, and λ are radar dependent, while G_r, S/N_{min}, NF, and B are receiver dependent factors.

Equation [3] relates the maximum detection range to bandwidth (B). The effects of the measurement bandwidth can significantly reduce the energy that can be measured from the peak power applied to the receiver input. Additional bandwidth details are provided in Sections 4-4, 4-7, and in other parts of this section

NOISE POWER, kT$_o$B

Thermal noise is spread more or less uniformly over the entire frequency spectrum. Therefore the amount of noise appearing in the output of an ideal receiver is proportional to the absolute temperature of the receiver input system (antenna etc) times the bandwidth of the receiver. The factor of proportionality is Boltzmann's Constant.

Mean noise power of ideal receiver = kT$_o$B = P$_N$ (Watts)
Mean noise power of a real receiver = (NF)kT$_o$B (Watts)

The convention for the temperature of T$_o$ is set by IEEE standard to be 290°K, which is close to ordinary room temperature. So, assuming T$_o$ = 290°K, and for a bandwidth B = 1 Hz, kT$_o$B = 4×10^{-21} W = -204 dBW = -174 dBm.

For any receiver bandwidth, multiply 4×10^{-21} W by the bandwidth in Hz, or if using dB;
10 log kT$_o$B = -174 dBm + 10 Log (actual BW in Hz)   or   -114 dBm + 10 Log (actual BW in MHz)

and so on, as shown by the values in Table 2.

Typical values for maximum sensitivity of receivers would be:

RWR	-65 dBm
Pulse Radar	-94 dBm
CW Missile Seeker	-138 dBm

Table 2. Sample Noise Power Values (kT$_o$B)				
Bandwidth	Bandwidth Ratio (dB)	Watts	dBW	dBm
1 Hz	0	4×10^{-21}	-204	-174
1 kHz	30	4×10^{-18}	-174	-144
1 MHz	60	4×10^{-15}	-144	-114
1 GHz	90	4×10^{-12}	-114	-84

If antenna contributions are ignored (see note in Table 4) for a CW receiver with a 4 GHz bandwidth, the ideal mean noise power would be -174 dBm + 10 Log(4×10^{9}) = -174 dBm + 96 dB = -78 dBm. A skilled operator might only be able to distinguish a signal 3 dB above the noise floor (S/N=3 dB), or -75 dBm. A typical radar receiver would require a S/N of 3 to 10 dB to distinguish the signal from noise, and would require 10 to 20 dB to track. Auto tracking might require a S/N of approximately 25 dB, thus, a receiver may only have sufficient sensitivity to be able to identify targets down to -53 dBm. Actual pulse receiver detection will be further reduced due to sin x/x frequency distribution and the effect of the measurement bandwidth as discussed in Sections 4-4 and 4-7. Integration will increase the S/N since the signal is coherent and the noise is not.

Noise Bandwidth

Equivalent Noise Bandwidth (B$_N$) - Set by minimum pulse width or maximum modulation bandwidth needed for the system requirements. A choice which is available to the designer is the relationship of pre- and post-detection bandwidth. Pre-detection bandwidth is denoted by B$_{IF}$, while post-detection is denoted B$_V$, where V stands for video. The most affordable approach is to set the post-detection filter equal to the reciprocal of the minimum pulse width, then choose the pre-detection passband to be as wide as the background interference environment will allow. Recent studies suggest that pre-detection bandwidths in excess of 100 MHz will allow significant loss of signals due to "pulse-on-pulse" conditions. Equations [4] and [5] provide B$_N$ relationships that don't follow the Table 3 rules of thumb.

Table 3. Rules of Thumb for B$_N$ a.k.a. B (Doesn't apply for S/N between 0 and 10 to 30 dB)		
S/N out	Linear Detector	Square Law Detector
High S/N (>15 to 20 dB)	B$_N$ = B$_V$ (> 20 to 30 dB)	B$_N$ = 4 B$_V$ (> 10 to 15 dB)
Low S/N (< 0 dB)	$B_N = \sqrt{(2 B_{IF} B_V - B_V^2)/ 4 (S/N)_{out}}$	$B_N = \sqrt{(2 B_{IF} B_V - B_V^2)/ (S/N)_{out}}$

For a square law detector: [1]

$$B_N = B_V \left[2 + \sqrt{4 + \frac{(2 B_{IF} / B_V) - 1}{(S/N)_{out}}} \right] \qquad [4]$$

At high $(S/N)_{out}$, the $1/(S/N_{out})$ term goes to zero and we have: $\quad B_N = B_V [2 + \sqrt{4}] = 4 B_V$

At low $(S/N)_{out}$, the $1/(S/N_{out})$ term dominates, and we have:

$$B_N = B_V \sqrt{\frac{(2 B_{IF} / B_V) - 1}{(S/N)_{out}}} = \sqrt{\frac{2 B_{IF} B_V - B_V^2}{(S/N)_{out}}}$$

For a linear detector: [1]

$$B_N = \frac{B_V}{2} + \frac{1}{4} \bullet \sqrt{B_V \left(4 B_V + \frac{H^2 (2 B_{IF} - B_V)}{(S/N)_{out}} \right)} \qquad [5]$$

H is a hypergeometric (statistical) function of $(S/N)_{in}$

$\quad H = 2$ for $(S/N)_{in} \ll 1$

$\quad H = 1$ for $(S/N)_{in} \gg 1$

At high $(S/N)_{out}$, the $1/(S/N_{out})$ term goes to zero and we have: $\quad B_N = \frac{B_V}{2} + \frac{1}{4}\sqrt{B_V (4 B_V)} = B_V$

At low $(S/N)_{out}$, the $1/(S/N_{out})$ term dominates, and we have:

$$B_N = \frac{1}{4} \bullet \sqrt{\frac{B_V H^2 (2 B_{IF} - B_V)}{(S/N)_{out}}} = \sqrt{\frac{2 B_{IF} B_V - B_V^2}{4 (S/N)_{out}}}$$

Note (1): From Klipper, *Sensitivity of crystal Video Receivers with RF Pre-amplification*, The Microwave Journal, August 1965.

TRADITIONAL "RULE OF THUMB" FOR NARROW BANDWIDTHS (Radar Receiver Applications)
Required IF Bandwidth For Matched Filter Applications:

$$B_{IF} = \frac{1}{PW_{min}} \quad \text{Where:} \quad \begin{aligned} & B_{IF} = Pre\text{-}detection\ RF\ or\ IF\ bandwidth \\ & PW_{min} = Specified\ minimum\ pulse\ width = \tau \end{aligned}$$

Matched filter performance gives maximum probability of detection for a given signal level, but: (1) Requires perfect centering of signal spectrum with filter bandwidth, (2) Time response of matched pulse does not stabilize at a final value, and (3) Out-of-band splatter impulse duration equals minimum pulse width. As a result, EW performance with pulses of unknown frequency and pulse width is poor.

Required Video Bandwidth Post-Detection
Traditional "Rule of Thumb" $\quad B_V = \frac{0.35}{PW_{min}} \quad Where: B_V = Post\text{-}detection\ bandwidth$

Some authors define B_V in terms of the minimum rise time of the underlined detected pulse, i.e., $B_V = (0.35$ to $0.5)/t_r$ min, where t_r = rise time.

REVISED "RULE OF THUMB" FOR WIDE BANDWIDTHS (Wideband Portion of RWRs)

$$B_{IF} = \frac{2\ to\ 3}{PW_{min}} \quad and \quad B_V = \frac{1}{PW_{min}}$$

The pre-detection bandwidth is chosen based upon interference and spurious generation concerns. The post-detection bandwidth is chosen to "match" the minimum pulse width. This allows (1) Half bandwidth mistuning between signal and filter, (2) Half of the minimum pulse width for final value stabilization, and (3) The noise bandwidth to be "matched" to the minimum pulse width. As a result, there is (1) Improved EW performance with pulses of unknown frequency and pulse width, (2) Measurement of in-band, but mistuned pulses, and (3) Rejection of out-of-band pulse splatter.

NOISE FIGURE / FACTOR (NF)

Electrical noise is defined as electrical energy of random amplitude, phase, and frequency. It is present in the output of every radio receiver. At the frequencies used by most radars, the noise is generated primarily within the input stages of the receiver system itself (Johnson Noise). These stages are not inherently noisier than others, but noise generated at the input and amplified by the receiver's full gain greatly exceeds the noise generated further along the receiver chain. The noise performance of a receiver is described by a figure of merit called the noise figure (NF). The term noise factor is synonymous, with some authors using the term "factor" for numeric and "figure" when using dB notation. (The notation "F_n" is also sometimes used instead of "NF".) The noise figure is defined as:

$$NF = \frac{Noise\ output\ of\ actual\ receiver}{Noise\ output\ of\ ideal\ receiver} = \frac{N_{out}}{GN_{in}} \quad or\ in\ dB: \quad 10\ Log\left[\frac{Noise\ output\ of\ actual\ receiver}{Noise\ output\ of\ ideal\ receiver}\right] = 10\log\frac{N_{out}}{GN_{in}}$$

A range of NF values is shown in Table 4.

Table 4. Typical Noise Figure / Factor Value	Decimal	dB
Passive lossy network (RF transmission line, attenuator, etc.) Example: 20 dB attenuator (gain = 0.01)	Same as reciprocal of gain value ex: 100	Same as dB value ex: 20
Solid State Amplifier (see manufacturers specifications)	4	6
Traveling Wave Tube (see manufacturers specifications)	10 to 100	10 to 20
Antennas (Below $\approx$ 100 MHz, values to 12 dB higher if pointed at the sun) Note: Unless the antenna is pointed at the sun, its negligible NF can be ignored. Antenna gain is not valid for NF calculations because the noise is received in the near field.	1.012 to 1.4	0.05 to 1.5

An ideal receiver generates no noise internally. The only noise in its output is received from external sources. That noise has the same characteristics as the noise resulting from thermal agitation in a conductor. Thermal agitation noise is caused by the continuous random motion of free electrons which are present in every conductor. The amount of motion is proportional to the conductor's temperature above absolute zero. For passive lossy networks, the noise factor equals the loss value for the passive element:

$$NF = \frac{N_{out}}{G\,N_{in}} = \frac{kTB}{\frac{1}{L}kTB} = L \quad \begin{array}{l} Where\ L = Ratio\ Value\ of\ Attenuation \\ i.e.\ For\ a\ 3\ dB\ attenuator, G = 0.5\ and\ L = 2 \\ \therefore NF = 2\ and\ 10\log NF = 3\ dB \end{array}$$

A typical series of cascaded amplifiers is shown in Figure 3.

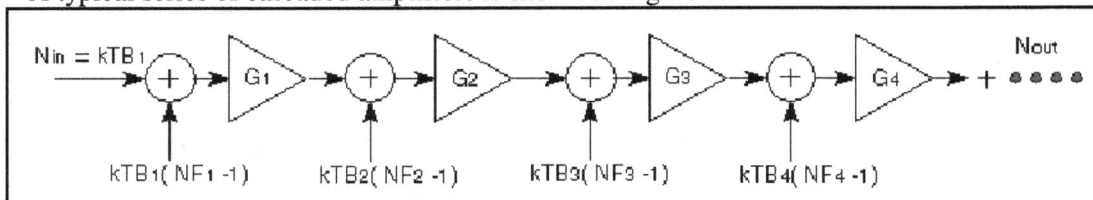

Figure 3. Noise Factors for Cascaded Amplifiers (NF$_{CA}$)

Loss (negative gain) can be used for the gain value of attenuators or transmission line loss, etc to calculate the noise out of the installation as shown in the following equation:

$$N_{out} = N_{in}\,G\,NF_{CA} = k\,T\,B_1\left(G_1G_2G_3\ldots\right)\left(NF_1 + \frac{B_2(NF_2-1)}{B_1G_1} + \frac{B_3(NF_3-1)}{B_1G_1G_2} + \frac{B_4(NF_4-1)}{B_1G_1G_2G_3} + \ldots\right) \quad (ratio\ form) \qquad [6]$$

If the bandwidths of the amplifiers are the same, equation [6] becomes:

$$N_{out} = N_{in}\,G\,NF_{CA} = k\,T\,B\left(G_1G_2G_3\ldots\right)\left(NF_1 + \frac{NF_2-1}{G_1} + \frac{NF_3-1}{G_1G_2} + \frac{NF_4-1}{G_1G_2G_3} + \ldots\right) \quad (ratio\ form) \qquad [7]$$

Pre-amplifier Location Affects Receiver Input Noise

As shown in Figure 4, if a 2 to 12 GHz receiver installation doesn't have enough sensitivity, it is best to install an additional amplifier closer to the antenna (case 1) instead of closer to the receiver (case 2). In both cases, the line loss (L) and the amplifier gain (G) are the same, so the signal level at the receiver is the same. For case 1, $S_1 = P_{in} + G - L$. In case 2, $S_2 = P_{in} - L + G$, so $S_1 = S_2$. The noise generated by the passive transmission line when measured at the receiver is the same in both cases. However, the noise generated inside the amplifier, when measured at the receiver input, is different.

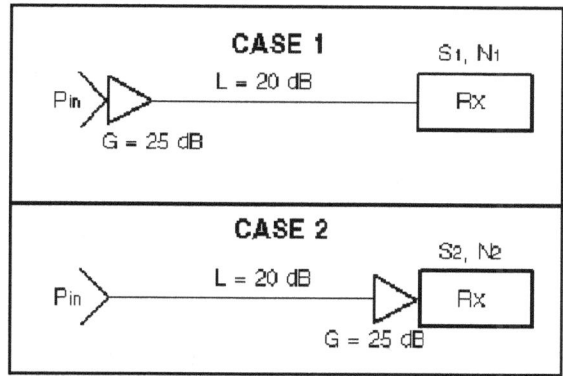

Figure 4. Pre-Amp S/N

For this example, case 2 has a noise level at the input to the receiver which is 19.7 dB higher than case 1 (calculations follow later).

Table	Case 1 Gain		Case 1 NF	
5a	Amp	L	Amp	L
dB	25	-20	6 *	20
ratio	316.2	0.01	4 *	100

Table	Case 2 Gain		Case 2 NF	
5b	L	Amp	L	Amp
dB	-20	25	20	6 *
ratio	0.01	316.2	100	4 *

* Amplifier NF value from Table 4.

Using equation [3] and the data in Tables 5a and 5b, the noise generated by the RF installation is shown in Tables 6a and 6b (the negligible noise contribution from the antenna is the same in both cases and is not included) (also see notes contained in Table 4):

Table 6a. Case 1	**Table 6b.** Case 2
$G(NF) = 316.2\,(0.01)\left(4 + \dfrac{100 - 1}{316.7}\right) = 13.64$	$G(NF) = 0.01\,(316.2)\left(100 + \dfrac{4 - 1}{0.01}\right) = 1264.8$
10 log G(NF) = 11.34 dB	10 log G(NF) = 31 dB
Noise at receiver:	
$N_{out\,1} = -74\ dBm + 11.34\ dB = -62.7\ dBm$	$N_{out\,2} = -74\ dBm + 31\ dB = -43\ dBm$

$N_{out\,2} - N_{out\,1} = 19.7$ dB. The input noise of -74 dBm was calculated using 10 log (kTB), where B = 10 GHz.

Note that other tradeoffs must be considered: (1) greater line loss between the antenna and amplifier improves (decreases) VSWR as shown in Section 6-2, and (2) the more input line loss, the higher the input signal can be before causing the pre-amplifier to become saturated (mixing of signals due to a saturated amplifier is addressed in Section 5-7).

Combining Receive Paths Can Reduce Sensitivity

If a single aircraft receiver processes both forward and aft signals as shown in Figure 5, it is desirable to be able to use the receiver's full dynamic range for both directions. Therefore, one needs to balance the gain, so that a signal applied to the aft antenna will reach the receiver at the same level as if it was applied to the forward antenna.

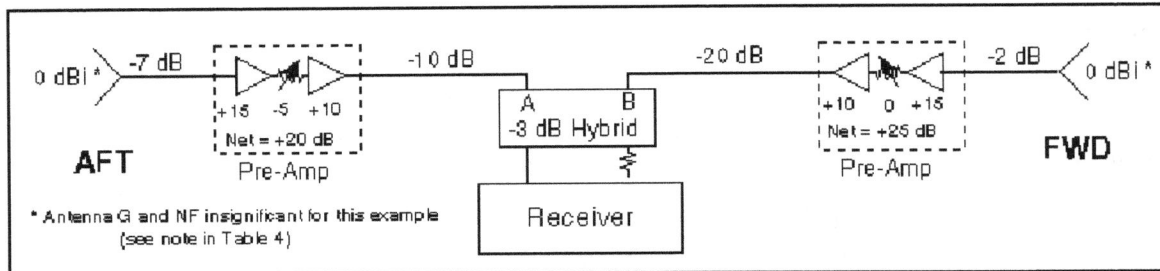

Figure 5. Example of Pre-Amplifier Affecting Overall Gain / Sensitivity

Common adjustable preamplifiers can be installed to account for the excessive transmission line loss. In this example, in the forward installation, the level of the signal at the receiver is the same as the level applied to the antenna. Since the aft transmission line has 5 dB less attenuation, that amount is added to the preamplifier attenuator to balance the gain. This works fine for strong signals, but not for weaker signals. Because there is less loss between the aft preamplifier and the receiver, the aft noise dominates and will limit forward sensitivity. If the bandwidth is 2-12 GHz, and if port A of the hybrid is terminated by a perfect 50Ω load, the forward noise level would be -65.3 dBm. If port B is terminated, the aft noise level would be -60.4 dBm. With both ports connected, the composite noise level would be -59.2 dBm (convert to mw, add, then convert back to dBm). For this example, if the aft preamplifier attenuation value is changed to 12 dB, the gain is no longer balanced (7 dB extra loss aft), but the noise is balanced, i.e. forward = -65.6 dBm, aft = -65.3 dBm, and composite -62.4 dBm. If there were a requirement to see the forward signals at the most sensitive level, extra attenuation could be inserted in the aft preamplifier. This would allow the forward noise level to predominate and result in greater forward sensitivity where it is needed. Calculations are provided in Tables 7 and 8.

		Aft					Fwd				
		RF Line	Amp	Attn	Amp	RF Line & hybrid	RF Line	Amp	Attn	Amp	RF Line & hybrid
Gain	dB	-7	15	-5	10	-13	-2	15	0	10	-23
	ratio	0.2	31.6	0.32	10	0.05	0.63	31.6	0	10	0.005
NF	dB	7	6	5	6	13	2	6	0	6	23
	ratio	5	4	3.16	4	20	1.585	4	0	4	200

Table 7. Summary of Gain and NF Values for Figure 5 Components

Aft NF = 22.79 therefore 10 log NF = 13.58 dB. Input noise level = -74 dBm + 13.58 dB = -60.42 dBm ≅ -60.4 dBm
Fwd NF = 7.495 therefore 10 log NF = 8.75 dB. Input noise level = -74 dBm + 8.75 dB = -65.25 dBm ≅ -65.3 dBm
The composite noise level at the receiver = -59.187 dBm ≅ -59.2 dBm

Table 8. Effect of Varying the Attenuation (shaded area) in the Aft Preamplifier Listed in Table 7.

Aft Attn NF	Aft Attn Gain	Aft Noise	Fwd Noise	Composite Noise	Min Signal Received ***	Aft Input	Fwd Input
0 dB	0 dB	-55.8 dBm	-65.3 dBm	-55.4 dBm	-43.4 dBm	-48.4 dBm	-43.4 dBm
5	-5	-60.4	-65.3	-59.2	-47.2 *	-47.2 *	-47.2 *
10	-10	-64.4	-65.3	-61.8	-49.8	-44.8	-49.8
12	-12	-65.6 **	-65.3 **	-62.4	-50.4	-43.4	-50.4
15	-15	-67.1	-65.3	-63.1	-51.1	-41.1	-51.1

* Gain Balanced	** Noise Balanced	*** S/N was set at 12 dB

TANGENTIAL SENSITIVITY

Tangential sensitivity (TSS) is the point where the top of the noise level with no signal applied is level with the bottom of the noise level on a pulse as shown in Figure 6. It can be determined in the laboratory by varying the amplitude of the input pulse until the stated criterion is reached, or by various approximation formulas.

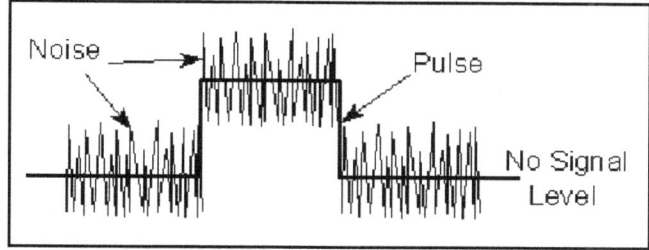

Figure 6. Tangential Sensitivity

The signal power is nominally 8±1 dB above the noise level at the TSS point. TSS depends on the RF bandwidth, the video bandwidth, the noise figure, and the detector characteristic.

TSS is generally a characteristic associated with receivers (or RWRs), however the TSS does not necessarily provide a criterion for properly setting the detection threshold. If the threshold is set to TSS, then the false alarm rate is rather high. Radars do not operate at TSS. Most require a more positive S/N for track (> 10 dB) to reduce false detection on noise spikes.

SENSITIVITY CONCLUSION

When all factors effecting system sensitivity are considered, the designer has little flexibility in the choice of receiver parameters. Rather, the performance requirements dictate the limit of sensitivity which can be implemented by the EW receiver.

1. Minimum Signal-to-Noise Ratio (S/N) - Set by the accuracy which you want to measure signal parameters and by the false alarm requirements.

2. Total Receiver Noise Figure (NF) - Set by available technology and system constraints for RF front end performance.

3. Equivalent Noise Bandwidth (B_N) - Set by minimum pulse width or maximum modulation bandwidth needed to accomplish the system requirements. A choice which is available to the designer is the relationship of pre- (B_{IF}) and post-detection (B_V) bandwidth. The most affordable approach is to set the post-detection filter equal to the reciprocal of the minimum pulse width, then choose the pre-detection passband to be as wide as the background interference environment will allow. Recent studies suggest that pre-detection bandwidths in excess of 100 MHz will allow significant loss of signals due to "pulse-on-pulse" conditions.

4. Antenna Gain (G) - Set by the needed instantaneous FOV needed to support the system time to intercept requirements.

This Page Blank

RECEIVER TYPES AND CHARACTERISTICS

Besides the considerations of noise and noise figure, the capabilities of receivers are highly dependant on the type of receiver design. Most receiver designs are trade-offs of several conflicting requirements. This is especially true of the Electronic Support Measures (ESM) receivers used in Electronic Warfare.

This section consists of a figure and tables that provide a brief comparison of various common ESM receiver types. Figures 1 and 2 show block diagrams of common ESM receivers. Table 1 is a comparison of major features of receivers. Table 2 shows the receiver types best suited for various types of signals and Tables 3 and 4 compare several direction of arrival (DOA) and emitter location techniques. Table 5 shows qualitative and quantitative comparisons of receiver characteristics.

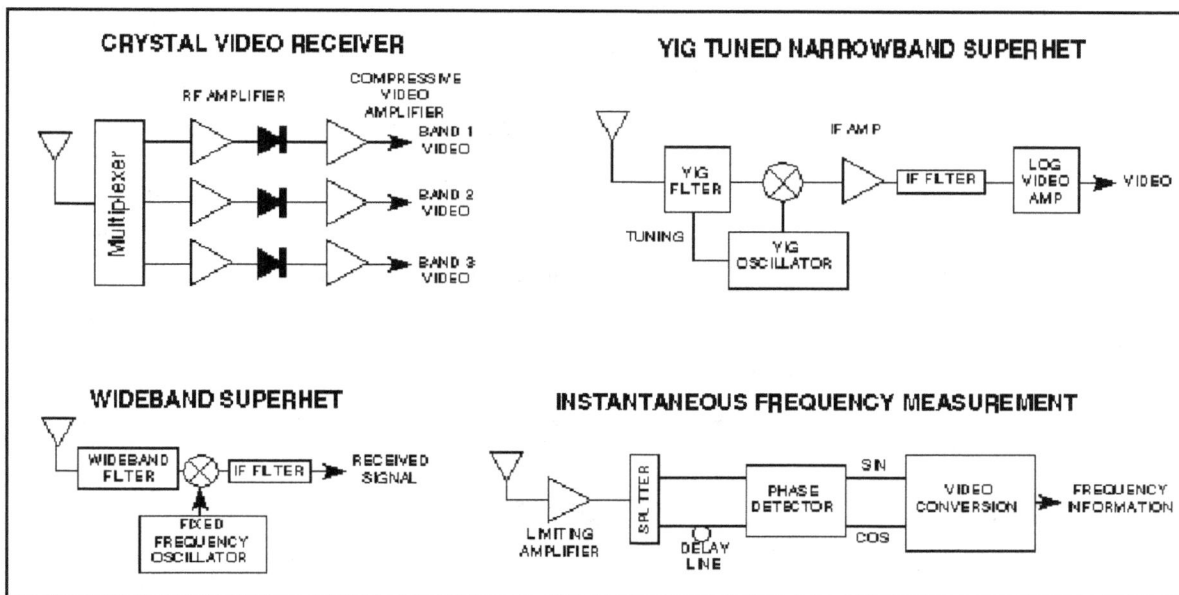

Figure 1. Common ESM Receiver Block Diagrams

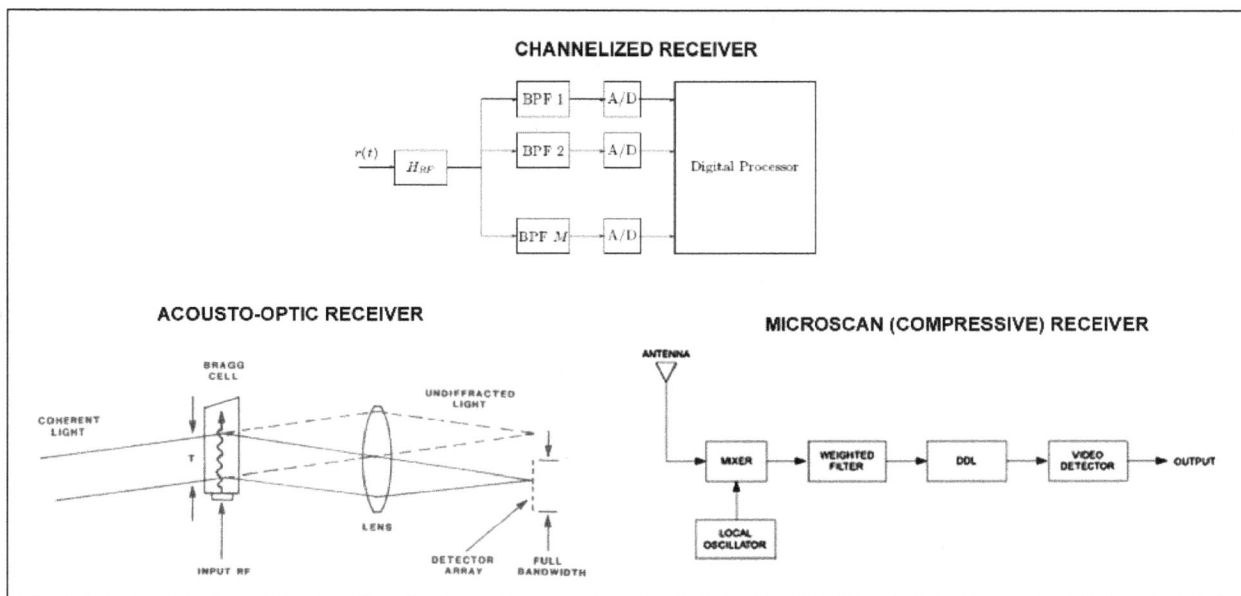

Figure 2. Common ESM Receiver Block Diagrams (Continued)

Table 1. Comparison of Major Features of Receivers

Receiver	Advantages	Disadvantages	Principal Applications
Wideband crystal video	Simple, inexpensive, instantaneous, High POI in frequency range	No frequency resolution Poor sensitivity and Poor simultaneous signal performance	RWR
Tuned RF Crystal Video	Simple, Frequency measurement Higher sensitivity than wideband	Slow response time Poor POI	Option in RWR, Frequency measurement in hybrid
IFM	Relatively simple Frequency resolution Instantaneous, high POI	Cannot sort simultaneous signals Relatively poor sensitivity	Shipboard ESM, Jammer power management, SIGINT equipment
Narrow-band scanning Superhet	High sensitivity Good frequency resolution Simultaneous signals don't interfere	Slow response time Poor POI Poor against frequency agility	SIGINT equipment Air and ship ESM Analysis part of hybrid
Wide-band Superhet	Better response time and POI	Spurious signals generated Poorer sensitivity	Shipboard ESM Tactical air warning
Channelized	Wide bandwidth, Near instantaneous, Moderate frequency resolution	High complexity, cost; Lower reliability; limited sensitivity	SIGINT equipment Jammer power management
Microscan	Near instantaneous, Good resolution and dynamic range, Good simultaneous signal capability	High complexity, Limited bandwidth No pulse modulation information Critical alignment	SIGINT equipment Applications for fine freq analysis over wide range
Acousto-optic	Near instantaneous, Good resolution, Good simultaneous signal capability Good POI	High complexity; new technology	

Note: The Microscan receiver is also known as a compressive receiver

Table 2. Receiver Types vs. Signal Types

Signal Type	Receiver Type							
	Wide-Band Crystal Video	TRF Crystal Video	IFM	Narrow-Band Superhet	Wide-Band Superhet	Channelized	Microscan	Acousto-optic
CW	Special design for CW	Special design for CW	Yes, but interferes with pulsed reception	Yes	Yes	Yes	Yes	Yes
Pulsed	Yes	Yes	Yes	Yes	Yes	Yes	Yes	Yes
Multiple Frequency	No	No	No	Yes, but won't recognize as same source	No	Yes	Yes	Yes
Frequency Agile	Yes, doesn't measure frequency	No	Yes	No	Yes (within passband)	Yes	Yes	No/Yes, depending on readout time
PRI Agile	Yes	Yes	Yes	No/Yes, depending on scan rate	Yes	Yes	No/Yes, imprecision in TOA	No/Yes, depending on readout time
Chirped	Yes, within acceptance BW	No	Yes	No/Yes, depending on BW	Yes	Yes (reduced sensitivity)	No/Yes, depending on scan rate	Yes (reduced sensitivity)
Spread Spectrum	Yes, within acceptance BW	No	Yes	No	No/Yes, depending on BW	Yes (reduced sensitivity)	Yes (reduced sensitivity)	Yes (reduced sensitivity)

Table 3. Direction of Arrival Measurement Techniques

	Amplitude Comparison	Phase Interferometer
Sensor Configuration	Typically 4 to 6 Equal Spaced Antenna Elements for 360° Coverage	2 or more RHC or LHC Spirals in Fixed Array
DF Accuracy	$$DF_{ACC} \approx \frac{\Theta_{bW}^2 \, \Delta C_{dB}}{24 \, S}$$ (Gaussian Antenna Shape)	$$DF_{ACC} \approx \frac{\lambda}{2 \, \pi \, d \cos \theta} \Delta \theta$$
DF Accuracy Improvement	Decrease Antenna BW; Decrease Amplitude Mistrack; Increase Squint Angle	Increase Spacing of Outer Antennas; Decrease Phase Mistrack
Typical DF Accuracy	3° to 10° rms	0.1° to 3° rms
Sensitivity to Multipath/Reflections	High Sensitivity; Mistrack of Several dB Can Cause Large DF Errors	Relatively Insensitive; Interferometer Can be Made to Tolerate Large Phase Errors
Platform Constraints	Locate in Reflection Free Area	Reflection Free Area; Real Estate for Array; Prefers Flat Radome
Applicable Receivers	Crystal Video; Channelizer; Acousto-Optic; Compressive; Superheterodyne	Superheterodyne
ΔC_{dB}= Amplitude Monopulse Ratio in dB S= Squint Angle in degrees θ_{BW}= Antenna Beamwidth in degrees		

Table 4. Emitter Location Techniques

Measurement Technique	Advantages	Disadvantages
Triangulation	Single Aircraft	Non-instantaneous location Inadequate accuracy for remote targeting Not forward looking
Azimuth/elevation	Single Aircraft Instantaneous location possible	Accuracy degrades rapidly at low altitude Function of range
Time Difference of Arrival (Pulsed signals)	Very high precision Can support weapon delivery position requirements Very rapid, can handle short on-time threat	Very complex, diverse systems required, at least 3 aircraft High quality receivers, DME (3 sites) very wideband data link Very high performance control processor; requires very high reliability subsystems

Table 5. Qualitative Comparison of Receivers　　From NRL Report 8737

Feature	Receiver Type							
	Wide-Band Crystal Video	TRF Crystal Video	IFM	Narrow-Band Superhet	Wide-Band Superhet	Channelized	Microscan	Acousto-optic
Instantaneous Analysis Bandwidth	Very wide	Narrow	Very wide	Narrow	Moderate	Wide	Wide	Moderate
Frequency Resolution	Very poor	Fair	Good	Very good	Poor	Fair	Good	Good
Sensitivity	Poor (No preamp) Fair (preamp)	Fair/ good	Poor (No preamp) Fair (preamp)	Very good	Fair	Fair/ good	Very good	Good
Dynamic Range	Fair	Fair/ good	Good	Very good	Fair	Good	Fair	Poor
Speed of Acquisition	Very Fast	Slow	Very Fast	Slow	Fast	Very Fast	Very Fast	Fast
Short pulse Width Capability	Good	Good	Good	Good	Very good	Good	Fair	Fair
Retention of Signal Character- istics	Fair	Fair	Poor	Good	Fair/ good	Good	Poor	Fair/ good
Applicability to Exotic Signals	Poor/ fair	Poor	Good	Poor	Fair/ good	Good	Fair/ good	Fair/ good
High signal Density Performance	Poor (high false alarm rate from background)	Fair/ good	Good	Poor	Fair (depending on BW)	Fair/good, depending on architecture & processing	Good	Poor
Simultaneous Signal Capability	Poor	Fair/ good	Poor	Good	Fair (depending on BW)	Good	Good	Good
Processing Complexity	Moderate depending on application	Moderate depending on application	Moderate	Moderate	Moderate	Low-high depending on architecture	Complex	Simple signal processing complex data processing
Immunity to Jamming	Poor	Fair	Poor/ Fair	Good	Poor/ Fair	Good	Good	Good
Power Requirements	Low	Low/ Moderate	Moderate	Moderate	Moderate	High	Moderate	Moderate/ High
RF Range (GHz)	Multi- octave (0.5-40)	0.15-18 separate	>0.5 to 40	<0.01 to 40	0.5 to 18	0.5 to 60	<0.5 to 8	0.5-4 (0.5-18 channelized and down conversion)
Max Instantane- ous Analysis Bandwidth	Multi- octave (to 17.5 GHz)	As high as desired with equivalent reduction in resolution	Multi- octave (1 octave per unit)	50 MHz	500 MHz	~2 GHz without degradation, 17.5 GHz with degradation	0.5 to 2 depending on PW limitation	1 GHz
Frequency Accuracy	Measurement accuracy no better than analysis BW	Measurement accuracy no better than analysis BW	5-10 MHz	0.5% to 1%	0.5 to 3 MHz	±1 MHz	10 KHz	±1 MHz
Pulse Width Range	CW to 50 ns	CW to 50 ns	CW to ~20 ns (depending on resolution)	CW to 100 ns with 20 MHz resolution	CW to 4 ns with 500 MHz resolution	CW to 30 ns (depending on resolution)	CW to 250 ns	CW to 0.5 μs

Feature	Receiver Type							
	Wide-Band Crystal Video	TRF Crystal Video	IFM	Narrow-Band Superhet	Wide-Band Superhet	Channelized	Microscan	Acousto-optic
Frequency Resolution	~400 MHz (no better than BW)	25 MHz	1 MHz	<0.1 MHz	100-500 MHz	10-125 MHz (less with freq vernier)	1 MHz	0.5 to 1 MHz
Sensitivity (dBm)	-40 to -50 (no preamp) -80 (with preamp)	Better than -80 with preamp	-40 (no preamp) -75 (preamp) 4 GHz BW	-90, 1 MHz BW	-80, 500 MHz BW	-70, 10-50 MHz BW	-90, 5-10 MHz BW	-70 to -80
Maximum Dynamic Range (dB)	70	70-80	80 (w/preamp) 100+ (saturated)	90	60	50-80	40-60	25-35
Tuning Time	-	50 ms	-	1.0 s (1 octave)	.12 s (200 MHz band)	-	0.3 μs LO scan time	0.5 ms (integration time)
Signal ID Time	100 ns	50 ms	2-10 ms	~0.1 s	-	2.10ms	~1 μs	-
Minimum Weight (lb)	20 (with processor)	30	<20 (octave unit) 65-75 (full coverage)	60-75	35 (tuner only)	1309-200 for 0.5 to 18 GHz coverage	25	29-55
Size / Minimum Volume (in³)	Small 300 (w/processor)	Small 375	Sm/Moderate 600-1000 ~100 miniaturized	Moderate 1500-3000	Moderate Several thousand	Large 4000-8000 (0.5-18 GHz coverage	Moderate 1200-2000	Small 800-1900
Minimum Power (W)	100 (with processor) <10 without processor	60 (without processor)	~50 (octave unit)	150	150 (tuner only)	350 to 1200 for 0.5 to 18 GHz coverage	70-80	200
Cost	Low	Low/ Moderate	Moderate	Moderate/ High	Moderate/ High	High	Moderate/ High	Low/ Moderate

This Page Blank

RADAR MODES

Typical Radar modes are listed below in the general functional category for which they were designed. Not all of these modes are applicable to all radars and certain radars have additional modes.

• NAVIGATION

Terrain avoidance - A mode in which the radar is set at a fixed depression angle and short range to continuously sweep the ground area directly in front of the aircraft in order to avoid mountains. This is particularly useful during flight into unfamiliar territory when clouds, haze, or darkness obscure visibility.

Ground mapping - A mode in which the radar uses a variety of techniques to enhance ground features, such as rivers, mountains and roads. The mode is unlike air-to-air modes where ground return is rejected from the display.

Precision velocity update / Doppler navigation - A mode in which the radar again tracks ground features, using Doppler techniques, in order to precisely predict aircraft ground speed and direction of motion. Wind influences are taken into account, such that the radar can also be used to update the aircraft inertial navigation system.

• FIGHTER MISSIONS

Pulse search - Traditional pulse techniques are used to accurately determine range, angle, and speed of the target. Limitations are easy deception by enemy jamming, and less range when compared to other modes.

Velocity search - A high PRF Pulse Doppler waveform is used for long range detection primarily against nose aspect targets, giving velocity and azimuth information. Although velocity search can work against tail-on targets, the Doppler return is weaker, consequently the maximum detection range is also much less. When the target is in the beam (flying perpendicular to the fighter), the closure (Doppler) is the same as ground return and target return is almost zero.

Track While Scan (TWS) - A system that maintains an actual track on several aircraft while still searching for others. Since the radar is sharing it's computing time between targets, the accuracy is less precise than for a single target track (STT) mode of operation.

Raid assessment - A mode in which the radar has an STT on a single target, but is routinely driven off by a small amount in order to determine if multiple aircraft exists in the immediate vicinity of the target aircraft.

Single-Target-Track (STT) (including air combat maneuvering modes) - Highly precise STT modes are used to provide the most accurate information to the fire control computer so that accurate missile or gun firing can be accom-plished. The fire control radar continuously directs energy at the target so that the fired missile locates and tracks on the reflected energy from the target. Air combat maneuvering modes are automatic modes in which the radar has several sweep patterns fixed about the aircraft axis, such that little or no work is required of the pilot in order to lock up a target.

- **AIR-TO-GROUND MISSIONS**

Weapons delivery - A mode in which ground features are tracked, and particular emphasis is placed on determining range to the ground target, angle of dive, weapons ballistic tables, and aircraft speed.

Surveillance/tracking of ground forces/targets - Similar to the above with emphasis on multiple ground features and less on weapons delivery data.

Reconnaissance - A specific navigational mode to aid in identifying specific targets.

- **AIR-TO-SURFACE MISSIONS**

ASW - Navigational techniques specializing in specific search patterns to aid in detection of enemy submarines.

- **TECHNIQUES USED FOR MULTIPLE APPLICATIONS**

Synthetic Aperture Radar (SAR) - A form of radar that uses the relative motion between an antenna and its target region, to provide coherent-signal variations, in order to obtain finer spatial resolution than is possible with conventional beam-scanning means. SAR is usually implemented by mounting a single beam-forming antenna on a moving platform such as an aircraft from which a target scene is repeatedly illuminated with pulses of radio waves at wavelengths anywhere from a meter down to millimeters. The many echo waveforms received successively at the different antenna positions are coherently detected and stored and then post-processed together to resolve elements in an image of the target region.

Over-The-Horizon Radar (OTHR) - uses the refraction of high frequency radiation through the ionosphere in order to detect targets beyond the line-of-sight. The complexities of the ionosphere can produce multipath propagation, which may result in multiple resolved detections for a single target. When there are multipath detections, an OTHR tracker will produce several spatially separated tracks for each target. Information conveying the state of the ionosphere is required in order to determine the true location of the target and is available in the form of a set of possible propagation paths, and a transformation from measured coordinates into ground coordinates for each path. Since may be no other information as to how many targets are in the surveillance region, or which propagation path gave rise to which track, there is a joint target and propagation path association ambiguity which must be resolved using the available track and ionospheric information.

GENERAL RADAR DISPLAY TYPES

There are two types of radar displays in common use today.

RAW VIDEO

Raw video displays are simply oscilloscopes that display the detected and amplified target return signal (and the receiver noise). Raw video displays require a human operator to interpret the various target noise and clutter signals.

On the left hand display of Figure 1, an operator could readily identify three targets and a ghost (a ghost is a phony target that usually fades in and out and could be caused by birds, weather, or odd temporary reflections - also referred to as an angel). Target 3 is a weak return and hidden in the noise - an operator can identify it as a target by the "mouse under the rug" effect of raising the noise base line.

SYNTHETIC VIDEO

Synthetic video displays use a computer to clean up the display by eliminating noise and clutter and creating it's own precise symbol for each target.

On the right hand display target 1 comes and goes because it is barely above the receiver noise level - notice that it is quite clear on the raw video. Target 3 wasn't recognized by the computer because it's to far down in the noise. The computer validated the ghost as a target. The ghost might be a real target with glint or ECM characteristics that were recognized by the computer but not the operator.

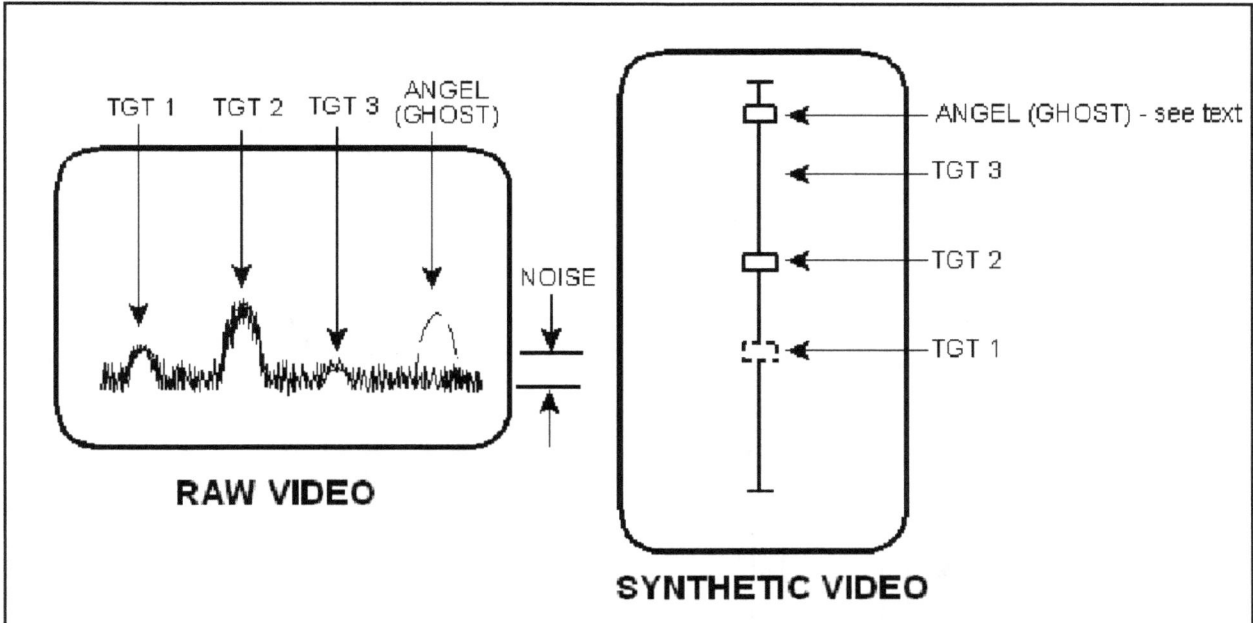

Figure 1. Radar Display Types

SEARCH AND ACQUISITION RADARS

They generally use either a PPI or a sector PPI display as shown in Figure 2. PPI displays can be either raw video or synthetic video.

PPI scope (plan position indicator).
  Polar plot of direction and distance.
  Displays all targets for 360 degrees.

Sector PPI scope.
  Polar plot of direction and distance.
  Displays all targets within a specific sector.
  Origin may be offset so that "your" radar position may be off the scope.

TRACKING RADARS

Usually use some combination of A, B, C, or E scope displays. There are many other types of displays that have been used at one time or another - including meters - but those listed here are the most common in use today.

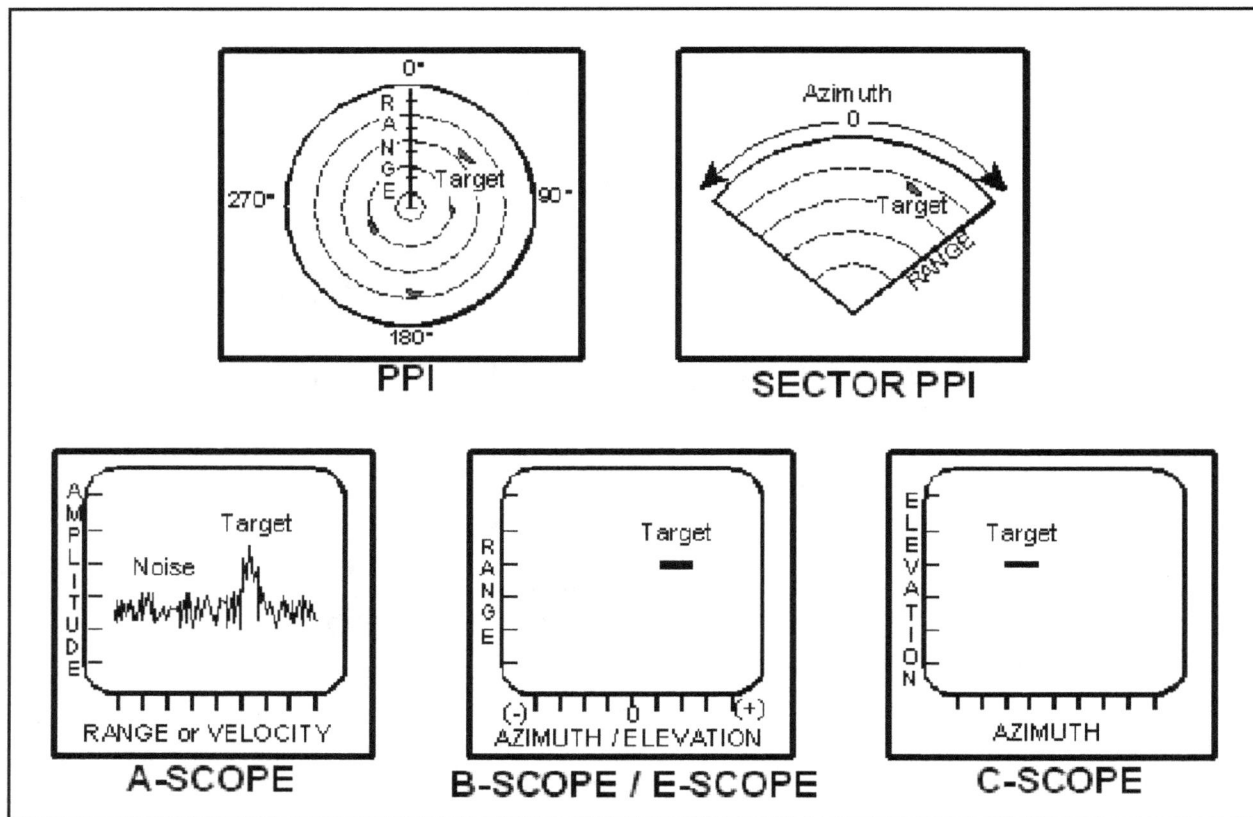

Figure 2. Common Radar Displays

A-SCOPE

Target signal amplitude vs range or velocity.

Displays all targets along pencil beam for selected range limits.

Displays tracking gate. Usually raw video. Some modern radars have raw video a-scopes as an adjunct to synthetic video displays.

Must be used with a separate azimuth and elevation display of some sort.

Also called a range scope (R-Scope).

B-SCOPE

Range vs azimuth or elevation. Displays targets within selected limits.

Displays tracking gate. May be raw or synthetic video.

Surface radars usually have two. One azimuth/one elevation which can result in confusion with multiple targets.

C-SCOPE

Azimuth vs elevation. Displays targets within selected limits of az and el.

Displays tracking gate. May display bull's-eye or aim dot.

May have range indicator inserted typically as a marker along one side. Usually synthetic video.

Pilots eye view and very common in modern fighter aircraft heads up displays for target being tracked.

Could be used in any application where radar operator needs an "aiming" or "cross hair" view like a rifle scope.

E-SCOPE

Elevation vs Range similar to a B-scope, with elevation replacing azimuth.

This Page Blank

IFF - IDENTIFICATION - FRIEND OR FOE

Originated in WWII for just that purpose - a way for our secondary radars to identify U.S. aircraft from enemy aircraft by assigning a unique identifier code to U.S. aircraft transponders.

The system is considered a secondary radar system since it operates completely differently and independently of the primary radar system that tracks aircraft skin returns only, although the same CRT display is frequently used for both.

The system was initially intended to distinguish between enemy and friend but has evolved such that the term "IFF" commonly refers to all modes of operation, including civil and foreign aircraft use.

There are five major modes of operation currently in use by military aircraft plus two sub-modes.
- Mode 1 is a non-secure low cost method used by ships to track aircraft and other ships.
- Mode 2 is used by aircraft to make carrier controlled approaches to ships during inclement weather.
- Mode 3 is the standard system also used by commercial aircraft to relay their position to ground controllers throughout the world for air traffic control (ATC).
- Mode 4 is secure encrypted IFF (the only true method of determining friend or foe) Military only
- Mode 5 – provides a cryptographically secured version of Mode S and ADS-B GPS position. (military only). Mode 5 is divided into two levels. Both are crypto-secure with Enhanced encryption, Spread Spectrum Modulation, and Time of Day Authentication. Level 1 is similar to Mode 4 information but enhanced with an Aircraft Unique PIN. Level 2 is the same as Mode 5 level one but includes additional information such as aircraft position and other attributes
- Mode "C" is the altitude encoder (military and civilian).
- Mode S – provides multiple information formats to a selective interrogation. Typically aircraft are assigned a unique 24-bit Mode S address. The Mode S address is partitioned and a group of address ranges are allocated to each country. Some countries change the assigned address for security reasons, and thus it might not be a unique address. (military and civilian)

The non-secure codes are manually set by the pilot but assigned by the air traffic controller.

A cross-band beacon is used, which simply means that the interrogation pulses are at one frequency and the reply pulses are at a different frequency. 1030 MHz and 1090 MHz is a popular frequency pair used in the U.S.

The secondary radar transmits a series of selectable coded pulses. The aircraft transponder receives and decodes the interrogation pulses. If the interrogation code is correct, the aircraft transponder transmits a different series of coded pulses as a reply.

The advantage of the transponder is that the coded pulses "squawked" by the aircraft transponders after being interrogated might typically be transmitted at a 10 watt ERP, which is much stronger than the microwatt skin return to the primary radar. Input power levels may be on the order of several hundred watts.

The transponder antenna is low gain so that it can receive and reply to a radar from any direction.

An adjunct to the IFF beacon is the altitude encoding transponder known as mode C - all commercial and military aircraft have them, but a fair percentage of general aviation light aircraft do not because of cost. The number of transponder installations rises around many large metropolitan areas where they are required for safety (easier identification of aircraft radar tracks).

Air traffic control primary radars are similar to the two-dimensional search radar (working in azimuth and range only) and cannot measure altitude.

The expanded display in figure 1 is typical of an air traffic control IFF response. The aircraft was told to squawk a four digit number such as "4732". The altitude encoded transponder provides the aircraft altitude readout to the ground controllers display along with the coded response identifying that particular aircraft.

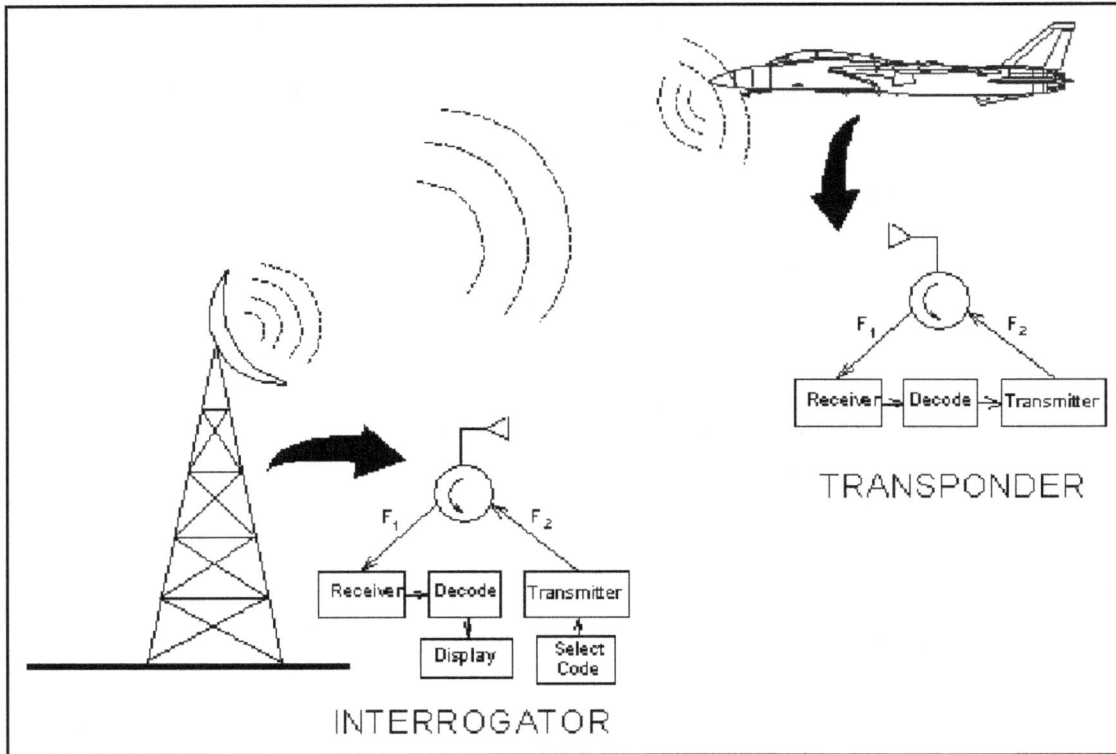

Figure 1. IFF Transponder

In addition to systems with active electronic data interchange between airborne and ground equipment, some military surveillance systems can provide targeting in tactical applications. The development of automated techniques for use against ground targets is typically referred to as Automatic Target Recognition (ATR). When used against air targets, it is typically referred to as Non-Cooperative Target Recognition (NCTR). The requirements for radar target recognition are complex since typical targets have background clutter and often multiple targets types exist.

RECEIVER TESTS

Two tone and spurious response (single signal) receiver tests should be performed on EW and radar receivers to evaluate their spurious free dynamic range. A receiver should have three ranges of performance: (1) protection from damage, (2) degraded performance permitted in the presence of a strong interfering signal(s) and no degradation when only a strong desired signal is present, and (3) full system performance.

The original MIL-STD-461A design requirement and its companion MIL-STD-462 test requirement specified four receiver tests. These standards allowed the interfering signal(s) to be both inband and out of band, which is meaningful for design and test of EW receivers, however inband testing generally is not meaningful for narrowband communications receivers. These standards were difficult to follow and had to be tailored to properly evaluate the EW and radar system. MIL-STD-461B/C still allowed the interfering signal(s) to be both inband and out of band but deleted the single signal interference test (CS08 Conducted Susceptibility test). MIL-STD-461D/-462D leave the pass/fail criteria entirely up to what is listed in the individual procurement specification. It also places all interfering signals <u>out of band</u>, redesignates each test number with a number "100" higher than previously used, and combines "CS08" as part of CS104. Therefore, to provide meaningful tests for EW and radar systems, the procurement specification must specify the three ranges of performance mentioned in the beginning of this section and that the tests are to be performed with the interfering signal(s) both inband and out of band. The four tests are as follows (listed in order of likelihood to cause problems):

Test Name	MIL-STD-461A	MIL-STD-461D
Undesired, Single signal interference test	CS08	Part of CS104
Desired with undesired, two signal interference tests	CS04	CS104
Two signal intermodulation test	CS03	CS103
Two signal cross modulation test	CS05	CS105

The rest of this section explains the application of these tests and uses the names of the original MIL-STD-461A tests to separate the tests by function.

TEST SETUP

A directional coupler used backwards (as shown here in Figure 1) is an easy way to perform two signal tests. The CW signal should be applied to the coupling arm (port B) since the maximum CW signal level is -10 dBm. The pulse signal should be applied to the straight-through path

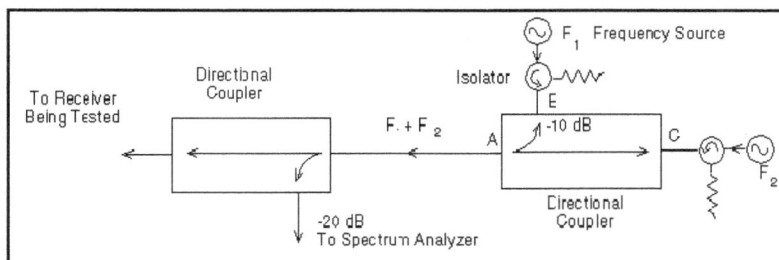

Figure 1. Receiver Test Setup When Antenna Can Be Removed

(port C) since the maximum pulse level is +10 dBm peak. These power levels are achievable with standard laboratory signal generators, therefore one doesn't have to resort to using amplifiers which may distort the signals. Always monitor the output signal to verify spectrally pure signals are being applied to the test unit. This can be accomplished by another directional coupler used in the standard configuration. Dissimilar joints or damaged or corroded microwave components can cause mixing. This can also result if the two signal generators are not isolated from one another. Therefore, even if a directional coupler is used to monitor the signal line, it is still advisable to directly measure the input to the receiver whenever there is a suspected receiver failure. This test does not need to be performed in an EMI shielded room and is more suitable for a radar or EW lab where the desired signals are readily available.

If the receiver's antenna is active or cannot be removed, a modified test as shown in Figure 2 should be performed. The monitoring antenna which is connected to the spectrum analyzer should be the same polarization as the antenna for the receiver being tested. Amplifiers may be required for the F_1 and F_2 signals. It is desirable to perform this test in an anechoic chamber or in free space.

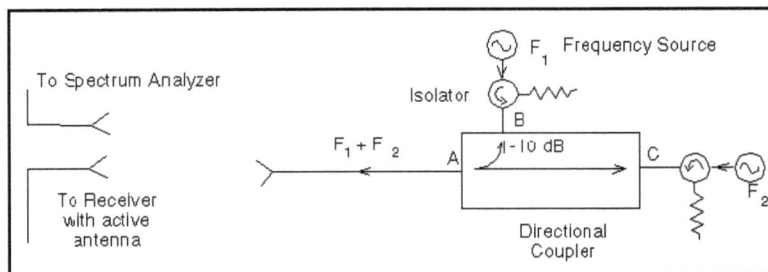

Figure 2. Receiver Test Setup When Antenna Is Active

In the following discussion of CS08, CS04, CS03, and CS05 tests, it is assumed that when the receive light illuminates, the receiver identifies a signal that matches parameters in the User Data File (UDF) or pre-programmed list of emitter identification parameters. If a receiver is different, the following procedures will have to be appropriately tailored. If the UDF does not have entries for very low level signals in the 10% and 90% regions of each band, complete testing is not possible. Most problems due to higher order mixing products and adjacent band leakage are only evident in these regions. In the following tests, the lowest level where the receive light is constantly on is used to identify the minimum receive level. If a receiver has a receive level hysteresis or other idiosyncracy, then using a 50% receive light blinking indicator may be more appropriate. Whatever technique is appropriate, it should be consistently used during the remainder of the test. The maximum frequency for testing is normally 20 GHz. If a millimeter wave receiver is being tested, the maximum frequency should be 110 GHz.

CS08 - UNDESIRED, SINGLE SIGNAL INTERFERENCE TEST

MIL-STD-461B/C (EMI design requirements) deleted this test. MIL-STD-461D allows a single signal test as part of CS104 (CS04) but specifies it as an out of band test. The original CS08 inband and out of band test is still needed and is the most meaningful test for wide band EW receivers which have a bandwidth close to an octave. This test will find false identification problems due to 1) lack of RF discrimination, 2) higher order mixing problems, 3) switch or adjacent channel/band leakage, and 4) cases where the absence of a desired signal causes the receiver to search and be more susceptible. In this latter case, a CS04 two signal test could pass because the receiver is captured by the desired signal, whereas a CS08 test could fail. Examples of the first three failures are as follows:

EXAMPLE 1

A 2 to 4 GHz receiver which uses video detection (e.g., crystal video) and doesn't measure RF is used for this example. This receiver assumes that if the correct Pulse Repetition Interval (PRI) is measured, it is from a signal in the frequency band of interest. Three cases can cause false identification. Refer to Figure 3.

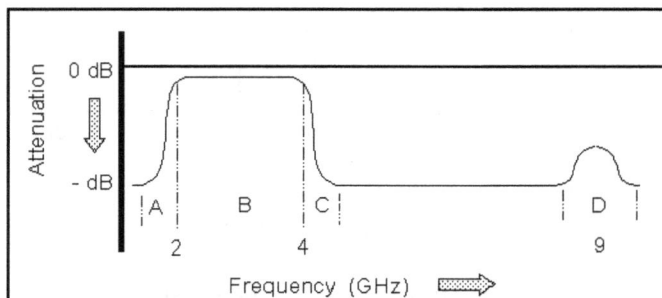

Figure 3. Frequency Areas in a Sample 2-4 GHz Receiver

(1) Region A&C. The 2 to 4 GHz band pass filter will pass strong signals in regions A&C. If they have the correct PRI, they will also be identified.

(2) Region B. Any other signal besides the desired signal in the 2 to 4 GHz region that has the correct PRI will also be identified as the signal of interest.

(3) Region D. Band pass filters with poor characteristics tend to pass signals with only limited attenuation at frequencies that are three times the center frequency of the band pass filter. If these signals have the correct PRI, they will be incorrectly identified.

High duty cycle signals (CW or pulse doppler) in regions A, B, C, and D may overload the processing of signals, saturate the receiver, or desensitize the receiver. This case is really a two signal CS04 test failure and will be addressed in the CS04 section.

EXAMPLE 2

A receiver measuring the carrier frequency of each pulse (i.e. instantaneous frequency measurement (IFM)) and the PRI is used for this example. False signal identification can occur due to higher order mixing products showing up in the receiver pass bands. These unwanted signals result from harmonics of the input RF mixing with harmonics of the Local Oscillator (LO). Refer to Figures 4 and 5.

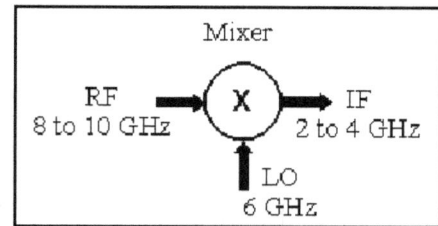

Figure 4. Low Side Mixing

Mixers are nonlinear devices and yield the sum, difference, and the original signals. Any subsequent amplifier that is saturated will provide additional mixing products.

If a 8.5 GHz signal with a 1 kHz PRI is programmed to be identified in the UDF, measurements are made at the 2.5 GHz Intermediate Frequency (IF), i.e., RF-LO = IF = 8.5-6 = 2.5 GHz.

The same 2.5 GHz signal can result from an RF signal of 9.5 GHz due to mixing with the second harmonic of the LO i.e., 2 X 6 - 9.5 = 2.5 GHz. This signal will be substantially attenuated (approximately 35 dB) when compared to the normal IF of 9.5 - 6 = 3.5 GHz. If the receiver has filters at the IF to reduce the signal density and a filter has minimum insertion loss at 2.5 GHz and maximum insertion loss at 3.5 GHz, then only the low level 2.5 GHz signal will be measured and assumed to be due to a 8.5 GHz input signal whereas the input is really at 9.5 GHz.

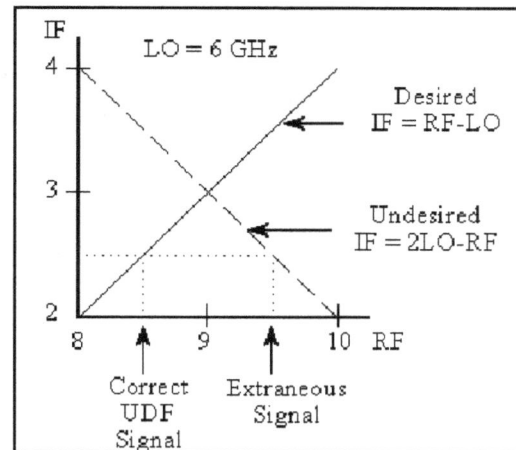

Figure 5. Low Side Mixing Results

Spurious intermodulation products can also result from high side mixing, but generally the suppression of undesired signals is greater. In this case, the LO is at a frequency higher than the RF input. This is shown in Figures 6 and 7.

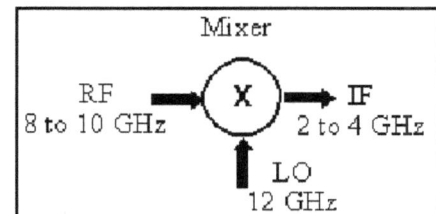

Figure 6. High Side Mixing

Table 1. Intermodulation Product Suppression

Harmonic of		Suppression
LO	RF	
1	1	0
1	2	$\Delta P-41$
1	3	$2\Delta P-28$
2	1	-35
2	2	$\Delta P-39$
2	3	$2\Delta P-44$
3	1	-10
3	2	$\Delta P-32$
3	3	$2\Delta P-18$
4	1	-35
4	2	$\Delta-39$
5	1	-14
5	3	$2\Delta P-14$
6	1	-35
6	2	$\Delta P-39$
7	1	-17
7	3	$2\Delta P-11$

Courtesy Watkins-Johnson

As previously mentioned, the amplitude of intermodulation products is greatly reduced from that of the original signals. Table 1 shows rule of thumb approximate suppression (reduction), where $\Delta P = P_{RF}(dBm) - P_{LO}(dBm)$. As can be seen, the strength of the LO is a factor. The higher the LO power, the more negative the suppression becomes.

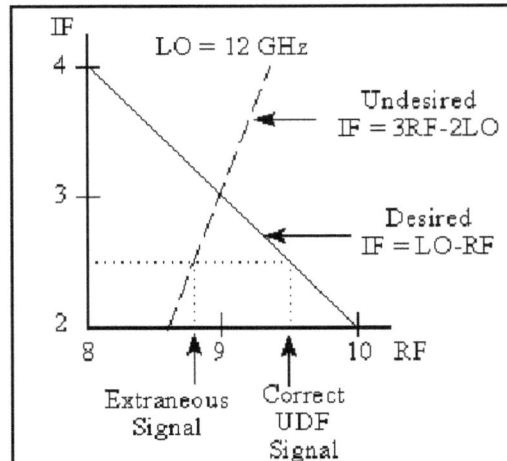

Figure 7. High Side Mixing Byproducts

If one assumes the maximum RF power for full system performance is +10 dBm and the LO power level is +20 dBm, then ΔP = -10 dB minimum. Therefore in this example, the 3RF-2LO mixing product would be $2\Delta P - 44 =$ -20 - 44 = -64 dB when compared to the desired mixing product.

The use of double mixing, as shown in Figure 8, can significantly reduce unwanted signals but it is more expensive. For a 8 GHz signal in, one still generates a 2 GHz IF but by mixing up, then down, unwanted signals are not generated or significantly suppressed.

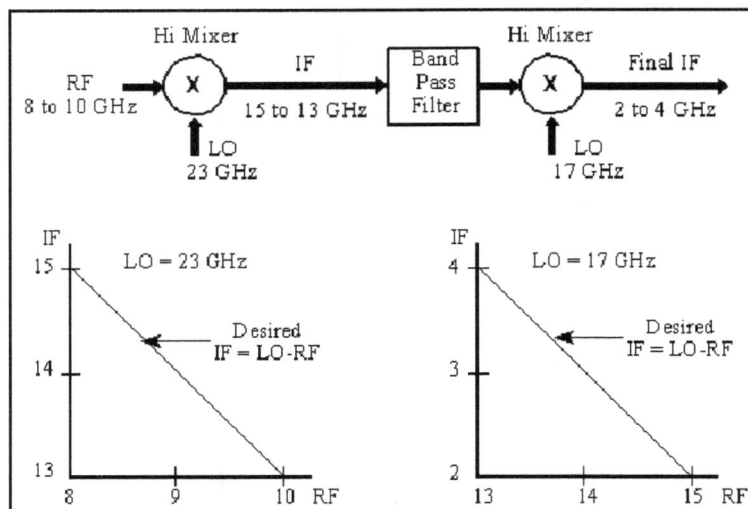

Figure 8. Double Mixing

Some of these problems can be corrected by :

(1) always having LOs on the high side versus low side of the input RF (but this is more expensive),

(2) using double mixing

(3) software programming the receiver to measure for the potential stronger signal when a weak signal is measured in a certain IF region, and

(4) improved filtering of the LO input to the mixer and the output from the mixer.

EXAMPLE 3

If the same receiver discussed in example 2 had additional bands (Figure 9) and used a switch at the IF to select individual bands, a strong signal in an adjacent band could be inadvertently measured because:

(1) the switch, which may have 80 dB of isolation when measured outside the circuit, may only have 35 dB isolation when installed in a circuit because of the close proximity of input and output lines,

(2) the strong signal in one band may have the same IF value that is being sought in an adjacent band, and

(3) the additional parameters such as PRI may be the same.

As shown in Figure 9, assume that in band 2 we are looking for a 4.5 GHz signal that has a PRI of 1 kHz. Measurements are made at an IF of 3.5 GHz since LO-RF = IF = 8-4.5 = 3.5 GHz. If a 6.5 GHz signal is applied to band 3, its IF also equals 3.5 since LO-RF = 10-6.5 = 3.5 GHz. If this is a strong signal, has a PRI of 1 kHz, and there is switch leakage, a weak signal will be measured and processed when the switch is pointed to band 2. The receiver measures an IF of 3.5 GHz and since the switch is pointed to band 2, it scales the measured IF using the LO of band 2 i.e., LO-IF = RF = 8-3.5 = 4.5 GHz. Therefore, a 4.5 GHz signal is assumed to be measured when a 6.5 GHz signal is applied. Similarly this 6.5 GHz signal would appear as a weak 3.5 GHz signal from band 1 or a 9.5 GHz signal from band 4.

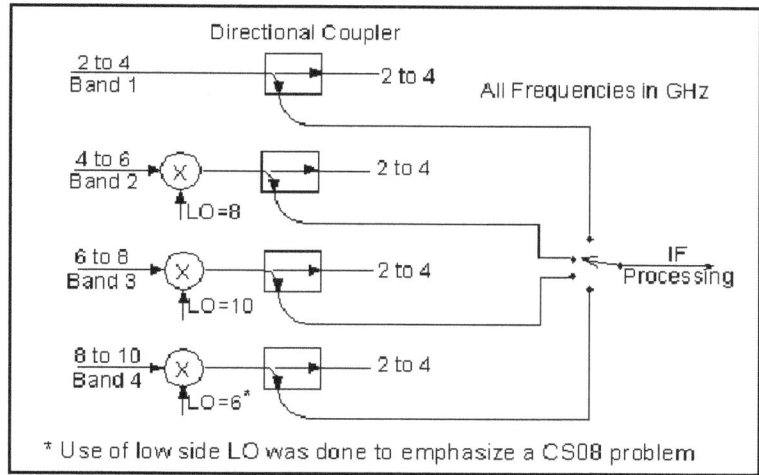

Figure 9. Multi Band Receiver With Common IF

In performing this test it is important to map the entries of the UDF for each band i.e., show each resulting IF, its PRI, and the sensitivity level that the receive light is supposed to illuminate, i.e., if a test in one band used a PRI corresponding to a PRI in another band where the receive threshold is programmed to not be sensitive this will negate the effectiveness of a cross coupling test. Mapping the UDF will facilitate applying a strong signal to one band using the PRI of a desired signal in an adjacent band.

CS08 TEST PROCEDURE

Assume that the receiver band is 2 to 4 GHz as shown in Figure 10. Pick the UDF entry that has the greatest sensitivity. UDF #1 entry is for a 3±.05 GHz signal with a PRI of 1 kHz. If the test signal is set for the UDF #1 PRI, a receive light will also occur at the frequencies of UDF #2 if it also has the same PRI (this is not a test failure). If adjacent bands don't also have entries with the same PRI, then the test should be repeated for the band being tested with at least one of the adjacent band PRI values.

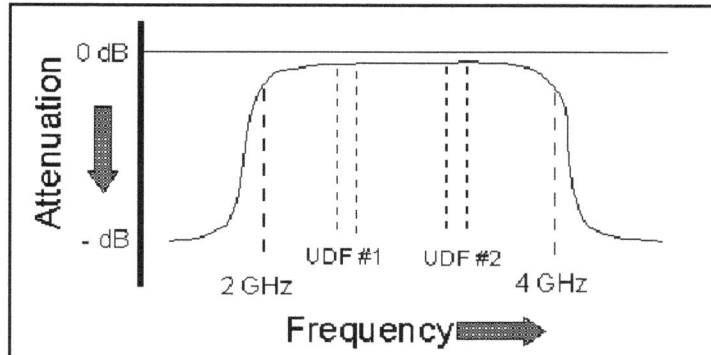

Figure 10. Receiver Band With Multiple UDF Entries

(1) Set the receiver or jammer to the receive mode, verify it is working for UDF #1 and record P_0, the minimum signal level where the receive light is constantly on.

(2) Raise this signal to its maximum specified level for full system performance. If a maximum level is not specified, use +10 dBm peak for a pulse signal or -10 dBm for a CW signal.

(3) Tune this strong RF signal outside the UDF #1 range and record any RF frequency where the receive light comes on. If another inband UDF has the same PRI, this is not a failure.

(4) This test is performed both inband and out of band. Out of band tests should be performed on the high end to five times the maximum inband frequency or 20 GHz, whichever is less, and on the low end to IF/5 or 0.05 F0, whichever is less, unless otherwise specified. The out of band power level is +10 dBm peak for a pulse signal or -10 dBm for a CW signal, unless otherwise specified.

(5) If a receive light comes on when it is not supposed to, record the RF and reduce the power level to where the receive light just stays on constantly. Record this level P1. The interference rejection level is P1-P0= PIR

(6) Repeat this test for each type of signal the receiver is supposed to process, i.e. pulse, PD, CW, etc.

CS04 - DESIRED WITH UNDESIRED, TWO SIGNAL INTERFERENCE TEST

The intent is for a weak desired signal to be received in the presence of an adjacent CW signal. The desired signal is kept tuned at minimal power level and a strong unmodulated signal is tuned outside the UDF region. Radar and EW receivers without preselectors are likely to experience interference when this test is performed inband. Receivers with nonlinear devices before their passive band pass filter, or filters that degrade out of band, are likely to experience susceptibility problems when this test is performed out of band.

Tests performed inband - An unmodulated CW signal is used. If the receiver is supposed to handle both pulsed and CW signals, this test is performed inband. If the pulse receiver is supposed to desensitize in order to only process pulse signals above the CW level, then only this limited function is tested inband i.e., normally the levels correspond, if a CW signal of -20 dBm is present, then the receiver should process pulse signals greater than -20 dBm.

CS04 TEST PROCEDURE

(1) As shown in Figure 11, initially the pulse signal is tuned to F_0 and the minimum receive level P_0 is recorded, i.e., minimum level where the receive light is constantly on.

(2) The pulse signal is raised to the maximum specified level for full system performance and tuned on either side of F_0 to find the frequencies on both sides (F_{High} and F_{Low}) where the receive light goes out. If a maximum pulse power level is not specified, then +10 dBm peak is used. In some receivers F_L and F_H are the band skirts.

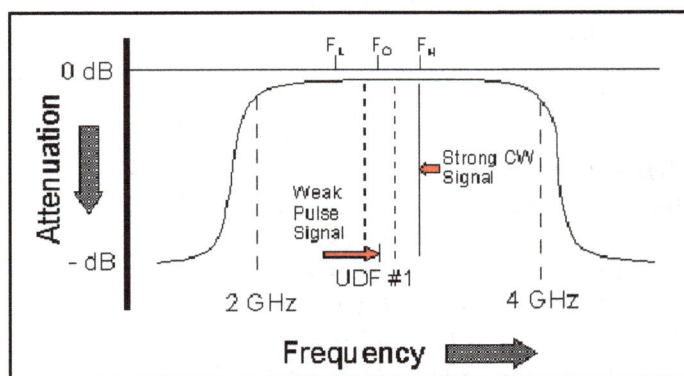

Figure 11. CS04 Test Signals

(3) The pulse signal is returned to the level found in step 1. A CW signal at the maximum specified CW power level for full system performance is tuned above F_H and below F_L. If a maximum CW power level is not specified, then -10 dBm is used. Anytime the receive light is lost, the tuned CW RF value is recorded. The CW signal should be turned off to verify that the pulse signal can still be received in the absence of interference. If the pulse signal is still being received, then the interfering CW signal should be reapplied and decreased to the lowest power level where the receive light stays on constantly. Record this level P_1. The interference rejection level is $P_1 - P_0 = P_{IR}$.

(4) Out of band tests should be performed to five times the maximum inband frequency or 20 GHz, whichever is less, and on the low end to IF/5 or 0.05 F_0, whichever is less, unless otherwise specified. The out of band CW power level is -10 dBm unless otherwise specified.

Failures - Out of band test

(1) If a non-linear device such as a limiter is placed before a band pass filter, a strong out of band signal can activate the limiter and cause interference with the inband signal. The solution is to place all non-linear or active devices after a passive band pass filter.

(2) Band pass filters with poor characteristics tend to pass signals with only limited attenuation at frequencies that are three times the center frequency of the band pass filter. Passage of a CW or high duty cycle signal that is out of band may desensitize or interfere with the processing of a weak inband signal.

CS03 INTERMODULATION TEST

This two signal interference test places a pulse signal far enough away (Δf) from the desired UDF frequency (F_0) that it won't be identified. A CW signal is initially placed $2\Delta f$ away. If an amplifier is operating in the saturated region, these two signals will mix and produce sum and difference signals. Subsequent mixing will result in a signal at the desired UDF frequency F_0 since $F_1 - (F_2 - F_1) = F_0$. These two signals are raised equally to strong power levels. If no problem occurs, the CW signal is tuned to the upper inband limit and then tuned out of band. A similar test is performed below F_0.

CS03 TEST PROCEDURES

(1) Set the receiver or jammer to the receive mode. Verify it is working at a desired signal frequency, (F_0), and record the minimum signal level i.e., lowest level where the receive light is constantly on (record this level P_0).

(2) The modulated signal is raised to the maximum specified level for full system performance and tuned on either side of F_0 to find the frequency F_1 on both sides where the receive light goes out. If a maximum power level is not specified, +10 dBm peak is used. The difference between F_1 and F_0 is Δf as shown in Figure 12.

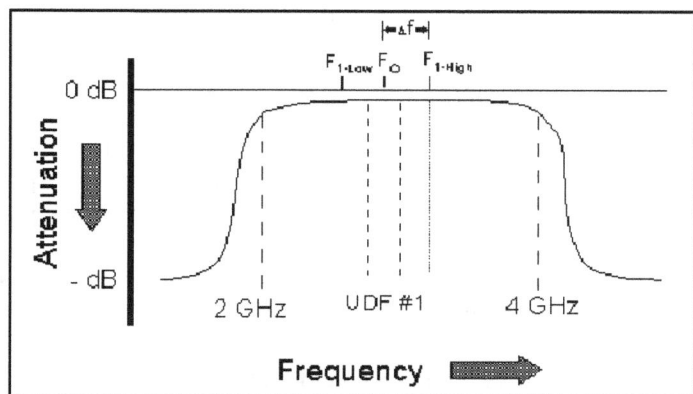

Figure 12. Initial CS03 Test Signal

(3) As shown in Figure 13, a pulse signal is tuned to F_1 and a CW signal is tuned to F_2 where $F_2 = F_1 + \Delta f$ on the high side. The power level of the two signals is initially set to P_0 and raised together until the maximum specified levels for full system performance are reached. If maximum power levels are not specified, then +10 dBm peak is used for the pulse signal and -10 dBm is used for the CW signal. Whenever the receive light comes on, the two signals should be turned off individually to verify that the failure is due to a combination of the two signals versus (1) a single signal (CS08) type failure or (2) another inband UDF value

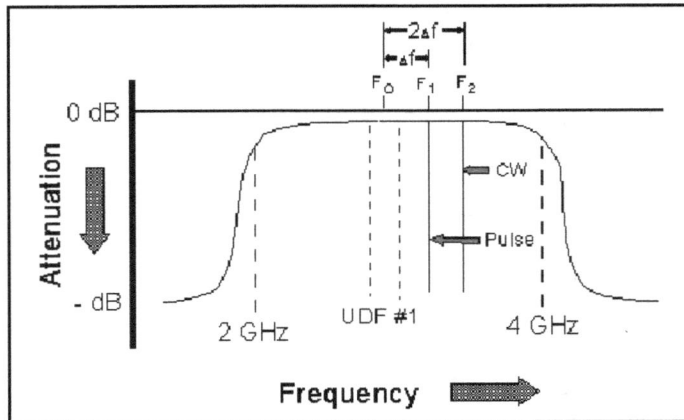

Figure 13. CS03 Testing Signal

has been matched. If the failure is due to the two signal operation, then the power level (P_1 and P_2) of F_1 and F_2 should be recorded. If $P_1=P_2$, the intermodulation rejection level is $P_1-P_0=P_{IM}$. If $P_1 \neq P_2$, it is desirable to readjust them to be equal when the receive light just comes on.

(4) Once the $F_1 + F_2$ signals are raised to the maximum power test levels described in step 3 without a failure, then F_2 is tuned to the upper limit of the band. F_2 should also be tuned out of band to five times the maximum inband frequency or 20 GHz whichever is less unless otherwise specified. The out of band power level is -10 dBm unless otherwise specified. Whenever the receive light comes on, F_2 should be turned off to verify that the failure is due to a two signal test. If it is, turn F_2 back on and equally drop the power levels of F_1 and F_2 to the lowest level where the receive light just comes on. Record the power levels (P_1 and P_2).

(5) Step 3 is repeated where F_1 is Δf below F_0 and $F_2=F_1-\Delta f$. Step 4 is repeated except F_2 is tuned to the lower limit of the band. F_2 should also be tuned out of band down to 0.1 F_0, unless otherwise specified.

(6) Normally if a failure is going to occur it will occur with the initial setting of F_1 and F_2. Care must be taken when performing this test to ensure that the initial placements of F_1 and F_2 do not result in either of the signals being identified directly.

As shown in Figure 14, if F_1 was placed at 3.2 GHz it would be identified directly and if F_2 was placed at 3.4 GHz it would be identified directly. Whereas, if F_1 was at 3.1 GHz and F_2 was at 3.2 GHz neither interfering signal would be identified directly but their intermodulation may result in an improper identification at F_0. Later when F_2 is tuned higher, the receive light will come on around 3.4 GHz and 3.6 GHz. This is not a test failure just a case of another inband UDF value being matched.

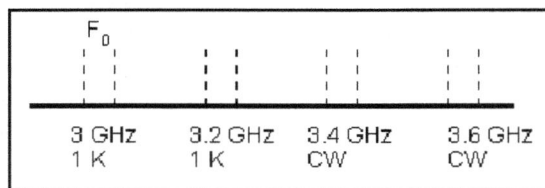

Figure 14. Sample UDF Entries

CS05 - CROSS MODULATION

This two signal interference test places a weak CW signal where the receiver is programmed for a pulse signal and tunes a strong pulse signal elsewhere. As shown in Figure 15, when an amplifier is saturated, lower level signals are suppressed. When an amplifier is operated in the linear region all signals receive the rated linear gain. In this test the pulse signal will cause the amplifier to kick in and out of saturation and modulate the weak CW signal. The receiver may measure the modulation on the CW signal and incorrectly identify it as a pulse signal.

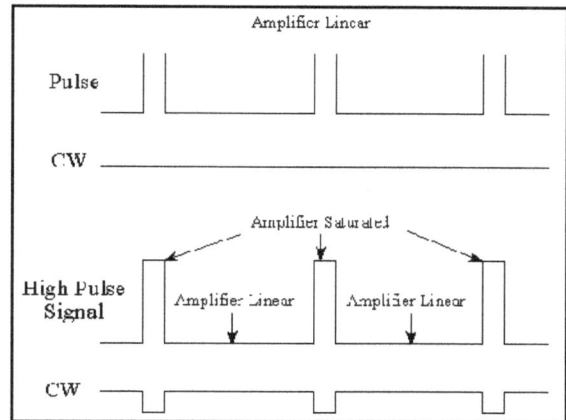

Figure 15. Cross Modulation Example

CS05 TEST PROCEDURE

(1) Initially the pulse signal is tuned to F_0 and the minimum power level P_0 where the receive light is constantly on is recorded.

(2) As shown in Figure 16, the signal is raised to the maximum specified level for full system performance for a pulse signal and tuned on either side of F_0 to find the frequencies on both sides, (F_{High} and F_{Low}) where the receive light goes out. If a maximum pulse power level is not specified, then +10 dBm peak is used.

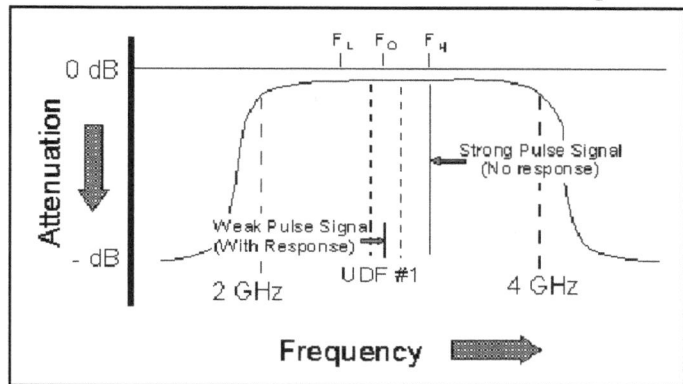

Figure 16. Initial CS05 Test Signals

(3) The pulse signal from step 2 is turned off and a second signal is placed at F_0. It is a CW signal that is 10 dB stronger than the peak power level (P_0) measured is step 1. The receive light should not come on.

(4) As shown in Figure 17, the strong pulse signal of step 2 is turned back on and tuned above F_H and then tuned below F_L. Out of band tests should be performed to the maximum RF of the system + maximum IF or 20 GHz whichever is less and on the low end to the minimum RF of the system minus the maximum IF, unless otherwise specified.

(5) If a receive light occurs, turn off the weak CW signal since the "failure" may be due to the tuned pulsed signal, i.e. a CS08 failure or another inband UDF value has been matched.

If the light extinguishes when the weak CW signal is turned off, then turn the signal back on, reduce the value of the high level pulse signal until the minimum level is reached where the light stays on constantly. Record this level as P_1. The cross modulation rejection level is P_1-P_0-10 dB = P_{CM}.

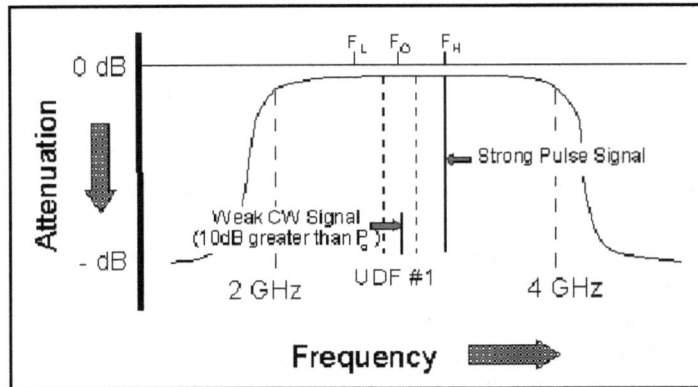

Figure 17. Final CS05 Test Signals

SIGNAL SORTING METHODS and DIRECTION FINDING

As shown in Figure 1, signal processing is basically a problem of signal detection, emitter parameter measurement and correlation, emitter sorting, identification, and operator notification.

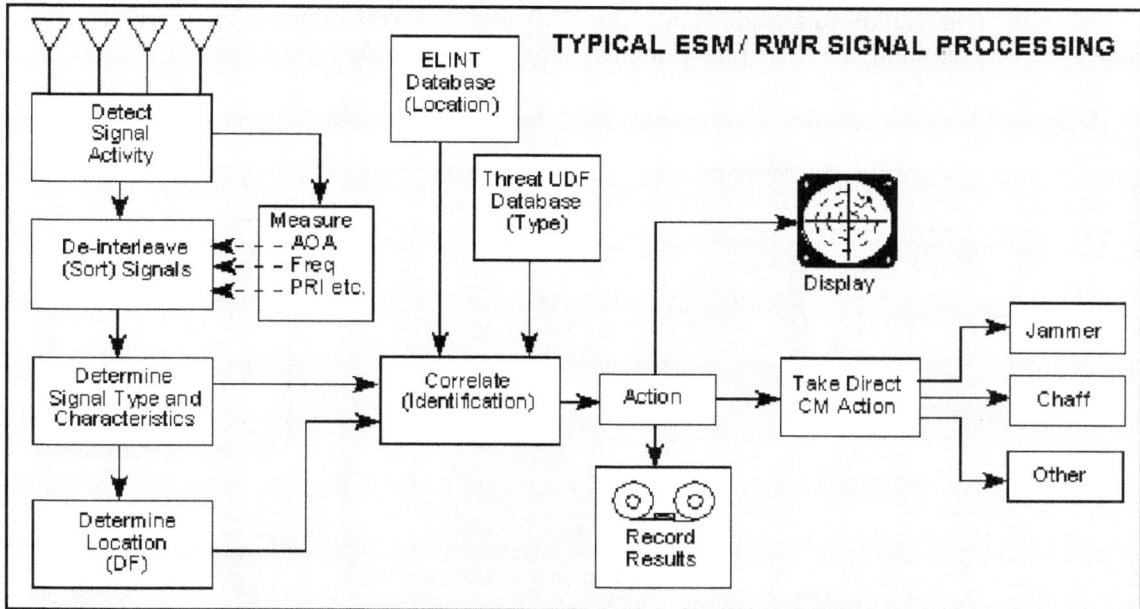

Figure 1. Signal Processing Steps

The ultimate goal of this processing is to classify radar signals by their unique characteristics and to use this data to identify enemy radars operating in the environment, determine their location or direction, assess their threat to friendly forces, and display this information to the operator.

While not all electronic support measures (ESM) or radar warning receiver (RWR) systems perform every step in this process, each completes some of them. For example, ESM systems seldom initiate direct CM action, while RWRs sometimes do. Also ESM systems frequently record electronic data for future use, but few RWRs do. ESM systems place more emphasis on accurate emitter location and hence direction finding capabilities, while RWRs usually give a rough estimate of position/distance.

The typical emitter characteristics that an ESM system can measure for a pulse radar include the following data:

1. Radio Frequency (RF)
2. Amplitude (power)
3. Direction of Arrival (DOA) - also called Angle of Arrival (AOA)
4. Time of Arrival (TOA)
5. Pulse Repetition Interval (PRI)
6. PRI type
7. Pulse Width (PW)
8. Scan type and rate
9. Lobe duration (beam width)

However, this list is not comprehensive. Other emitter parameters are available which may be necessary to characterize the threat system.

More sophisticated ESM systems can measure additional parameters, such as PRI modulation characteristics, inter-and intra-pulse Frequency Modulation (FM), missile guidance characteristics (e.g., pattern of pulse spacing within a pulse group), and Continuous Wave (CW) signals.

Still other parameters which can describe an electromagnetic wave but are currently not commonly used for identification include polarization and phase. However, as threat emitters begin to use this data more frequently to avoid jamming the more important they may become in identifying signals.

Some of the emitter characteristics which describe an electromagnetic wave are shown in Figure 2.

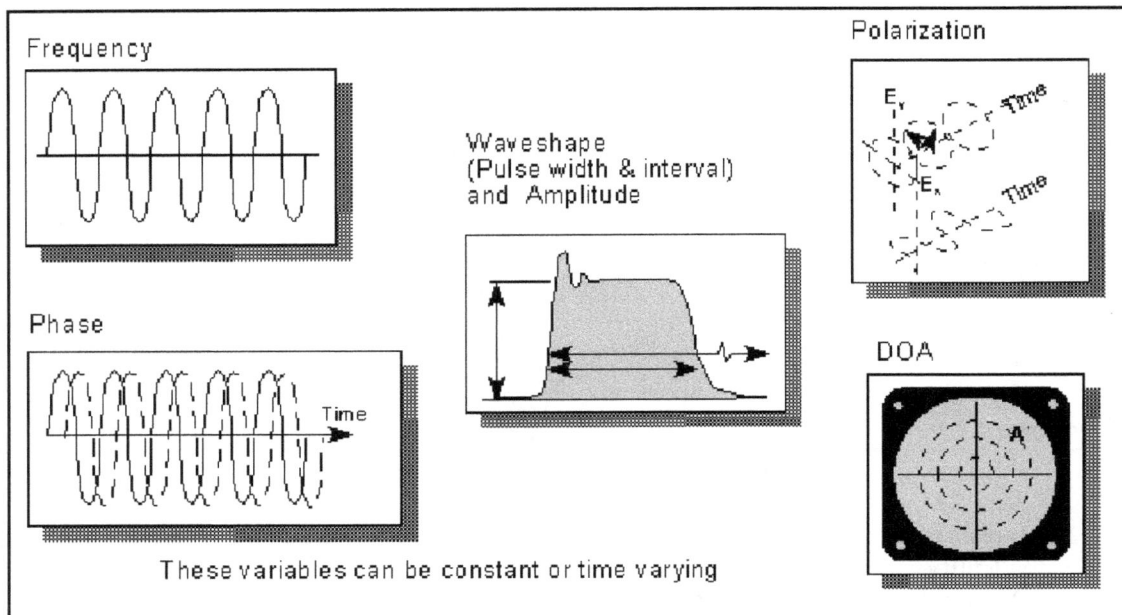

Figure 2. Information Content of an Electromagnetic Wave

Table 1 illustrates the relative importance of several measured parameters during various stages of signal processing.

Table 1. Importance of Emitter Parameters During Signal Processing

Parameter	Pulse Train De-interleavement	Emitter Identification	Intercept Correlation
Frequency	2	2	2
Amplitude	1	0	1
Angle of Arrival	2	0	2
TOA	0	0	1
PRI	2	2	2
PRI type	2	2	2
PW	2	1	1
Scan rate and type	0	2	1
Lobe Duration	0	1	1
0 Not Useful	1 Some Use	2 Very Useful	

Some emitter parameters can be measured using a single pulse; these parameters are referred to as monopulse parameters. The monopulse parameters include RF, PW, DOA, amplitude and TOA. RF can be determined on a pulse-by-pulse basis by receivers that can measure frequency. Frequency is very useful for emitter identification since most radars operate at a single frequency. Most real-time systems measure pulse width instead of pulse shape because the latter is much more difficult to characterize mathematically. Unfortunately, the apparent pulse width can be severely distorted by reflections, and consequently, its usefulness for emitter identification is limited. DOA cannot be used for emitter identification, but is excellent for sorting signals. A number of ESM systems use both frequency and DOA information to distinguish the new signals from the old (that is, known) ones. Amplitude also cannot be used for emitter identification. However, it can be used for sorting and for gross distance estimation using precompiled emitter's effective radiated power. Moreover, amplitude in conjunction with TOA can be used to determine the emitter's scan characteristics.

Other emitter parameters such as PRI, guidance and scan characteristics can be determined only by analyzing a group of pulses. All these parameters are useful for emitter identification; unfortunately, they require time for data collection and analysis, and call for sophisticated signal processing algorithms.

The problem of signal recognition in real-time is complicated by two factors: modulation of the signals and the very high pulse densities expected in the environment. Complex modulations (for example, inter-pulse RF modulation, intra-pulse RF modulation and agile Pulse Repetition Frequencies (PRFs)) present a significant pattern recognition problem for a number of ESM systems. It is expected that during some missions, hundreds of emitters will be transmitting simultaneously in the same vicinity. Wide-open antenna/receiver combination systems may have to cope with up to a million PPS. Even narrow-band receivers can expect data rates up to 100,000 PPS. At these rates, a single modern computer cannot be expected to process all the pulses, derive the characteristics for all emitters and identify the emitters in real-time. Other factors which encumber signal recognition include missing pulses, atmospheric noise and multiple reflections of pulses.

Present RWRs are designed primarily to cope with stable emitters. A stable emitter is one whose frequency and pulse repetition interval (PRI) remain relatively constant from pulse to pulse. The future threat will move steadily away from the stable emitter towards agile emitters which vary their frequency and PRI characteristics. The first change in this direction is towards the patterned agile emitter which varies its pulse and frequency parameters in accordance with a specific pattern. Examples of patterned agile emitters are MTI radars which use staggered PRFs, pulse Doppler radars which change frequency and PRF on a block-to-block basis, and certain frequency-agile radars whose transmitter frequency is mechanically modulated in a systematic pattern (e.g., spin-tuned magnetron). The next step in this evolution is towards truly agile emitters which change their frequency and PRF in a random manner on a pulse-to-pulse basis. One tempering factor in this evolution is that radars which process Doppler must maintain a constant frequency for at least two consecutive pulses.

In addition to agile frequency and PRI parameters, the future threat will be composed of a number of high-PRF pulsed Doppler, burst-frequency, CW, pulse-compression, agile-beam, and LPI radars, which use pseudo-noise waveforms. This conglomeration of radar types will cause a high signal density which must be segmented into a manageable data stream by the use of both frequency and spatial filtering in the RWR. While frequency and PRI are good parameters for sorting present-day non-agile emitters, they are poor or useless parameters for sorting agile emitters.

> Angle of arrival is generally regarded as the best initial sorting parameter because it cannot be varied by the emitter from pulse to pulse.

PASSIVE DIRECTION FINDING AND EMITTER LOCATION

Direction finding (DF) systems provide several important functions in modern EW systems. We have already discussed the importance of measuring the emitter's bearing, or angle of arrival (AOA), as an invariant sorting parameter in the deinterleaving of radar signals and in separating closely spaced communication emitters. In addition, the conservation of jamming power in power-managed ECM systems depends on the ability of the associated ESM system to measure the direction to the victim emitter. A function which is becoming increasingly important in defense suppression and weapon delivery systems involves locating the emitter's position passively. This can be accomplished from a single moving platform through successive measurements of the emitter's angular direction, or from multiple platforms which make simultaneous angular measurements.

The emitter identification function requires identifying and associating consecutive pulses produced by the same emitter in angle of arrival (AOA) and frequency. The AOA is a parameter which a hostile emitter cannot change on a pulse-to-pulse basis. However, to measure the AOA of pulses which overlap in the time domain first requires them to be separated in the frequency domain. The advanced ESM receivers which accomplish this function must operate over several octaves of bandwidth while providing RMS bearing accuracies on the order of at least 2 degrees with high POI and fast reaction time in dense signal environments.

There are basically three methods, depicted in Figure 3, which allow the passive location of stationary ground-based emitters from airborne platforms.

These are:

1. The azimuth triangulation method where the intersection of successive spatially displaced bearing measurements provides the emitter location.

2. The azimuth/elevation location technique, which provides a single-pulse instantaneous emitter location from the intersection of the measured azimuth/elevation line with the earth's surface.

3. The time difference of arrival (TDOA), or precision emitter location system (PELS) method, which measures the difference in time of arrival of a single pulse at three spatially remote locations.

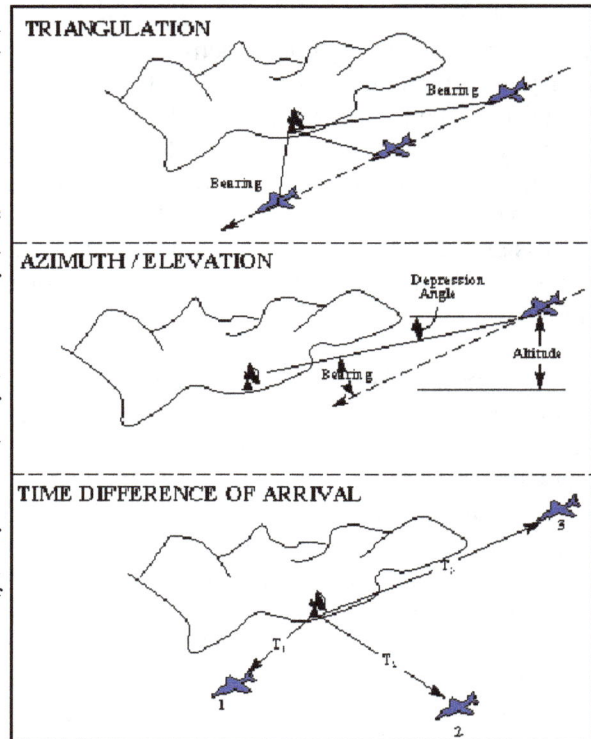

Figure 3. Passive Emitter Location Techniques

Additional methods include:

1. Phase rate of change, which is similar to triangulation, except it makes calculations using the phase derivative.

2. Angle distance techniques, where the distance from the emitter is derived from the signal strength (with known "threat" characteristics).

3. RF Doppler processing, which measures Doppler changes as the aircraft varies direction with respect to the "target" radar.

The relative advantages and disadvantages of each are given in Table 2.

Table 2. Emitter Location Techniques

Measurement Technique	Advantages	Disadvantages
Triangulation	Single Aircraft	Non-Instantaneous Location; Inadequate Accuracy for Remote Targeting; Not Forward Looking
Azimuth/Elevation	Single Aircraft Instantaneous Location Possible	Accuracy Degrades Rapidly at Low Altitude; Function of Range
Time Difference of Arrival (Pulsed Signals)	Very High Precision Can Support Weapon Delivery Position Requirements Very Rapid, Can Handle Short On-Time Threat	Very Complex, At Least 3 Aircraft; High Quality Receivers; DME (3 Sites); Very Wideband Data Link; Very High Performance Control Processor; Requires Very High Reliability Subsystems. Requires common time reference and correlation operation for non-pulse signals.

The triangulation method has the advantage of using a single aircraft, and its accuracy is greatest for a long baseline and the broadside geometry. The accuracy degenerates as the aircraft heading line approaches the boresight to the emitter.

The azimuth/elevation technique also has the advantage of using a single aircraft, but suffers from the difficultness of making an accurate elevation measurement with limited vertical aperture and in the presence of multipath effects.

The TDOA technique requires multiple aircraft and is complex, but has high potential accuracy. The determination of the location of the site involves the solution of at least two simultaneous second order equations for the intersection of two hyperbolas which represent $T_2 - T_1$ = Constant #1 and $T_3 - T_2$ = Constant #2. This method can be used to obtain a fix for an emitter which radiates only a single pulse.

ANGLE-OF-ARRIVAL (AOA) MEASUREMENTS

Several of the above DF measurements require AOA determination. Threat AOA measurements are also required to inform the aircrew in order to position the aircraft for optimal defense.

As shown in Figure 4, angle-of-arrival measuring systems fall into three main system categories of:
1. Scanning beam
2. Amplitude comparison or Simultaneous-multiple-beam
3. Phased Interferometer techniques

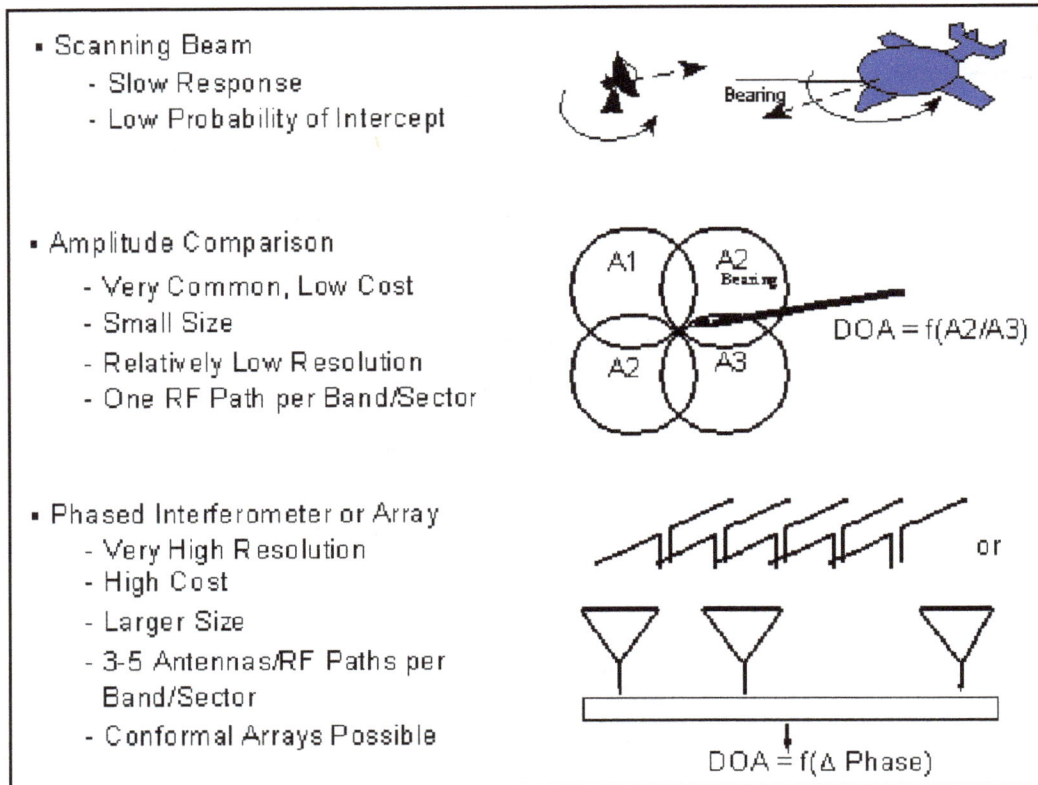

Figure 4. Angle-of-Arrival Measurement Techniques

Scanning Beam

The mechanically scanning beam, or "spinner," requires only a single receiver and also exhibits high sensitivity due to the use of a directive antenna. The disadvantage is that the "spinner" usually exhibits slow response because it must rotate through the coverage angle (e.g., 360 degrees) to ensure that it intercepts an emitter. Also, if the emitter uses a scanning directional antenna, both beams must point at each other for maximum sensitivity, which is a low probability occurrence. Both of these effects cause the mechanically scanning beam technique to have a low probability of intercept (POI).

Amplitude Comparison

The two primary techniques used for direction finding are the amplitude-comparison method and the interferometer or phase-comparison method. The phase-comparison method generally has the advantage of greater accuracy, but the amplitude-comparison method is used extensively due to its lower complexity and cost. Regardless of which technique is used, it should be emphasized that the ultimate rms angular accuracy is given by:

$$\Delta\theta = \frac{k\,\theta_B}{\sqrt{SNR}}$$
where θ_B is the antenna's angular beamwidth, or interferometer lobe width, and SNR is the signal-to-noise ratio.

Thus, phase interferometers that typically use very widebeam antennas require high signal-to-noise ratios to achieve accurate angle-of-arrival measurements. Alternately, a multi-element array antenna can be used to provide relatively narrow interferometer lobes, which require modest signal-to-noise ratios.

Virtually all currently deployed radar warning receiving (RWR) systems use amplitude-comparison direction finding (DF). A basic amplitude-comparison receiver derives a ratio, and ultimately angle-of-arrival or bearing, from a pair of independent receiving channels, which utilize squinted antenna elements that are usually equidistantly spaced to provide an instantaneous 360° coverage. Typically, four or six antenna elements and receiver channels are used in such systems, and wideband logarithmic video detectors provide the signals for bearing-angle determination. The monopulse ratio is obtained by subtraction of the detected logarithmic signals, and the bearing is computed from the value of the ratio.

Amplitude comparison RWRs typically use broadband cavity-backed spiral antenna elements whose patterns can be approximated by Gaussian-shaped beams. Gaussian-shaped beams have the property that the logarithmic output ratio slope in dB is linear as a function of angle of arrival. Thus, a digital look-up table can be used to determine the angle directly. However, both the antenna beamwidth and squint angle vary with frequency over the multi-octave bands used in RWRs. Pattern shape variations cause a larger pattern crossover loss for high frequencies and a reduced slope sensitivity at low frequencies. Partial compensation of these effects, including antenna squint, can be implemented using a look-up table if frequency information is available in the RWR. Otherwise, gross compensation can be made, depending upon the RF octave band utilized.

Typical accuracies can be expected to range from 3 to 10 degrees rms for multi-octave frequency band amplitude-comparison systems which cover 360 degrees with four to six antennas.

The four-quadrant amplitude-comparison DF systems employed in RWRs have the advantage of simplicity, reliability, and low cost. Usually, only one antenna per quadrant is employed which covers the 2 to 18 GHz band. The disadvantages are poor accuracy and sensitivity, which result from the broad-beam antennas employed. Both accuracy and sensitivity can be improved by expanding the number of antennas employed. For example, expanding to eight antennas would double the accuracy and provide 3 dB more gain. As the number of antennas increases, it becomes appropriate to consider multiple-beam-forming antennas rather than just increasing the number of individual antennas. The geometry of multiple-beam-forming antennas is such that a conformal installation aboard an aircraft is difficult. Therefore, this type of installation is typically found on naval vessels or ground vehicles where the space is available to accommodate this type of antenna.

Simultaneous-multiple-beam (amplitude comparison)

The simultaneous-multiple-beam system uses an antenna, or several antennas, forming a number of simultaneous beams (e.g., Butler matrix or Rotman lens), thereby retaining the high sensitivity of the scanning antenna approach while providing fast response. However, it requires many parallel receiving channels, each with full frequency coverage. This approach is compatible with amplitude-monopulse angular measuring techniques which are capable of providing high angular accuracy.

A typical example of a multiple-beam antenna is a 16-element circular array developed as part of a digital ESM receiver. This system covers the range from 2 to 18 GHz with two antenna arrays (2 to 7.5 GHz and 7.5 to 18 GHz), has a sensitivity of -55 to -60 dBm and provides an rms bearing accuracy of better than 1.7 degrees on pulsewidths down to 100 ns.

Phased Interferometer Techniques

The term interferometer generally refers to an array type antenna in which large element spacing occurs and grating lobes appear.

Phase interferometer DF systems are utilized when accurate angle-of-arrival information is required. They have the advantage of fast response, but require relatively complex microwave circuitry, which must maintain a precise phase match over a wide frequency band under extreme environmental conditions. When high accuracy is required (on the order of 0.1 to 1°), wide baseline interferometers are utilized with ambiguity resolving circuitry. The basic geometry is depicted in Figure 5, whereby a plane wave arriving at an angle is received by one antenna earlier than the other due to the difference in path length.

The time difference can be expressed as a phase difference:

$$\varphi = \omega\Delta\tau = 2\pi a(f/c) = 2\pi\,(d\sin\theta)/\lambda,$$

where θ is the angle of arrival,
d is the antenna separation, and
λ is the wavelength in compatible units.

The unambiguous field of view (FOV) is given by $\theta = 2\sin^{-1}(\pi/2d)$, which for $\lambda/2$ spacing results in 180° coverage. This spacing must be established for the highest frequency to be received.

Interferometer elements typically use broad antenna beams with beamwidths on the order of 90°. This lack of directivity produces several adverse effects. First, it limits system sensitivity due to the reduced antenna gain. Secondly, it opens the system to interference signals from within the antenna's broad angular coverage. The interference signals often include multipath from strong signals which can limit the accuracy of the interferometer.

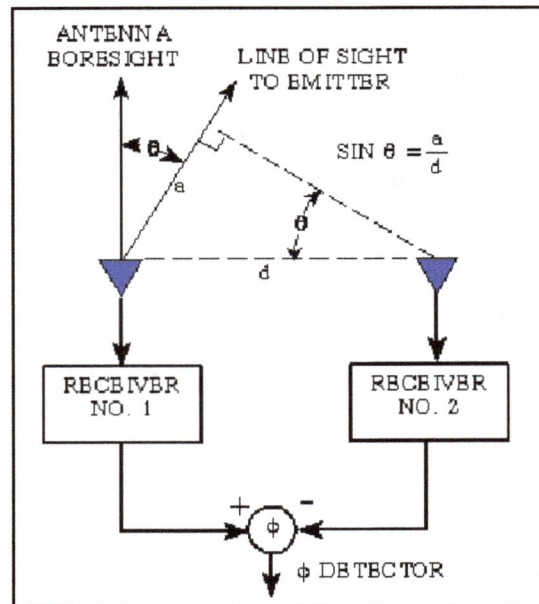

Figure 5. Phase Interferometer Principle

In an interferometer, the locus of points that produce the same time or phase delay forms a cone. The indicated angle is the true azimuth angle multiplied by the cosine of the elevation angle. The error in assuming the incident angle to be the azimuth angle is negligible for signals near the antenna's boresight. At 45° azimuth and 10° elevation, the error is less than 1°, increasing to 15° for both at 45°. Two orthogonal arrays, one measuring the azimuth angle and the other the elevation angle can eliminate this error. For targets near the horizon, the depression angle is small, thereby requiring only horizontal arrays.

The rms angular accuracy of an interferometer in radians is given by:

$$\sigma_\theta = \Delta\alpha\,/\,(\,\pi\bullet\sqrt{SNR}\,),\quad \text{where } \Delta\alpha = \lambda/(d\cdot\cos\theta) \text{ is the separation between adjacent nulls.}$$

For a two-element interferometer, the spacing (d) must be $\lambda/2$ or less to provide unambiguous, or single lobe ± 90°, coverage. This, in effect, sets a wide interferometer (or grating) lobe which must be split by a large factor to achieve high accuracy. This, in turn, imposes a requirement for high SNR to achieve the large beam-splitting factor. For example, if 0.1° accuracy is required from an unambiguous two-element interferometer, then a SNR of about 50 dB is required to achieve this accuracy. This may be difficult to achieve considering the inherently low sensitivity of an interferometer system.

When high accuracy is required from an interferometer system, it is usual to employ separations greater than $\lambda/2$. The increased separation sets up a multi-grating-lobe structure through the coverage angle which requires less SNR to achieve a specified accuracy. For example, a two-element interferometer with 16λ spacing would set up a 33-grating-lobe structure (including the central lobe) throughout the $\pm 90°$ coverage angle. Within each of the 33 grating lobes, it would only require a SNR on the order of 20 dB to achieve $0.1°$ accuracy. However, there would be 33 ambiguous regions within the $\pm 90°$ angular coverage and also 32 nulls (where the phase detector output is zero), about which the system would be insensitive to an input signal. The ambiguities could be resolved by employing a third antenna element with $\lambda/2$ spacing, which would provide an accuracy on the order of $3°$ with 20 dB SNR. This accuracy is sufficient to identify which of the 33 lobes contains the signal. Providing coverage in the null regions requires additional antenna elements.

Interferometers employing multiple antenna elements are called multiple-baseline interferometers. In a typical design, the receiver consists of a reference antenna and a series of companion antennas. The spacing between the reference element and the first companion antenna is $\lambda/2$; other secondary elements are placed to form pairs separated by 1, 2, 4, and 8 wavelengths. The initial AOA is measured unambiguously by the shortest-spaced antenna pair. The next greatest spaced pair has a phase rate of change which is twice that of the first, but the information is ambiguous due to there being twice as many lobes as in the preceding pair. A greater phase rate of change permits higher angular accuracy while the ambiguity is resolved by the previous pair. Thus, the described multiple-baseline interferometer provides a binary AOA measurement where each bit of the measurement supplies a more accurate estimate of the emitter's AOA.

Harmonic multiple-baseline interferometers use elements which are spaced at $2^n \cdot \lambda/2$, with $n = 0, 1, 2, 3$. In nonharmonic interferometers, no pair of antennas provides a completely unambiguous reading over the complete field of view. For example, the initial spacing in the nonharmonic interferometer might be λ, while the next companion element spacing is $3\lambda/2$. Ambiguities are resolved by truth tables, and hence the accuracy is set by the spacing of the widest baseline antenna pair. Nonharmonic interferometers have been implemented over 9:1 bandwidths (2 to 18 GHz) with rms accuracies from 0.1 to $1°$ and with no ambiguities over $\pm 90°$. The principal advantage of the nonharmonic over the harmonic interferometer is the increased bandwidth for unambiguous coverage.

Interferometer DF accuracy is determined by the widest baseline pair. Typical cavity-backed spirals, track to 6 electrical degrees, and associated receivers track to $9°$, resulting in an rms total of $11°$. At a typical 16 dB SNR, the rms phase noise is approximately 9 electrical degrees. For these errors and an emitter angle of $45°$, a spacing of 25λ is required for $0.1°$ rms accuracy while a spacing of 2.5λ is needed for $1°$ accuracy. For high accuracy, interferometer spacings of many feet are required. In airborne applications, this usually involves mounting interferometer antennas in the aircraft's wingtips.

The characteristics of typical airborne amplitude comparison and phase interferometer DF systems are summarized in Table 3. The phase interferometer system generally uses superheterodyne receivers which provide the necessary selectivity and sensitivity for precise phase measurements.

Table 3. Direction Of Arrival Measurement Techniques

	Amplitude Comparison	Phase Interferometer
Sensor Configuration	Typically 4 to 6 Equispaced Antenna Elements for 360° Coverage	2 or more RHC or LHC Spirals in Fixed Array
DF Accuracy	$DF_{ACC} \approx \dfrac{\theta_{BW}^2 \, \Delta C_{dB}}{24S}$ (Gaussian Shape)	$DF_{ACC} = \dfrac{\lambda}{2\pi d \cos\theta} \Delta\theta$
DF Accuracy Improvement	Decrease Antenna BW Decrease Amplitude Mistrack Increase Squint Angle	Increase Spacing of Outer Antennas; Decrease Phase Mistrack
Typical DF Accuracy	3° to 10° rms	0.1° to 3° rms
Sensitivity to Multipath/ Reflections	High Sensitivity Mistrack of Several dB Can Cause Large DF Errors	Relatively Insensitive; Interferometer Can Be Made to Tolerate Large Phase Errors
Platform Constraints	Locate in Reflection Free Area	Reflection Free Area; Real Estate For Array; Prefers Flat Radome
Applicable Receivers	Crystal Video; Channelizer; Acousto-Optic; Compressive; Superheterodyne	Superheterodyne

ΔC_{dB} = Amplitude Monopulse Ratio in dB
S = Squint Angle in degrees
θ_{BW} = Antenna Beamwidth in degrees

MICROWAVE / RF COMPONENTS

MICROWAVE WAVEGUIDES and COAXIAL CABLE

In general, a waveguide consists of a hollow metallic tube of arbitrary cross section uniform in extent in the direction of propagation. Common waveguide shapes are rectangular, circular, and ridged. The rectangular waveguide has a width a and height b as shown in figure 1. Commonly used rectangular waveguides have an aspect ratio b/a of approximately 0.5. Such an aspect ratio is used to preclude generation of field variations with height and their attendant unwanted modes. Waveguides are used principally at frequencies in the microwave range; inconveniently large guides would be required to transmit radio-frequency power at

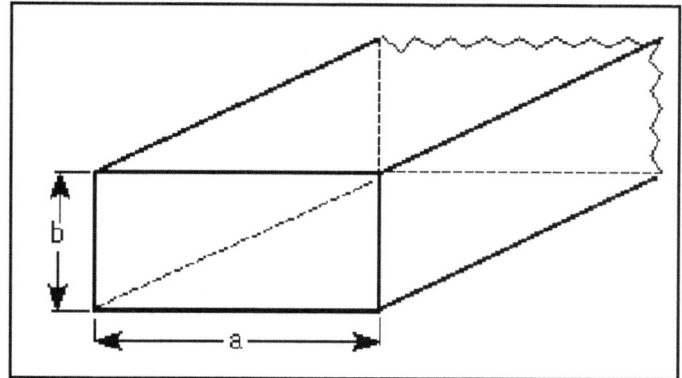

Figure 1. The Rectangular Waveguide

longer wavelengths. In the X-Band frequency range of 8.2 to 12.4 GHz, for example, the U.S. standard rectangular waveguide, WR-90, has an inner width of 2.286 cm (0.9 in.) and an inner height of 1.016 cm (0.4 in.).

In waveguides the electric and magnetic fields are confined to the space within the guides. Thus no power is lost to radiation. Since the guides are normally filled with air, dielectric losses are negligible. However, there is some I^2R power lost to heat in the walls of the guides, but this loss is usually very small.

It is possible to propagate several modes of electromagnetic waves within a waveguide. The physical dimensions of a waveguide determine the cutoff frequency for each mode. If the frequency of the impressed signal is <u>above</u> the cutoff frequency for a given mode, the electromagnetic energy can be transmitted through the guide for that particular mode with minimal attenuation. Otherwise the electromagnetic energy with a frequency <u>below</u> cutoff for that particular mode will be attenuated to a negligible value in a relatively short distance. This grammatical use of cutoff frequency is opposite that used for coaxial cable, where cutoff frequency is for the highest useable frequency. The dominant mode in a particular waveguide is the mode having the lowest cutoff frequency. For rectangular waveguide this is the TE_{10} mode. The TE (transverse electric) signifies that all electric fields are transverse to the direction of propagation and that no longitudinal electric field is present. There is a longitudinal component of magnetic field and for this reason the TE_{mn}

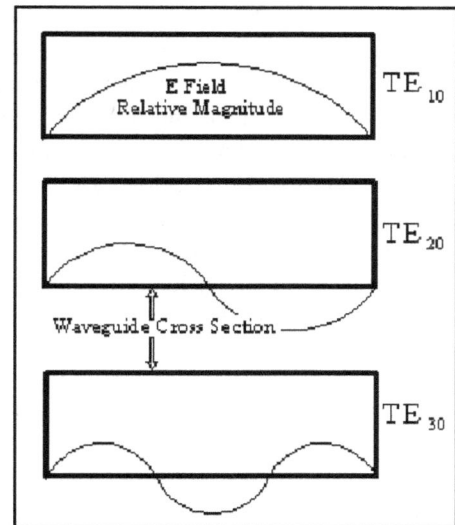

Figure 2. TE Modes

waves are also called H_{mn} waves. The TE designation is usually preferred. Figure 2 shows a graphical depiction of the E field variation in a waveguide for the TE_{10}, TE_{20}, and TE_{30} modes. As can be seen, the first index indicates the number of half wave loops across the width of the guide and the second index, the number of loops across the height of the guide - which in this case is zero. It is advisable to choose the dimensions of a guide in such a way that, for a given input signal, only the energy of the dominant mode can be transmitted through the guide. For example, if for a particular frequency, the width of a rectangular guide is too large, then the TE_{20} mode can propagate causing a myriad of problems. For rectangular guides of low aspect ratio the TE_{20} mode is the next higher order mode and is harmonically related to the cutoff frequency of the TE_{10} mode.

It is this relationship together with attenuation and propagation considerations that determine the normal operating range of rectangular waveguide.

The discussion on circular waveguides will not be included because they are rarely used in the EW area. Information regarding circular waveguides can be found in numerous textbooks on microwaves.

CHARACTERISTICS OF STANDARD RECTANGULAR WAVEGUIDES

Rectangular waveguides are commonly used for power transmission at microwave frequencies. Their physical dimensions are regulated by the frequency of the signal being transmitted. Table 1 tabulates the characteristics of the standard rectangular waveguides. It may be noted that the number following the EIA prefix "WR" is in inside dimension of the widest part of the waveguide, i.e. WR90 has an inner dimension of 0.90".

DOUBLE RIDGE RECTANGULAR WAVEGUIDE

Another type of waveguide commonly used in EW systems is the double ridge rectangular waveguide. The ridges in this waveguide increase the bandwidth of the guide at the expense of higher attenuation and lower power-handling capability. The bandwidth can easily exceed that of two contiguous standard waveguides. Introduction of the ridges mainly lowers the cutoff frequency of the TE_{10} mode from that of the unloaded guide, which is predicated on width alone. The reason for this can easily be explained when the field configuration in the guide at cutoff is investigated. At cutoff there is no longitudinal propagation down the guide. The waves simply travel back and forth between the side walls of the guide. In fact the guide can be viewed as a composite parallel plate waveguide of infinite width where the width corresponds to the direction of propagation of the normal guide. The TE_{10} mode cutoff occurs where this composite guide has its lowest-order resonant frequency. This occurs when there is only one E field maximum across the guide which occurs at the center for a symmetrical ridge. Because of the reduced height of the guide under the ridge, the effective TE_{10} mode resonator is heavily loaded as though a shunt capacitor were placed across it. The cutoff frequency is thus lowered considerably. For the TE_{20} mode the fields in the center of the guide will be at a minimum. Therefore the loading will have a negligible effect. For guides of proper aspect ratio, ridge height, and ridge width, an exact analysis shows that the TE_{10} mode cutoff can be lowered substantially at the same time the TE_{20} and TE_{30} mode cutoffs are raised slightly. Figure 3 shows a typical double ridged waveguide shape and Table 2 shows double ridged waveguide specifications. In the case of ridged waveguides, in the EIA designation, (WRD350 D36) the first "D" stands for double ridged ("S" for single ridged), the 350 is the starting frequency (3.5 GHz), and the "D36" indicates a bandwidth of 3.6:1. The physical dimensions and characteristics of a WRD350 D24 and WRD350 D36 are radically different. A waveguide with a MIL-W-23351 dash number beginning in 2 (i.e. 2-025) is a double ridge 3.6:1 bandwidth waveguide. Likewise a 1- is a single ridge 3.6:1, a 3- is a single ridge 2.4:1, and a 4- is a double ridge 2.4:1 waveguide.

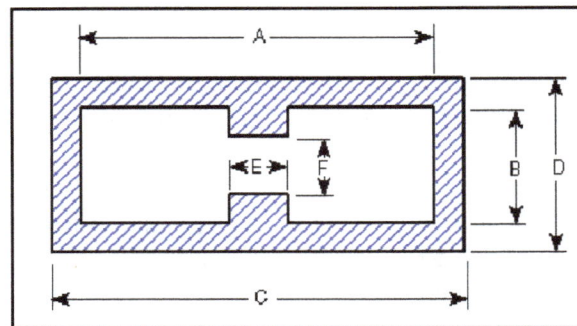

Figure 3. Double Ridge Waveguide

Figure 4 shows a comparison of the frequency /attenuation characteristics of various waveguides. The attenuation is based on real waveguides which is higher than the theoretical values listed in Tables 1 and 2. Figure 5 shows photographs of waveguides with some common connectors. Figure 6 shows attenuation characteristics of various RF coaxial cables.

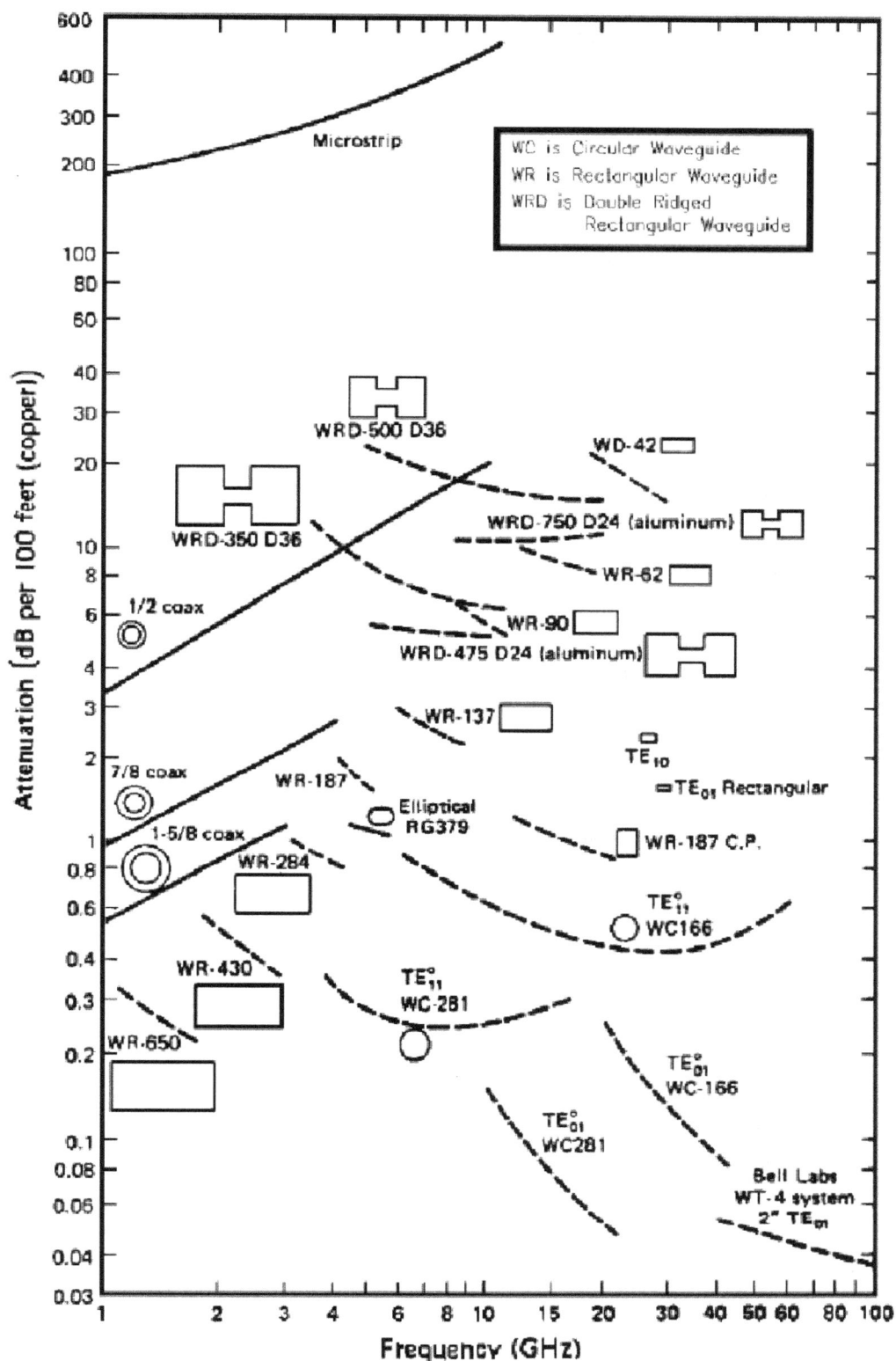

Figure 4. Attenuation vs Frequency for a Variety of Waveguides and Cables

Table 1. Rectangular Waveguide Specifications

Waveguide Size	JAN WG Desig	MIL-W-85 Dash #	Material	Freq Range (GHz)	Freq Cutoff (GHz)	Power (at 1 Atm)		Insertion Loss (dB/100ft)	Dimensions (Inches)	
						CW	Peak		Outside	Wall Thickness
WR284	RG48/U RG75/U	1-039 1-042	Copper Aluminum	2.60 - 3.95	2.08	45 36	7650	.742-.508 1.116-.764	3.000x1.500	0.08
WR229	RG340/U RG341/U	1-045 1-048	Copper Aluminum	3.30 - 4.90	2.577	30 24	5480	.946-.671 1.422-1.009	2.418x1.273	0.064
WR187	RG49/U RG95/U	1-051 1-054	Copper Aluminum	3.95 - 5.85	3.156	18 14.5	3300	1.395-.967 2.097-1.454	1.000x1.000	0.064
WR159	RG343/U RG344/U	1-057 1-060	Copper Aluminum	4.90 - 7.05	3.705	15 12	2790	1.533-1.160 2.334-1.744	1.718x0.923	0.064
WR137	RG50/U RG106/U	1-063 1-066	Copper Aluminum	5.85 - 8.20	4.285	10 8	1980	1.987-1.562 2.955-2.348	1.500x0.750	0.064
WR112	RG51/U RG68/U	1-069 1-072	Copper Aluminum	7.05 - 10.0	5.26	6 4.8	1280	2.776-2.154 4.173-3.238	1.250x0.625	0.064
WR90	RG52/U RG67/U	1-075 1-078	Copper Aluminum	8.2 - 12.4	6.56	3 2.4	760	4.238-2.995 6.506-4.502	1.000x0.500	0.05
WR75	RG346/U RG347/U	1-081 1-084	Copper Aluminum	10.0 - 15.0	7.847	2.8 2.2	620	5.121-3.577 7.698-5.377	0.850x0.475	0.05
WR62	RG91/U RG349/U	1-087 1-091	Copper Aluminum	12.4 - 18.0	9.49	1.8 1.4	460	6.451-4.743 9.700-7.131	0.702x0.391	0.04
WR51	RG352/U RG351/U	1-094 1-098	Copper Aluminum	15.0 - 22.0	11.54	1.2 1	310	8.812-6.384 13.250-9.598	0.590x0.335	0.04
WR42	RG53/U	1-100	Copper	18.0 - 26.5	14.08	0.8	170	13.80-10.13	0.500x0.250	0.04
WR34	RG354/U	1-107	Copper	2.0 - 33.0	17.28	0.6	140	16.86-11.73	0.420x0.250	0.04
WR28	RG271/U	3-007	Copper	26.5 - 40.0	21.1	0.5	100	23.02-15.77	0.360x0.220	0.04

Standard waveguide - 7 mm

Double ridge waveguide - SMA jack

Figure 5. Waveguides With Some Common Connections

Table 2. Double Ridge Rectangular Waveguide Specifications

Waveguide Size	MIL-W-23351 Dash #	Material	Freq Range (GHz)	Freq Cutoff (GHz)	Power (at 1 Atm) CW	Power (at 1 Atm) Peak	Insertion Loss (dB/ft)	Dimensions (Inches) A	B	C	D	E	F
WRD250		Alum Brass Copper Silver Al	2.60 - 7.80	2.093	24	120	0.025 0.025 0.018 0.019	1.655	0.715	2	1	0.44	0.15
WRD350 D24	4-029 4-303 4-031	Alum Brass Copper	3.50 - 8.20	2.915	18	150	0.0307 0.0303 0.0204	1.48	0.688	1.608	0.816	0.37	0.292
WRD475 D24	4-033 4-034 4-035	Alum Brass Copper	4.75 - 11.00	3.961	8	85	0.0487 0.0481 0.0324	1.09	0.506	1.19	0.606	0.272	0.215
WRD500 D36	2-025 2-026 2-027	Alum Brass Copper	5.00 - 18.00	4.222	4	15	0.146 0.141 0.095	0.752	0.323	0.852	0.423	0.188	0.063
WRD650		Alum Brass Copper	6.50 - 18.00	5.348	4	25	0.106 0.105 0.07	0.720	0.321	0.820	0.421	0.173	0.101
WRD750 D24	4-037 4-038 4-039	Alum Brass Copper	7.50 - 18.00	6.239	4.8	35	0.0964 0.0951 0.0641	0.691	0.321	0.791	0.421	0.173	0.136
WRD110 D24	4-041 4-042 4-043	Alum Brass Copper	11.00 - 26.50	9.363	1.4	15	0.171 0.169 0.144	0.471	0.219	0.551	0.299	0.118	0.093
WRD180 D24	4-045 4-046 4-047	Alum Brass Copper	18.00 - 40.00	14.995	0.8	5	0.358 0.353 0.238	0.288	0.134	0.368	0.214	0.072	0.057

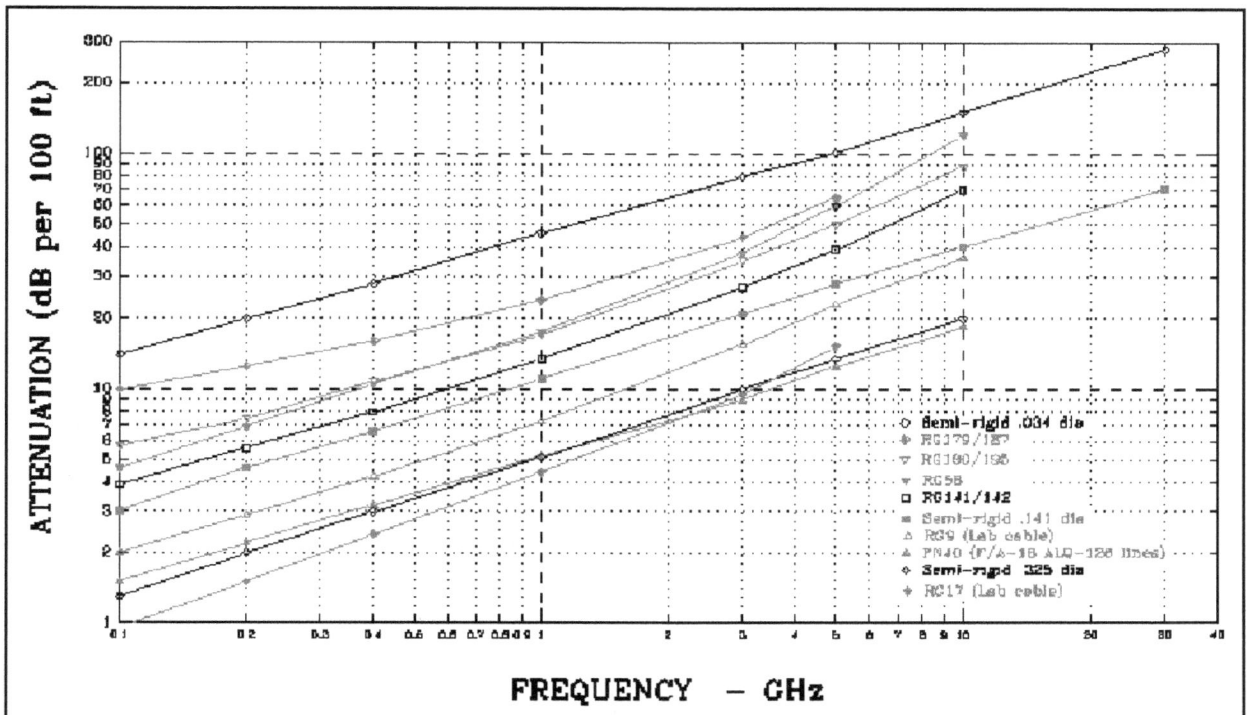

Figure 6. Attenuation vs Frequency for a Variety of Coaxial Cables

This Page Blank

VOLTAGE STANDING WAVE RATIO (VSWR) / REFLECTION COEFFICIENT
RETURN LOSS / MISMATCH LOSS

When a transmission line is terminated with an impedance, Z_L, that is not equal to the characteristic impedance of the transmission line, Z_O, not all of the incident power is absorbed by the termination. Part of the power is reflected back so that phase addition and subtraction of the incident and reflected waves creates a voltage standing wave pattern on the transmission line. The ratio of the maximum to minimum voltage is known as the Voltage Standing Wave Ratio (VSWR) and successive maxima and minima are spaced by 180° ($\lambda/2$).

$$VSWR = \frac{E_{max}}{E_{min}} = \frac{E_i + E_r}{E_i - E_r}$$

where
E_{max} = maximum voltage on the standing wave
E_{min} = minimum voltage on the standing wave
E_i = incident voltage wave amplitude
E_r = reflected voltage wave amplitude

The reflection coefficient, ρ, is defined as E_r/E_i and in general, the termination is complex in value, so that ρ will be a complex number.

Additionally we define: $\Gamma = \dfrac{Z_L - Z_O}{Z_L + Z_O}$ The refection coefficient, ρ, is the absolute value of the magnitude of Γ. If the equation for VSWR is solved for the reflection coefficient, it is found that:

$$\frac{Reflection}{Coefficient} = \rho = |\Gamma| = \frac{VSWR - 1}{VSWR + 1} \quad \text{Consequently, } VSWR = \frac{1 + \rho}{1 - \rho}$$

The return loss is related through the following equations:

$$\frac{Return}{Loss} = 10 \log\left[\frac{P_i}{P_r}\right] = -20 \log\left[\frac{E_r}{E_i}\right] = -20 \log\left[\frac{VSWR - 1}{VSWR + 1}\right] = -20 \log \rho$$

Return loss is a measure in dB of the ratio of power in the incident wave to that in the reflected wave, and as defined above always has a positive value. For example if a load has a Return Loss of 10 dB, then 1/10 of the incident power is reflected. The higher the return loss, the less power is actually lost.

Also of considerable interest is the Mismatch Loss. This is a measure of how much the transmitted power is attenuated due to reflection. It is given by the following equation:

Mismatch Loss = $-10 \log (1 - \rho^2)$

For example, an antenna with a VSWR of 2:1 would have a reflection coefficient of 0.333, a mismatch loss of 0.51 dB, and a return loss of 9.54 dB (11% of your transmitter power is reflected back). In some systems this is not a trivial amount and points to the need for components with low VSWR.

VSWR	Return Loss (dB)	% Power / Voltage Loss	Reflection Coefficient	Mismatch Loss (dB)
1	∞	0 / 0	0	0.000
1.15	23.1	0.49 / 7.0	0.07	.021
1.25	19.1	1.2 / 11.1	0.111	.054
1.5	14.0	4.0 / 20.0	0.200	.177
1.75	11.3	7.4 / 27.3	0.273	.336
1.9	10.0	9.6 / 31.6	0.316	.458
2.0	9.5	11.1 / 33.3	0.333	.512
2.5	7.4	18.2 / 42.9	0.429	.880
3.0	6.0	25.1 / 50.0	0.500	1.25
3.5	5.1	30.9 / 55.5	0.555	1.6
4.0	4.4	36.3 / 60.0	0.600	1.94
4.5	3.9	40.7 / 63.6	0.636	2.25
5.0	3.5	44.7 / 66.6	0.666	2.55
10	1.7	67.6 / 81.8	0.818	4.81
20	0.87	81.9 / 90.5	0.905	7.4
100	0.17	96.2 / 98.0	0.980	14.1
∞	.000	100 / 100	1.00	∞

* Divide % Voltage loss by 100 to obtain ρ
(reflection coefficient)

If 1000 watts (60 dBm/30 dBW) is applied to this antenna, the return loss would be 9.54 dB. Therefore, 111.1 watts would be reflected and 888.9 watts (59.488 dBm/29.488 dBW) would be transmitted, so the mismatch loss would be 0.512 dB.

Transmission line attenuation improves the VSWR of a load or antenna. For example, a transmitting antenna with a VSWR of 10:1 (poor) and a line loss of 6 dB would measure 1.5:1 (okay) if measured at the transmitter. Figure 1 shows this effect.

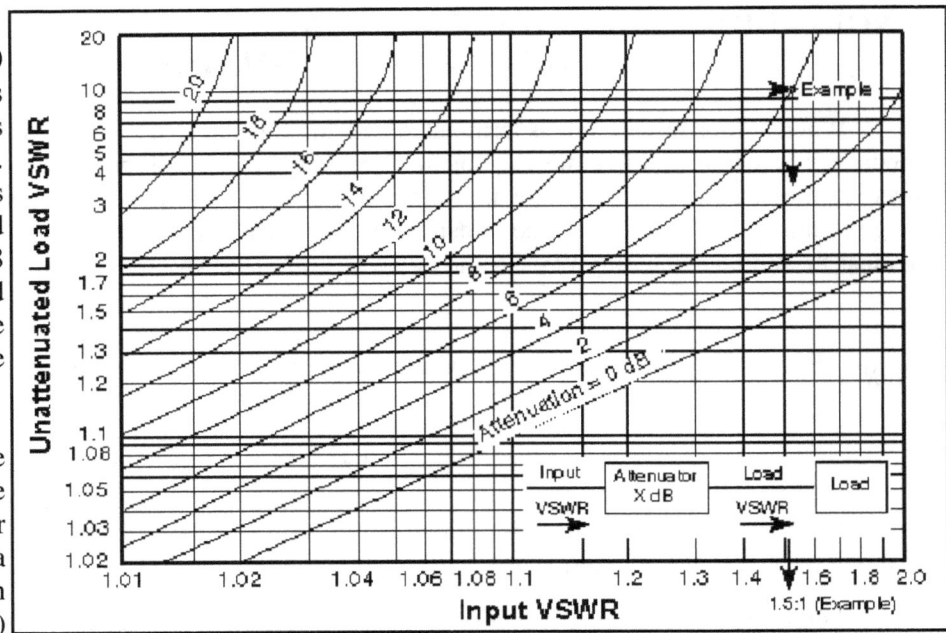

Figure 1. Reduction of VSWR by Attenuation

Therefore, if you are interested in determining the performance of antennas, the VSWR should always be measured at the antenna connector itself rather than at the output of the transmitter. Transmit cabling will load the line and create an illusion of having a better antenna VSWR. Transmission lines should have their insertion loss (attenuation) measured in lieu of VSWR, but VSWR measurements of transmission lines are still important because connection problems usually show up as VSWR spikes.

Historically VSWR was measured by probing the transmission line. From the ratio of the maximum to minimum voltage, the reflection coefficient and terminating impedance could be calculated. This was a time consuming process since the measurement was at a single frequency and mechanical adjustments had to be made to minimize coupling into circuits. Problems with detector characteristics also made the process less accurate. The modern network analyzer system sweeps very large frequency bandwidths and measures the incident power, P_i, and the reflected power, P_r. Because of the considerable computing power in the network analyzer, the return loss is calculated from the equation given previously, and displayed in real time. Optionally, the VSWR can also be calculated from the return loss and displayed real time.

If a filter is needed on the output of a jammer, it is desirable to place it approximately half way between the jammer and antenna. This may allow the use of a less expensive filter, or a reflective filter vs an absorptive filter.

Special cases exist when comparing open and shorted circuits. These two conditions result in the same ∞ VSWR and zero dB return loss even though there is a 180° phase difference between the reflection coefficients. These two conditions are used to calibrate a network analyzer.

MICROWAVE COAXIAL CONNECTORS

For high-frequency operation, the average circumference of a coaxial cable must be limited to about one wavelength in order to reduce multimodal propagation and eliminate erratic reflection coefficients, power losses, and signal distortion.

Except for the sexless APC-7 connector, all other connectors are identified as either male (plugs) which have a center conductor that is a probe or female (jacks) which have a center conductor that is a receptacle. Sometimes it is hard to distinguish them as some female jacks may have a hollow center "pin" which appears to be male, yet accepts a smaller male contact.

An adapter is an approximately zero loss interface between two connectors and is called a barrel when both connectors are identical. A number of common of coaxial connectors are described below, however other special purpose connectors exist, including blind mate connectors where spring fingers are used in place of threads to obtain shielding (desired connector shielding should be at least 90 dB).

APC-2.4 (2.4mm) - The 50 Ω APC-2.4 (Amphenol Precision Connector-2.4 mm) is also known as an OS-50 connector. It was designed to operate at extremely high microwave frequencies (up to 50 GHz).	
APC-3.5 (3.5mm) - The APC-3.5 connector provides repeatable connections and has a very low VSWR. Either the male or female end of this 50 Ω connector can mate with the opposite type of SMA connector. The APC-3.5 connector can work at frequencies up to 34 GHz.	2.4 mm jack - 3.5 mm jack
APC-7 (7mm) - The APC-7 was developed by HP, but has been improved and is now manufactured by Amphenol. The connector provides a coupling mechanism without male or female distinction and is the most repeatable connecting device used for very accurate 50 Ω measurement applications. Its VSWR is extremely low up to 18 GHz. Other companies have 7mm series available.	7mm - 3.5mm plug
BNC (OSB) - The BNC (Bayonet Neill-Concelman or Bayonet Navy Connector) was originally designed for military system applications during World War II. The connector operates best at frequencies up to about 4 GHz; beyond that it tends to radiate electromagnetic energy. The BNC can accept flexible cables with diameters of up to 6.35 mm (0.25 in.) and characteristic impedance of 50 to 75 Ω. It is now the most commonly used connector for frequencies under 1 GHz. Other names the BNC has picked up over the years include: "Baby Neill-Concelman", "Baby N connector", "British Naval Connector", and "Bayonet Nut Connector"	BNC male plug (above) BNC female jack

	C Plug (top)
C - The C (Concelman) coaxial connector is a medium size, older type constant 50 Ω impedance. It has a bayonet (twist and lock) connection. It is larger than the BNC, but about the same as Type N. It has a frequency range of 0-11 GHz	
SC (OSSC) - The SC type connector is a screw version of the "C" connector.	
	C jack (bottom)
SMA (OSM/3mm) - The SMA (Sub-Miniature A) connector was originally designed by Bendix Scintilla Corporation, but it has been manufactured by the Omni-Spectra division of M/ACOM (as the OSM connector) and many other electronic companies. It is a 50 Ω threaded connector. The main application of SMA connectors is on components for microwave systems. The connector normally has a frequency range to 18 GHz, but high performance varieties can be used to 26.5 GHz.	SMA Plug (above)
	SMA Jack
SSMA (OSSM) - The SSMA is a microminiature version of the SMA. It is also 50 Ω and operates to 26.5 GHz with flexible cable or 40 GHz with semi-rigid cable.	
	SSMA jack - BNC jack

SMC (OSMC) - The SMC (Sub-Miniature C) is a 50 Ω or 75 Ω connector that is smaller than the SMA. The connector can accept flexible cables with diameters of up to 3.17 mm (0.125 in.) for a frequency range of up to 7-10 GHz.	SMC Plug - SMA Jack (Some call this a SMC Jack, even though it has a female connector)
SMB (OSMB) - The SMB is like the SMC except it uses quick disconnect instead of threaded fittings. It is a 50 / 75 Ω connector which operates to 4 GHz with a low reflection coefficient and is useable to 10 GHz.	SMB Plug (above) SMB Jack
TNC (OST) - The TNC (Threaded Neill-Concelman or threaded Navy Connector) is merely a threaded BNC. The function of the thread is to stop radiation at higher frequencies, so that the connector can work at frequencies up to 12 GHz (to 18 GHz when using semi-rigid cable). It can be 50 or 75 Ω. For size comparison here is a SMA plug - TNC plug	TNC Plug (above) TNC Jack

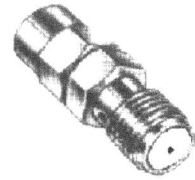

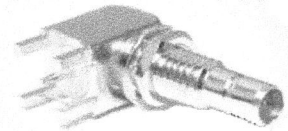

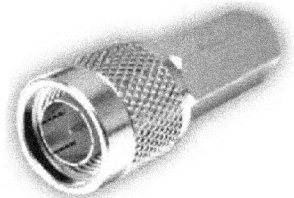

Type N (OSN) - The 50 or 75 Ω Type N (Navy) connector was originally designed for military systems during World War II and is the most popular measurement connector for the frequency range of 1 to 11 GHz. The precision 50 Ω APC-N and other manufacturers high frequency versions operate to 18 GHz.

For size comparison here is a picture of Type N plug - TNC jack

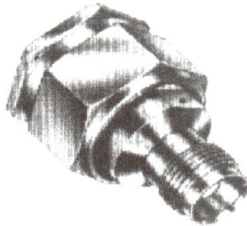

N Plug (above)

N Jack

Note: Always rotate the movable coupling nut of the plug, not the cable or fixed connector, when mating connectors. Since the center pin is stationary with respect to the jack, rotating the jack puts torque on the center pin. With TNC and smaller connectors, the center pin will eventually break off.

An approximate size comparison of these connectors is depicted below (not to scale).

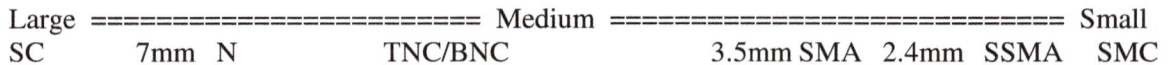

Large ========================= Medium ============================= Small
SC   7mm N   TNC/BNC   3.5mm SMA 2.4mm SSMA   SMC

Figure 1 shows the frequency range of several connectors.

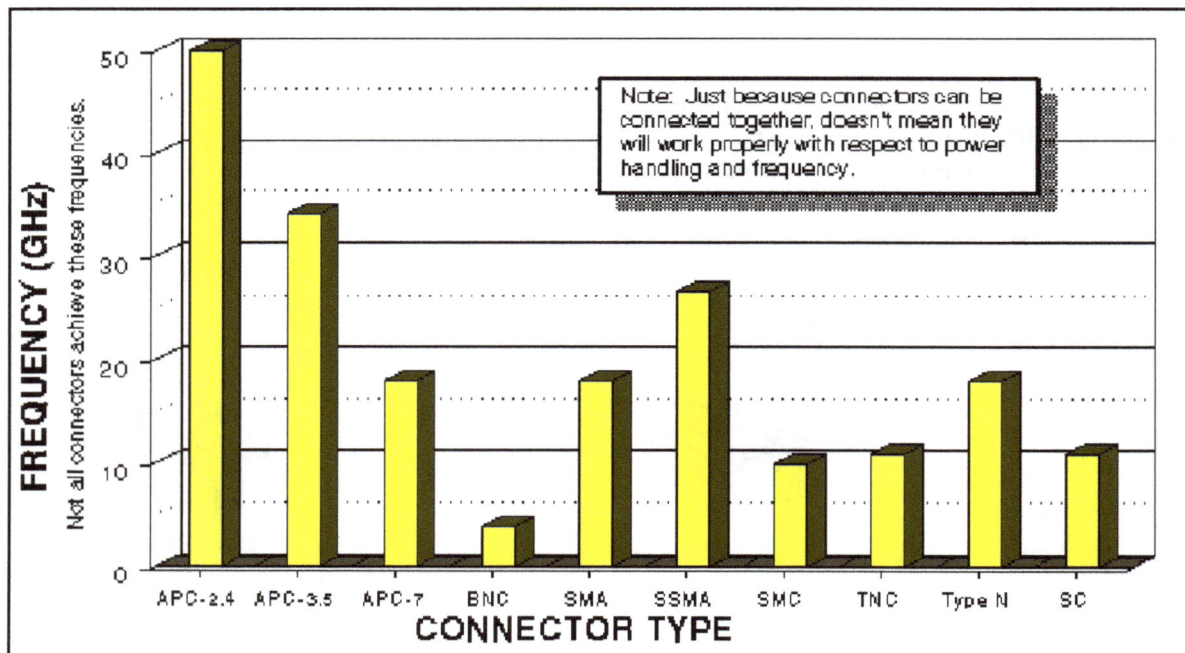

Figure 1. Frequency Range of Microwave Connectors

POWER DIVIDERS AND DIRECTIONAL COUPLERS

A directional coupler is a passive device which couples part of the transmission power by a known amount out through another port, often by using two transmission lines set close enough together such that energy passing through one is coupled to the other. As shown in Figure 1, the device has four ports: input, transmitted, coupled, and isolated. The term "main line" refers to the section between ports 1 and 2. On some

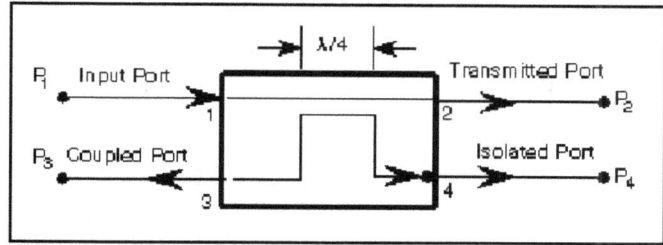

Figure 1. Directional Coupler

directional couplers, the main line is designed for high power operation (large connectors), while the coupled port may use a small SMA connector. Often the isolated port is terminated with an internal or external matched load (typically 50 ohms). It should be pointed out that since the directional coupler is a linear device, the notations on Figure 1 are arbitrary. Any port can be the input, (as in Figure 3) which will result in the directly connected port being the transmitted port, adjacent port being the coupled port, and the diagonal port being the isolated port.

Physical considerations such as internal load on the isolated port will limit port operation. The coupled output from the directional coupler can be used to obtain the information (i.e., frequency and power level) on the signal without interrupting the main power flow in the system (except for a power reduction - see Figure 2). When the power coupled out to port three is half the input power (i.e. 3 dB below the input power level), the power on the main transmission line is also 3 dB below the input power and equals the coupled power. Such a coupler is referred to as a 90 degree hybrid, hybrid, or 3 dB coupler. The frequency range for coaxial couplers specified by manufacturers is that of the coupling arm. The main arm response is much wider (i.e. if the spec is 2-4 GHz, the main arm could operate at 1 or 5 GHz - see Figure 3). However it should be recognized that the coupled response is periodic with frequency. For example, a $\lambda/4$ coupled line coupler will have responses at $n\lambda/4$ where n is an odd integer.

Common properties desired for all directional couplers are wide operational bandwidth, high directivity, and a good impedance match at all ports when the other ports are terminated in matched loads. These performance characteristics of hybrid or non-hybrid directional couplers are self-explanatory. Some other general characteristics will be discussed below.

COUPLING FACTOR

The coupling factor is defined as: $Coupling\ factor\ (dB) = -10 \log \dfrac{P_3}{P_1}$

where P_1 is the input power at port 1 and P_3 is the output power from the coupled port (see Figure 1).

The coupling factor represents the primary property of a directional coupler. Coupling is not constant, but varies with frequency. While different designs may reduce the variance, a perfectly flat coupler theoretically cannot be built. Directional couplers are specified in terms of the coupling accuracy at the frequency band center. For example, a 10 dB coupling ± 0.5 dB means that the directional coupler can have 9.5 dB to 10.5 dB coupling at the frequency band center. The accuracy is due to dimensional tolerances that can be held for the spacing of the two coupled lines. Another coupling specification is frequency sensitivity. A larger frequency sensitivity will allow a larger frequency band of operation. Multiple quarter-wavelength coupling sections are used to obtain wide frequency bandwidth directional couplers. Typically this type of directional coupler is designed to a frequency bandwidth ratio and a maximum coupling ripple within the frequency band. For example a typical 2:1 frequency bandwidth coupler design that produces a 10 dB coupling with a ±0.1 dB ripple would, using the previous accuracy specification, be said to have 9.6 ± 0.1 dB to 10.4 ± 0.1 dB of coupling across the frequency range.

LOSS

In an ideal directional coupler, the main line loss port 1 to port 2 ($P_1 - P_2$) due to power coupled to the coupled output port is:

$$Insertion\ loss\ (dB) = 10 \log \left[1 - \frac{P_3}{P_1} \right]$$

Coupling dB	Insertion Loss - dB
3	3.00
6	1.25
10	0.458
20	0.0436
30	0.00435

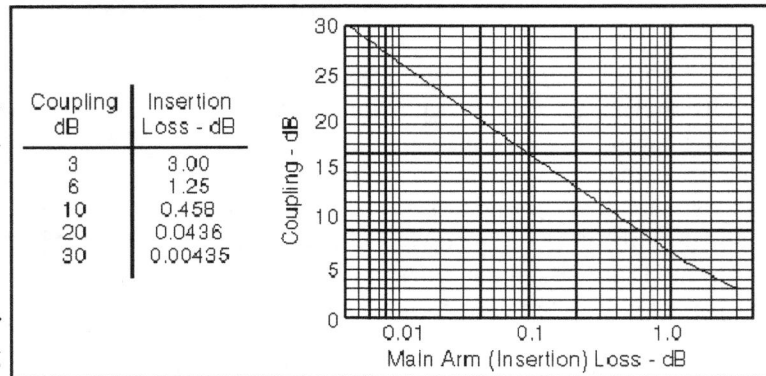

Figure 2. Coupling Insertion Loss

The actual directional coupler loss will be a combination of coupling loss, dielectric loss, conductor loss, and VSWR loss. Depending on the frequency range, coupling loss becomes less significant above 15 dB coupling where the other losses constitute the majority of the total loss. A graph of the theoretical insertion loss (dB) vs coupling (dB) for a dissipationless coupler is shown in Figure 2.

ISOLATION

Isolation of a directional coupler can be defined as the difference in signal levels in dB between the input port and the isolated port when the two output ports are terminated by matched loads, or:

$$Isolation\ (dB) = -10 \log \frac{P_4}{P_1}$$

Isolation can also be defined between the two output ports. In this case, one of the output ports is used as the input; the other is considered the output port while the other two ports (input and isolated) are terminated by matched loads.

Consequently: $Isolation\ (dB) = -10 \log \dfrac{P_3}{P_2}$

The isolation between the input and the isolated ports may be different from the isolation between the two output ports. For example, the isolation between ports 1 and 4 can be 30 dB while the isolation between ports 2 and 3 can be a different value such as 25 dB. If both isolation measurements are not available, they can assumed to be equal. If neither are available, an estimate of the isolation is the coupling plus return loss (see VSWR section). The isolation should be as high as possible. In actual couplers the isolated port is never completely isolated. Some RF power will always be present. Waveguide directional couplers will have the best isolation.

If isolation is high, directional couplers are excellent for combining signals to feed a single line to a receiver for two-tone receiver tests. In Figure 3, one signal enters port P_3 and one enters port P_2, while both exit port P_1. The signal from port P_3 to port P_1 will experience 10 dB of loss, and the signal from port P_2 to port P_1 will have 0.5 dB loss. The internal load on the isolated port will dissipate the signal losses from port P_3 and port P_2. If the isolators in Figure 3 are neglected, the isolation measurement (port P_2 to port P_3) determines the amount of power from the signal

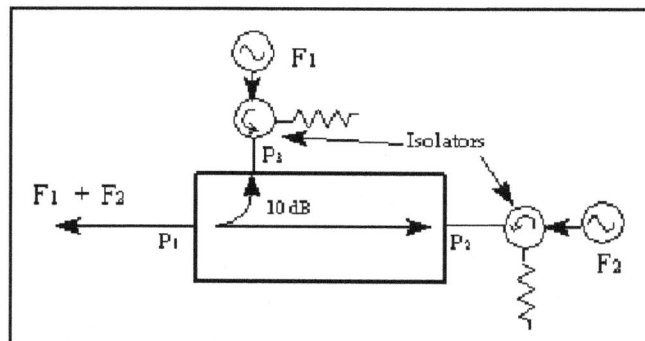

Figure 3. Two-Tone Receiver Tests

generator F_2 that will be injected into the signal generator F_1. As the injection level increases, it may cause modulation of signal generator F_1, or even injection phase locking. Because of the symmetry of the directional coupler, the reverse injection will happen with the same possible modulation problems of signal generator F_2 by F_1. Therefore the isolators are used in Figure 3 to effectively increase the isolation (or directivity) of the directional coupler. Consequently the injection loss will be the isolation of the directional coupler plus the reverse isolation of the isolator.

DIRECTIVITY

Directivity is directly related to Isolation. It is defined as:

$$Directivity\ (dB) = -10 \log \frac{P_4}{P_3} = -10 \log \frac{P_4}{P_1} + 10 \log \frac{P_3}{P_1}$$

where: P_3 is the output power from the coupled port and P_4 is the power output from the isolated port. The directivity should be as high as possible. Waveguide directional couplers will have the best directivity. Directivity is not directly measurable, and is calculated from the isolation and coupling measurements as:

Directivity (dB) = Isolation (dB) - Coupling (dB)

HYBRIDS

The hybrid coupler, or 3 dB directional coupler, in which the two outputs are of equal amplitude takes many forms. Not too long ago the quadrature (90 degree) 3 dB coupler with outputs 90 degrees out of phase was what came to mind when a hybrid coupler was mentioned. Now any matched 4-port with isolated arms and equal power division is called a hybrid or hybrid coupler. Today the characterizing feature is the phase difference of the outputs. If 90 degrees, it is a 90 degree hybrid. If 180 degrees, it is a 180 degree hybrid. Even the Wilkinson power divider which has 0 degrees phase difference is actually a hybrid although the fourth arm is normally imbedded.

Applications of the hybrid include monopulse comparators, mixers, power combiners, dividers, modulators, and phased array radar antenna systems.

AMPLITUDE BALANCE

This terminology defines the power difference in dB between the two output ports of a 3 dB hybrid. In an ideal hybrid circuit, the difference should be 0 dB. However, in a practical device the amplitude balance is frequency dependent and departs from the ideal 0 dB difference.

PHASE BALANCE

The phase difference between the two output ports of a hybrid coupler should be 0, 90, or 180 degrees depending on the type used. However, like amplitude balance, the phase difference is sensitive to the input frequency and typically will vary a few degrees.

The phase properties of a 90 degree hybrid coupler can be used to great advantage in microwave circuits. For example in a balanced microwave amplifier the two input stages are fed through a hybrid coupler. The FET device normally has a very poor match and reflects much of the incident energy. However, since the devices are essentially identical the reflection coefficients from each device are equal. The reflected voltage from the FETs are in phase at the isolated port and are 180° different at the input port. Therefore, all of the reflected power from the FETs goes to the load at the isolated port and no power goes to the input port. This results in a good input match (low VSWR).

If phase matched lines are used for an antenna input to a 180° hybrid coupler as shown in Figure 4, a null will occur directly between the antennas. If you want to receive a signal in that position, you would have to either change the hybrid type or line length. If you want to reject a signal from a given direction, or create the difference pattern for a monopulse radar, this is a good approach.

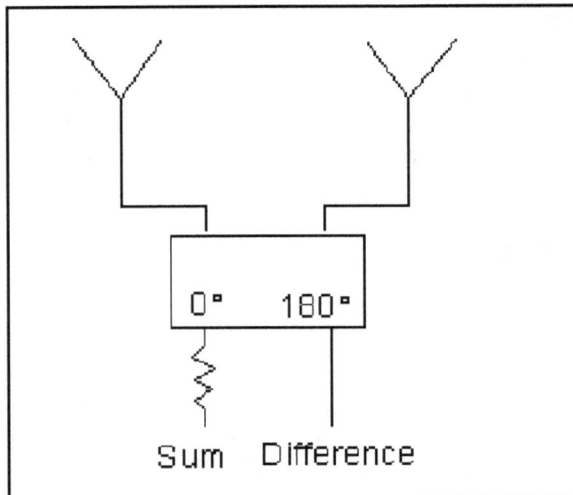

Figure 4. Balanced Antenna Input

OTHER POWER DIVIDERS

Both in-phase (Wilkinson) and quadrature (90°) hybrid couplers may be used for coherent power divider applications. The Wilkinson's power divider has low VSWR at all ports and high isolation between output ports. The input and output impedances at each port is designed to be equal to the characteristic impedance of the microwave system. A typical power divider is shown in Figure 5. Ideally, input power would be divided equally between the output ports. Dividers are made up of multiple couplers, and like couplers, may be reversed and used as multiplexers. The drawback is that for a four channel multiplexer, the output consists of only 1/4 the power from each, and is relatively inefficient. Lossless multiplexing can only be done with filter networks.

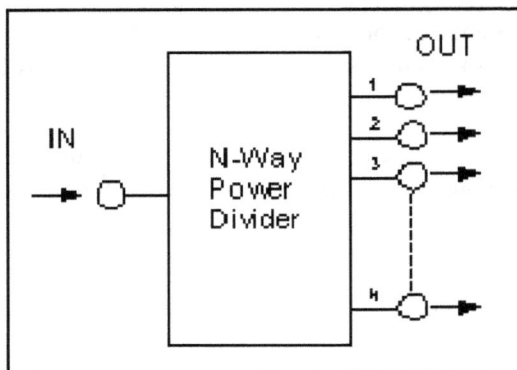

Figure 5. Power Divider

Coherent power division was first accomplished by means of simple Tee junctions. At microwave frequencies, waveguide tees have two possible forms - the H-Plane or the E-Plane. These two junctions split power equally, but because of the different field configurations at the junction, the electric fields at the output arms are in-phase for the H-Plane tee and are anti-phase for the E-Plane tee. The combination of these two tees to form a hybrid tee allowed the realization of a four-port component which could perform the vector sum (Σ) and difference (Δ) of two coherent microwave signals. This device is known as the magic tee.

POWER COMBINERS

Since hybrid circuits are bi-directional, they can be used to split up a signal to feed multiple low power amplifiers, then recombine to feed a single antenna with high power as shown in Figure 6. This approach allows the use of numerous less expensive and lower power amplifiers in the circuitry instead of a single high power TWT. Yet another approach is to have each solid state amplifier (SSA) feed an antenna and let the power be combined in space or be used to feed a lens which is attached to an antenna. (See Section 3-4)

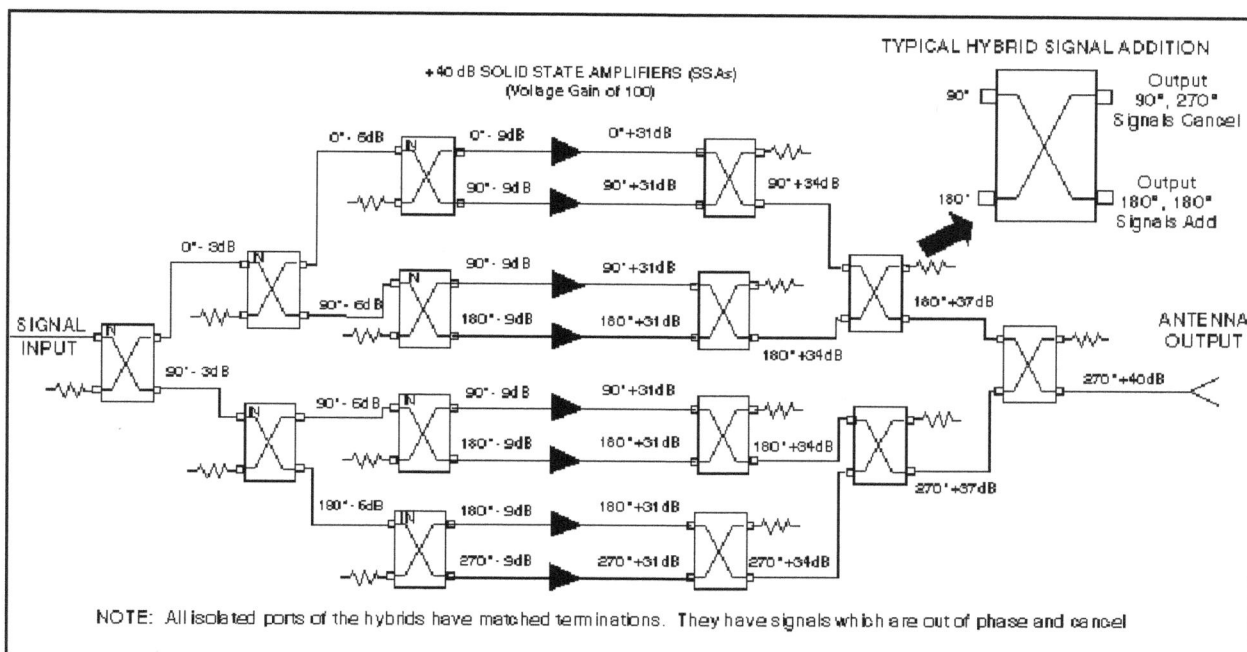

Figure 6. Combiner Network

Sample Problem:

If two 1 watt peak unmodulated RF carrier signals at 10 GHz are received, how much peak power could one measure?

A. 0 watts

B. 0.5 watts

C. 1 watt

D. 2 watts

E. All of these

The answer is all of these as shown in Figure 7.

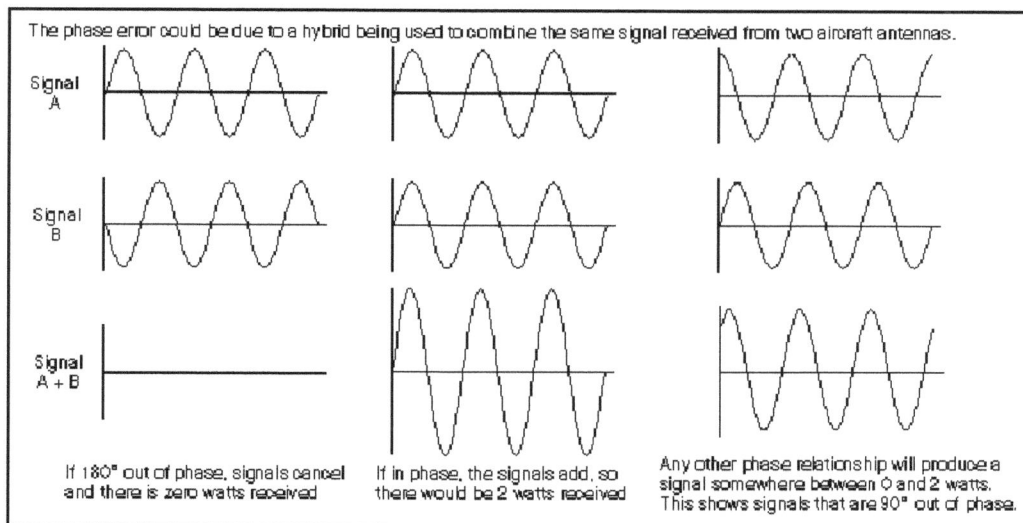

Figure 7. Sine Waves Combined Using Various Phase Relationships

This Page Blank

ATTENUATORS / FILTERS / DC BLOCKS

ATTENUATORS

An attenuator is a passive microwave component which, when inserted in the signal path of a system, reduces the signal by a specified amount. They normally possess a low VSWR which makes them ideal for reducing load VSWR in order to reduce measurement uncertainties. They are sometimes used simply to absorb power, either to reduce it to a measurable level, or in the case of receivers to establish an exact level to prevent overload of following stages.

Attenuators are classified as either fixed or variable and either reflective or non-reflective. The fixed and variable attenuators are available in both waveguide and coaxial systems. Most of the receivers under 20 GHz use coaxial type attenuators.

FIXED

The performance characteristics of a fixed attenuator are:
1. input and output impedances
2. flatness with frequency
3. average and peak power handling capability
4. temperature dependence

VARIABLE

The variable attenuator can be subdivided into two kinds: step attenuator and continuously variable attenuator. In a step attenuator, the attenuation is changed in steps such as 10 dB, 1 dB or 0.5 dB. In a continuously variable attenuator, the attenuation is changed continuously and a dial is usually available to read the attenuation either directly or indirectly from a calibration chart.

For a variable attenuator, additional characteristics should be considered, such as:
1. amount or range of attenuations
2. insertion loss in the minimum attenuation position
3. incremental attenuation for step attenuator
4. accuracy of attenuation versus attenuator setting
5. attenuator switching speed and switching noise.

REFLECTIVE

A reflective attenuator reflects some portion of the input power back to the driving source. The amount reflected is a function of the attenuation level. When PIN diodes are zero or reverse biased, they appear as open circuits when shunting a transmission line. This permits most of the RF input power to travel to the RF output. When they are forward biased, they absorb some input, but simultaneously reflect some back to the input port. At high bias current, most RF will be reflected back to the input resulting in a high input VSWR and high attenuation.

ABSORPTIVE

The VSWR of a non-reflective (absorptive) PIN diode attenuator remains good at any attenuation level (bias state). This is accomplished by configuring the diodes in the form of a Pi network that remains matched for any bias state or by use of a 90° hybrid coupler to cancel the waves reflected to the input connector.

MICROWAVE FILTERS

INTRODUCTION

Microwave filters are one of the most important components in receivers. The main functions of the filters are: (1) to reject undesirable signals outside the filter pass band and (2) to separate or combine signals according to their frequency. A good example for the latter application is the channelized receiver in which banks of filters are used to separate input signals. Sometimes filters are also used for impedance matching. Filters are almost always used before and after a mixer to reduce spurious signals due to image frequencies, local oscillator feedthrough, and out-of-frequency band noise and signals. There are many books which are devoted to filter designs. There are many kinds of filters used in microwave receivers, so it is impossible to cover all of them.

If a filter is needed on the output of a jammer, it is desirable to place it approximately half way between the jammer and antenna vs adjacent to either. The transmission line attenuation improves the VSWR of the filter at the transmitter. This may allow use of a less expensive filter, or use of a reflective filter vs an absorptive filter.

A filter is a two-port network which will pass and reject signals according to their frequencies. There are four kinds of filters according to their frequency selectivities. In the examples that follow, f_L = low frequency, f_M = medium frequency, and f_H = high frequency. Their names reflect their characteristics, and they are:

1. A low-pass filter which passes the low frequency signals below a predetermined value as shown in Figure 1.

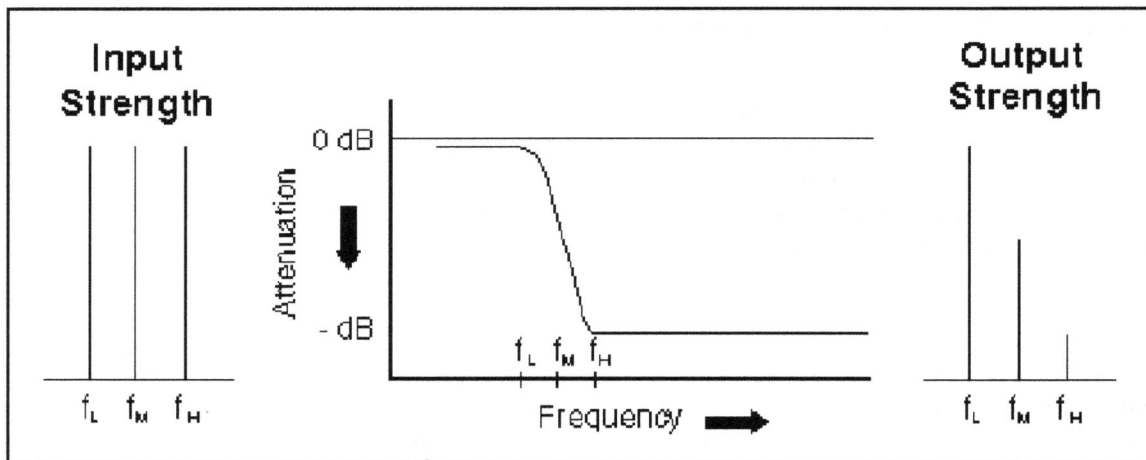

Figure 1. Low-Pass Filter

2. A high-pass filter which passes the high frequency signals above a predetermined value as in Figure 2.

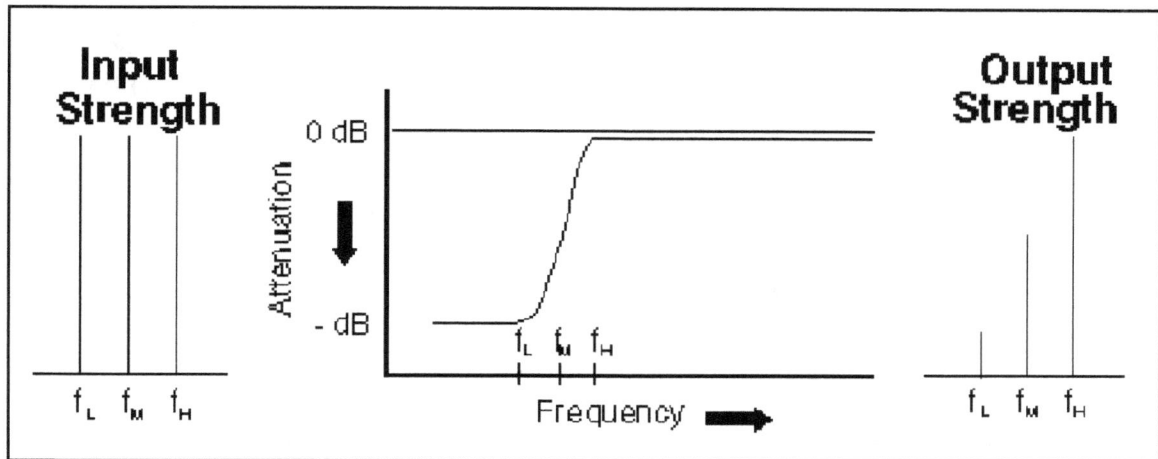

Figure 2. High-Pass Filter

3. A band-pass filter which passes signals between two predetermined frequencies as shown in Figure 3.

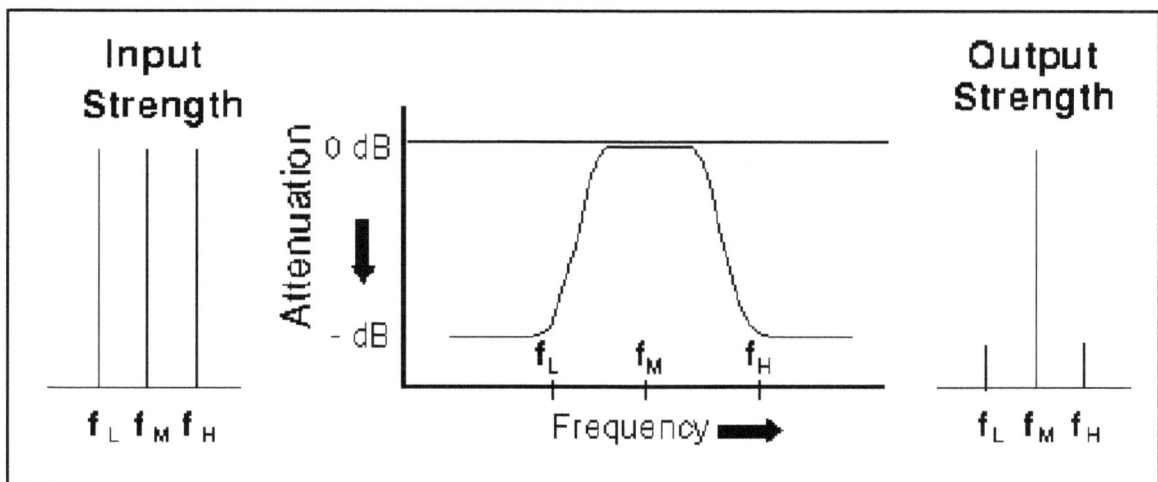

Figure 3. Band-Pass Filter

A band-pass filter with different skirt slopes on the two sides of the pass band is sometimes referred to as an asymmetrical filter. In this filter the sharpness of the rejection band attenuation is significantly different above and below the center frequency. One additional note regarding band-pass filters or filters in general, their performance should always be checked in the out-of-band regions to determine whether or not they posses spurious responses. In particular they should be checked at harmonics of the operating frequency.

4. A band reject filter (sometimes referred to as a bandstop or notch filter) which rejects signals between two predetermined frequencies such as high power signals from the aircraft's own radar as shown in Figure 4.

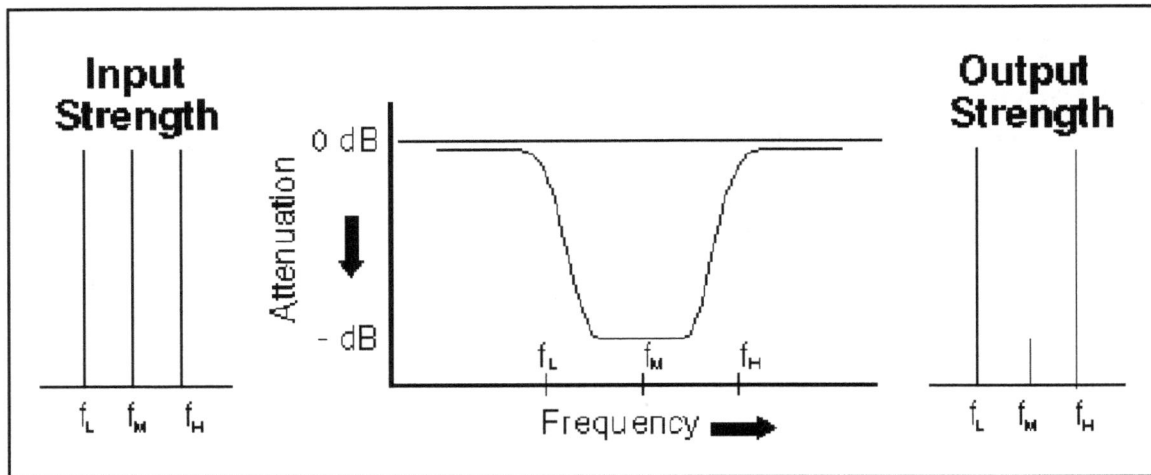

Figure 4. Band-Reject Filter

In general, filters at microwave frequencies are composed of resonate transmission lines or waveguide cavities that, when combined, reflect the signal power outside the filter frequency pass band and provide a good VSWR and low loss within the frequency pass band. As such, specifications for filters are maximum frequency, pass band loss, VSWR, and rejection level at a frequency outside of the pass band. The trade-offs for filters are a higher rejection for a fixed frequency pass band or a larger frequency pass band for a fixed rejection, which requires a filter with more resonators, which produce higher loss, more complexity, and larger size.

DC BLOCKS

DC Blocks are special connectors which have a capacitor (high pass filter) built into the device. There are three basic types:

1. INSIDE - The high pass filter is in series with the center conductor as shown in Figure 5. DC is blocked on the center conductor.

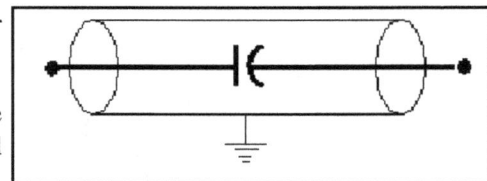

Figure 5. Inside DC Block

2. OUTSIDE - The high pass filter is in series with the cable shield as shown in Figure 6.

3. INSIDE/OUTSIDE - A high pass arrangement is connected to both the inner and outer conductors.

DC Blocks are ideal for filtering DC, 60 Hz, and 400 Hz from the RF line.

In general, capacitors with a large value of capacitance do not have the least loss at microwave frequencies. Also, since capacitance is proportional to size, a large size produces more capacitance with more inductance. Because of these reasons, D.C. blocks are typically available with a high pass frequency band starting in the region of 0.1 to 1 GHz.

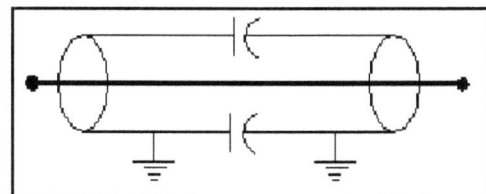

Figure 6. Outside DC Block

TERMINATIONS / DUMMY LOADS

A termination is a one-port device with an impedance that matches the characteristic impedance of a given transmission line. It is attached to a certain terminal or port of a device to absorb the power transmitted to that terminal or to establish a reference impedance at that terminal. Important parameters of a termination are its VSWR and power handling capacity. In a receiver, terminations are usually placed at various unconnected ports of components such as hybrid and power dividers to keep the VSWR of the signal path low. It is extremely important that the isolated port in a directional coupler and the unused port of a power divider (i.e., only three ports of a four-way power divider are used) be properly terminated. All of the design considerations of directional couplers and power dividers are based on the fact that all ports are terminated with matched loads. If an unused port is not properly terminated, then the isolation between the output ports will be reduced which may severely degrade the performance of the receiver.

A termination is the terminology used to refer to a low power, single terminal device intended to terminate a transmission line. Similar devices designed to accommodate high power are generally termed dummy loads.

TERMINATIONS:

Terminations are employed to terminate unconnected ports on devices when measurements are being performed. They are useful as dummy antennas and as terminal loads for impedance measurements of transmission devices such as filters and attenuators.

The resistive elements in most terminations are especially fabricated for use at microwave frequencies. Two types are commonly employed: (1) resistive film elements, and (2) molded resistive tapers. The resistive film is very thin compared to the skin depth and normally very short relative to wavelength at the highest operating frequency. The molded taper consists of a dissipative material evenly dispersed in a properly cured dielectric medium. Both forms of resistive elements provide compact, rugged terminations suitable for the most severe environmental conditions with laboratory stability and accuracy.

Terminations should be properly matched to the characteristic impedance of a transmission line. The termination characteristics of primary concern are:

a. operating frequency range   d. VSWR
b. average power handling capability   e. size
c. operating temperature range   f. weight

Many microwave systems employ directional couplers which require terminations on at least one port, and most have various modes of operation or test where terminations are needed on certain terminals.

A matched termination of a generalized transmission line is ideally represented by an infinite length of that line having small, but non-zero loss per unit length so that all incident energy is absorbed and none is reflected.

Standard mismatches are useful as standards of reflection in calibrating reflectometer setups and other impedance measuring equipment. They are also used during testing to simulate specific mismatches which would be encountered on the terminals of components once the component is installed in the actual system.

The following table shows common mismatches with the impedance that can provide the mismatch.

Common Mismatches (Z_O = 50 Ω)		
Ratio	Z_L (higher)	Z_L (lower)
1.0 : 1	50 Ω (matched)	50 Ω (matched)
1.25 : 1	62.5 Ω	40 Ω
1.50 : 1	75 Ω	33.3 Ω
2.00 : 1	100 Ω	25 Ω

DUMMY LOADS

A dummy load is a high power one port device intended to terminate a transmission line. They are primarily employed to test high power microwave systems at full power capacity. Low power coaxial loads are generally termed terminations and typically handle one watt or less.

Most radars or communications systems have a dummy load integrated into them to provide a non-radiating or EMCON mode of operation, or for testing (maintenance).

Three types of dissipative material are frequently employed in dummy loads: (1) lossy plastic, (2) refractory, and (3) water.

The lossy plastic consists of particles of lossy material suspended in plastic medium. This material may be designed to provide various attenuations per unit length but is limited as to operating temperature. It is employed primarily for low power applications.

The refractory material is a rugged substance that may be operated at temperatures up to 1600°F. It is virtually incapable of being machined by ordinary means but is often fabricated through diamond wheel grinding processes. Otherwise material must be fired in finished form. Such material is employed in most high power applications.

The dissipative properties of water are also employed for dummy load applications. Energy from the guide is coupled through a leaky wall to the water which flows alongside the main guide. Water loads are employed for extremely high power and calorimetric applications.

While dummy loads can operate over full waveguide bands, generally a more economical unit can be manufactured for use over narrower frequency ranges.

The power rating of a dummy load is a complex function dependent upon many parameters, including average and peak power, guide pressure, external temperature, guide size, air flow, and availability of auxiliary coolant. The average and peak powers are interrelated in that the peak power capacity is a function of the operating temperature which in turn is a function of the average power.

CIRCULATORS AND DIPLEXERS

A microwave circulator is a nonreciprocal ferrite device which contains three or more ports. The input from port n will come out at port n + 1 but not out at any other port. A three-port ferrite junction circulator, usually called the Y-junction circulator, is most commonly used. They are available in either rectangular waveguide or strip- line forms. The signal flow in the three-port circulator is assumed as 1→2, 2→3, and 3→1 as shown in Figure 1.

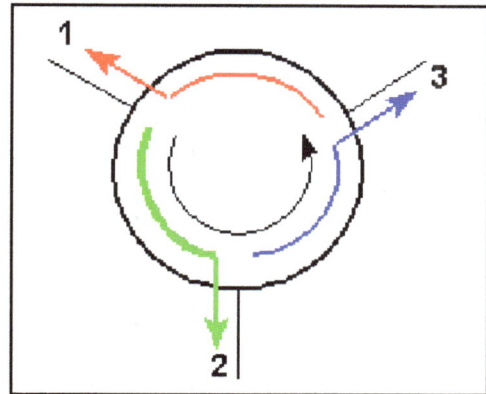

Figure 1. Symbolic Expression for a Y-Junction Circulator

If port 1 is the input, then the signal will come out of port 2; in an ideal situation, no signal should come out of port 3 which is called the isolated port. The insertion loss of the circulator is the loss from 1 to 2, while the loss from 1 to 3 is referred to as isolation. A typical circulator will have a few tenths of a dB insertion loss from port 1 to 2 and 20 dB of isolation from port 1 to 3 for coaxial circulators (30 dB or more for waveguide circulators). When the input is port 2, the signal will come out of port 3 and port 1 is the isolated port. Similar discussions can be applied to port 3.

> Since circulators contain magnets, they should not be mounted near ferrous metals since the close proximity of metals like iron can change the frequency response.

As shown in Figure 2, if one port of a circulator is loaded, it becomes an isolator, i.e. power will pass from ports one to two, but power reflected back from port two will go to the load at port three versus going back to port one.

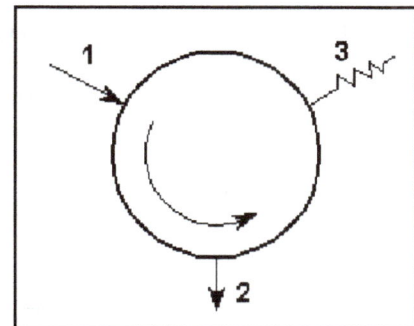

Figure 2. Isolator From A Circulator

As shown in Figure 3 this circulator is made into a diplexer by adding a high pass filter to port two. Frequencies from port one that are below 10 GHz will be reflected by port two. Frequencies above 10 GHz will pass through port two. At the 10 GHz crossover frequency of the diplexer, a 10 GHz signal will be passed to both ports two and three but will be half power at each port. Diplexers or triplexers (one input and three output bands), must be specifically designed for the application.

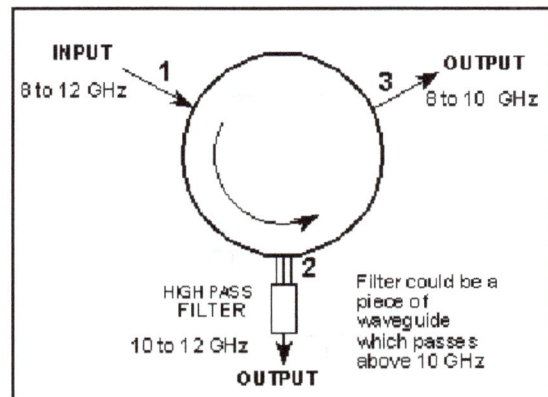

Figure 3. Diplexer From A Circulator

Another useful device is the 4-port Faraday Rotator Circulator shown symbolically in Figure 4. These waveguide devices handle very high power and provide excellent isolation properties. It is useful when measurements must be made during high power application as shown. A water load is used to absorb the high power reflections so that a reasonable power level is reflected to the receiver or measurement port.

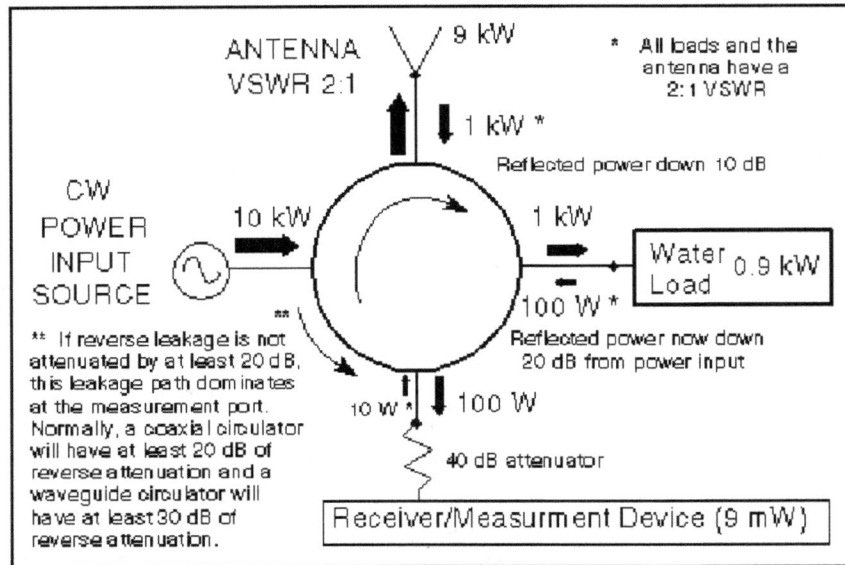

Figure 4. Faraday Rotator Circulator

The Maximum Input Power to a Measurement Device - The ideal input to a measurement device is in the 0 to 10 dBm (1 to 10 mW) range. Check manufacturer's specification for specific maximum value.

If the RF transmission lines and their components (antenna, hybrid, etc.) can support the wider frequency range, circulators could be used to increase the number of interconnecting RF ports from two as shown in Figure 5, to four as shown in Figure 6. Figure 7 shows an alternate configuration using diplexers which could actually be made from circulators as shown previously in Figure 3.

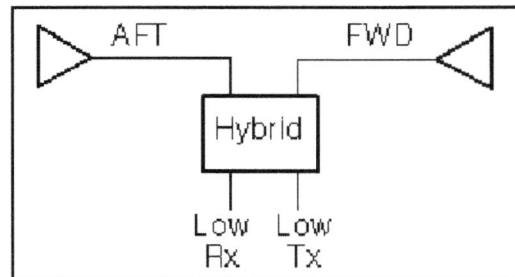

Figure 5. Low Band Configuration

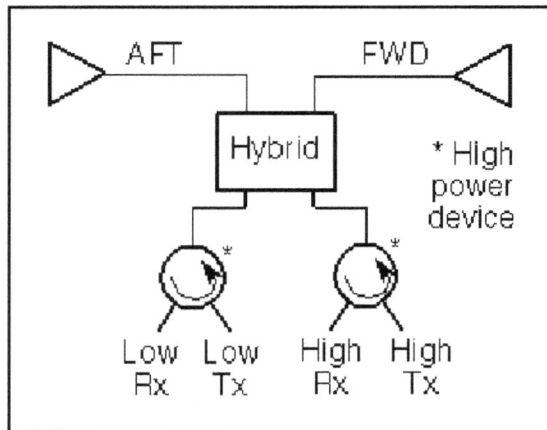

Figure 6. Low/High Band Configuration

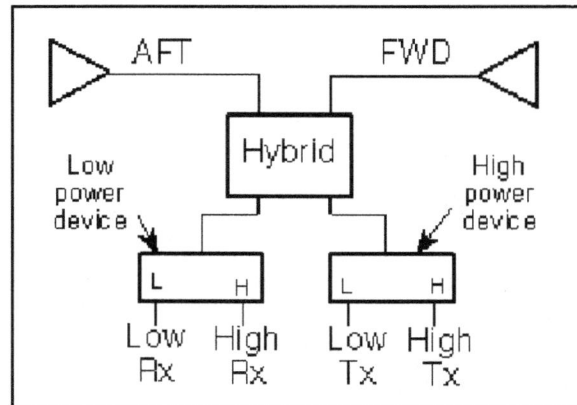

Figure 7. Alternate Low/High Band Configuration

MIXERS AND FREQUENCY DISCRIMINATORS

Mixers are used to convert a signal from one frequency to another. This is done by combining the original RF signal with a local oscillator (LO) signal in a non-linear device such as a Schottky-barrier diode.

The output spectrum includes:
- The original inputs, LO and RF
- All higher order harmonics of LO and RF
- The two primary sidebands , LO ± RF (m,n = 1)
- All higher order products of mLO ± nRF (where m,n are integers)
- A DC output level

The desired output frequency, commonly called the intermediate frequency (IF), can be either the lower (LO-RF) or upper (LO+RF) sideband. When a mixer is used as a down converter, the lower sideband is the sideband of interest.

A microwave balanced mixer makes use of the 3 dB hybrid to divide and recombine the RF and LO inputs to two mixing diodes. The 3 dB hybrid can be either the 90° or 180° type. Each has certain advantages which will be covered later. The critical requirement is that the LO and RF signals be distributed uniformly (balanced) to each mixer diode.

Figure 1 is a typical balanced mixer block diagram. The mixer diodes are reversed relative to each other; the desired frequency (IF) components of each diode are then in-phase while the DC outputs are positive and negative respectively.

The two diode outputs are summed in a tee where the DC terms cancel and only the desired IF component exists at the IF port.

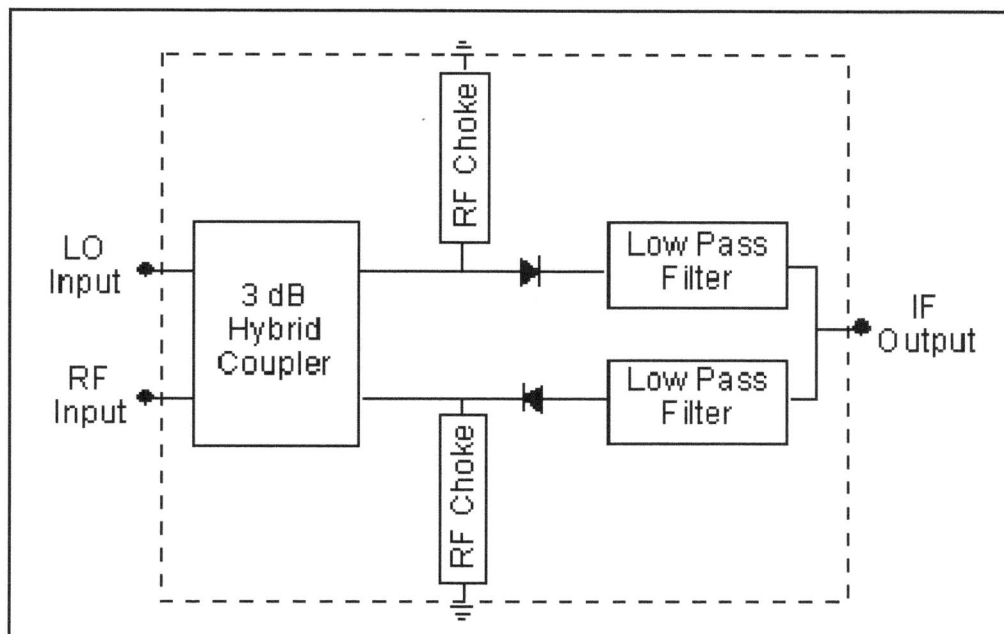

Figure 1. Mixer Block Diagram

Other types of mixers exist, including the double-balanced mixer, and the Ortho-Quad® (quadrature fed dual) mixer. The relative advantages and disadvantages of each of the four types are summarized in Table 1.

Table 1. Mixer Comparison

Mixer Type	VSWR [1]	Conversion Loss [2]	LO/RF Isolation [3]	Harmonic Suppression [4]	Dynamic Range	IF Bandwidth
90° Hybrid	good	lowest	poor	poor-fair	high	wide
180° Hybrid	poor	low	good	good	high	wide
Double-Balanced	poor	low	Very good -excellent	very good	high	extremely wide
Ortho Quad	good	low	very good	fair	high	wide

NOTES:
(1) Poor = 2.5:1 typical ; Good = 1.3:1 typical
(2) Conversion loss: lowest: 5-7 dB typical; Low 7-9 dB typical
(3) Poor: 10 dB typical ; Good: 20 dB typical ; Very Good: 25-30 dB typical ; Excellent: 35-40 dB typical
(4)   Poor: partial rejection of LO/RF even harmonics
         Fair: slightly better
         Good: can reject all LO even harmonics
         Very Good: can reject all LO and RF even harmonics

Used in various circuits, mixers can act as modulators, phase detectors, and frequency discriminators.

The phase discriminators can serve as a signal processing network for systems designed to monitor bearing, polarization, and frequency of AM or FM radiated signals.

A frequency discriminator uses a phase discriminator and adds a power divider and delay line at the RF input as shown in Figure 2. The unknown RF signal "A" is divided between a reference and delay path. The differential delay (T) creates a phase difference (θ) between the two signals which is a linear function of frequency (f) and is given by $\theta = 2\pi f T$.

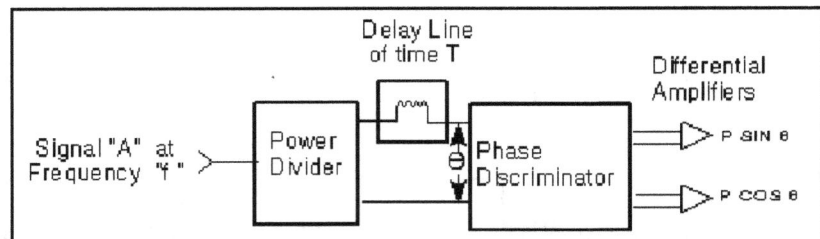

Figure 2. Frequency Discriminator

When the two output signals are fed to the horizontal and vertical input of an oscilloscope, the resultant display angle will be a direct function of frequency.

DETECTORS

A detector is used in receiver circuits to recognize the presence of signals. Typically a diode or similar device is used as a detector. Since this type of detector is unable to distinguish frequency, they may be preceded by a narrow band-pass filter.

A typical simplistic circuit is shown in Figure 1.

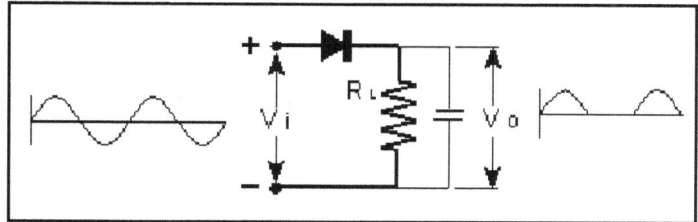

Figure 1. Typical Diode Detector Circuit

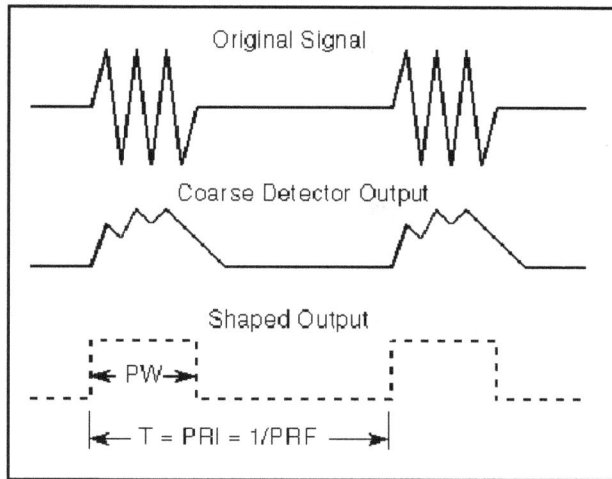

Figure 2. Demodulated Envelope Output

When the diode is reverse biased, very little current passes through unless the reverse breakdown voltage is exceeded. When forward biased and after exceeding the cut-in voltage, the diode begins to conduct as shown in Figure 3. At low voltages, it first operates in a square law region. Detectors operating in this region are known as small signal type. If the voltage is higher, the detector operates in a linear region, and is known as the large signal type.

The power/voltage characteristics for a typical diode detector is shown in Figure 4.

Square Law Detector

In the square law region, the output voltage V_o is proportional to the square of the input voltage V_i, thus V_o is proportional to the input power.

$$V_o = nV_i^2 = nP_i \quad \text{or} \quad P_i \propto V_o$$

Where n is the constant of proportionality

To integrate a pulse radar signal, we can add capacitance to the circuit in parallel with the output load R_L to store energy and decrease the bleed rate. Figure 2 shows a typical input/output waveform which detects the envelope of the pulse radar signal. From this information pulse width and PRF characteristics can be determined for the RWR UDF comparison.

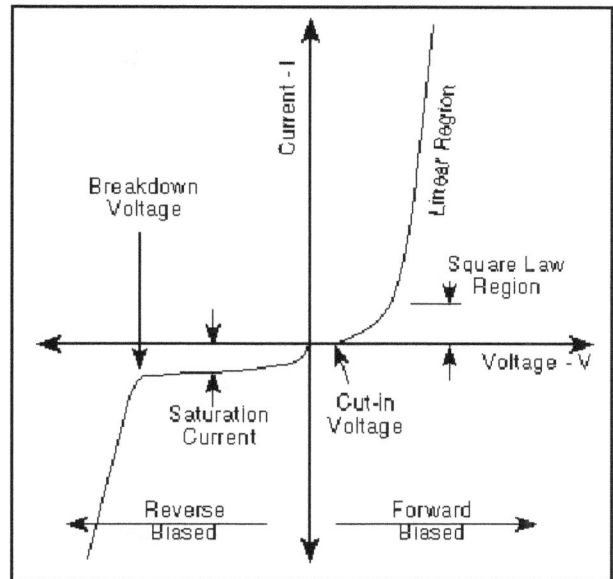

Figure 3. Diode Electrical Characteristics

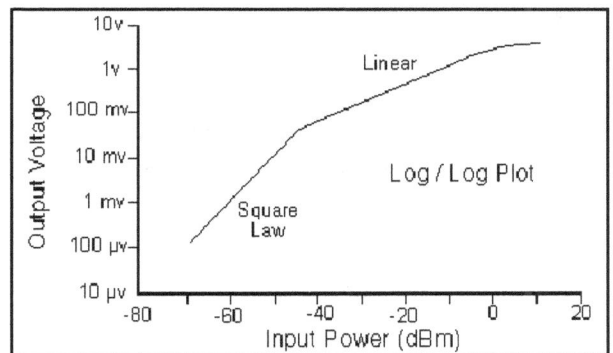

Figure 4. Diode Power/Voltage Characteristic

Linear Detector

In the linear detection region, the output voltage is given by:

$$V_o = mV_i \quad \text{and since } P=V^2/R, \quad P_i \propto V_o^2$$
Where m is the constant of proportionality

Log Detector Amplifier

Another type of detector arrangement is the Log detector amplifier circuit shown in Figure 5. It is formed by using a series of amplifiers and diode detectors. Due to the nature of the amplifier/diode characteristics, the output voltage is related to the power by:

$$P_i \propto 10^{pV_o + q}$$
Where p and q are constants of proportionality

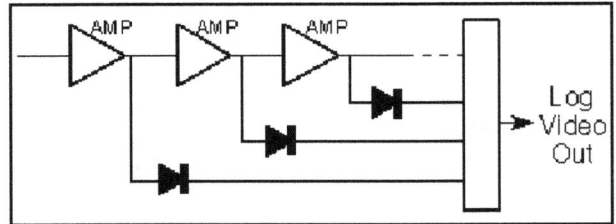

Figure 5. Log Detector

The Log detector has good range, but is hampered by large size when compared to a single diode detector.

Pulse Width Measurements

If the pulse width of a signal was specified at the one-half power point, the measurements of the detected signal on an oscilloscope would vary according to the region of diode operation. If the region of operation is unknown, a 3 dB attenuator should be inserted in the measurement line. This will cause the power to decrease by one-half. The (temporary) peak amplitude on the oscilloscope becomes the amplitude reference point for measuring the pulse width when the external 3 dB attenuator is removed.

These voltage levels for half power using the three types of detectors are shown in Table 1.

Table 1. Detector Characteristics

	Square Law	Linear	Log
Output Voltage When Input Power is reduced by Half (3 dB)	$0.5\ V_{in}$	$0.707\ V_{in}$	A very small value. ~ $0.15\ V_{in}$ for typical 5 stage log amplifier
Sensitivity & Dynamic Range	Good sensitivity Small dynamic range	Less sensitivity Greater dynamic range	Poorest sensitivity Greatest dynamic range (to 80 dB)

Also see the Microwave Measurements section subsection entitled "Half Power or 3 dB Measurement Point".

RF / MICROWAVE AMPLIFIERS

An amplifier is one of the most essential and ubiquitous elements in modern RF and microwave systems. Fundamentally, an amplifier is a type of electronic circuit used to convert low-power signals into ones of significant power. The specific requirements for amplification are as varied as the systems where they are used. Amplifiers are realized using a wide range of different technologies, and are available in many form factors. While the performance of an amplifier can be measured by wide range of attributes, several important ones include: gain, power added efficiency (PAE), input and output return loss, 1-dB compression point (P_{1dB}), stability, linearity, intermodulation distortion, and noise figure.

Traveling Wave Tubes (TWTs)

A traveling-wave tube (TWT) is a specialized vacuum tube used in electronics to amplify radio frequency (RF) signals to very high power. A TWT is a component of an electronic assembly known as a traveling-wave tube amplifier (TWTA), often pronounced "Tweet-uh". While trades between these parameters exist, modern TWTAs are capable of providing very high gains (>50 dB), multiple octaves of bandwidth, high efficiency, and output powers that range from tens to thousands of watts. Pulsed TWTAs can reach even higher output powers. Figure 1 presents a simplified diagram of a helix-type TWT.

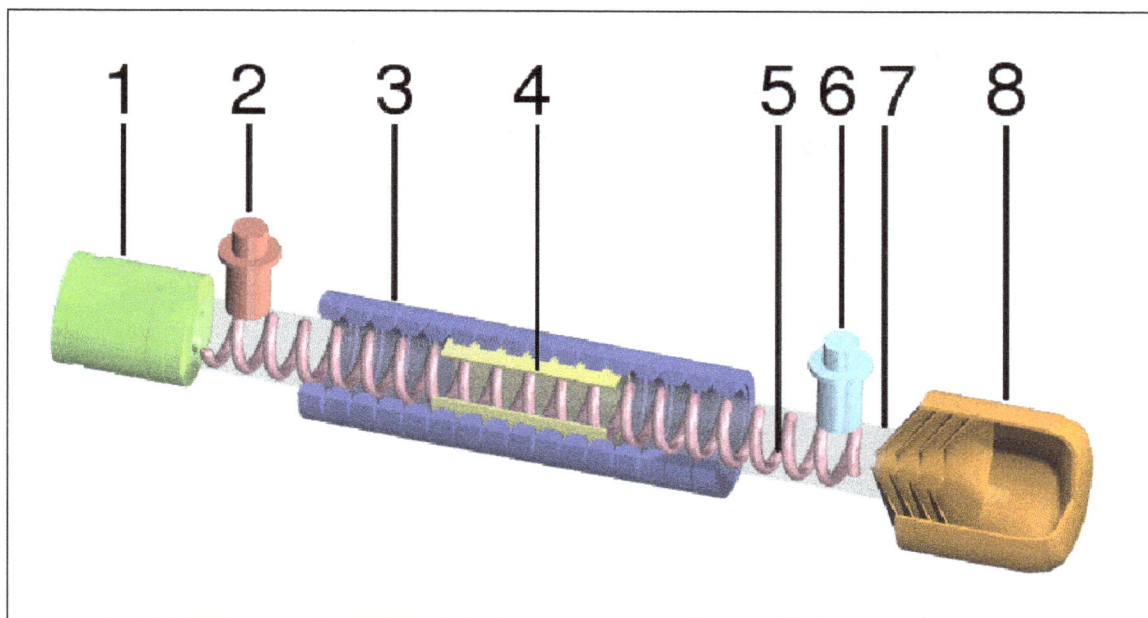

Figure 1: Cutaway view of a helix TWT.
(1) Electron gun; (2) RF input; (3) Magnets; (4) Attenuator;
(5) Helix coil; (6) RF output; (7) Vacuum tube; (8) Collector.
Reproduced from Wikipedia.

The device is an elongated vacuum tube with an electron gun (a heated cathode that emits electrons) at one end. A solenoid coil wrapped around the tube creates a magnetic field which focuses the electrons into a beam, which then passes down the middle of a wire helix that stretches the length of the tube, finally striking a collector at the other end. An input signal is coupled into the helix near the emitter, is amplified by the electron beam as it travels down the length of the tube, and the amplified signal is coupled to an external port near the collector.

The helix, which acts as a RF transmission line, delays the signal to near the same propagation speed as the electron beam. The speed at which the electromagnetic wave travels down the tube can be varied by changing the number or diameter of the turns in the helix. While propagating along the tube, the EM wave interacts with the electron beam. Since the electromagnetic wave effectively propagates slower than the electron beam, the electrons "bunch up" and modulate the input signal, giving up energy in the process - an effect known as velocity modulation. Thus, the traveling wave progressively grows in amplitude as it propagates down the tube towards the collector.

Figure 2 features a photo of a high voltage power supply and TWT.

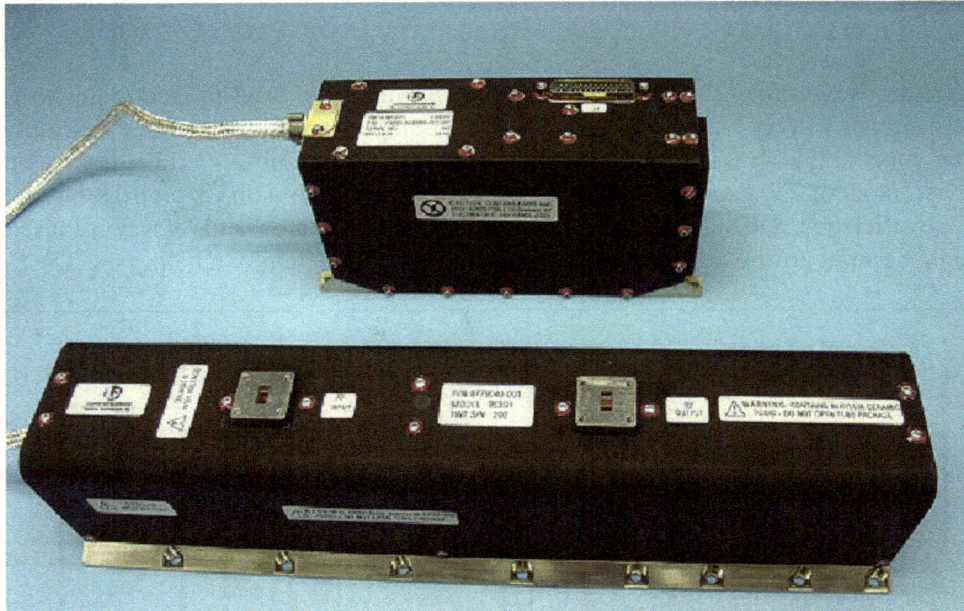

Figure 2: TWTA including high voltage power supply (top) and TWT (bottom)

Historically, TWTAs have been used in satellite transponders, radars, and in electronic warfare and self-protection systems. Recently, with the advent of wideband, high power solid-state amplifier solutions, however, TWTAs are slowly being replaced due to the higher reliability of their solid state counterparts.

Microwave Power Modules (MPMs)

A Microwave Power Module (MPM) is a hybrid solution between solid-state and vacuum tube electronics, which aims to take advantage of the best features of both technologies. They feature a solid-state pre-amplifier, miniaturized TWTA, and a high-density power supply, all integrated into a unit much more compact and lightweight than the traditional TWTA. While MPMs generally don't provide as much power as their larger TWTA counterparts, their lighter weight, compact form factor, and relaxed power supply requirements (often 28 VDC or 270 VDC) enable use in applications where a TWTA would not be possible.

Similar to TWTAs, MPMs are capable of providing very high gains (>50 dB), multiple octaves of bandwidth, high efficiency. Typical power levels range from tens to hundreds of watts, with ~1kW capability for pulsed MPMs.

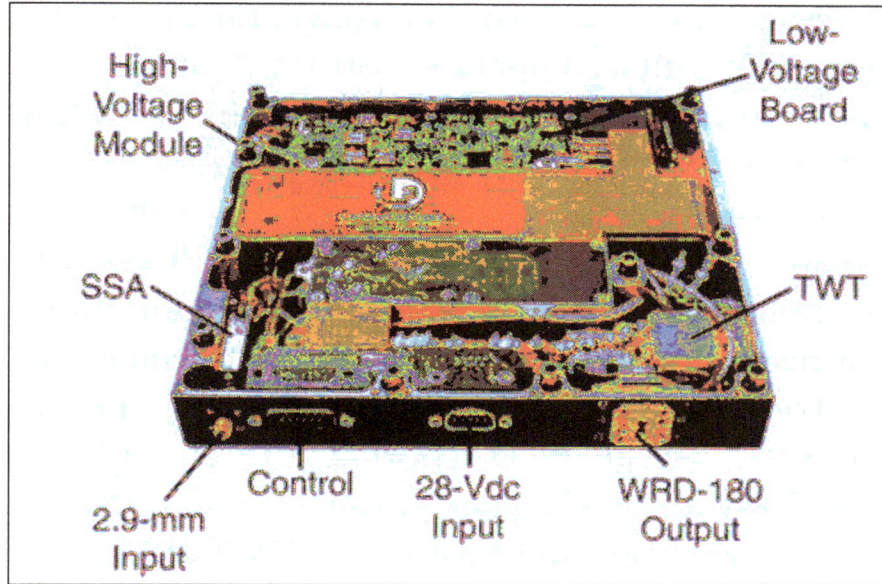

Figure 3: 40W Ka-band MPM with components identified.
Image from IEEE Magazine, December 2009, page 42

They have found applications in phased array antennas, lower-power radar transmitters, satellite communications, EW systems, and UAVs.

Solid State

Solid-state electronics, including amplifiers, are built entirely from solid materials and in which the electrons, or other charge carriers, are confined entirely within the solid material. The building material is most often a crystalline semiconductor. Solid-state power amplifiers, or SSPAs, are fabricated on many different semiconductor technologies, some of which include GaAs, GaN, InP, SiGe, CMOS. A photo of a GaN SSPA is presented in Figure 4.

Figure 4: Typical SSPA. Note the GaN die and external impedance matching networks.

In general, SSPAs are heralded as having a higher reliability than TWTAs. While high power SSPAs are available on the market, achieving high efficiency has been a challenge as many amplifier stages often need to be combined to meet to achieve power levels in the tens to hundreds of watts. Mutli-stage designs increase size, weight, and power (SWAP) and decrease PAE (due to ohmic losses in the combining networks), which is one reason why TWTs and MPMs still provide excellent alternatives for medium to high power applications.

Below is a table that attempts to compare and contrast different solid-state technologies in the context of microwave networks and amplifiers. It is by no means complete, but provides a general overview.

Table 1: Comparison between different semiconductor process technologies

Technology	Advantages	Disadvantages
Silicon variants (CMOS, SiGe)	• Cheapest substrate due to CPU industry • Can be fabricated with nanometer accuracy	• Not a good overall microwave substrate • Results in lossy, high noise figure, low power components • Crystal is fragile • Junction temperatures limited to ~110C
Indium Phosphide Variants (InP/InGaP)	• High performance at low voltage • Good thermal stability • Results in high efficiency PAs, particularly at lower operating voltage • Extremely low noise figure • Useful through W-band and beyond	• Less mature, "niche" fabrication houses • Brittle, fragile material • Higher cost than GaAs • Low breakdown voltage – not good for high power
Gallium Arsenide variants (GaAs)	• Most mature, widespread technology • Many transistor variants (MESFET, PHEMT, MHEMT, HBT, etc) • High reliability • Fairly low cost (but more than silicon) • Great microwave substrate • Low loss, high ε_r. • 16-20V breakdown possible • Junction temperatures up to 150C	• High noise figure, Noise figure and power performance • Difficult to summarize - depends on transistor type used
Gallium Nitride (GaN)	• Up to 10x the power density of GaAs • High breakdown voltage (100V possible) • Results in high efficiency, high frequency, wide bandwidth PAs • Can operate hotter than GaAs, Si, or SiGe	• Currently more expensive than GaAs, but costs are decreasing • Difficult to fabricate • High power density leads to thermal challenges

SIGNAL GENERATION

Signal generators, also known variously as function generators, RF and microwave signal generators, pitch generators, arbitrary waveform generators, digital pattern generators or frequency generators are electronic devices that generate repeating or non-repeating electronic signals (in either the analog or digital domains). They are generally used in designing, testing, troubleshooting, and repairing electronic or electro-acoustic devices; though they often have artistic uses as well.

There are many different types of signal generators, with different purposes and applications (and at varying levels of expense); in general, no device is suitable for all possible applications.

Analog Signal Generators

RF signal generators are capable of producing CW (continuous wave) tones. The output frequency can usually be tuned anywhere in their frequency range. Many models offer various types of analog modulation, either as standard equipment or as an optional capability to the base unit. This could include AM, FM, ΦM (phase modulation) and pulse modulation. Another common feature is a built-in attenuator which makes it possible to vary the signal's output power. Depending on the manufacturer and model, output powers can range from -135 to +30 dBm. A wide range of output power is desirable, since different applications require different amounts of signal power. For example, if a signal has to travel through a very long cable out to an antenna, a high output signal may be needed to overcome the losses through the cable and still have sufficient power at the antenna. But when testing receiver sensitivity, a low signal level is required to see how the receiver behaves under low signal-to-noise conditions.

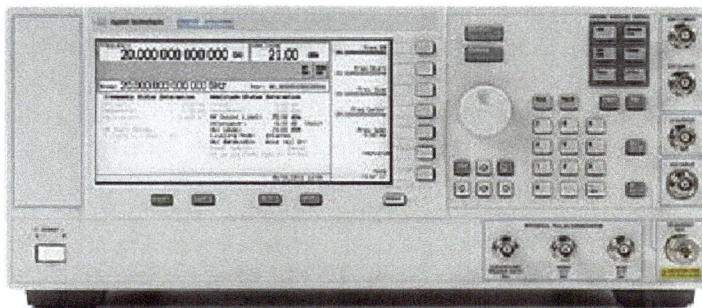

RF signal generators are required for servicing and setting up analog radio receivers, and are used for professional RF applications.

Arbitrary Waveform Generator

An arbitrary waveform generator (AWG) is a piece of electronic test equipment used to generate electrical waveforms. These waveforms can be either repetitive or single-shot (once only) in which case some kind of triggering source is required (internal or external). The resulting waveforms can be injected into a device under test and analyzed as they progress through the device, confirming the proper operation of the device or pinpointing a fault in the device.

Unlike function generators, AWGs can generate any arbitrarily defined wave shape as their output. The waveform is usually defined as a series of "waypoints" (specific voltage targets occurring at specific times along the waveform) and the AWG can either jump to those levels or use any of several methods to interpolate between those levels.

Because AWGs synthesize the waveforms using (baseband) digital signal processing techniques, their maximum frequency is usually limited to no more than a few gigahertz (~10 GHz being the latest state-of-the-art in 2012).

A major difficulty in generating non-repetitive waveforms at higher frequencies with AWGs is the large amount of data required to describe high-frequency baseband signals. For example, a 20 gigasample/sec arbitrary waveform with 8 bits of resolution requires 20 GB of data to represent every 1 second of signal, regardless of that signal's nature. Generating such large digital data streams and delivering them to the DAC in the AWG is an increasingly difficult problem as DAC upper frequencies continue to grow.

AWGs, like most signal generators, may also contain an attenuator, various means of modulating the output waveform, and often contain the ability to automatically and repetitively "sweep" the frequency of the output waveform (by means of a voltage-controlled oscillator) between two operator-determined limits. This capability makes it very easy to evaluate the frequency response of a given electronic circuit.

AWGs can operate as conventional function generators. These would include standard waveforms such as sine, square, ramp, triangle, noise and pulse. Some units include additional built-in waveforms such as exponential rise and fall times, sinx/x, cardiac. Some AWGs allow users to retrieve waveforms from a number of digital and mixed-signal oscilloscopes. Some AWGs may display a graph of the waveform on their screen - a graph mode. Some AWGs have the ability to generate a pattern of words from multiple bit output connector to simulate data transmission, combining the properties of both AWGs and digital pattern generators.

One feature of direct digital synthesizer (DDS) based arbitrary waveform generators is that their digital nature allows multiple channels to be operated with precisely controlled phase offsets or ratio-related frequencies. This allows the generation of polyphase sine waves, I-Q constellations, or simulation of signals from geared mechanical systems such as jet engines. Complex channel-channel modulations are also possible.

Vector Signal Generators

Modern vector signal generators can be seen as a hybrid of the arbitrary waveform generator and the analog signal generator, combining a lower-bandwidth AWG with analog upconversion, allowing a moderate-bandwidth (~100 MHz, circa 2012) digital signal to be upconverted to any frequency (~50 GHz typically upper frequency typically).

With the advent of digital communications systems, it is no longer possible to adequately test these systems with traditional analog signal generators. This has led to the development of vector signal generators, also known as digital signal generators. These signal generators are capable of generating digitally-modulated radio signals that may use any of a large number of digital modulation formats such as QAM, QPSK, FSK, BPSK, and OFDM. In addition, since modern commercial digital communication systems are almost all based on well-defined industry standards, many vector signal generators can generate signals based on these standards. Examples include GSM, W-CDMA (UMTS), CDMA2000, LTE, Wi-Fi (IEEE 802.11), and WiMAX (IEEE 802.16). In contrast, military communication systems such as JTRS, which place a great deal of importance on robustness and information security, typically use very proprietary methods. To test these types of communication systems, users will often create their own custom waveforms and download them into the vector signal generator to create the desired test signal.

DIGITAL SIGNAL PROCESSING COMPONENTS

The goal of DSP is usually to measure, filter and/or compress continuous real-world analog signals. The first step is usually to convert the signal from an analog to a digital form, by sampling and then digitizing it using an analog-to-digital converter (ADC), which turns the analog signal into a stream of numbers. However, often, the required output signal is another analog output signal, which requires a digital-to-analog converter (DAC). Even if this process is more complex than analog processing and has a discrete value range, the application of computational power to digital signal processing allows for many advantages over analog processing in many applications, such as error detection and correction in transmission as well as data compression. With the increasing bandwidth and dynamic range of ADC and digital components, common analog RF signal processing operations such as filtering, threshold detection, and pulse compression are being carried out in the digital domain. Additionally new capabilities such as synthetic aperture radar (SAR), digital RF memory (DRFM), and space-time adaptive processing (STAP) are wholly enabled by the increasing power of digital processing components.

Analog-to-Digital Converter

An analog-to-digital converter (abbreviated ADC, A/D or A to D) is a device that converts a continuous quantity to a discrete time digital representation. An ADC may also provide an isolated measurement. The reverse operation is performed by a digital-to-analog converter (DAC).

Typically, an ADC is an electronic device that converts an input analog voltage or current to a digital number proportional to the magnitude of the voltage or current. However, some non-electronic or only partially electronic devices, such as rotary encoders, can also be considered ADCs.

Resolution

The resolution of the converter indicates the number of discrete values it can produce over the range of analog values. The values are usually stored electronically in binary form, so the resolution is usually expressed in bits. In consequence, the number of discrete values available, or "levels", is a power of two. For example, an ADC with a resolution of 8 bits can encode an analog input to one in 256 different levels, since $2^8 = 256$. The values can represent the ranges from 0 to 255 (i.e. unsigned integer) or from -128 to 127 (i.e. signed integer), depending on the application.

Resolution can also be defined electrically, and expressed in volts. The minimum change in voltage required to guarantee a change in the output code level is called the least significant bit (LSB) voltage. The resolution Q of the ADC is equal to the LSB voltage. The voltage resolution of an ADC is equal to its overall voltage measurement range divided by the number of discrete values:

$$Q = \frac{E_{FSR}}{2^M}$$

where M is the ADC's resolution in bits and E_{FSR} is the full scale voltage range (also called 'span').

E_{FSR} is given by

$$E_{FSR} = V_{RefHi} - V_{RefLow}$$

where V_{RefHi} and V_{RefLow} are the upper and lower extremes, respectively, of the voltages that can be coded.

Normally, the number of voltage intervals is given by

$$N = 2^M - 1$$

where M is the ADC's resolution in bits.

That is, one voltage interval is assigned in between two consecutive code levels.

A typical 3 bit (2^8 = 8-level) coding scheme is depicted in Figure 1.

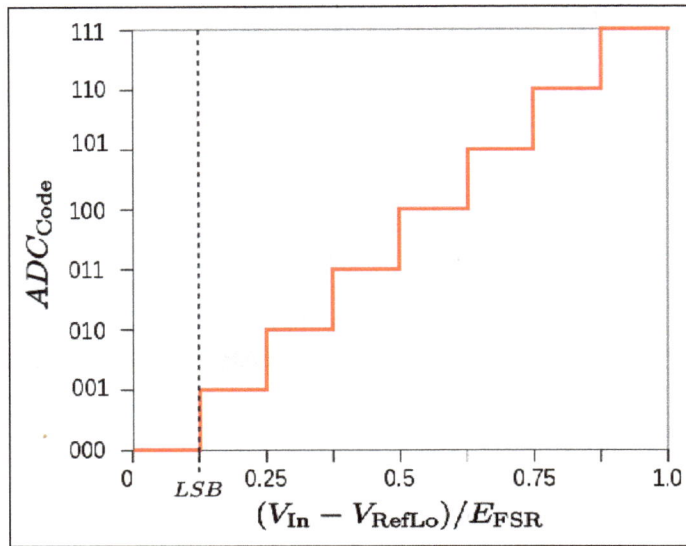

Figure 1. An 8-level (3 bit) ADC coding scheme

Example:

Assume input signal x(t) = A cos(t), A = 5V
Full scale measurement range = -5 to 5 volts
ADC resolution is 8 bits: 2^8 - 1 = 256 - 1 = 255 quantization levels (codes)
ADC voltage resolution, Q = (5 V − [-5] V) / 255 = 10 V / 255 ≈ 0.039 V ≈ 39 mV.

In practice, the useful resolution of a converter is limited by the best signal-to-noise ratio (SNR) that can be achieved for a digitized signal. An ADC can resolve a signal to only a certain number of bits of resolution, called the effective number of bits (ENOB). One effective bit of resolution changes the signal-to-noise ratio of the digitized signal by 6 dB, if the resolution is limited by the ADC. If a preamplifier has been used prior to A/D conversion, the noise introduced by the amplifier can be an important contributing factor towards the overall SNR.

Accuracy

An ADC has several sources of errors. Quantization error and (assuming the ADC is intended to be linear) non-linearity are intrinsic to any analog-to-digital conversion. There is also a so-called aperture error which is due to a clock jitter and is revealed when digitizing a time-variant signal (not a constant value).

These errors are measured in a unit called the least significant bit (LSB). In the above example of an eight-bit ADC, an error of one LSB is 1/256 of the full signal range, or about 0.4%.

Sampling Rate

The analog signal is continuous in time and it is necessary to convert this to a flow of digital values. It is therefore required to define the rate at which new digital values are sampled from the analog signal. The rate of new values is called the sampling rate or sampling frequency of the converter.

A continuously varying band limited signal can be sampled (that is, the signal values at intervals of time T, the sampling time, are measured and stored) and then the original signal can be exactly reproduced from the discrete-time values by an interpolation formula. The accuracy is limited by quantization error. However, this faithful reproduction is only possible if the sampling rate is higher than twice the highest frequency of the signal. This is essentially what is embodied in the Shannon-Nyquist sampling theorem.

Oversampling

Usually, for economy, signals are sampled at the minimum rate required, with the result that the quantization noise introduced is white noise spread over the whole pass band of the converter. If a signal is sampled at a rate much higher than the Nyquist frequency and then digitally filtered to limit it to the signal bandwidth there are the following advantages:

- Digital filters can have better properties (sharper roll-off, phase) than analog filters, so a sharper anti-aliasing filter can be realized and then the signal can be down-sampled giving a better result
- A 20-bit ADC can be made to act as a 24-bit ADC with 256× oversampling
- The signal-to-noise ratio due to quantization noise will be higher than if the whole available band had been used. With this technique, it is possible to obtain an effective resolution larger than that provided by the converter alone.
- The improvement in SNR is 3 dB (equivalent to 0.5 bits) per octave of oversampling which is not sufficient for many applications. Therefore, oversampling is usually coupled with noise shaping. With noise shaping, the improvement is 6L+3 dB per octave where L is the order of loop filter used for noise shaping. e.g. - a 2nd order loop filter will provide an improvement of 15 dB/octave.

Digital to Analog Converter (DAC)

Digital-to-analog converters (DACs) perform the inverse function of the ADC. As such, the abovementioned principles of accuracy, resolution, sampling and oversampling, etc. apply equivalently.

Field-Programmable Gate Array (FPGA)

A field-programmable gate array (FPGA) is an integrated circuit designed to be configured by the customer or designer after manufacturing—hence "field-programmable". The FPGA configuration is generally specified using a hardware description language (HDL), similar to that used for an application-specific integrated circuit (ASIC). Circuit diagrams were previously used to specify the configuration, as they were for ASICs, but this is increasingly rare. FPGAs can be used to implement any logical function that an ASIC could perform. The ability to update the functionality after shipping, partial re-configuration of a portion of the design and the low non-recurring engineering costs relative to an ASIC design (notwithstanding the generally higher unit cost), offer advantages for many applications.

FPGAs contain programmable logic components called "logic blocks", and a hierarchy of reconfigurable interconnects that allow the blocks to be "wired together"—somewhat like many (changeable) logic gates that can be inter-wired in (many) different configurations. Logic blocks can be configured to perform complex combinational functions, or merely simple logic gates like AND and XOR. In most FPGAs, the logic blocks also include memory elements, which may be simple flip-flops or more complete blocks of memory.

FPGAs are becoming increasingly more popular for implementing RF signal processing functions such as signal compression (matched filtering), channel selection, modulation / demodulation, etc.

MICROWAVE MEASUREMENTS

Measurement Procedures

Calculate your estimated power losses before attempting to perform a measurement. The ideal input to a measurement device is in the 0 to 10 dBm (1 to 10 mW) range.

Linearity Check

To verify that a spectrum measurement is accurate and signals are not due to mixing inside the receiver, a linearity check should be performed, i.e. externally insert a 10 dB attenuator - if measurements are in the linear region of the receiver, all measurements will decrease by 10 dB. If the measurements decrease by less than 10 dB , the receiver is saturated. If the measurements disappear, you are at the noise floor.

Half-Power or 3 dB Measurement Point

To verify the half power point of a pulse width measurement on an oscilloscope, externally insert a 3 dB attenuator in the measurement line, and the level that the peak power decreases to is the 3 dB measurement point (Note: you cannot just divide the peak voltage by one-half on the vertical scale of the oscilloscope).

VSWR Effect on Measurement

Try to measure VSWR (or reflection coefficient) at the antenna terminals. Measuring VSWR of an antenna through it's transmission line can result in errors. Transmission lines should be measured for insertion loss not VSWR.

High Power Pulsed Transmitter Measurements

When making power measurements on a high power pulsed transmitter using a typical 40 dB directional coupler, an additional attenuator may be required in the power meter takeoff line, or the power sensor may be burnt out.

For example, assume we have a 1 megawatt transmitter, with PRF = 430 pps, and PW = 13 µs. Further assume we use a 40 dB directional coupler to tap off for the power measurements. The power at the tap would be:

$$10 \log(P_p) - 10 \log(DC) - \text{Coupler reduction} =$$

$$10 \log(10^9 \text{mW}) - 10 \log(13 \times 10^{-6})(430) - 40 \text{ dB} =$$

$$90 \text{ dBm} - 22.5 \text{ dB} - 40 \text{ dB} = 27.5 \text{ dBm (too high for a power meter)}$$

Adding a 20 dB static attenuator to the power meter input would give us a value of 7.5 dBm or 5.6 mW, a good level for the power meter.

High Power Measurements With Small Devices

When testing in the presence of a high power radar, it is normally necessary to measure the actual field intensity. The technique shown in Figure 4, in Section 6-7, may not be practical if the measurement device must be small. An alternate approach is the use of a rectangular waveguide below its cutoff frequency. In this manner, the "antenna" waveguide provides sufficient attenuation to the frequency being measured so it can be coupled directly to the measurement device or further attenuated by a low power attenuator. The attenuation of the waveguide must be accurately measured since attenuation varies significantly with frequency.

This Page Blank

ELECTRO-OPTICS AND IR

ELECTRO-OPTICAL SYSTEMS AND EW COUNTERMEASURES

INTRODUCTION

The development of infrared countermeasures and the evaluation of their performance against threat missiles is a broad, complex technical field. Much of the detailed information about the threats to be countered and the characteristics of the countermeasures themselves is understandably sensitive, beyond the limitations of this document. More detailed information can be requested, provided the requesting organization has the necessary clearance and need to know. Such requests will be considered on a case-by-case basis. Information will be provided only upon written concurrence of the controlling Government organization.

There are many electro-optical (EO) electronic warfare (EW) systems, which are analogous to radio frequency (RF) EW systems. These EO EW systems operate in the optical portion of the electromagnetic spectrum. Electro-optics (EO), as the name implies, is a combination of electronics and optics. By one definition EO is the science and technology of the generation, modulation, detection and measurement, or display of optical radiation by electrical means. Most infrared (IR) sensors, for example, are EO systems. In the popularly used term "EO/IR," the EO is typically used to mean visible or laser systems. The use of EO in this context is a misnomer. Actually, almost all "EO/IR" systems are EO systems as defined above. Another often-used misnomer is referring to an EO spectrum. EO systems operate in the optical spectrum, which is from 0.01 to 1000 micrometers. EO systems include, but are not limited to, lasers, photometry, infrared, and other types of visible, and UV imaging systems.

Within the broad field of Electronic Warfare, electro-optical systems are prevalent for communication systems and offensive and defensive applications. Lasers have been used extensively for weapons guidance purposes, warhead fuzing applications, targeting systems and other offensive weapons related purposes. Understanding electro-optical and radiometric principles and sensors is critical to the development of vehicle survivability systems. These principles range from signature reduction and camouflage to active countermeasure systems such as lamp-based and laser jammers to passive threat warning systems and expendable flare decoys.

Although military systems operate in many portions of the electro-optical spectrum, the infrared is of paramount importance for remote detection systems and weapons applications. Missile seekers, Forward Looking Infrared (FLIR) systems, and Infrared Search and Track Systems all operate in the infrared portion of the spectrum.

OPTICAL SPECTRUM

The optical spectrum is that portion of the electromagnetic spectrum from the extreme ultraviolet (UV) through the visible to the extreme IR. Figure 1 shows the optical spectrum in detail. The end points of the optical spectrum are somewhat arbitrary.

IR spectrum terminology has also varied through the years, with near (near visible), or short-wavelength infrared (SWIR) being on the high frequency end. Then as frequency decreases the spectrum is followed by intermediate or mid-wavelength infrared (MWIR), then far or long-wavelength IR (LWIR), and finally extreme IR.

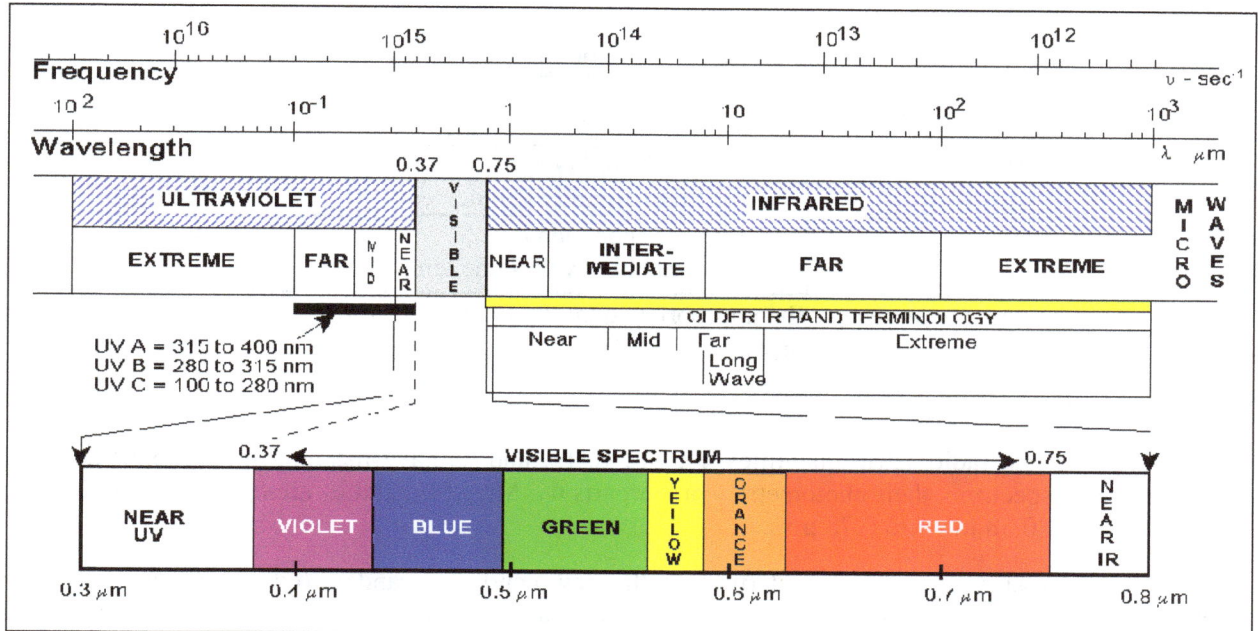

Figure 1. Optical Spectrum

RADIOMETRIC QUANTITIES AND TERMINOLOGY

The common terms used to describe optical radiation are the source parameters of power, radiant emittance (older term) or radiant exitance (newer term), radiance, and radiant intensity. They refer to how much radiation is given off by a body. The parameter measured by the detector (or collecting object/surface) is the irradiance. Any of these quantities can be expressed per unit wavelength in which case the subscript is changed from e (meaning energy derived units) to λ and the term is then called "Spectral ...X...", i.e. I_e is radiant intensity, while I_λ is spectral radiant intensity. These quantities in terms of currently preferred "Système International d'Unités" (SI units) are defined in Table 1.

Table 1. Radiometric SI Units.

Symbol	Name	Description	Units
Q	Radiant Energy		J (joules)
Φ_e	Radiant Power (or flux)	Rate of transfer of radiant energy	W (watts)
M_e	Radiant Exitance	Radiant power per unit area emitted from a surface	$W\,m^{-2}$
L_e	Radiance	Radiant power per unit solid angle per unit projected area	$W\,m^{-2}sr^{-1}$
I_e	Radiant Intensity	Radiant power per unit solid angle from a point source	$W\,sr^{-1}$
E_e	Irradiance	Radiant power per unit area incident upon a surface	$W\,m^{-2}$
X_λ	Spectral ...X..	(Quantity) per unit wavelength interval	(Units) nm^{-1} or μm^{-1}

Where X_λ is generalized for each unit on a per wavelength basis; for example, L_λ would be called "spectral radiance" instead of radiance.

In common usage, irradiance is expressed in units of Watts per square centimeter and wavelengths are in μm instead of nanometers (nm). Other radiometric definitions are shown in Table 2.

Table 2. Other Radiometric Definitions

Symbol	Name	Description	Units
α	Absorptance[1]	α = (*) absorbed / (*) incident	numeric
ρ	Reflectance	ρ = (*) reflected / (*) incident	numeric
τ	Transmittance	τ = (*) transmitted / (*) incident	numeric
$\in$	Emissivity	$\in$ = (*) of specimen / (*) of blackbody @ same temperature	numeric

Where (*) represents the appropriate quantity Q, Φ, M, E, or L

Note (1) Radiant absorptance should not be confused with absorption coefficient.

PHOTOMETRIC QUANTITIES

Whereas the radiometric quantities Φ_e, M_e, I_e, L_e, and E_e have meaning throughout the entire electromagnetic spectrum, their photometric counterparts Φ_v, M_v, I_v, L_v, and E_v are meaningful only in the visible spectrum (0.38 μm thru 0.78 μm).

The "standard candle" was redefined as the new candle or candela (cd). One candela is the luminous intensity of 1/60th of 1 cm^2 of the projected area of a blackbody radiator operating at the temperature of the solidification of platinum (2045 °K). By definition, the candela emits one lumen (lm) per steradian.

Table 3 displays the photometric quantities and units. These are used in dealing with optical systems such as aircraft television camera systems, optical trackers, or video recording.

Table 3. Photometric SI Units.

Symbol	Name	Description	Units
Q_v	Luminous energy		lumen sec (lm s)
F or Φ_v	Luminous flux or Luminous Power	Rate of transfer of luminant energy	lumen (lm)
M_v	Luminous Exitance or flux density (formerly luminous emittance)	Luminant power per unit area emitted from a surface	lm m^{-2}
L_v	Luminance (formerly brightness)	Luminous flux per unit solid angle per unit projected area	nit (nt) or candela / m^2 or lm/sr·m^2
I_v	Luminous Intensity (formerly candlepower)	Luminous power per unit solid angle from a point source	candela (cd) or lm/sr
E_v	Illuminance (formerly illumination)	Luminous power per unit area incident upon a surface	lux or lx or lm/m^2
K	Luminous efficacy	K= Φ_v / Φ_e	lm / w

Table 4 displays conversion factors for commonly used illuminance quantities.

Table 4. Illuminance Conversion Units

		Lux (lx)	Footcandle (fc)	Phot (ph)
1 lux (lm m^{-2})	=	1	0.0929	1 x 10^{-4}
1 footcandle (lm ft^{-2})	=	10.764	1	0.001076
1 phot (lm cm^{-2})	=	1 x 10^4	929	1

THE BASIC PRINCIPLES

The processes of absorption, reflection (including scattering), and transmission account for all incident radiation in any particular situation, and the total must add up to one in order that energy be conserved:

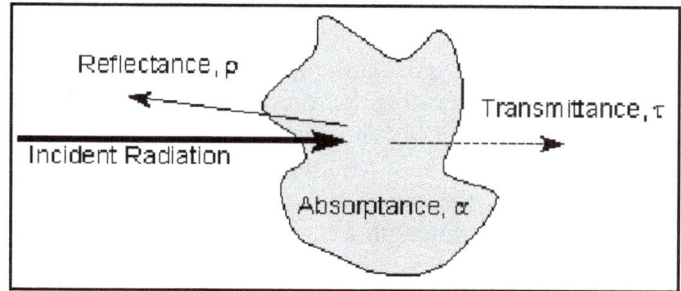

Figure 2. Radiation Incident on a body

$$\alpha + \rho + \tau = 1, \quad \text{as shown in Figure 2.}$$

If a material is opaque (no transmission), then: absorption + reflection = 1

In addition to the above processes, optical (including IR) radiation interacts with matter the same as radiation in any other part of the spectrum, including:

1. **Diffraction** – around edges
2. **Emission** – from matter by conversion from another form of energy
3. **Interference** – constructive and destructive
4. **Refraction** – bends when passing between two media with different propagation speeds (Snell's Law)
5. **Scattering** – when interacts with particles whose size approaches length of the wave
6. **Polarized** - electric field is partially polarized by reflection from dielectric

STERADIAN – SOLID ANGLE

Of significance to many terms and units in radiometric calculations is the *solid angle*. Figure 3 is a pictorial depicting the relationship of area, distance, and solid angle. By definition, the ratio of area on the surface of a sphere to the square of distance (the radius) is the unit less parameter *solid angle*, or steradian in the SI system of units. Solid angle is usually abbreviated as "sr" or given the Greek letter, Ω. The steradian is a dimensionless quantity in radiometric calculations.

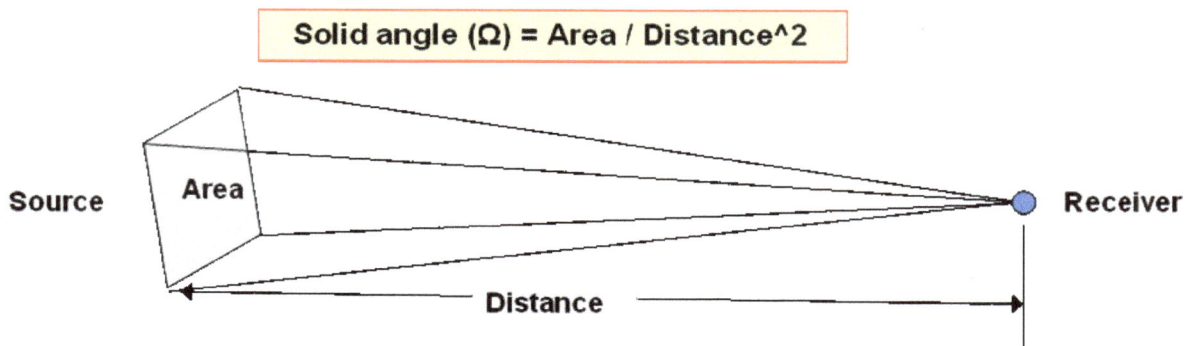

Figure 3. Solid Angle

A sphere contains a solid angle of 4π steradians; a hemisphere contains 2π steradians, and so on. The area is a curved surface, but in most applications, the solid angles are sufficiently small that the area can be approximated as a plane. Also, for small angles, the solid angle in steradians is approximately equal to the product of two plane angles in radians.

CONVERSIONS

IR wavelengths are typically expressed in μm, visible wavelengths in μm or nm, and UV wavelengths in nm or angstroms. Table 5 lists conversion factors for converting from one unit of wavelength to another. The conversion is from column to row. For example, to convert from μm to nm, multiply the value expressed in μm by 10^3. IR wavelengths are also sometimes expressed in a frequency-like unit called wavenumbers or inverse centimeters. A wavenumber value can be found by dividing 10,000 by the wavelength expressed in μm. For example, 2.5 μm converts to a wavenumber of 4000 or 4000 inverse centimeters (cm^{-1}).

Table 5. Wavelength Conversion Units

From ->	Angstroms - Å	Nanometers - nm	Micrometers - μm
To get ↓		Multiply by	
Angstroms - Å	1	10	10^4
Nanometers - nm	10^{-1}	1	10^3
Micrometers - μm	10^{-4}	10^{-3}	1

BASIC RADIANT POWER RELATIONSHIPS

Radiant intensity is the most commonly used term to describe the radiant power of a source per unit solid angle. Radiant Intensity offers the advantage of being a source term, like radiance, that is not related to the size of the radiating source. In practice, radiant intensity is a derived term and is not directly measurable. If the Instantaneous Field of View (IFOV), which represents the smallest optical resolution element of a remote sensor, subtends an angle smaller than the size of the radiating source, the sensor responds directly to *radiance*. If the IFOV of the sensor subtends an angle larger than the radiating source, the sensor responds to *irradiance*. The relationship among radiant intensity, radiance, and irradiance is shown in the following equation:

$$I = ED^2 = E\frac{A}{\Omega} = LA = L\Omega D^2$$

Where:

I = Intensity. Radiant intensity is the target source power per unit solid angle, in Watts/steradian.

E = Irradiance. Irradiance is the received power density in Watts/cm2 incident on a distant sensor. Irradiance is the quantity measured or detected by a distant sensor where the target is not spatially resolved (i.e., where the target subtends an angle smaller than the resolution or field of view of the instrument.)

L = Radiance. Radiance is the intensity per unit area of source, in W/cm^2/steradians. Radiance is the quantity measured by a camera or imaging system, where the target is optically resolved (i.e., where the target subtends an angle much greater than the resolution or instantaneous field of view of the imager.)

A = Area. Area is the cross-sectional or projected area of the target in cm^2.

D = Distance. Distance between the target and sensor in cm.

Ω = Solid angle subtended by the target in steradians. Solid angle appears directly or indirectly in many infrared quantities. The solid angle subtended by a source is the ratio of the source area to the square of distance.

All remote sensors receive energy from a source through an atmospheric path. The atmosphere attenuates the propagation of energy due to scattering, absorption, and molecular rotation and vibration, depending on the wavelengths involved. The atmosphere itself also radiates. All incident energy on a remote sensor, except as received under vacuum conditions, is detected through the atmosphere. The term *apparent* is applied to a radiometric quantity to acknowledge the presence of atmospheric effects. A remote sensor, responding the radiance of a source over an atmospheric path, responds to *apparent radiance*.

Real sensors do not respond uniformly to energy. Sensors have non-uniform spectral and spatial response due to many factors, including detector and optical characteristics. All energy received by the detector is spectrally weighted by the spectral response of the instrument. The term *effective* is applied to acknowledge the non-uniform response of the instrument.

Outside of a vacuum, all radiation incident on a remote sensor is attenuated by atmospheric effects such as scattering and absorption. Where the atmosphere absorbs, it also emits. Path radiance is the term applied to the contribution of the atmospheric path to the received radiation. The term *apparent* is applied to radiometric quantities to acknowledge the influence of the atmosphere on the received radiation.

DOMAINS, DISTRIBUTIONS, AND SENSOR RESPONSE

Radiant power that is emitted or reflected from a source is distributed across multiple dimensions or domains. The three domains that are most important to radiometric applications are spectral (Figure 4), spatial (Figure 5), and temporal (Figure 6). Knowing how the radiant power from a source is distributed across each domain is important to understanding performance of an EW system, whether it be a sensor, a weapon, or a countermeasure.

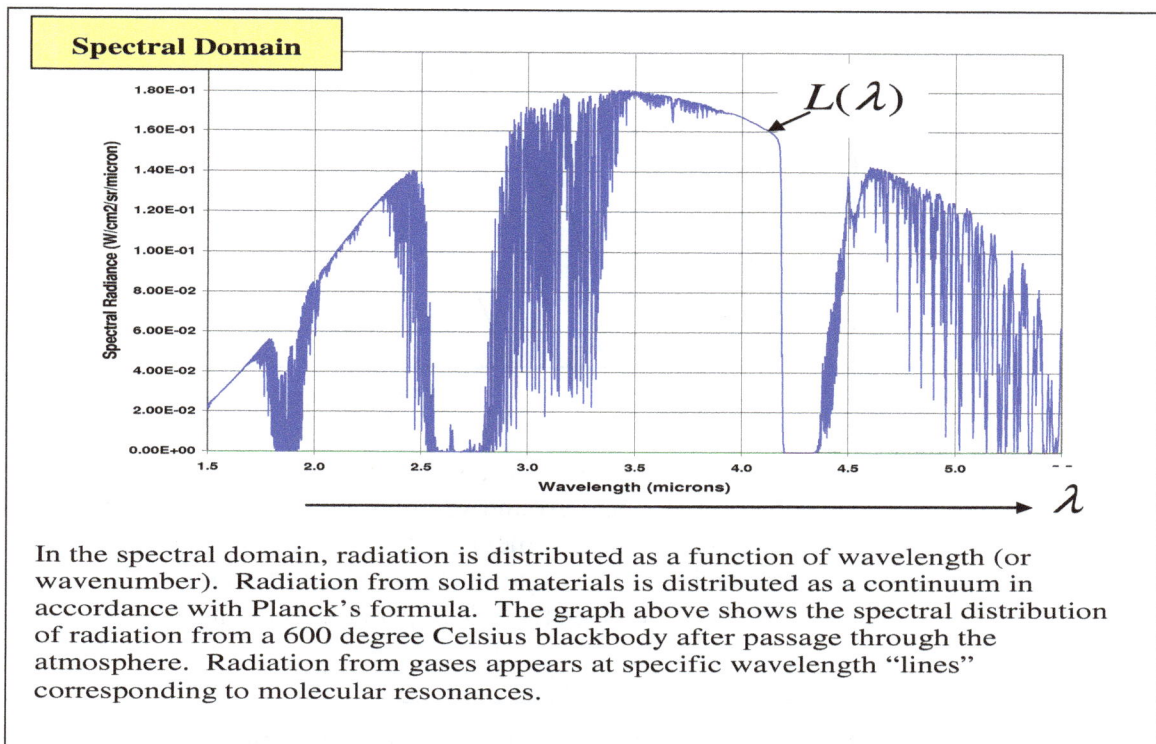

In the spectral domain, radiation is distributed as a function of wavelength (or wavenumber). Radiation from solid materials is distributed as a continuum in accordance with Planck's formula. The graph above shows the spectral distribution of radiation from a 600 degree Celsius blackbody after passage through the atmosphere. Radiation from gases appears at specific wavelength "lines" corresponding to molecular resonances.

Figure 4. Sample Spectral Distribution of a Solid Material

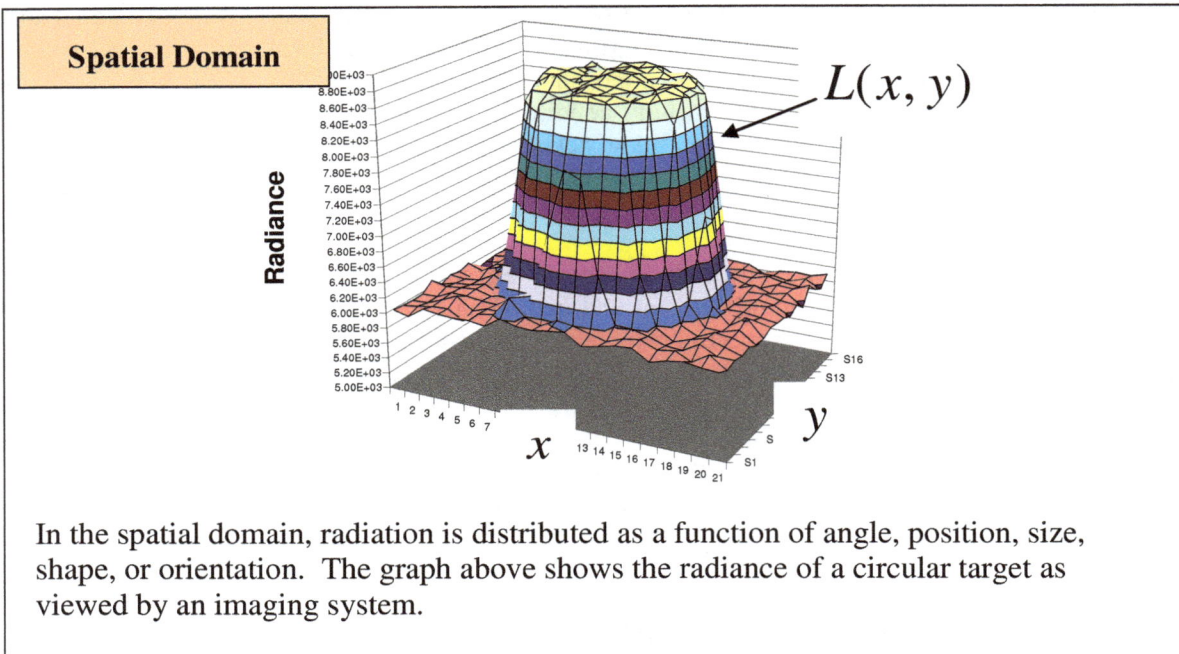

Spatial Domain

In the spatial domain, radiation is distributed as a function of angle, position, size, shape, or orientation. The graph above shows the radiance of a circular target as viewed by an imaging system.

Figure 5. Sample Spatial Distribution of a Circular Target Image

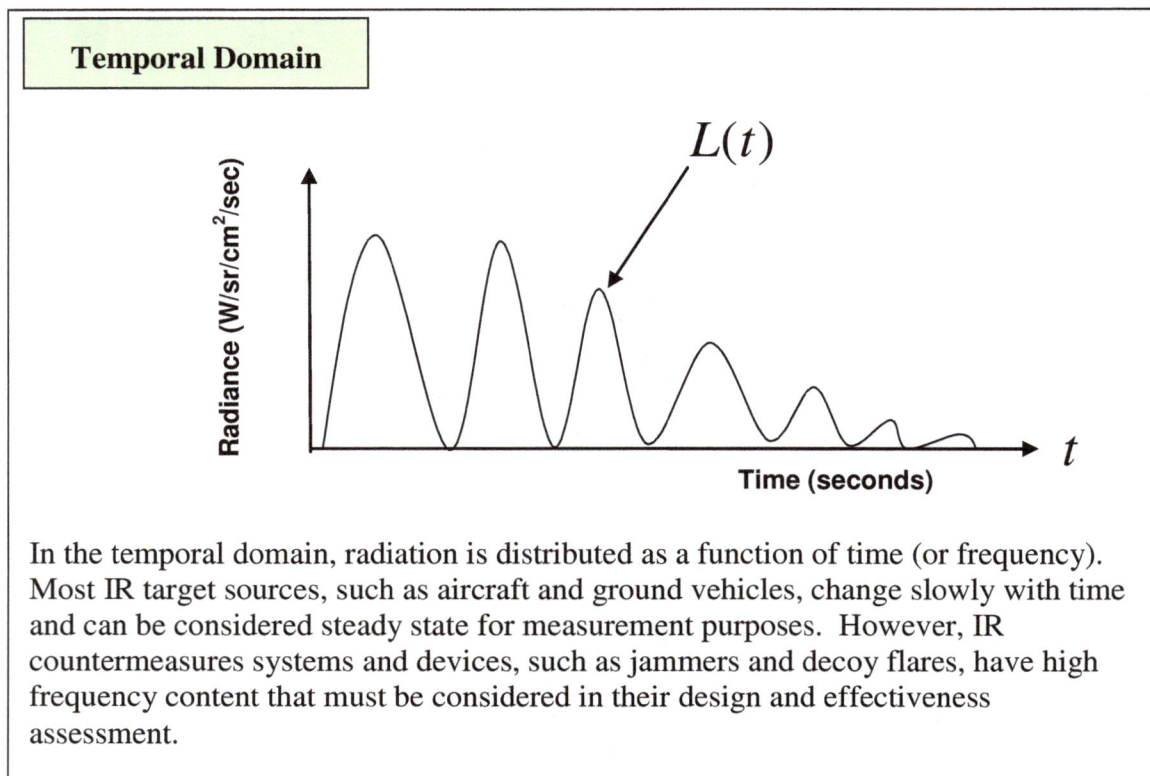

Temporal Domain

In the temporal domain, radiation is distributed as a function of time (or frequency). Most IR target sources, such as aircraft and ground vehicles, change slowly with time and can be considered steady state for measurement purposes. However, IR countermeasures systems and devices, such as jammers and decoy flares, have high frequency content that must be considered in their design and effectiveness assessment.

Figure 6. Sample temporal distribution of a modulated source

All electro-optical sensors have some sensitivity to how radiation is distributed across the different domains. This sensitivity or response takes the following forms:

1. **Spectral response**:
   a. May be broad in the case of a radiometer or imager, or narrow in the case of a spectrometer where response takes the form of resolution.
   b. Spectral response is chosen to exploit particular spectral features of interest in a target.
   c. Non-uniformity of spectral response can have a significant effect on the performance of sensor.

2. **Spatial response**:
   a. May be broad in the case of a radiometer or spectrometer, where spatial response takes the form of sensitivity across the sensor's FOV. In the case of an imager, the spatial response may be narrow because spatial response takes the form of the resolution or IFOV.
   b. Required spatial resolution is defined by the size and radiant power of the source, the distance between the source and the target, and by the sensitivity of the sensor.
   c. Non-uniformity of spatial response can have a significant effect on the performance of sensor.
   d. Response to radiation from the background within the FOV is primarily a spatial response issue and can be significant to sensor performance.

3. **Temporal response**:
   a. Takes the form of bandwidth or, in the case of a digital system, of sampling frequency.
   b. Unlike spectral and spatial response, temporal response includes the response time of the detector and all of the electronics, data transfer time, and processing time for a sensor.
   c. Non-uniformity across a region is usually not a concern. Temporal response is usually uniform for target frequencies that fall well below the low-pass or Nyquist breakpoints.

Sensors respond to received radiant power by integrating the power distribution in each domain under the corresponding response curve. Sensors respond in multiple domains, so output is proportional to a multi-dimensional, weighted integral. Figure 7 illustrates response-weighting, or convolution, of radiation with instrument response in the spectral domain. The concept applies equally in the spatial domain.

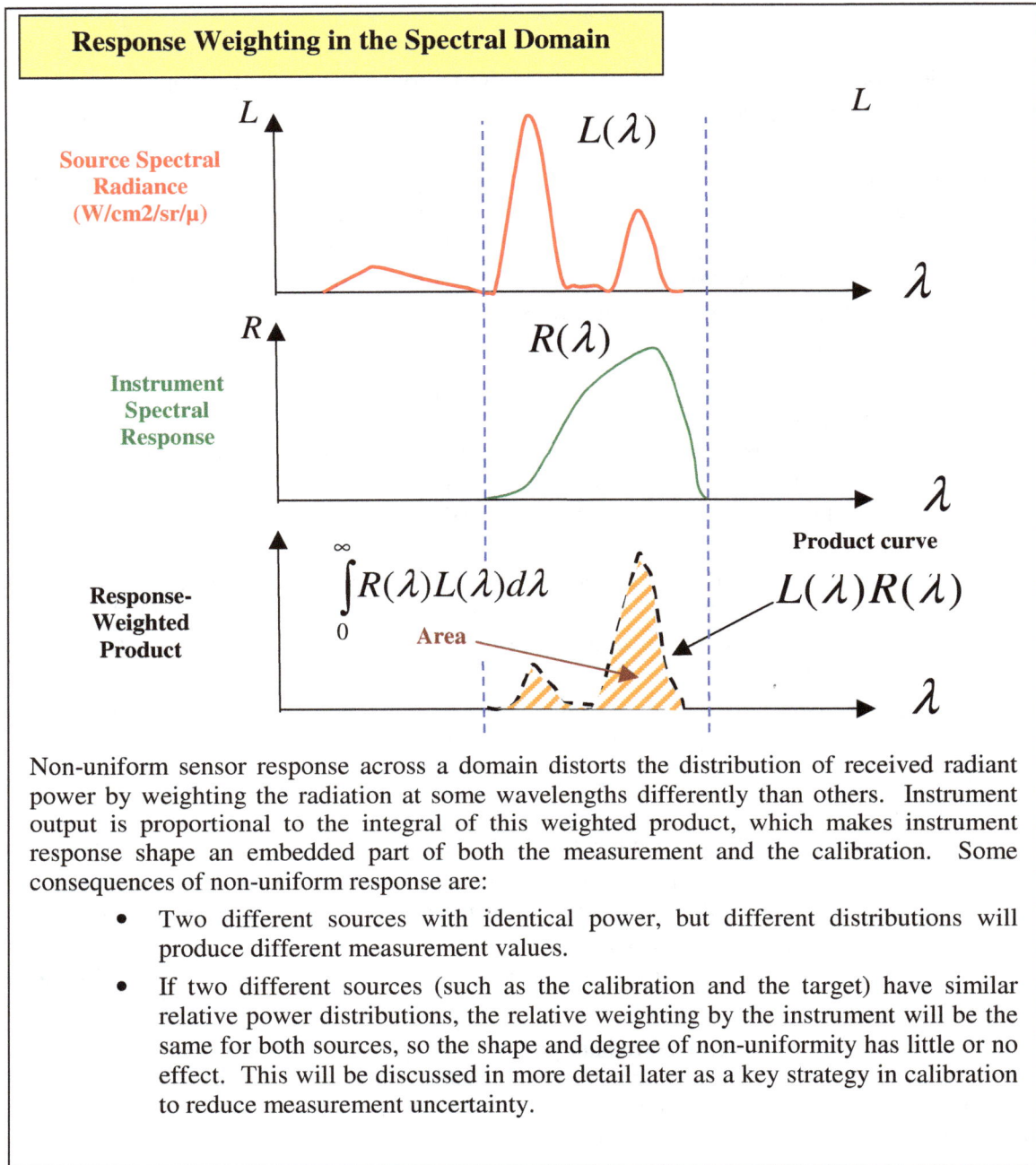

Response Weighting in the Spectral Domain

Source Spectral Radiance (W/cm2/sr/µ)

$L(\lambda)$

Instrument Spectral Response

$R(\lambda)$

Response-Weighted Product

$$\int_0^\infty R(\lambda)L(\lambda)d\lambda$$

Area

Product curve

$L(\lambda)R(\lambda)$

Non-uniform sensor response across a domain distorts the distribution of received radiant power by weighting the radiation at some wavelengths differently than others. Instrument output is proportional to the integral of this weighted product, which makes instrument response shape an embedded part of both the measurement and the calibration. Some consequences of non-uniform response are:

- Two different sources with identical power, but different distributions will produce different measurement values.

- If two different sources (such as the calibration and the target) have similar relative power distributions, the relative weighting by the instrument will be the same for both sources, so the shape and degree of non-uniformity has little or no effect. This will be discussed in more detail later as a key strategy in calibration to reduce measurement uncertainty.

Figure 7. Response-Weighting in the Spectral Domain.

The consequence of instrument output being proportional to multiple weighted integrals is that the characteristics of the instrument affect the measured value in ways that cannot be easily extracted or corrected. Understanding non-uniformity in sensor response is critical to understanding performance and evaluating the effectiveness of the system.

INFRARED SOURCE CHARACTERISTICS

Reflectance

Reflectance is generally categorized as either *specular* (mirror-like) or *diffuse* (scattered by reflection from a rough surface). Most surfaces exhibit both types of reflection, but one typically dominates. Reflections from smooth surfaces are specular. Reflectance is a unitless quantity between zero and one that is the ratio of reflected power to the incident power. Reflectance varies with angle from the normal and with wavelength.

Diffuse reflectors scatter incident radiation broadly. Johann Heinrich Lambert described an ideal diffuse reflector in which the intensity of reflected rays is distributed as a cosine of the angle from the normal, regardless of the angles of the arriving incident rays. Such a surface is described as *Lambertian*. The same relationship applies to a diffuse source. If the source is perfectly diffuse, the radiance is independent of the viewing angle because the projected area of the source also varies as a cosine function of the angle from normal.

While no surface is perfectly diffuse, standard paint on most aircraft, for example, is near enough for at least a first approximation. Significant effort has gone into measuring and describing surface properties of military vehicles to support modeling and simulation activities. Bidirectional Reflectance Distribution Function (BRDF) is defined as the ratio of the reflected radiance to the incident irradiance at a wavelength, λ.

The BRDF provides a complete description of the reflectance properties of a surface. Measurements of BRDF are made in specially equipped optical laboratories. For aircraft, a typical use of BRDF data is in computer models that predict the reflections of the earth and sun from the aircraft fuselage and wings.

Emissivity

Emissivity is a unitless quantity that is a measure of the efficiency of a surface as an absorber or emitter. Emissivity is expressed as a number between 0.0 and 1.0. According to Kirchoff's Law, for an opaque object in thermal equilibrium, i.e., no net heat transfer, the emissivity equals the absorptance. In other words, a perfect emitter is also a perfect absorber. This ideal emitter is known as a *blackbody*. A surface with an emissivity of 1.0 emits the maximum radiation that Planck's Law allows. A body for which the emissivity is constant with wavelength but less than 1.0 is commonly known as a *graybody*. Planck's equation is typically expressed for an ideal blackbody emitter, but multiplying the blackbody expression by the emissivity term expresses the spectral distribution of power for a graybody.

Electromagnetic (EM) radiation, including infrared radiation, can be characterized by amplitude, frequency (or wavelength), coherency, and polarization properties. Amplitude refers to the magnitude of the electromagnetic wave. Coherency refers to the degree in which the electromagnetic waves maintain a constant phase difference, both spatially and temporally. Polarization is a description of the orientation of the wave's propagation perpendicular to the direction of travel. Like all electromagnetic radiation, infrared radiation travels as an EM field with the propagation velocity of the speed of light in a vacuum, and slower through air and other dielectrics. , Infrared radiation also exhibits both wave and particle properties. Which property is used depends on the application: Waves are used for applications involving propagation and geometrical optics, while particles (photons) are used for most detector applications. The energy of a photon is inversely proportional to its wavelength (hc/λ) where h is the Planck's constant, and c is the speed of light.

Infrared radiation also interacts with matter in a variety of ways:

- Reflects – wave is reflected from a surface.
- Refracts – direction of wave bends when passing between two transparent media with different propagation speeds.
- Scatters – when interacts with particles whose size approaches the wavelength of the radiation.
- Diffracts – around edges of an obstruction.
- Interferes – constructively and destructively.
- Absorbs – when absorbed by matter, radiation is (usually) converted into heat energy.
- Emits – radiation is emitted from matter by conversion from another form of energy.
- Transmitted – propagates through a transparent medium (or vacuum).

Plank's Law:

Infrared is directly related to the heat radiated from matter. Anything with a temperature above absolute zero (-273.15 degrees Celsius) radiates in the infrared. Planck's Law, discovered by German Physicist Max Planck, mathematically describes the distribution of radiant power across the spectrum for a given temperature. The form presented below calculates source radiance in Watts/sr/cm^2/ μ:

$$L_\lambda = \frac{C_1}{\pi \lambda^5 \left[e^{\left(\frac{C_2}{\lambda T}\right)} - 1 \right]}$$

Where: $C_1 = 2\pi c^2 h = 3.7415 \times 10E{+}04$ W cm^{-2} μ^4
$C_2 = ch/k = 1.4389E{+}04$ μ K
c = speed of light; h = Plank's constant; k = Boltzmann's constant
With λ in μ and T in Kelvin (= °C + 273.15)

The behavior of Planck's curve with temperature is fundamental to every infrared detection scenario. Figure 8 shows Planck's equation for a single temperature. The shape of this curve resembles a wave of water, with a steep rise in power on the short wavelength side of the peak and a tail off on the long wavelength side.

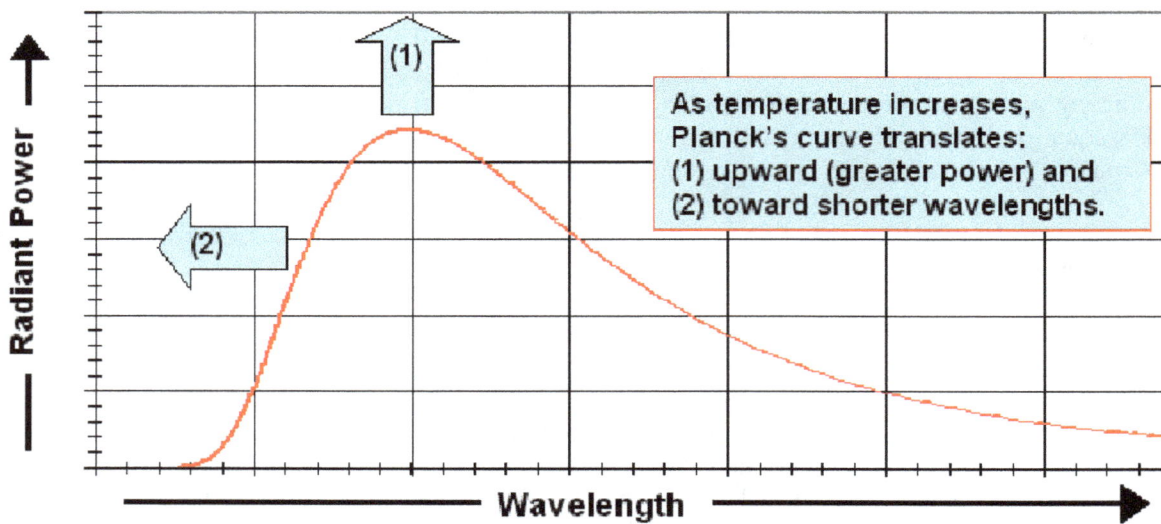

As temperature increases, Planck's curve translates:
(1) upward (greater power) and
(2) toward shorter wavelengths.

Figure 8. Temperature Effects and Power Distribution

When temperature increases, two significant changes in power distribution occur and are governed by the following two relationships:

Stefan-Boltzmann Law:

The total power under the curve increases proportionally to the fourth power of the temperature. According to the Stefan-Boltzmann law, the total radiant emittance (power) of a blackbody is:

$$M = \sigma T^4 \qquad \text{Where:} \quad \sigma = \frac{2\pi^5 k^4}{15 c^2 h^3} = 5.67 \times 10^{-12} \, Watts \; cm^{-2} \, {}^\circ K^{-4}$$

$$\text{and} \qquad T \text{ is in Kelvin}$$

This is Plank's radiation law integrated over all values of λ.

Wein's Displacement Law

The peak emission translates toward shorter wavelengths.

Wein's displacement law takes the derivative of the Plank's law equation to find the wavelength for maximum spectral exitance (emittance) at any given temperature:

$$\lambda_{max} = 2897.8 \,/\, T \qquad \text{where } T = \text{temperature is in Kelvin}$$

The surface of the sun radiates with a spectral distribution like that of a 5900 Kelvin source. In accordance with Wein's Law, the wavelength at which the radiation for a 5900 Kelvin source peaks is approximately 490 nanometers. The maximum sensitivity of the day-adapted human eye (photopic) occurs at about 555 nanometers, which happens to be in a highly transparent region of the atmosphere. In other words, the temperature of our sun and the transmission of our atmosphere are conveniently matched to the response of the human eye. Figure 9 shows the relationship between the response of the human eye and the spectral distribution of solar irradiance.

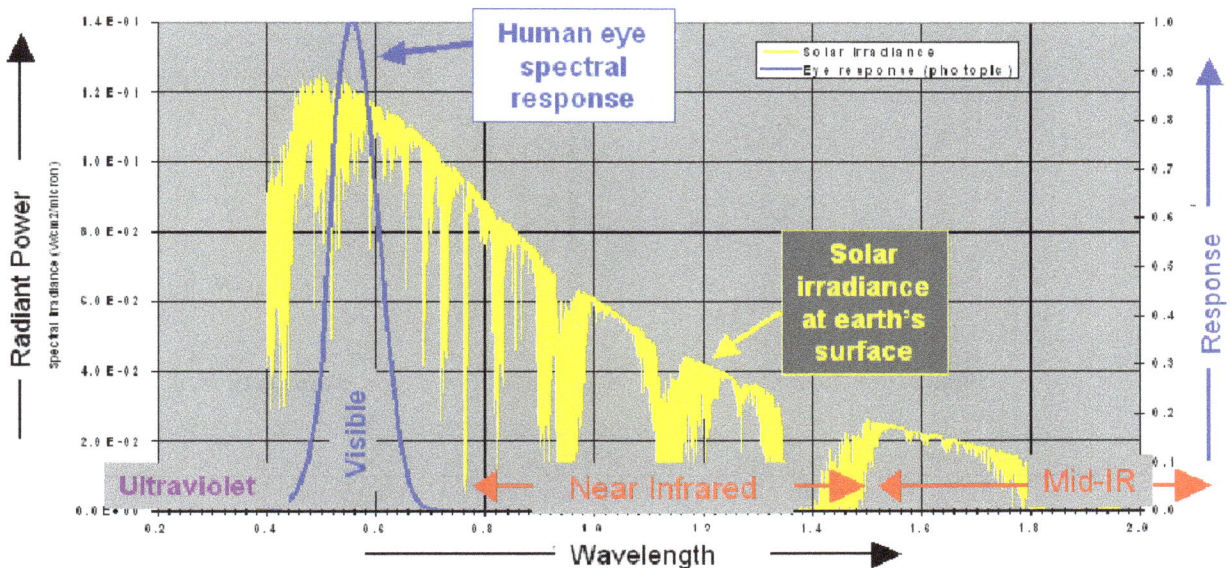

Figure 9. Power Distribution of the Sun Related to the Human Eye

Unlike other electromagnetic radiation, the peak radiation from objects at the temperature of the earth, and of humans, vehicles, aircraft, etc., lies at infrared wavelengths. This makes the infrared portion of the spectrum militarily important.

For detection of an object, whether it is with the human eye or with an electro-optical sensor, it is not where the maximum emissions occur in the spectrum that is important. It is where the maximum difference between the target and background lies that is most significant.

The *absolute* signature of a source or target is the power radiated and reflected by the target without influence from background radiation. Absolute signature can be measured with an instrument such as an imaging camera at close range, producing highly resolved IR imagery. The difference between radiation from a target and its background is its *contrast*. Contrast is the target signature relative to a specific background. Contrast irradiance is the quantity detected by any remote sensor or missile. Contrast varies greatly with background conditions.

Figure 10 shows several different types of source distributions encountered by infrared sensors. Most targets of military interest are a complex combination of spectral contributions from a variety of sources. For example, the signature of an aircraft can typically be described by the contributions from airframe, hot engine parts, and plume.

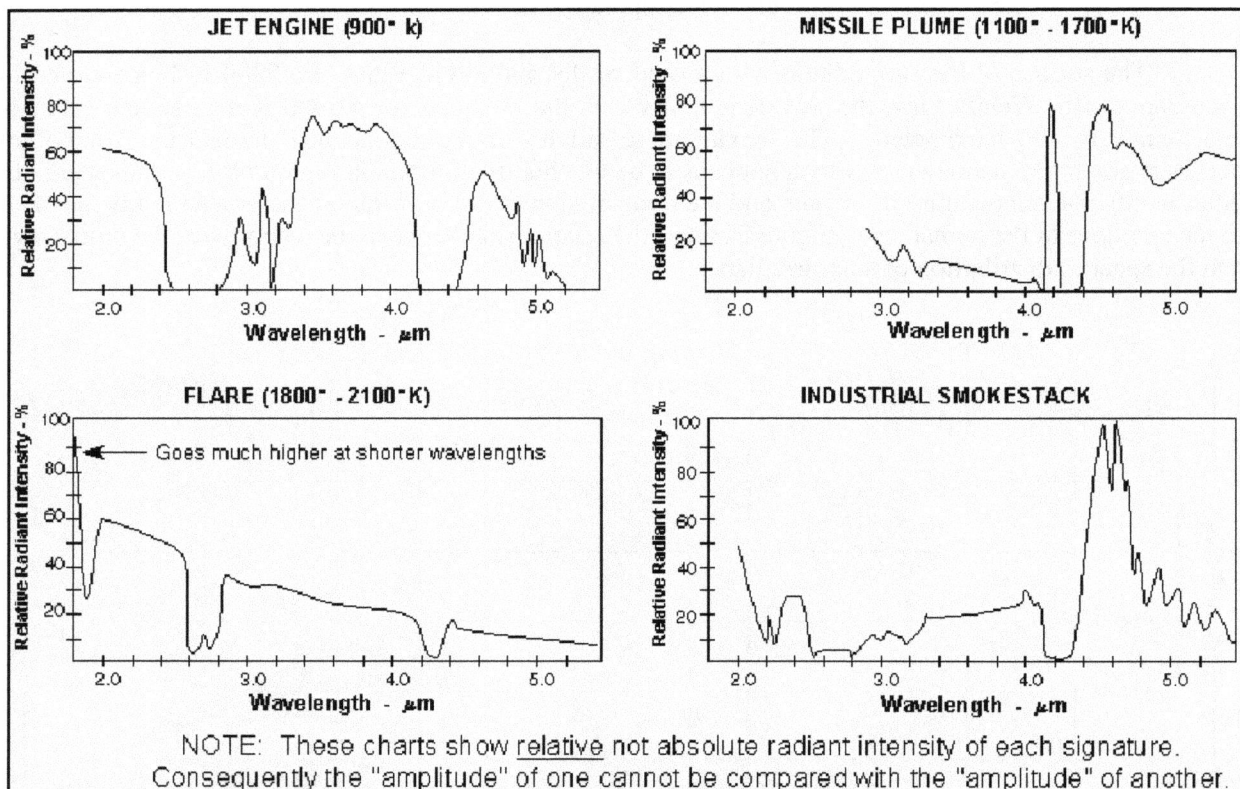

Figure 10. Spectral Distribution of Various Targets

ATMOSPHERIC TRANSMISSION

The radiation emitted or reflected from the targets and backgrounds must pass through the intervening atmosphere before reaching the detection system. The radiation is absorbed and re-emitted by molecular constituents of the atmosphere and scattered into and out of the path by various aerosol components.

Figure 11 reveals the presence of atmospheric windows, i.e. regions of reduced atmospheric attenuation. IR detection systems are designed to operate in these windows. Combinations of detectors and spectral bandpass filters are selected to define the operating region to conform to a window to maximize performance and minimize background contributions.

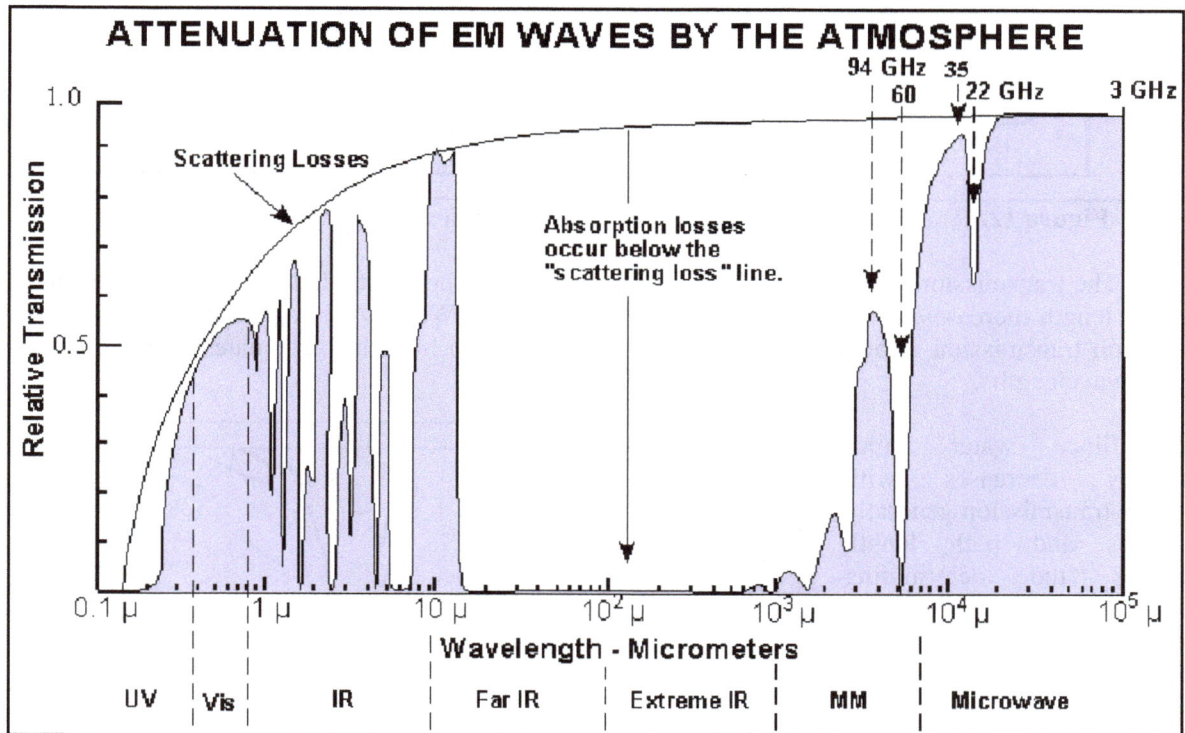

Figure 11. Atmospheric Transmission Over 1 NM Sea Level Path

The molecules that account for most of the absorption in the IR region are water, carbon dioxide, nitrous oxide, ozone, carbon monoxide, and methane. Figure 12 shows an expanded view of the infrared portion of the spectrum.

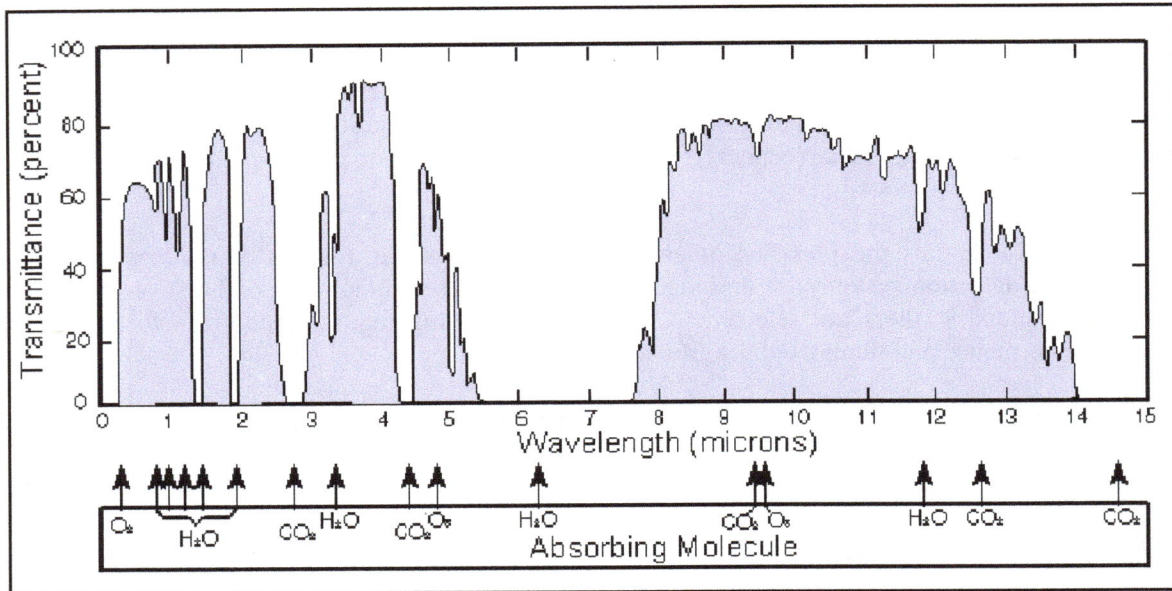

Figure 12. Transmittance of Atmosphere Over 1 NM Sea Level Path (Infrared Region)

The transmission in a window is greatly dependent on the length and characteristics of the path. As path length increases, absorption gets deeper and broader. Water vapor also has a significant effect overall on transmission through the atmosphere. High relative humidity attenuates transmission at all optical wavelengths.

Since water vapor generally decreases with altitude, transmission generally increases and path length becomes the determining factor. However, path length does not affect transmission of all wavelengths the same.

The altitude effects on transmittance are shown in Figure 13.

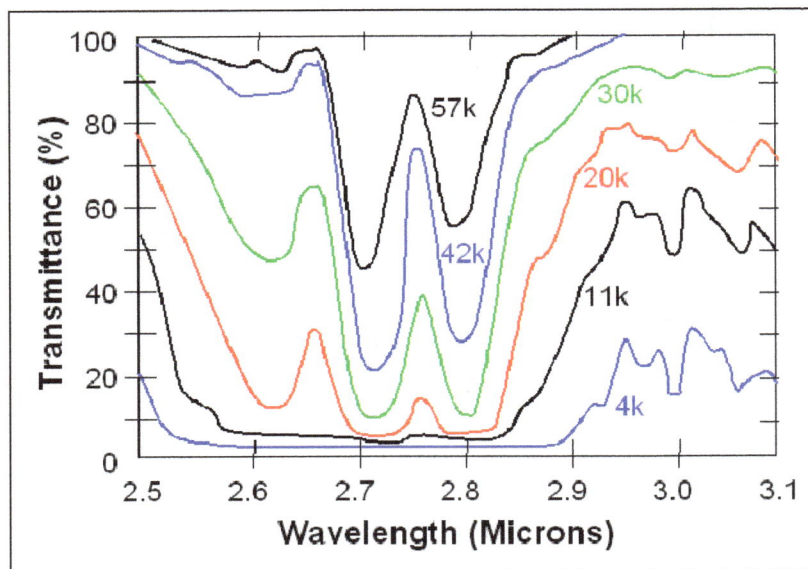

Figure 13. Transmittance at Various Altitudes

A great deal of work has gone into developing high fidelity atmospheric models. One of the most commonly used tools is MODTRAN™ (MODerate Resolution Atmospheric TRANsmission). MODTRAN™ models atmospheric propagation of EM radiation from 0.2 to 100 um. MODTRAN was developed by Spectral Sciences Inc. and the United States Air Force Research Laboratory.

ELECTRO-OPTICAL COMPONENTS AND SENSORS

Almost all IR instruments, missiles, search systems, etc. have similar functional components. Basic components typically include:

- Optics - Reflective or refractive lenses to:
    - Collect radiation. Irradiance (power density) is increased by collecting radiation over a large area and focusing down to a small area.
    - Form or focus an image that will be used to extract information about the target.

- Filter(s) - Spectral and spatial filters to distinguish target from background and to extract
    - Target information.
    - Spectral filters restrict sensitive wavelength range.
    - A spatial filter separates image information by features such as size or position.

- Detector - A transducer to convert received radiation to an electrical signal for processing.

- Electronics - Used to amplify and condition the detector signal and perform some action, such as controlling a servo for tracking and guidance or recording the received information. In addition to the above, the optical head may also contain a window to protect the electronics and an output unit consisting of indicators or displays.

<u>**Windows / Domes / Lens Materials**</u>

For most applications of EO systems in EW the detection system is protected from the environment by a window or dome of optically transmissive material. The window operates both as a weather seal and, in some cases, helps to define the spectral response region of the system. In some cases the window functions as a lens.

IR energy interacts with matter in ways we associate with light (reflection, refraction, and transmission). Lower energy of IR photons results in different optical properties than light. For example:

- Glass and water are not transparent to wavelengths longer than about 3 microns.
- Silicon and germanium are highly transparent to infrared radiation, but are opaque to visible light.

Transmission bands of representative window or lens materials are shown in Figure 14. The end points depicted are for the 10% transmission wavelengths. Not shown in Figure 14 are the various UV transmissive glasses such as Pyrex, Corex, and Vycor or Amorphous Material Transmitting IR radiation (AMTIR) which are various combinations of Ge-As-Se (AMTIR-1), As-Se (AMTIR-2, 4, & 5), Ge-Sb-Se (AMTIR-3), As-S (AMTIR-6).

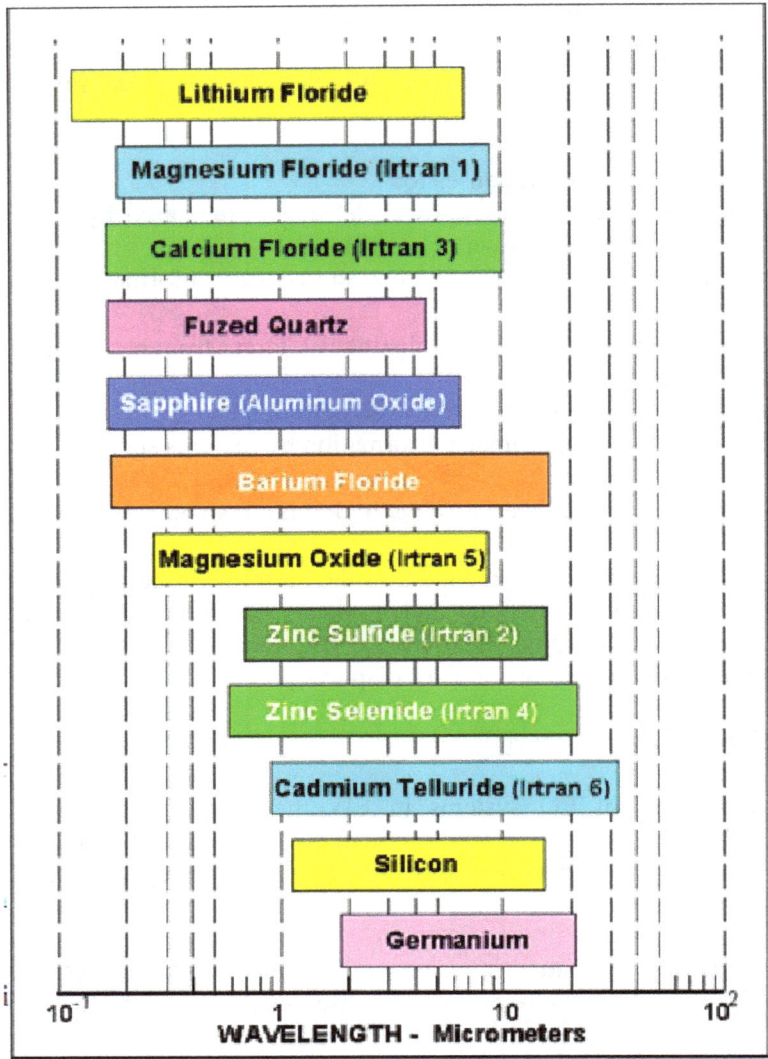

Figure 14. Transmission of Selected Window / Lens Materials

Objective Lens

The objective lens is the first optical element in a lens in a typical sensor or missile seeker. The objective lens serves two main functions:

- Collect radiation (i.e., multiply irradiance (power density) by collecting over a large area and focusing onto a small area)
- Form an image of the target scene onto a filter and detector array.

Lens Types

Lenses for the IR can be either refractive or reflective. Refractive optics are "straight through" lenses, with the light never making large bends as shown in Figure 15.

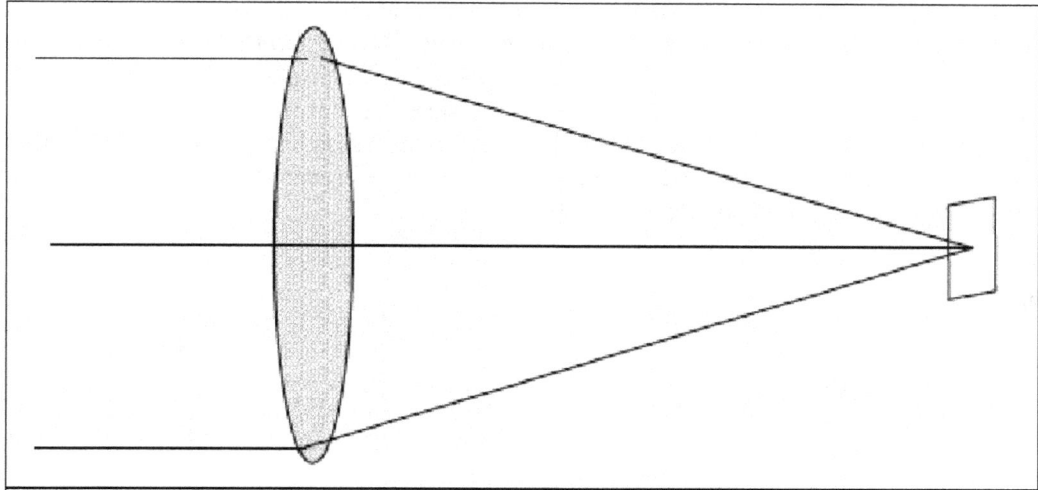

Figure 15. Refractive Lens

A common reflective design used in many missiles is the Cassegrain. The Cassegrain design shown in Figure 16 is compact in size for its focal length.

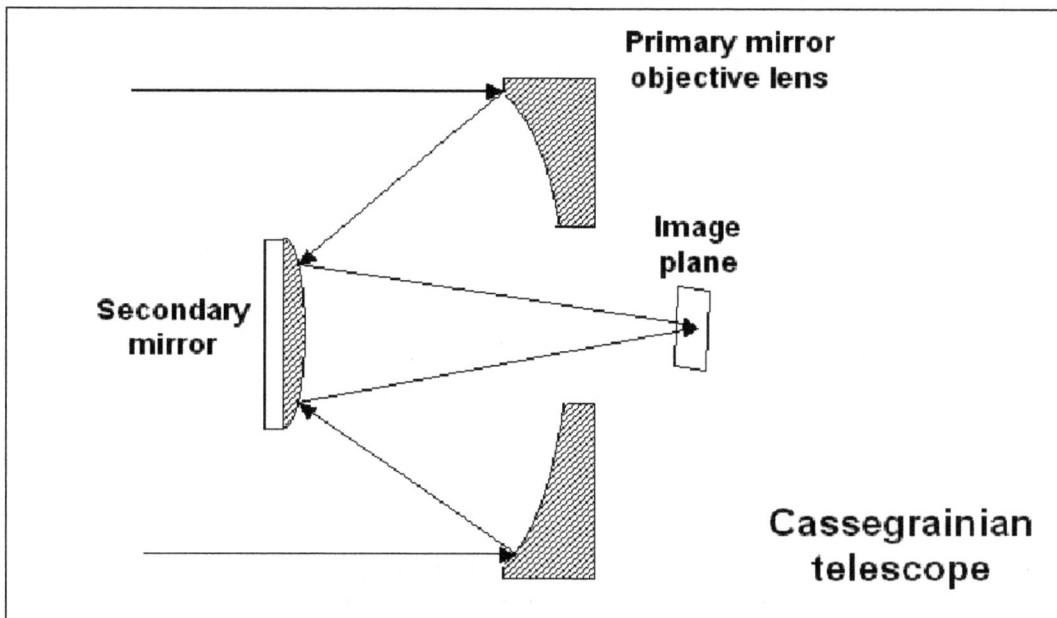

Figure 16. Cassegrain Lens

Optical Filters

Filters may be divided into two major categories: (1) spectral and (2) spatial.

Most optical radiation detectors have a wider sensitivity band than desired for the particular application. Spectral filters restrict sensitive wavelength range. Reasons for filtering include: enhancement of target-to-background contrast, avoidance of unwanted plume emissions and/or atmospheric absorption regions, and extraction and measurement of target spectral features.

To further define the system sensitivity, band interference filters or absorption filters are used. An absorption filter is a bulk material with a sharp cut-on or cut-off in its transmission characteristic. A cut-on and a cut-off filter can be combined to make a bandpass filter. By selecting absorption characteristics of absorption filters combined with the response of a detector, the desired system response can be obtained. An interference filter is composed of dielectric coatings on an appropriate substrate combined in such a way to produce cut-on, cut-off, or bandpass filters. Interference filters allow more control of the final response characteristics and smaller elements. See Figure 17.

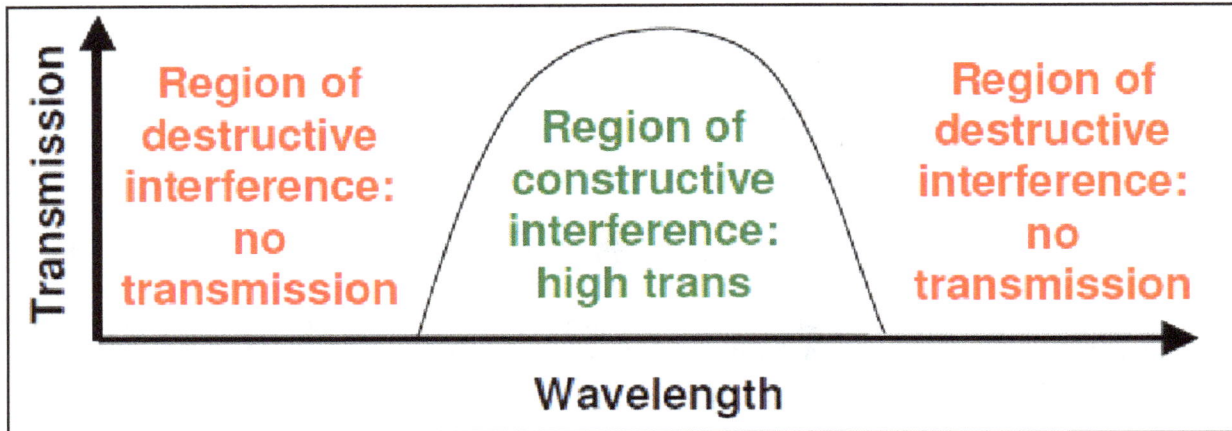

Figure 17. Spectral Bandpass Filter

Most spectral filters are of the thin-film interference type. Layers of dielectric material are vacuum deposited on a substrate window material. Typical substrate materials in IR are sapphire, silicon, and germanium. Thickness of deposited layers designed to have constructive interference to pass desired radiation at desired wavelengths and destructive interference to block undesired wavelengths Besides spectral filters, EO system optics often have antireflection (or AR) coatings to eliminate or greatly reduce unwanted reflections between optical elements.

A spatial filter separates information in a scene image by features such as size or position. Spatial filters take a variety of forms. Some common types and their functions include:

- Field stop: limits an instrument's field of view. Blocks unwanted sources (such as sun) outside nominal field of view.
- Mechanical modulator or "chopper."
- Reticle: A mechanical modulator used in many missile designs. Usually discriminates against extended sources (such as background) in favor of "point" target sources and provides target directional information from modulation phase.

Detector Coolers

Many IR detectors have to be cooled for proper operation. Most systems use closed-cycle coolers or thermoelectric coolers. Thermoelectric coolers use the Peltier effect, which produces a reduced temperature by passing a d-c current through a thermoelectric junction. Multi-stage coolers can cool a detector down to below 200°K. Closed-cycle coolers typically are of the Stirling cycle design and utilize the expansion of a gas (helium) to cool a cold finger attached to the detector. These generally operate at liquid nitrogen temperature (77°K).

Detectors

A detector is a transducer that transforms electromagnetic radiation into a form, which can be more easily detected. In the detectors of interest to EW the electromagnetic radiation is converted into an electrical signal. In some systems the signal is processed entirely within the system to perform its function. In others the signal is converted to a form to allow the human eye to be used for the final detection and signal analysis.

Detectors are transducers than convert optical radiation into electrons. The physical effects by which electromagnetic radiation is converted to electrical energy are divided into two categories: photon effects and thermal effects. EW systems primarily use detectors dependent on photon effects. These effects can be divided into internal photo effects and external photo effects.

The external photo effect is known as photoemission. In the photoemissive effect, photons impinging on a photocathode drive electrons from its surface. These electrons may then be collected by an external electrode and the photocurrent thus obtained is a measure of the intensity of the received radiation.

Internal photoeffects of interest are the photoconductive effect and the photovoltaic effect. In the photoconductive effect, absorbed photons cause an increase in the conductivity of a semiconductor. The change is detected as a decrease in the resistance in an electrical circuit. In the photovoltaic effect, absorbed photons excite electrons to produce a small potential difference across a p-n junction in the semiconductor. The photovoltage thus produced may be amplified by suitable electronics and measured directly.

Thermal detectors respond directly to heat. Examples of these devices include bolometers, thermopiles, and pyroelectric detectors. The pyroelectric effect is an example of the thermal effect. The pyroelectric effect is a change in polarization in a crystal due to changes in temperature. Radiation falling on such a crystal is detected by observing the change in polarization as a build up of surface charge due to local heating. When coated with a good black absorber, the crystal will be sensitive to a wide band of wavelengths.

Figure 18 shows the spectral sensitivity range of typical detectors using these effects.

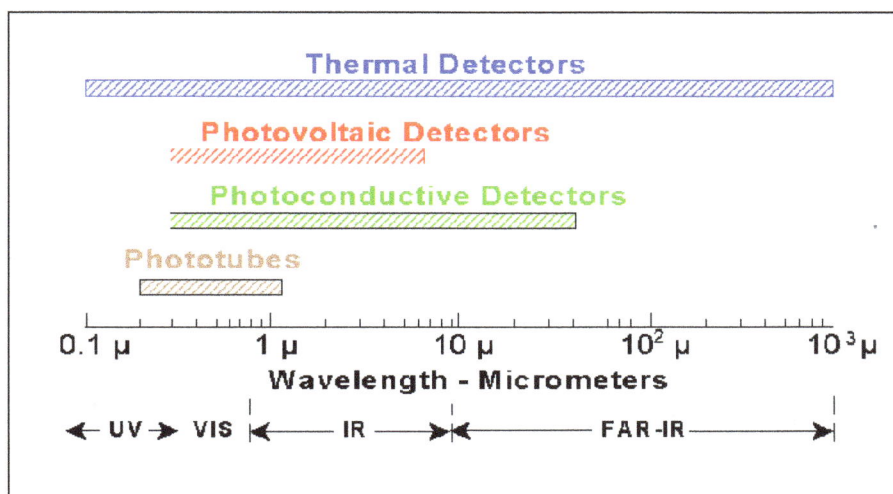

Figure 18. Spectral Range of Various Detectors

Detector Types

Photoconductive detectors operate as resistors in a circuit. The resistance of the detector changes as the radiation incident on its surface changes. For EW applications, the most photoconductive detector types include: Indium Antimonide (InSb), which can also be operated in photovoltaic mode; Gallium Arsenide (GaAs); Lead Sulfide (PbS); and Lead Selenide (PbSe).

Photovoltaic detectors, the most common detectors used in modern EW and military sensor applications, produce a voltage that is proportional to the incident radiation. Common examples of photovoltaic detectors are Indium Antimonide (InSb) and Mercury Cadmium Telluride (HgCdTe). Both of these detector types offer high sensitivity when cryogenically cooled.

Diode phototubes and photomultipliers are commonly used detectors for UV systems including many operational missile-warning systems. These types of tubes offer the advantage of operating uncooled which can significantly reduce the complexity of a sensor system and offer increased reliability. Most of the modern IR sensors require cooled detectors. InSb, for example, requires cooling to 77 Kelvin to achieve the necessary sensitivity. Depending on the application, HgCdTe can be operated at somewhat higher temperature conditions.

A Photoelectromagnetic (PEM) detector has a junction that generates a current when exposed to light in a magnetic field.

Some detectors (such as InSb) have multiple modes of operation, including: Photoconductive (PC), Photovoltaic (PV), or Photoelectromagnetic (PEM) modes of operation.

Detector Parameters and Figures of Merit

The important parameters in evaluating a detector are the spectral response, time constant, the sensitivity, and the noise figure.

The spectral response determines the portion of the spectrum to which the detector is sensitive.

The time constant is a measure of the speed of response of the detector. It is also indicative of the ability of the detector to respond to modulated radiation. When the modulation frequency is equal to one over the time constant, the response has fallen to 70.7 % of the maximum value. The time constant is related to the lifetime of free carriers in photoconductive and photovoltaic detectors and to the thermal coefficient of thermal detectors. The time constant in photoemissive devices is proportional to the transit time of photoelectrons between the photocathode and anode.

The sensitivity of a detector is related to its responsivity. The responsivity is the ratio of the detected signal output to the radiant power input. For photoconductive and photovoltaic detectors the responsivity is usually measured in volts per watt -- more correctly, RMS volts per RMS watt. However, the sensitivity of a detector is limited by detector noise. Responsivity, by itself, is not a measure of sensitivity. Detector sensitivity is indicated by various figures of merit, which are analogous to the minimum detectable signal in radar. Such a quantity is the noise equivalent power (NEP). The NEP is a measure of the minimum power that can be detected. It is the incident power in unit bandwidth, which will produce a signal voltage equal to the noise voltage. That is, it is the power required to produce a signal-to-noise ratio of one when detector noise is referred to unit bandwidth. The units of NEP are usually given as watts, but more correctly, are watts/Hz$^{1/2}$ or watts·sec$^{1/2}$.

Another figure of merit is the noise equivalent irradiance (NEI). The NEI is defined as the radiant power per unit area of the detector required to produce a signal-to-noise ratio of one. The units of NEI are watts per square centimeter.

Noise equivalent power (NEP) is the radiant power required to produce a signal to noise ratio of one for a detector. Detectivity (D) of a detector is defined as the reciprocal of the NEP. The units of D are watts $^{-1} \cdot$sec$^{-\frac{1}{2}}$. A higher value of detectivity indicates an improvement in detection capability. Since D depends on detector area, an alternate figure of merit, known as D-star (D*). D* is the detectivity measured with a bandwidth of one hertz and reduced to a responsive area of one square centimeter. The units of D* are cm$\cdot$watts $^{-1} \cdot$sec$^{-\frac{1}{2}}$. D* is the detectivity usually given in detector specification sheets. Typical spectral detectivity characteristics for various detectors are shown in Figure 19.

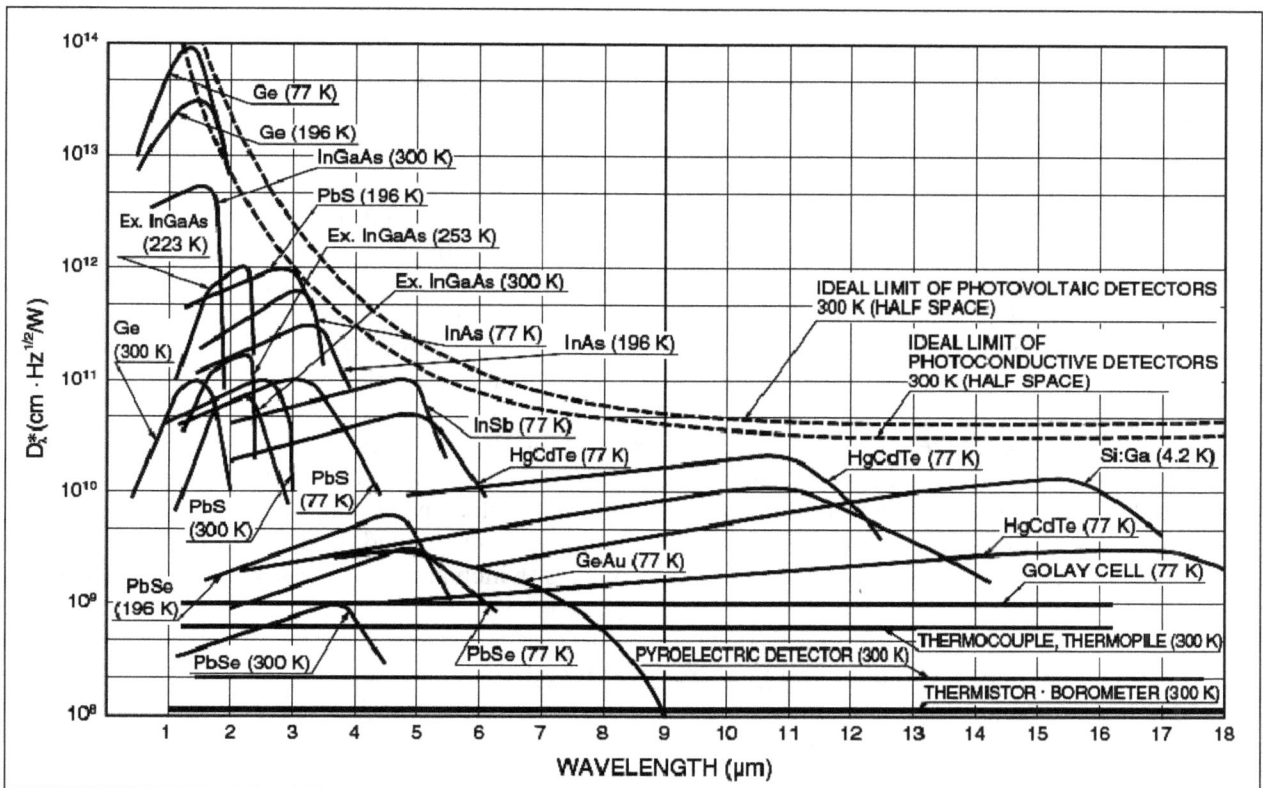

Figure 19. Spectral Detectivity of Various Detectors

Besides the NEI mentioned above, the quantum efficiency of the photocathode is also a figure of merit for photoemissive devices. Quantum efficiency is expressed as a percent -- the ratio of the number of photoelectrons emitted per quantum of received energy expressed as a percent. A quantum efficiency of 100 percent means that one photoelectron is emitted for each incident photon.

There are other figures of merit for television cameras. The picture resolution is usually described as the ability to distinguish parallel black and white lines and is expressed as the number of line pairs per millimeter or TV lines per picture height. The number of pixels in the scene also defines the quality of an image. A pixel, or picture element, is a spatial resolution element and is the smallest distinguishable and resolvable area in an image. CCD cameras with 512 x 512 elements are common. Another resolution quantity is the gray scale, which is the number of brightness levels between black and white a pixel can have.

Noise in Detectors

The performance of a detector is limited by noise. The noise is the random currents and voltages that compete with or obscure the signal or information content of the radiation. Five types of noise are most prominent in detectors: (1) thermal, (2) temperature, (3) shot, (4) generation-recombination, and (5) 1/f noise.

Thermal noise, also known as Johnson noise or Nyquist noise, is electrical noise due to random motions of charge carriers in a resistive material.

Temperature noise arises from radiative or conductive exchange between the detector and its surroundings, the noise being produced by fluctuations in the temperature of the surroundings. Temperature noise is prominent in thermal detectors.

Shot noise occurs due to the discreteness of the electronic charge. In a photoemissive detector shot noise is due to thermionic emission from the photocathode. Shot noise also occurs in photodiodes and is due to fluctuations in the current through the junction.

Generation-recombination noise is due to the random generation and recombination of charge carriers (holes and electrons) in semiconductors. When the fluctuations are caused by the random arrival of photons impinging upon the detector, it is called photon noise. When it is due to interactions with phonons (quantized lattice vibrations), it is called generation-recombination noise. Johnson noise is predominant at high frequencies, shot noise predominates at low frequencies, and generation-recombination and photon noise are predominant at intermediate frequencies.

As the name implies, 1/f noise has a power spectrum that is inversely proportional to frequency. It is dominant at very low frequencies. In photoemissive detectors it is called flicker noise and has been attributed to variation in the emission from patches of the photocathode surface due to variation in the work function of the surface. In semiconductors 1/f noise is also called modulation noise. Here it is apparently due to surface imperfections and ohmic contacts (which are a form of surface imperfection).

Infrared Spectral Region and Features of Interest

Different portions of the infrared spectrum are common for particular applications. The reasons for the selection of a specific window are often sensitive and beyond the scope of this document, but selections are typically based on several key considerations:

- Target characteristics such as size and spectral distribution of signature.
- Background radiance and clutter.
- Atmospheric effects (transmission, path radiance, scintillation, etc.).
- Distinguishing characteristics between natural and man-made sources.

Table 6 describes some of the types of characteristics that are prevalent in the short-wavelength (SWIR) infrared (0.7 to 3.0 microns), mid-wavelength (MWIR) infrared (3.0 to 6.0 microns) and long-wavelength (LWIR) infrared (7.0 to 14.0 microns) along with some types of systems that operate in these regions.

Table 6. Infrared Features, Regions, and Types of Systems

Near or Short Wave IR (SWIR)	**Dominant natural source**:	**Sun**
	Atmospheric:	
	Transmission:	High
	Path radiance:	Scattered sunlight
	Dominant aircraft IR component:	Sunlit airframe
	Anti-aircraft threat:	Vehicle-launched SAM

Mid-wave IR (MWIR)	**Dominant natural source**:	**Sun**
	Atmospheric:	
	Transmission:	High transmission "windows" between H_2O and CO_2 absorption
	Path radiance:	Scattered sunlight below 3 microns Thermal at longer than 3 microns
	Dominant aircraft IR component:	Engine hot parts and plume
	Anti-aircraft threat:	All AAMs and SAMs

Far or Long-wave IR (LWIR)	**Dominant natural source**:	**Earth**
	Atmospheric:	
	Transmission:	High
	Path radiance:	Low: small thermal emission from ozone
	Dominant aircraft IR component:	Airframe direct emission and terrestrial illumination
	Anti-aircraft threat:	Airborne IRST. No anti-aircraft missiles

Sensors and Detection

Figure 20 shows a generalized detection problem. On the left of the diagram are the radiation sources - the sun, background, and the target of interest. In the middle is the intervening atmosphere, which attenuates the radiation as it travels to the detection system shown on the right of the diagram.

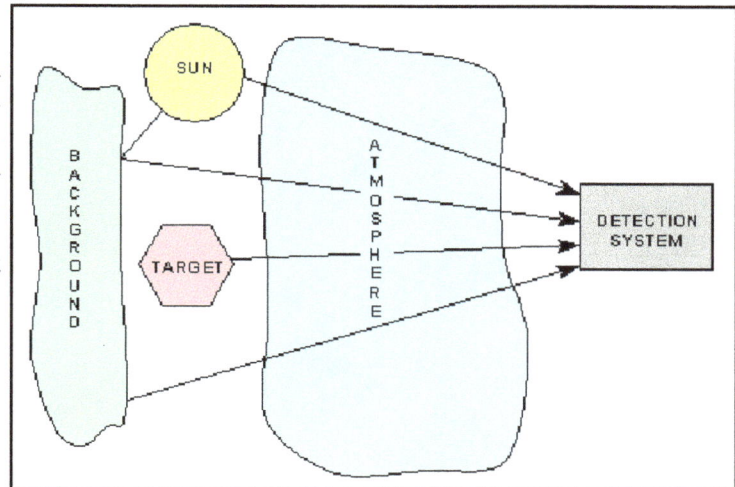

Figure 20. Generalized Detection Problem

Figure 21 shows the basic relationships that are critical to detection of a target in the infrared. The figure is based on a generalized aircraft, but the principles apply whether the target is in fact an aircraft against a sky background or a ground vehicle being viewed from above against a terrain background. At detection, most targets are unresolved. The sensor's ability to detect the target against background in this case is driven primarily by noise equivalent irradiance.

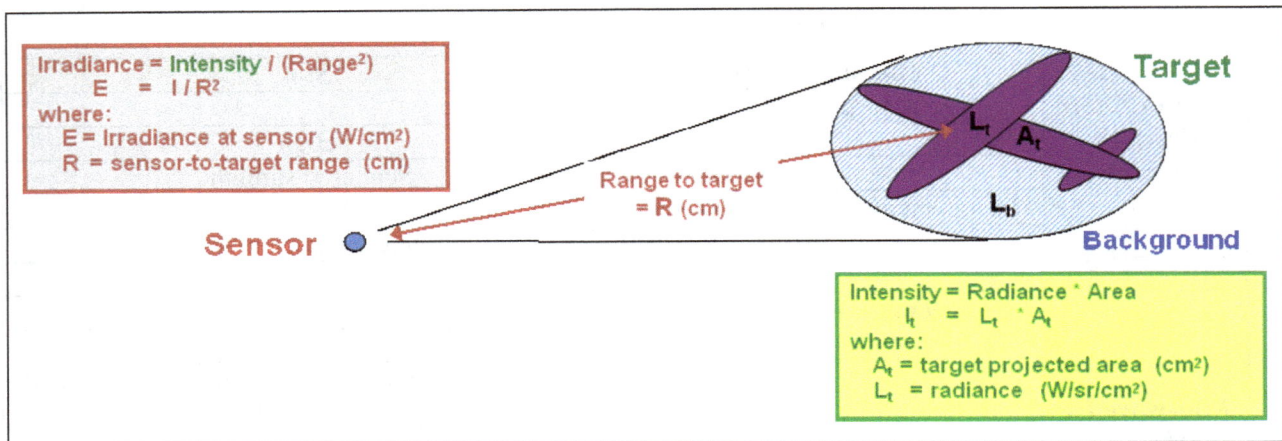

Irradiance = Intensity / (Range^2)
$E = I / R^2$
where:
E = Irradiance at sensor (W/cm^2)
R = sensor-to-target range (cm)

Range to target
= R (cm)

Sensor

Target

L_t A_t

L_b

Background

Intensity = Radiance * Area
$I_t = L_t * A_t$
where:
A_t = target projected area (cm^2)
L_t = radiance (W/sr/cm^2)

Figure 21. Detection of a Target with a Remote Sensor

Each of the equations shown in Figure 21 in reality has atmospheric effects and attenuation due to transmission losses and contributes to path radiance. Just as the power distribution of the target and background vary with wavelength, atmospheric effects are also spectrally selective.

Figure 22 shows the roll off of irradiance as a function of range for two different aircraft. Detection occurs at the point of intersection with the sensors noise equivalent irradiance. In the case of threat missiles, there is often a signal-to-noise threshold required for launch of the missile to ensure target quality prior to launch. The product of noise equivalent irradiance and the threshold for these systems is known typically as the minimum trackable irradiance (MTI). This is the figure used to calculate detection range for such systems.

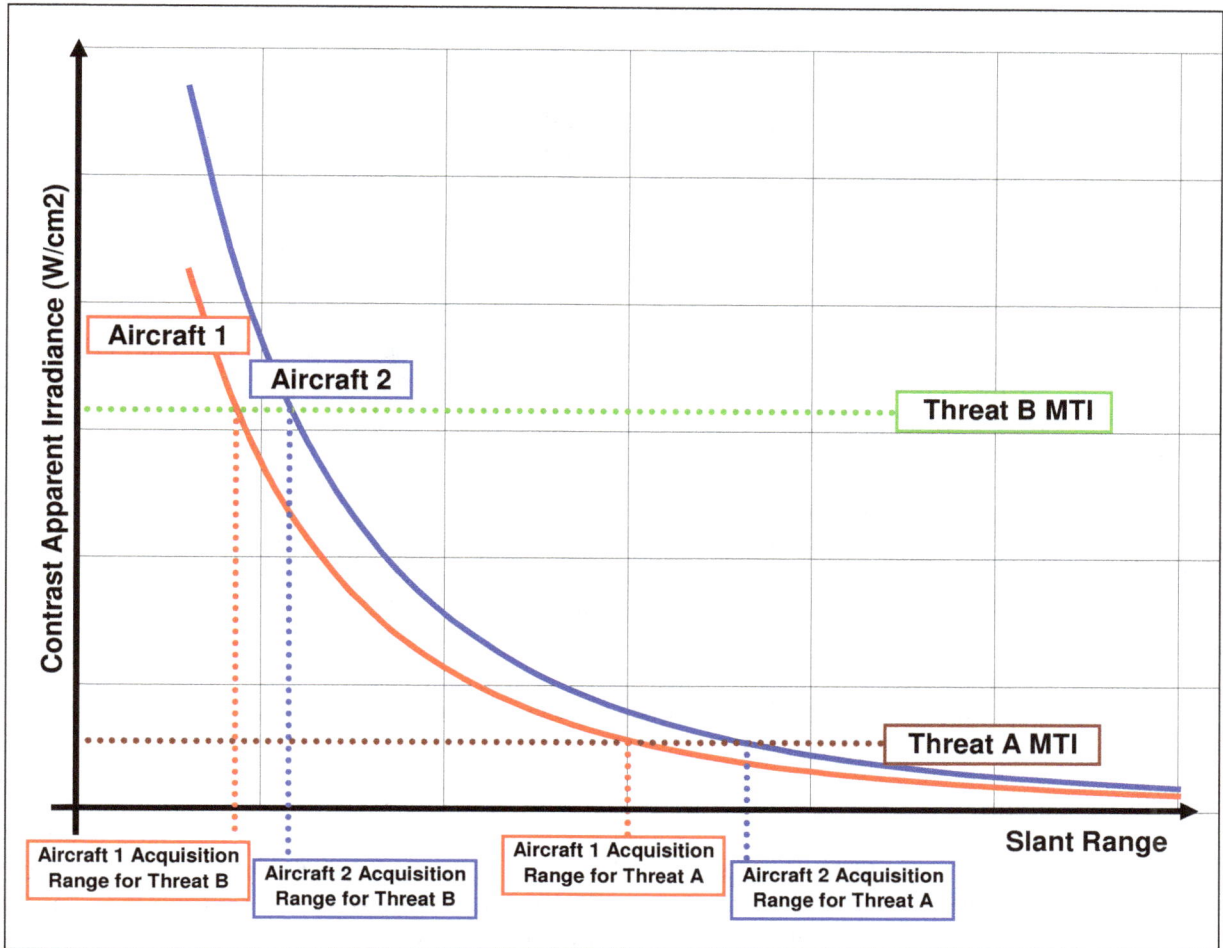

Figure 22. Detection Range Calculation

In an effort to simplify calculations, band average atmospheric transmission values are often applied during analysis of detection scenarios. The target itself is non-uniformly distributed as a function of wavelength, and the atmospheric effect is non-uniform, so this approach is mathematically incorrect since it pulls a non-constant term out of an integral. The degree of error introduced by the band average approach depends on the spectral distribution of the source and the overall transmission of the band in question, but caution should be applied when applying band averages. All calculations involving atmospheric propagation should be done spectrally and then integrated to provide the in-band value.

Sensor Characterization

As described in a previous section, the output of every sensor is proportional to an integral of the received radiation weighted by the instrument's response function in that domain. Every sensor responds to radiation in accordance with the characteristics of the sensor and its components.

Characterization quantifies the sensor's response shape. Knowledge of response shape is essential to the design of a sensor and understanding its performance in with changing ambient conditions and against various types of targets in real environments.

Normalization

Calibration of a sensor, which describes its response to known input sources, and characterization of the sensor would ideally be the same process. Ideally, the absolute instrument response would be mapped over a domain with a traceable standard laboratory source that was tunable across the range of interest. In the spectral domain, for example, this would require a tunable monochromater whose output beam provided a level of spectral irradiance traceable to a radiation standard and that also had a cryogenically-cooled background source. The result would be an absolute spectral response function in units of output reading per unit radiance (or irradiance) as a function of wavelength.

In practice, it is sufficient and more practical to separate the characterization and calibration processes, so characterization determines only the relative response shape rather than absolute. Calibration then incorporates the results of the shape characterization to determine the absolute instrument response. Both, however, are important to understanding performance of a sensor.

Characterization uses a variety of different methods and sources to map relative response shapes. The response curve is then normalized and this normalized curve is used in the later derivation of the calibration coefficients. Different normalizations, such as normalization to an average value, are possible, but the convention throughout most of the measurement community today is to normalize response curves to unity at the peak.

When the contributions of all the components are combined into one curve, the result is then peak normalized and this, now unitless, curve is used in the calibration calculations.

Calibration

Sensor calibration, which is the process of relating the known input power to the output of a sensor, requires the use of standard sources, typically National Institute of Standards and Technology (NIST) traceable blackbodies and various other laboratory equipment such as collimators which make all of the rays coming from the source parallel to each other, thus representing a source at infinity. Figure 23 is a pictorial illustration of the calibration of a sensor.

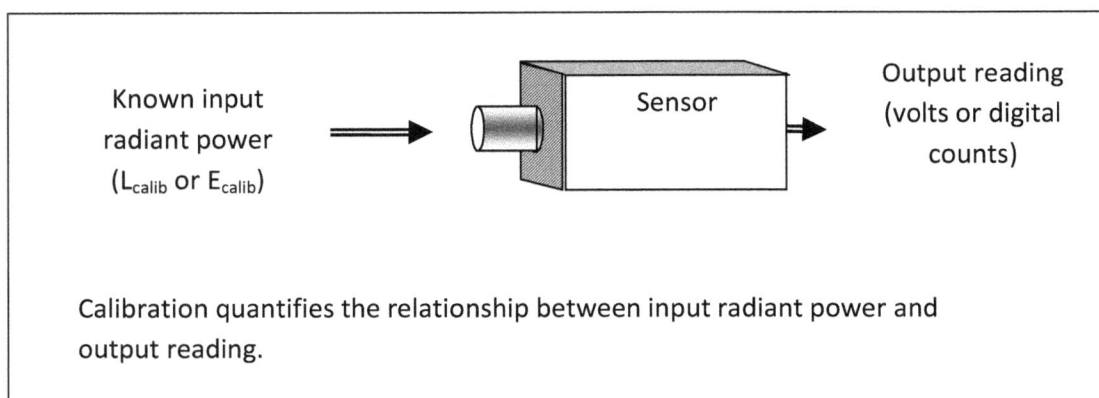

Known input radiant power (L_{calib} or E_{calib}) → Sensor → Output reading (volts or digital counts)

Calibration quantifies the relationship between input radiant power and output reading.

Figure 23. Sensor Calibration Relates Input Power to Output

Calibration of a sensor usually involves two major steps. Responsivity is the change in output of the sensor to changing input. For a sensor that responds linearly, for example, responsivity represents the slope of the curve when source radiance or incident irradiance is plotted against output voltage or counts for a digitized system. Figure 24 represents a calibration curve for a sensor that has a linear change in output over its dynamic range with changing input power. The slope of the curve is the "m" in the linear equation. Not all sensors respond linearly with power. Higher order response coefficients are common, especially for bolometers and infrared focal plane array sensors operated at short integration times. For these the process of determining the response of the instrument is the same, the curve just yields higher order terms.

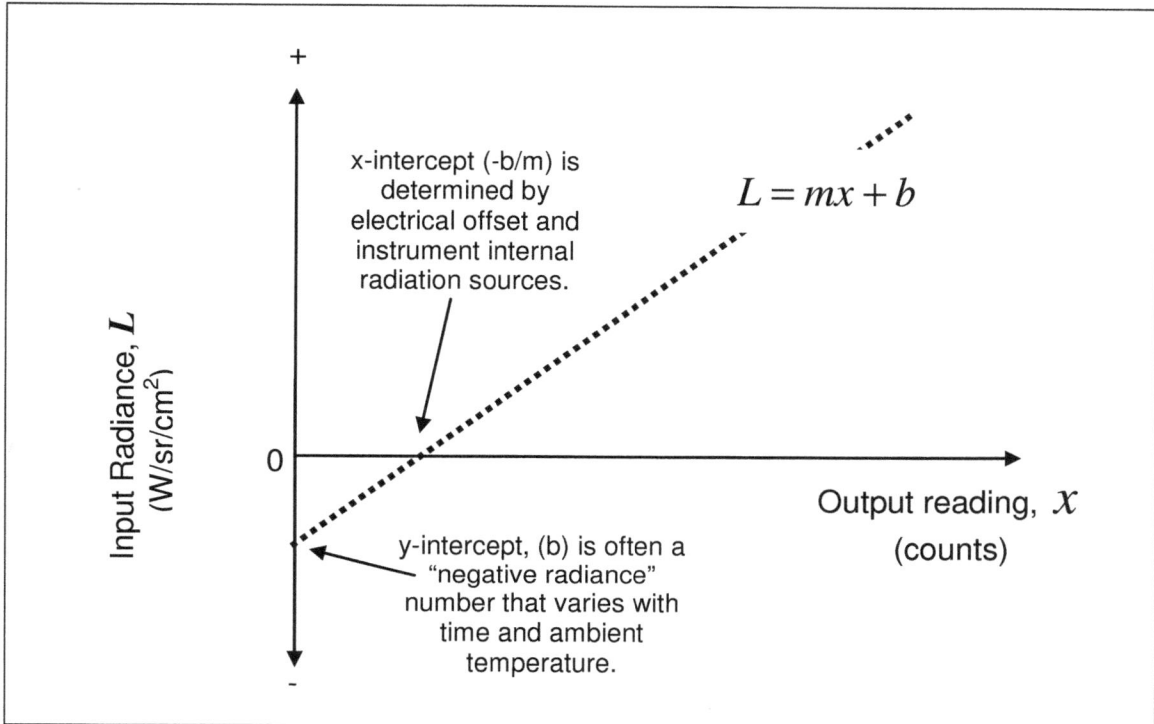

Figure 24. Calibration of a Sensor to Determine Responsivity and Offset

For most electro-optical sensors, the responsivity does not change with ambient temperature. In other words, the non-constant terms in the calibration equation, whether it is linear or higher order, do not change with changes in temperature. Over time, however, as detectors decay, responsivity decreases. This would show itself on the Figure 24 as an increase in the slope value, i.e., higher input power is required for the same output as the detector becomes less responsive.

Offset

Offset is another important parameter for instrument calibration. For any real (non-ideal) instrument, the response curve does not pass through zero. There are several reasons for this; one being that except in a complete vacuum, zero radiance does not exist. Additionally, contributions from detector noise and radiation from the optical elements in the lens, which cause the offset to drift with changes in ambient temperature, contribute to the offset term. Some amount of offset is designed into the system as well. All electronic circuits have some amount of DC drift. To prevent clipping of the signal if this drift should go below the lower limit of the analog-to-digital converter, the "bottom end" level is adjusted up to some offset level. The consequence of this offset voltage is the addition of a y-intercept term (b), which also must be quantified by the calibration if the sensor is a laboratory instrument. For a sensor that is used for contrast detection, the intercept value is unimportant since it subtracts out in the contrast calculation.

Sensitivity

Sensitivity for a sensor is determined to a large extent by the noise level in the detector output. For focal plane array detector, pixel-to-pixel non-uniformity also limits the sensitivity of the system since detection is determined by contrast with surrounding pixels. In practice, at least for cooled infrared sensors, detection is typically limited by background and not noise limits.

Instrument Response Uniformity and Non-Uniformity Correction

In reality, all sensors exhibit non-uniform response in all of the domains referenced previously. For example, in the spatial domain, the raw output of an infrared focal plane array detector exhibits pixel-to-pixel offset differences and response differences across the field-of-view. The response change is the result of two primary factors. Since each pixel is essentially a unique detector, it exhibits unique response because of manufacturing tolerances, slight differences in crystal structure, etc. Additionally, most electro-optical sensors implement an aperture or "field stop" in the case of infrared sensors, that limits the radiation that can reach the detector outside of the sensor's desired field-of-view. Radiation entering at angles off of normal to the detector shows a cosine roll-off in incident power. The result is a reduction in responsivity for pixels that are radially separated from the center of the detector. For the majority of systems, an optical gain correction can be applied to compensate for the change in response. The typical method involves using an extended blackbody source that fills the FOV of the sensor. Reference images are collected with the source at two temperatures that are well separated across the sensor's dynamic range. This process is typically called a "2-point" correction. Actual temperature is unimportant. Slope corrections can be determined for each pixel. The result is a gain map that can be stored in the sensor electronics that can be applied to each image to correct for the non-uniform spatial response across the detector array. Pixel-to-pixel offset maps can be determined using one of the same reference images. Pixel slope and offset corrections are typically derived as normalized quantities relative to a center pixel, average of center pixels, or maximum value. The application of the correction maps to the images is commonly referred to as "non-uniformity correction." Figure 25 shows the transition from a raw image to a non-uniformity corrected image.

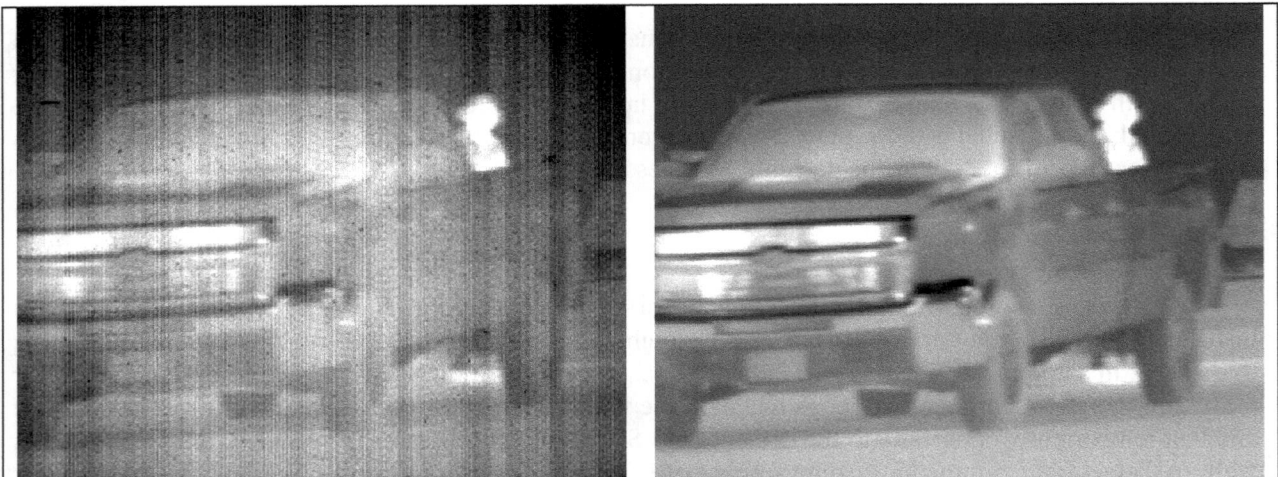

Figure 25. Non-Uniformity Correction of a Mid-IR Image

Bad Pixel Replacement

Focal plane arrays have pixels that are either unresponsive or responsive outside of useful limits. Figure 25 shows some of the bad pixels that appear as small black spots in an image from an InSb IRFPA imager. Bad pixels can be identified during laboratory calibration or with a sensor mounted reference source. There are many approaches to replacement of bad pixels and the best approach often depends on the sensor characteristics and its application. One common approach is simply to replace the pixel with the average of its nearest neighbors.

INFRARED THREATS TO AIRCRAFT AND THEIR COUNTERMEASURES

IR guided missiles are the largest single cause for aircraft losses since the start of the 1991 Gulf war. All missiles designed within the last 20 years have counter-countermeasures circuitry. Every missile can be defeated with IR countermeasures given time to develop and test devices and techniques, but many missiles have not been exploited and the variety and complexity of the different designs present formidable challenges to the US countermeasure community.

The IR "signature" of any aircraft has three main components:
- Engine exhaust plumes
- Engine hot parts (tailpipe, etc)
- Airframe (aerodynamic heating & reflection from sun, earth, etc)

Infrared guided missiles modulate the signal produced by the aircraft in contrast with its background. Previous generations of missiles used reticles to produce signal waveforms that would provide spatial and temporal information from which signal processing could produce trackable information.

IR Missile Operation

Aircraft (or any other object) can be intercepted using several different types of guidance. The simplest type is pure pursuit, where the missile is always pointed directly at the target location. This is not aerodynamically efficient since the missile would follow a longer (curving) flight path when following a crossing target.

Most missile guidance systems are designed to lead the target so that intercept occurs at the point where the target will be at the time the missile arrives. This requires that the missile fly a course so the relative bearing to the target stays constant (constant "line of sight" angle). The LOS angle is determined by missile speed relative to the target (higher closing speed = smaller angle). The size of the angle isn't important; only that it be constant (zero line of sight rate) as shown in Figure 26.

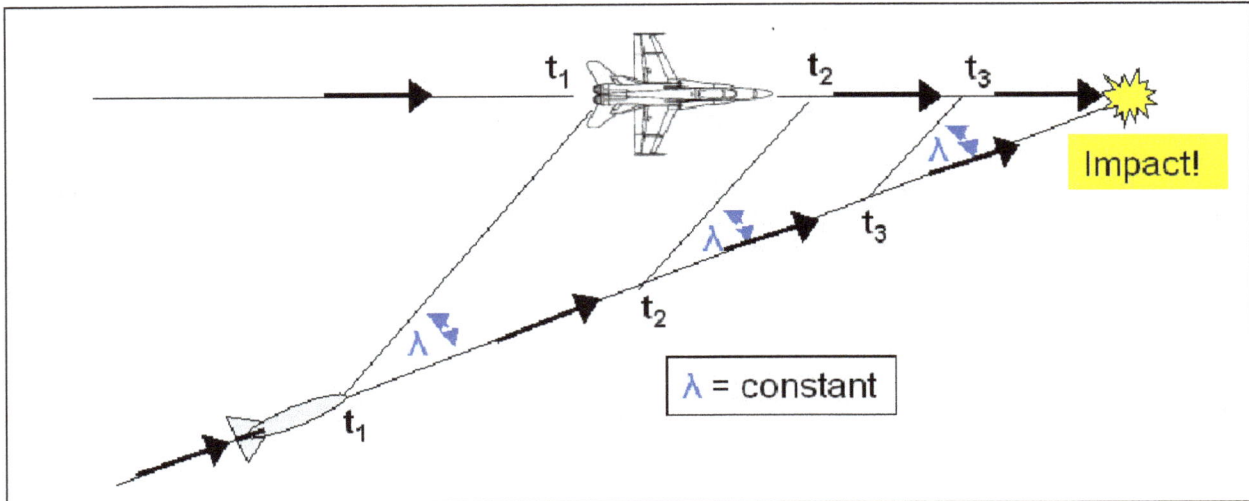

Figure 26. Missile Proportional Navigation

This intercept course ("proportional navigation") requires that the missile have two separate servo loops: (1) a target tracker and (2) a wing control servo to control direction of flight

For a missile to guide to its intended target, it needs a tracker, which contains the following elements:

- Optics to collect and focus IR from target.
- Gimbals to allow movement to point the optics.
- Gyro stabilization to isolate optics from missile body.
- Detector to convert the received IR to electrical signal.
- Stabilization (gyro) to isolate from missile body.
- A method to determine target direction to enable closed-loop tracking.
- A method to distinguish the target from natural background.

The target tracker is the "window" into the missile's guidance through which it can be deceived by countermeasures.

The problem of determining target direction with a single detector was solved by forming an image of the target scene onto the center of a reticle disk that spun with the optics. Unlike, for example, the reticle in a rifle telescope that superimposes cross hairs, the reticle in a missile acts as a kind of shutter that blocks the passage of IR through part of the reticle and allows IR to pass through the other part. A target image falling on the opaque portion is blocked and produces no detector signal. A target image falling on the transparent portion is passed on to the detector. When the reticle is spun, IR from a target off center is alternately passed and blocked, resulting in amplitude modulation (AM). The phase of this modulation relative to a spin reference is used to tell target direction from center. A closed servo loop moves the optics to keep the target centered on the reticle. This is depicted in Figure 27.

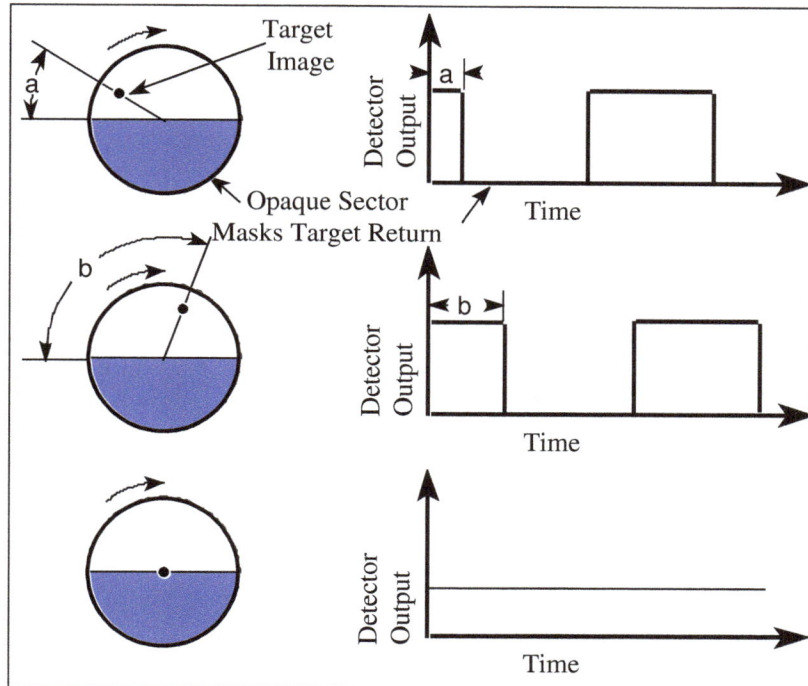

Figure 27. Basic Reticle Design

Target trackers have another problem: The aircraft target must be distinguished from natural background sources, such as sunlit clouds and terrain. To solve this, they look for features in the spectral, spatial, and temporal domains where the target is different from background.

- Temporal: There is no difference. Neither clouds nor aircraft signature are time varying.
- Spectral: Some difference. Choice of wavelength band yields helpful differences between target and background, but this is not sufficient by itself.
- Spatial: The most viable option. Aircraft are smaller than clouds and terrain. Background radiation can be greatly reduced by spatial filtering.

If half of the reticle is made with opaque "spokes," then some irradiance from targets with small images (such as aircraft) will be modulated more completely and generate a stronger signal at a faster modulating rate than large images (clouds) as shown in Figure 28.

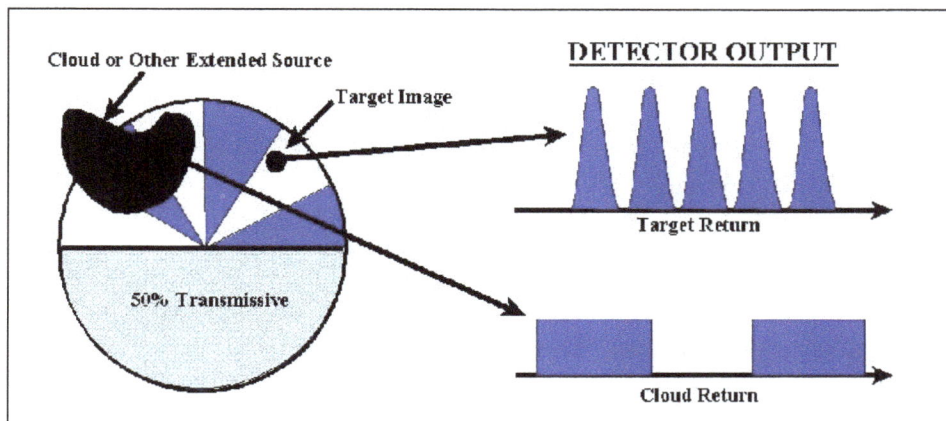

Figure 28. IR Seeker Design for Background Discrimination

The past figure and the following two figures (29 and 30) depict a spin-scan reticle used on the early Sidewinder designs. After the detector preamp, signal goes through a narrow bandpass filter to improve S/N. The AM waveform is then rectified and filtered. Target direction is determined from AM envelope phase.

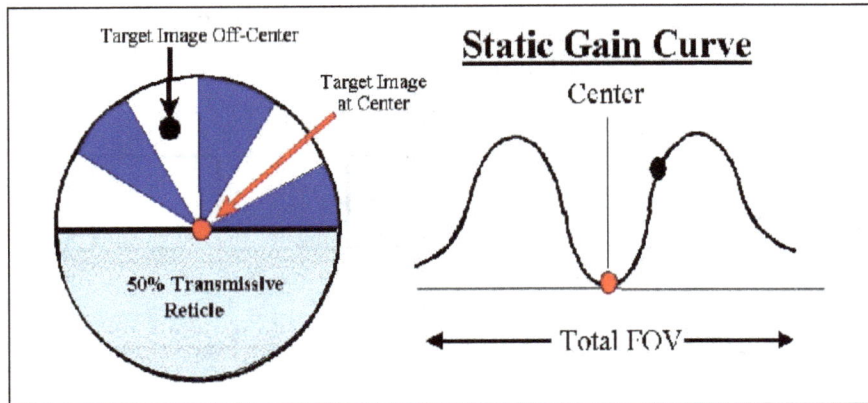

Figure 29. Spin Scan Seeker

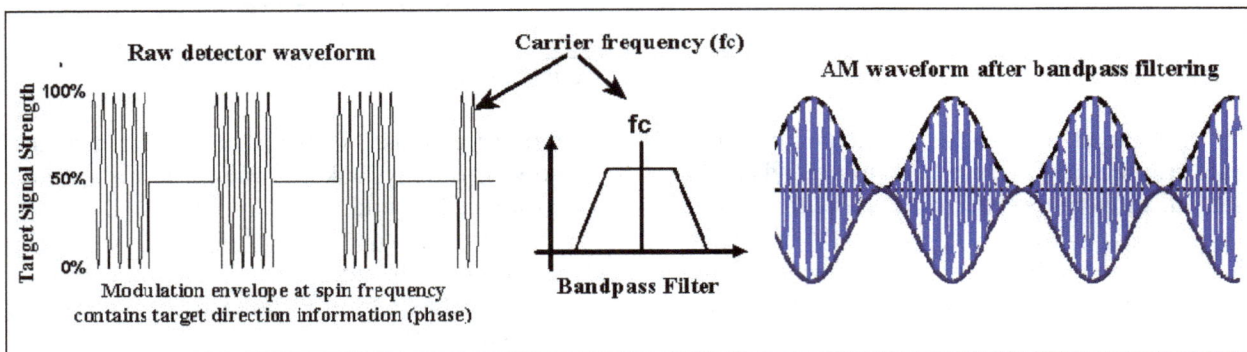

Figure 30. Spin Scan Waveforms for Off-Center Target

Spin-scan has the following characteristics that are important to countermeasures:
- The tracker loop drives to null the signal to zero. This occurs when the target is on the optical axis and the target image is at the center of the reticle.
- If the target is off-center, an AM carrier "error signal" is generated, where the phase of the modulation envelope indicates the target direction.

With spin scan, the missile is always looking at the target. This vulnerability to jammers led to the next evolution in target trackers: conical scan.

Conical scan borrows concept from early fire-control radars, which used a nutating feed horn. A con scan tracker is shown in Figure 31. With con scan:
- The secondary mirror of the Cassegrain is canted so the field of view seen by the detector sweeps out a pattern of overlapping circles.
- A target image at boresight falls near the edge of the reticle instead of center.
- Reticle pattern is same all the way around. (Usually tapered spokes.)
- Modulation of target near boresight is FM rather than AM. This allows tighter tracking.
- For larger angles off boresight, the target image falls outside the FOV of the detector for part of the scan. The modulation then becomes AM.

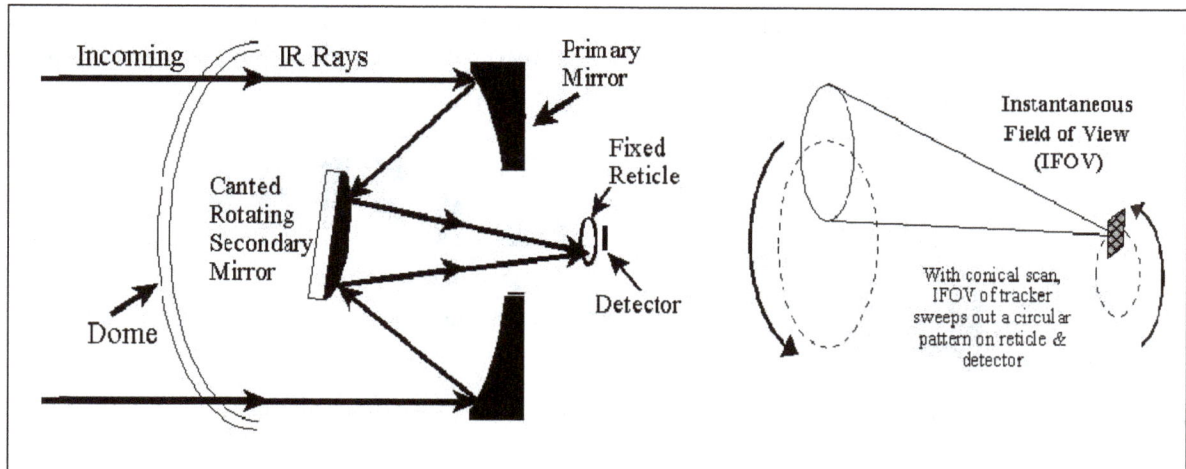

Figure 31. Conical Scan Tracker

In the con scan tracker, as the missile instantaneous field of view nutates about a target on boresite, (moving through positions at t_1 through t_5 shown in Figure 32), the apparent position of a target image on the reticle sweeps out the circular pattern shown.

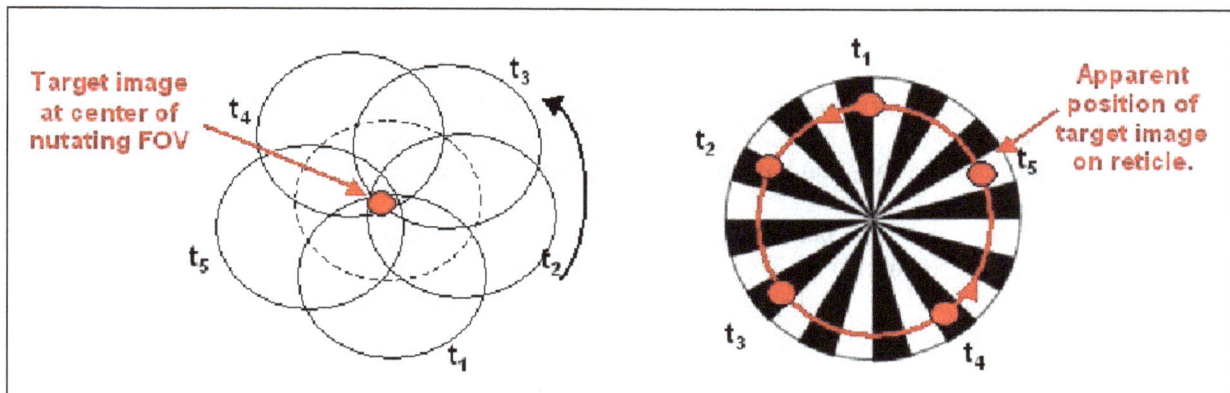

Figure 32. Image on a Con Scan Reticle: Target at Boresite

If the target is off boresight as shown in Figure 33, the detector receives a signal of varying pulse widths.

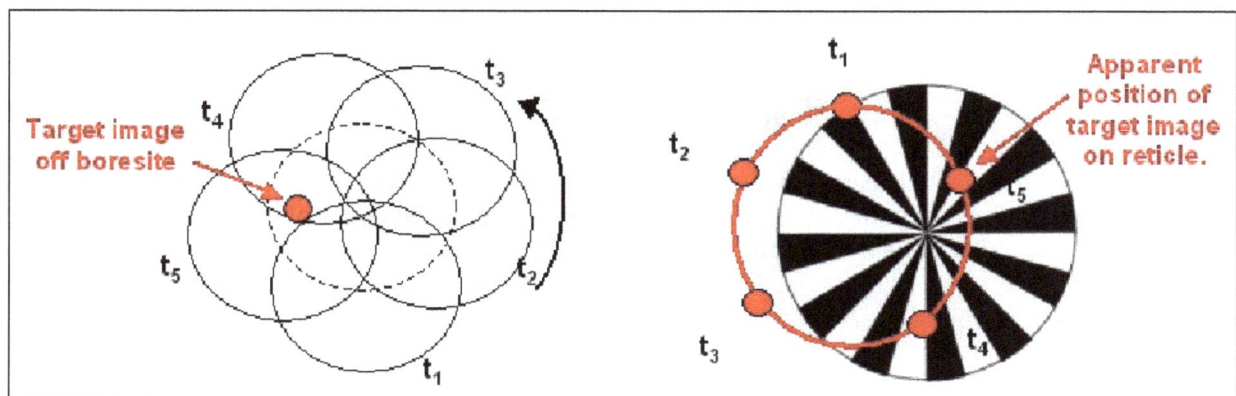

Figure 33. Image on a Con Scan Reticle: Target off Boresite

The waveform produced by a target on boresite is a constant amplitude carrier at the reticle chopping frequency as shown in Figure 34. A target slightly off boresite produces a constant amplitude carrier that is frequency modulated at spin frequency.

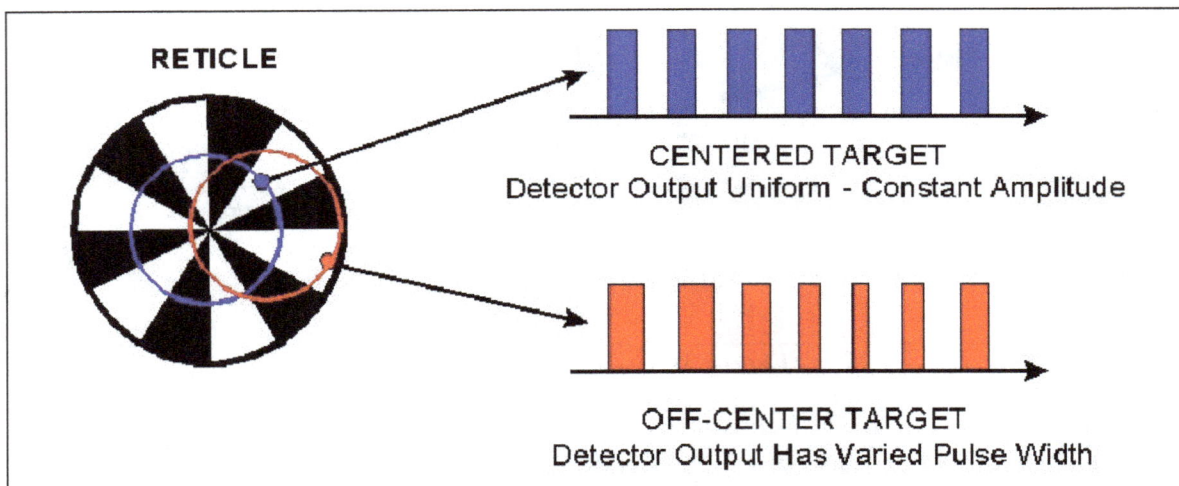

Figure 34. Conical Scan Seeker Output

A target further off boresite leaves the missile field of view during part of the scan, producing an amplitude-modulated waveform similar to that of a spin scan tracker. The important difference is that with a spin scan tracker, the target never leaves the missile field of view. With con scan, the target may fall outside the missile FOV at certain times during the scan. Because con-scan trackers do not necessarily view the target continuously, they can have high resistance to jammers.

Other types of seeker scan patterns now exist. The Rosette scan pattern shown in Figure 35 is one such example. It has an even higher resistance to countermeasures.

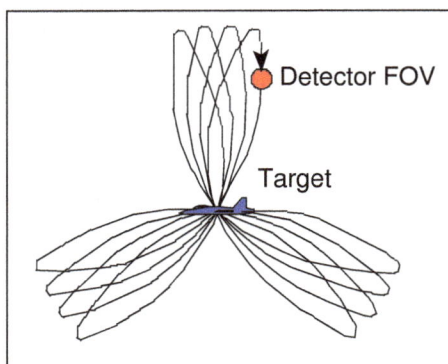

Figure 35. Rosette Scan Pattern

Imaging arrays of detectors without reticles are newer yet. They may be classified as either staring (every pixel sees the entire scene), or scanning arrays, where the optics plays a role in determining which "pixels" are exposed to optical / IR radiation.

INFRARED COUNTERMEASURE (IRCM)

Flares

Figure 36 shows a Navy F/A-18E Super Hornet aircraft dispensing IRCM flares from its internal flare dispensers. IRCM have been the staple of countermeasures protection for military aircraft more than four decades. Flares are designed to transfer the track of an attacking infrared missile by exhibiting characteristics that confuse the tracking and guidance algorithms built into the missile. Modern missiles incorporate Counter-Countermeasures (CCM) capabilities including hardware configurations, circuitry, and logic to help identify countermeasures and reject them from processing. CCM capabilities can be based on spectral, spatial, and temporal features of the target scene. As missiles continue to improve in their sensitivity, range, maneuverability, and CCM capabilities, flares continue to evolve in order to keep pace with the evolving threat.

Figure 36. Navy F/A-18E Aircraft Dispensing IRCM Flares

Over the years the Navy has fielded many flare types including the MJU-2/B, MJU-8, MJU-32, MJU-38, MJU-27, MJU-49, and many other improved versions of these flares and other types as well. IRCM flares continue to be the prevailing countermeasure for military aircraft protection by offering cost effective and robust protection.

Dispenser Systems

Most Navy fixed wing and rotary wing aircraft are equipped with countermeasures dispenser systems. These systems are critical to the survivability of the aircraft in a hostile threat environment. Modern dispensers such as the AN/ALE-47 offer high reliability and substantial programming capability that allows flare dispenses to be tailored for maximum protection of the host aircraft type. AN/ALE-47 is highly integrated into the aircraft over the 1553 data buses. The dispenser can incorporate information from several aircraft systems including missile-warning systems to improve its responses to threats and provide vital situational awareness to the aircrew.

Impulse Cartridges

IRCM flares are dispensed from the aircraft flare dispensers with electrically initiated impulse cartridges. Impulse cartridges incorporate energetic materials within a small confined canister. Upon application of a voltage to the electrical leads, a bridge wire in contact with the energetic materials burns through, igniting the propellant materials. The expanding gases push the flare from its case, held captive in the aircraft dispenser. Impulse cartridges for Navy use have been designed to withstand the extreme electro-magnetic environments encountered around aircraft carriers and other combat ships. Examples of impulse cartridges include the CCU-63 and CCU-136.

Infrared Jammers

Several lamp-based and mechanically modulated jammers have been developed over the years for protection of aircraft. Examples include the AN/ALQ-144 and AN/ALQ-157, predominantly used on helicopters and cargo aircraft. These jammers offer some level of protection over a broad field-of-regard and offer the advantage of continuous operation.

In principle, these jammers produce a modulated signal in the track band of the threat that corrupts the target tracking pulses in the missile seeker.

Passive Missile Warning Systems

Infrared-guided weapons provide passive attack capabilities against military vehicles. Unlike a radar-guided weapon that actively emits radiation and tracks the reflected pulses from the target, infrared weapons track radiation already being emitted from the target. Attacking missiles fly at very high speeds, and they are exceptionally maneuverable. Missile warning systems must be capable of detecting the threat, alerting the aircrew, and cueing a countermeasures response within sufficient time to counter the attacking missile. The time from launch to impact can be very short, making timely detection critical. Active warning systems have been developed in the past that use Doppler Radar capabilities to detect missiles, but passive missile warning systems have been preferred because of the desire to minimize emissions from the aircraft under attack.

Several passive missile-warning systems have been developed over the years for military aircraft. These systems operate in a variety of different parts of the electro-optical spectrum, but the most common are ultraviolet and infrared sensor systems. Examples of passive missile warning systems include AN/AAR-47, AN/AAR-54, AN/AAR-57, AN/AAQ-24 (both passive and active components), and the Joint and Allied Threat Awareness System (JATAS), currently under development by the Navy. Passive missile warning sensors continue to improve with advances in detector technologies, particularly with imaging detectors. These sensors provide excellent angle-of-arrival information, necessary to support cueing of laser based countermeasures, and advanced processing to detect and declare threat missiles in cluttered environments.

Laser Countermeasures

Laser-based infrared countermeasures have been in development for many years. Several systems have been fielded over the past fifteen years including the AN/AAQ-24 system on Air Force cargo aircraft and helicopters and a derivative system for Marine Corps helicopters.

Although configurations vary, most of these systems incorporate a single multi-band laser or several single-band lasers that produce modulated waveforms designed to corrupt a missile's guidance target tracking. The laser optics are located in a tracking gimbal that provides agile and rapid pointing over a broad field-of-regard. Laser based countermeasures require a relatively high angle of arrival accuracy from the host aircraft's missile warning sensor. Upon declaration of the threat, the missile warning system hands-off track to the tracking gimbal and cues the lasers to lase. A tracking camera in the tracking gimbal with high optical resolution helps to maintain track on the threat missile through the engagement period.

LASERS

The word laser comes from <u>L</u>ight <u>A</u>mplification by <u>S</u>timulated <u>E</u>mission of <u>R</u>adiation. A laser system emits light that is generated through a process of stimulated emission. The radiation produced by a laser exhibits high temporal and spatial coherence. In order to begin the process of stimulated emission, the lasing medium absorbs the energy from a pump source. The atoms in the lasing medium are excited to a higher energy state. These atoms will eventually return to their ground state. A large number of atoms that are excited to higher states create a population inversion. Population inversion describes the number of atoms in excited state versus the number of atoms in the ground state. In order for the atoms to return to their ground state, they must release energy. This energy is released in the form of photons. Energy of a photon is expressed as

$$E = \frac{hc}{\lambda}$$

Where
  E = Energy, generally electron volts (eV)
  h = Planck's constant = 4.136 x 10^{-15} (eV·s)
  c = speed of light = 2.998 x 10^8 (m/s)
  λ = wavelength of light in meters

The energy that must be released by the atom to return to the ground state will direct the wavelength of the photon emitted since h and c are constants. If all the excited atoms released the same amount of energy to return to their ground state, the released photons would all have the same wavelength and would be considered fully monochromatic. Most lasers do not emit a single wavelength but a range of slightly differing wavelengths ($\Delta\lambda$).

The lasing medium may be a solid, a gas, liquid or plasma. Some laser types include gas, chemical, dye, fiber-based, solid-state and semiconductor lasers. The laser radiation can be output in a continuous wave (CW) or in a pulsed wave. A continuous wave laser emits light that maintains a steady amplitude and frequency. A pulse wave will vary in amplitude and is also characterized by the systems pulse repetition frequency (PRF). The PRF is defined as the number of pulses emitted during a unit of time. Figure 37 shows the spectral output of several laser types.

The first laser was constructed by Theodore Maiman at Hughes Research Laboratories in Malibu, California. This laser was a pulsed, solid-state ruby laser. The ruby laser uses a synthetic ruby crystal as the lasing medium. A xenon flash lamp is used to excite the atoms in a ruby rod to higher energy levels. The highly polished and mirrored ends of the rod form a resonant cavity. One end of the rod has a slightly lower reflectivity. The lamp excitation produces an inverted population of excited atoms, which are stimulated to relax to lower energy levels releasing their extra energy as photons. Repeated reflections off the mirrored ends of the rod causes the photons to bounce back and forth through the rod stimulating further emissions at the same wavelength and phase producing a highly coherent beam, which finally passes through the lower reflectivity end.

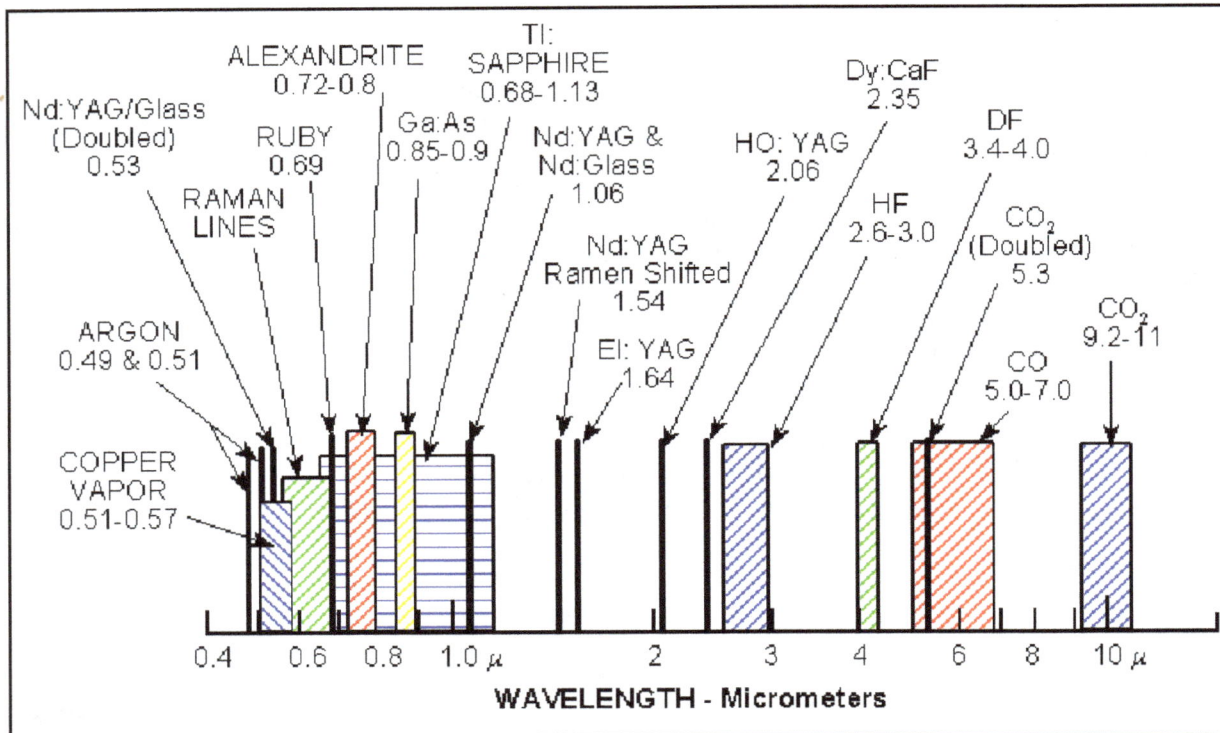

Figure 37. Spectral Lines / Ranges of Available Lasers

The typical laser rangefinder uses a solid-state laser with a neodymium-YAG crystal lasing at 1.06 μm.

Gas lasers can be pulsed or CW. The gas dynamic laser obtains its inverted population through a rapid temperature rise produced by accelerating the gas through a supersonic nozzle. In chemical lasers the inversion is produced by a chemical reaction. In the electric discharge laser the lasing medium is electrically pumped. The gas can also be optically pumped. In an optically pumped gas laser the lasing medium is contained in a transparent cylinder. The cylinder is in a resonant cavity formed by two highly reflective mirrors.

Many gas lasers use carbon dioxide as the lasing medium (actually a mixture of CO_2 and other gases). These are the basis for most high energy or high power lasers. The first gas laser was an optically pumped CW helium-neon laser. The common laser pointer is a helium-neon laser operating at 0.6328 μm. The lasing medium is a mixture of helium and neon gas in a gas discharge or plasma tube.

The dye laser is an example of a laser using a liquid for the lasing medium. The lasing medium is an organic dye dissolved in a solvent such as ethyl alcohol. Dye lasers operate from the near UV to the near IR, are optically pumped, and are tunable over a fairly wide wavelength range.

Another type of laser is the semiconductor or injection laser, also known as a laser diode. The junctions of most semiconductor diodes will emit some radiation if the devices are forward biased. This radiation is the result of energy released when electrons and holes recombine in the junction. There are two kinds of semiconductor diode emitters: (1) the light emitting diode (LED), which produces incoherent spontaneous emission when forward biased and which has a broad (800 angstrom) spectral output, and (2) the laser diode, which maintains a coherent emission when pulsed beyond a threshold current and which has a narrow spectral width (< 10 angstrom). In the laser diode the end faces of the junction region are polished to form mirror surfaces. They can operate CW at room temperatures, but pulsed operation is more common. Figure 38 shows a typical diode laser structure.

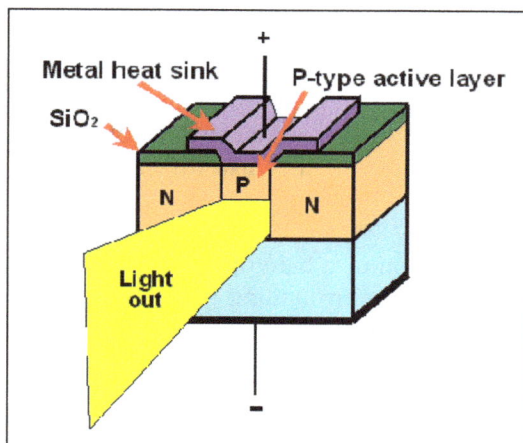

Figure 38. Diode Laser Construction

Fiber lasers use fibers that are doped with rare-earth elements as the pumping medium. These rare-earth elements include elements such as Erbium (most common), Ytterbium, and Neodymium. There are other elements such as Thulium that are used for doping purposes. Erbium doped fiber lasers can emit in the 1.5 to 1.6 micron wavelength, which is important due to eye safety concerns in this part of the spectrum. Other wavelength emissions for Erbium include 2.7 and 0.55 microns.

Fiber based laser systems are beneficial in many ways. The fiber gain medium is compact compared to many other types of gain medium and is highly efficient. The fiber gain medium can also be physically manipulated to save space. Fiber based lasers are able to achieve high output powers. The gain medium of a fiber laser can extend for several kilometers to achieve these higher power outputs. The fact that the light is already propagating in a flexible fiber can also allow for system designs that implement a gain cavity in one location and then deliver the output in another location.

Q-switching is a common means of obtaining short intense pulses from lasers. The Q-switch inhibits lasing until a very large inverted population builds up. The switch can be active or passive. A passive Q-switch switches at a predetermined level. An active Q-switch is controlled by external timing circuits or mechanical motion. The switch is placed between the rod (or lasing medium) and the 100 percent mirror. Figure 39 shows an arrangement using a Pockels cell as an active Q-switch.

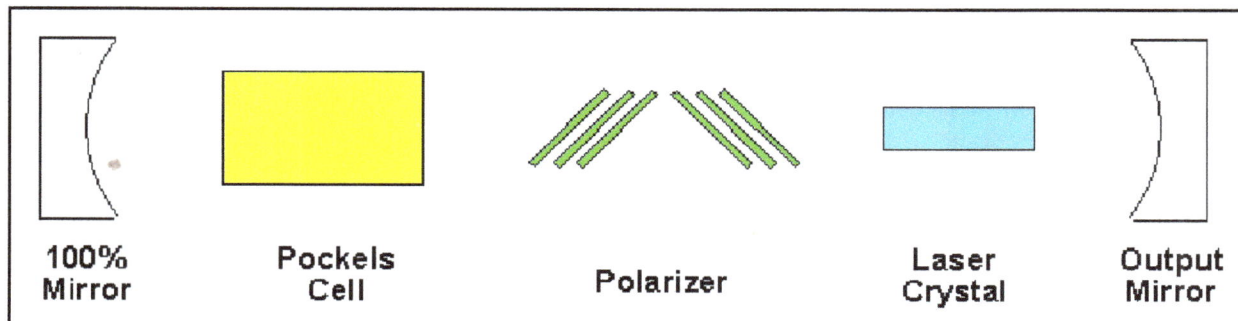

Figure 39. Q-Switch Arrangement

Other methods of obtaining pulsed operation include using pump sources that are pulsed and mode-locking.

FIBER OPTICS

Fiber optic cables are the optical analog of RF waveguides. Fiber optic cables are made from transparent dielectrics. The fiber optic cable acts as an optical waveguide allowing light to propagate along the length of the fiber by using the principle of total internal reflection. This phenomenon can only occur under certain conditions relating to the material indices of refraction and the light ray's angle of incidence. Some benefits of fiber optic fiber include low losses, bandwidth, electromagnetic interference immunity, size and weight.

Consider the physical construction of a bare optical fiber, depicted in Figure 40. A bare optical fiber is simply the inner glass core and the surrounding glass sleeve. The core must have a higher index of refraction than the cladding, $n_1 > n_2$. When $n_1 > n_2$, light impinging the boundary between the core and the cladding will totally internally reflect if the incident angle at each reflection is greater than the critical angle, θ_c. $\sin \theta_c = (n_2/ n_1)$

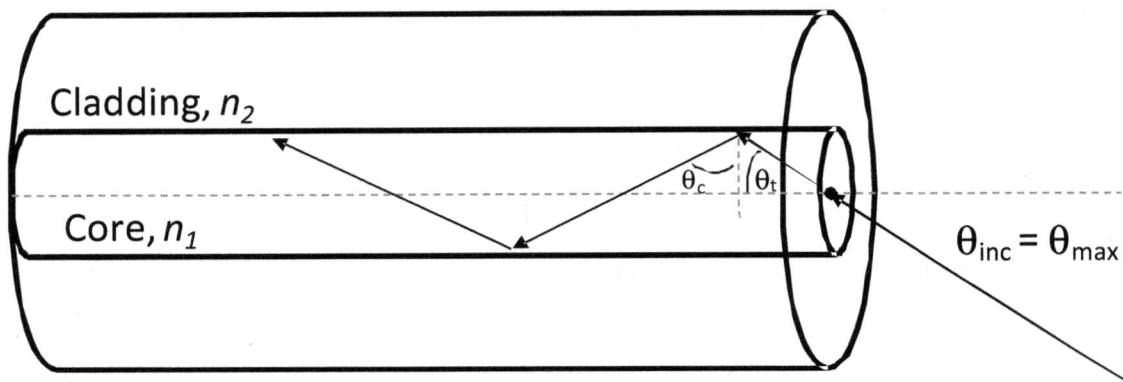

Figure 40. Bare Fiber Optic Cable

Incident rays on the face of the fiber must intersect at angles less than θ_{max} for the internal ray, θ_t, to intersect at θ_c. When rays intersect the front face of the fiber at angles greater than θ_c, they are only partially reflected in the core and will leak out.

There are many varieties of optical fibers. Optical fibers can either be single mode or multimode. Single mode fibers are fibers which propogate a single mode down the length of the fiber while multimode fibers can propogate many modes. Single mode fibers typically have a much smaller core diameter, typically around 8 to 10 μm. Their cladding is usually 125 μm. Multimode fibers typically have core diameters around 62.5 μm with 125 μm claddings. These diameters can vary depending on the application. Loss in multimode fibers over a 1 kilometer distance is typically around 1 dB at 1310 nm. This value will vary some with changes in wavelength. Single mode fibers can maintain the quality of a light pulse over longer distances than multimode fibers due to modal dispersion effects that occur in multimode fibers. Typical losses for a single mode fiber over 1 kilometer is approximately .3 dB at 1310 nm. Again, this value will vary some with changes in wavelength. However, multimode fiber is much less expensive than single-mode and can have a lower connection loss due to the larger core diameter. Multimode fiber is commonly used in communications.

In addition to single mode or multimode, a fiber can have a step index profile or a graded index profile. Figure 41 depicts the two profiles. The step index profile maintains a uniform index of refraction within the core. A graded index profile has a peak index of refraction at the center of the core. The index of refraction value rolls off from the center to lower values closer to the cladding interface. This profile assists with the modal dispersion issue found in multimode fiber.

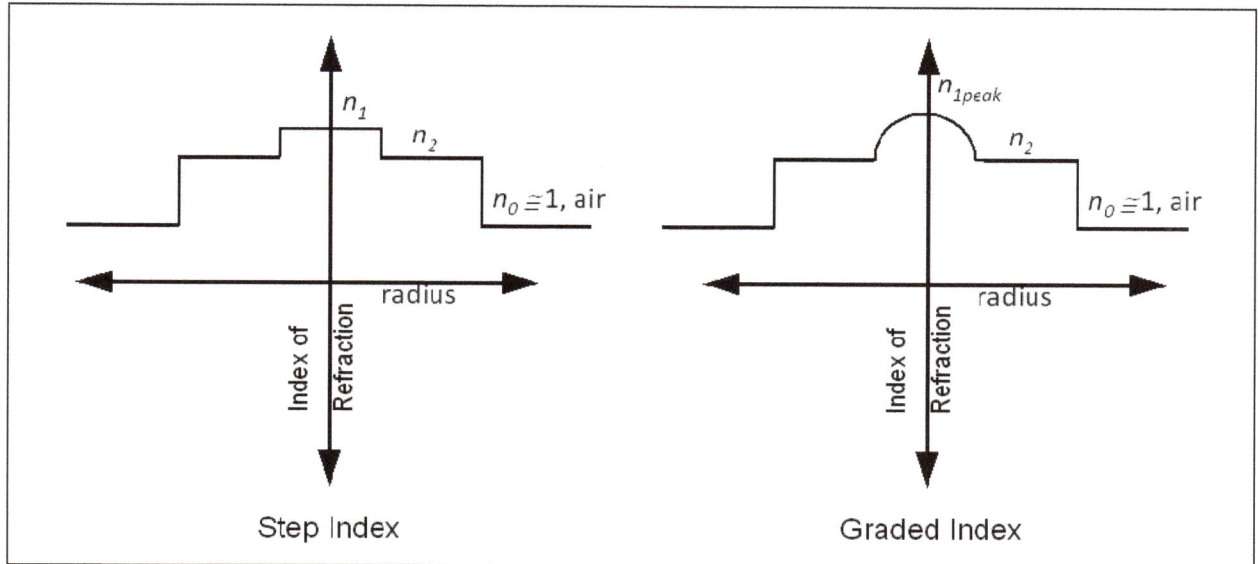

Figure 41. Fiber Profiles for a Single Mode Step Index and a Multimode Graded-Index Fiber

Most fiber is not used in a bare form and has some additional layers of protection around the cladding. These layers can include a 250 μm buffer with a 900 μm PVC tight buffer. Some fiber will also contain aramid yarn followed by a 3 mm PVC furcation tube. Buffer tubes are often used to assist with identification and provide damage protection. The outer layers can provide additional isolation from environmental factors and lower optical crosstalk.

There are also more specialized types of fiber that include polarization maintaining and photonic crystal fibers. Polarization maintaining fibers are not constructed with a cylindrical core but instead use elliptical, bow-tie styled cores or stress rods located in the cladding (PANDA style). These are shown in Figure 42.

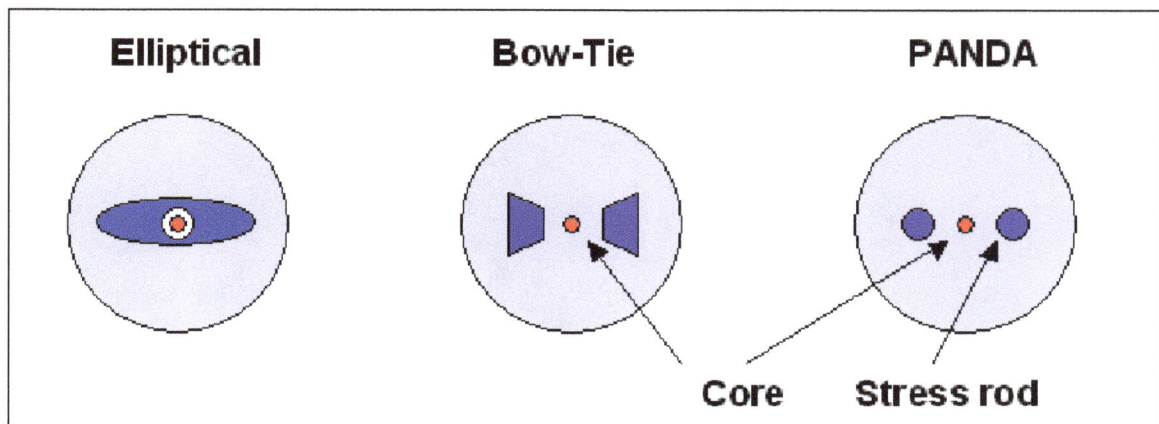

Figure 42. Polarization Maintaining Optical Fiber Types

Polarization maintaining fibers maintain the state of the linearly polarized light propogating through the fiber. This type of fiber is used when the polarization state of the light can not vary within a system. Photonic crystals propogate light by an arrangement of very small and closely spaced air holes that are maintained throughout the length of the fiber. Applications of photonic crystal fibers are varying and can be used in fiber lasers, amplifiers, sensors and telecom.

It is well known that fiber optics has many communication applications; however, improvements in fiber optic technology have lent themselves to many EO applications. Many EO components are now fiber based and can interface with the tremendous advancements in fiber-based laser systems as well as other EO systems. An example of the use of fiber optics in an EW system is the AN/ALE-50 and 55 Fiber-Optic Towed Decoy (FOTD). The FOTD uses fiber optic cabling to communicate with the jammer.

LASER SAFETY

Lasers are divided into the following classes:

Class 1	Low power / non-hazardous
Class 2/2a	Low power / minor controls necessary
	Emit less than 1 mW visible CW radiation. Not considered hazardous for momentary (<0.25 sec) unintentional exposure. Class 2a lasers are those class 2 lasers not intended to be viewed, i.e. supermarket scanners.
Class 3a/3b	Medium power / direct viewing hazard / little diffuse reflection hazard.
Class 3a	Visible lasers with 1-5 mW power output, invisible lasers, and those having 1-5 times the Accessible Emission Limit (AEL) of class 1 lasers.
Class 3b	All other class 3 lasers at all wavelengths which have a power output less than 500 mW.
Class 4	High power / eye & skin hazard / potential diffuse reflection hazard or fire hazard

There are several pertinent instructions and guidelines regarding laser use. They are:
- OPNAVINST 5100.27B Navy Laser Hazards Control Program
  (which replaced OPNAVINST 5100.27A and SPAWARINST 5100.12B)
- MIL-HDBK-828B, Range Laser Safety
- ANSI Z136.1-2007, American National Standard for the Safe Use of Lasers (Parent)

Every Navy command which uses lasers must have a Laser System Safety Officer (LSSO). All LSSOs must attend a Navy LSSO course.

There are four categories of LSSOs.
- Administrative Laser Safety Officer (ALSO)
- Technical Laser Safety Officer (TLSO)
- Laser Safety Specialist (LSS)
- Range Laser Safety Specialist (RLSS)

See OPNAVINST 5100.27B for details of their qualifications and responsibilities.

The hazard ranges of interest are the NOHD for direct viewing of a beam and the $r_{1(safe)}$ or $r_{2(safe)}$ for viewing a beam reflected off an object such as a wall. These are depicted in figure 1. The Maximum Permissible Exposure (MPE) values present laser safety levels as a function of exposure time, laser PRF, pulse duration, and wavelength. Different tables are used for eye safety while directly viewing a beam, for viewing a diffusely reflected beam, and for skin exposure.

For repeated pulses the following equation is used to calculate the maximum permissible exposure (MPE).

$$\text{MPE (repeated pulse)} = \frac{MPE(single\ pulse)}{(\ PRF\ x\ t_e\ )^{1/4}} \quad [1]$$

Where PRF is the pulse repetition frequency of the laser and t_e is the exposure duration.
For visible lasers t_e is usually taken as 1/4 second and for non-visible lasers a value of 10 seconds is used.

Figure 1 depicts some laser hazard distances.

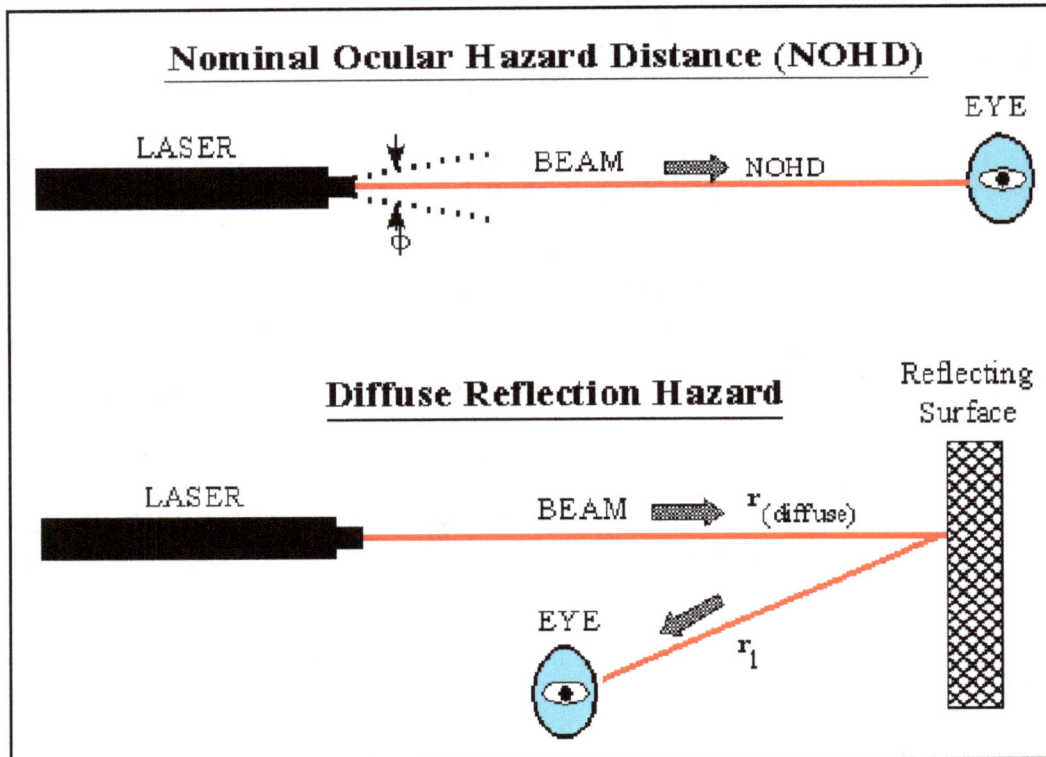

Figure 1. Laser Hazard Distances

Range laser safety specialistss shall be designated for external operations. Range test plans shall specify:

- Permissible aircraft flight paths, and ship or vehicle headings.
- Hazard areas to be cleared.
- Operational personnel locations.
- Types of surveillance to be used to ensure a clear range.
- Radio / communications procedures.

During laser operations no portion of the laser beam may extend beyond the controlled target area unless adequate surveillance can prevent radiation of unprotected areas. Class 3 and class 4 lasers shall not be directed above the horizon unless coordinated with those responsible for the given airspace (FAA, Navy, Air Force, etc).

In an industrial environment, warning and hazard signs and lights will be posted, a hazard zone shall be designated when lasers are in operation, and training shall be provided to operators in the proper eye and body (skin) protection required. Interlocks to laser operation shall be provided when there is the possibility of unauthorized personnel entering the hazard area.

Fiber optic cables usually have laser power sources so appropriate warnings or labels need to be applied to connections or possible breakage points.

AIRCRAFT DYNAMICS CONSIDERATIONS

FREE FALL / AIRCRAFT DRAG

The purpose of this section is to get an awareness of the distance traveled by a flare or other object such as a bomb, which is jettisoned or dropped by an aircraft. This will give the reader an appreciation for the significance of aircraft tactical altitude.

From Newton's second law of motion:

$$F = m_o a$$

where: F = Force
m_o = Mass of object
a = Acceleration

and the law of gravitation:

$$F = G \frac{m_o m_e}{r^2}$$

	English Units	SI Units
where: F = Force of attraction	lb_f	Newton
G = universal gravitational constant	3.44×10^{-8} ft^4/lb-sec^4	6.67×10^{-11} m^3/kg-sec^2
m_o, m_e= Masses (not weight) of object & earth	slug	kg
r = distance between center of gravity of objects	feet	meter

Combining the two equations and solving for "a" :

$$a = \frac{G m_e}{r^2} = g$$, the familiar constant acceleration due to gravity.

Since G and m_e are fixed and the variation in r (the distance from the earth's center) is small except for satellites, "g" is considered fixed at 32.2 ft/sec^2.

For objects with a constant acceleration (g), it can be shown that:

$$d = v_i t + \frac{1}{2} g t^2$$ where

d = distance traveled
v_i = Initial velocity
t = time
g = acceleration

For a falling object, Figure 1 on the following page may be used to estimate time/distance values.

- The upper curve is for an object shot upward with an initial velocity of 50 ft/sec.

- The middle curve is for an object shot horizontally with an initial velocity of 50 ft/sec or one that is a free-falling object dropped with no initial vertical velocity.

- The lower curve is for an object with a downward initial velocity of 50 ft/sec.

Notes:
1) 50 ft/sec is the typical cartridge ejection velocity of a flare/chaff expendable.
2) The top curve actually goes up 39 feet before starting back down, but this is difficult to see due to the graph scale.
3) This simplification ignores the effects of air drag or tumbling effects on a falling object which will result in a maximum terminal velocity, with resultant curve straightening.

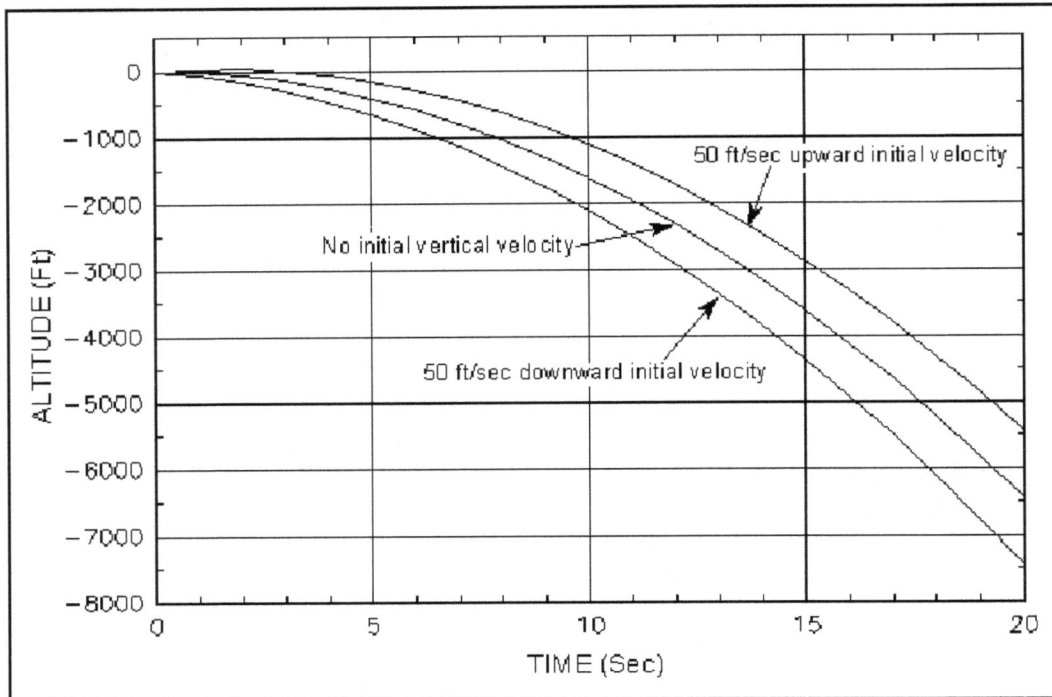

Figure 1. Object Fall Rate

SAMPLE CALCULATIONS:

Let us assume that we want to know how far a bomb or other object has fallen after 13 seconds if it had been dropped from an aircraft traveling at 450 kts which was in a 40° dive.

Our initial vertical velocity is: 450 kts (Sin 40°) (1.69 ft/sec per knot) = 489 ft/sec downward

$$d = V_it + \tfrac{1}{2}gt^2 = -489(13) + \tfrac{1}{2}(-32.2)(13)^2 = -6355 - 2721 = -9,076 \text{ ft.}$$

Remember to keep the signs (+/-) of your calculations in agreement with whatever convention you are using. Gravity pulls downward, so we used a minus sign for acceleration. Also the initial velocity was downward.

In reality, any object may well have reached terminal velocity before the time indicated using the above formula or Figure 1. In this example, the actual distance determined from ballistics tables would have been 8,000 ft, which is about 13% less than the above calculation would indicate. The drag characteristics of the object determine how much shorter the distance will be. In any case, it will not have dropped farther.

AIRCRAFT DRAG INDEX POINTS - Tactical aircraft carry stores in various combinations depending upon the mission. Each store has a different drag load which affects range. The pilot needs to know the total drag load in order to determine his aircraft range on a particular mission. Adding up the total drag in pounds of force for wind resistance would be cumbersome. Therefore, the drag of the stores is compared to a known reference drag (usually the aircraft), and expressed as a percentage of aircraft drag multiplied by some constant. This ratio is variously called drag count, drag index, or drag points. For instance, if a missile has 100 pounds of drag and the reference aircraft drag is 50,000 pounds, the ratio is 100/50,000 = 0.002. Multiply this by a constant of 100 (for example) and the drag index point is 0.2. The pilot only needs to look on a chart to see what the drag index points are for his stores, add up the drag points, and look on a chart to see what his aircraft range and best range (or endurance) speed will be.

MACH NUMBER and AIRSPEED vs ALTITUDE

MACH NUMBER is defined as a speed ratio, referenced to the speed of sound, i.e.

$$MACH\ NUMBER = \frac{Velocity\ of\ Interest}{Velocity\ of\ Sound} \quad \textit{(at the given atmospheric conditions)} \quad [1]$$

Since the temperature and density of air decreases with altitude, so does the speed of sound, hence a given true velocity results in a higher MACH number at higher altitudes.

AIRSPEED is a term that can be easily confused. The unqualified term airspeed can mean any of the following:

a. Indicated airspeed (IAS) - the airspeed shown by an airspeed indicator in an aircraft. Indicated airspeed is expressed in knots and is abbreviated KIAS.

b. Calibrated airspeed (CAS) - indicated airspeed corrected for static source error due to location of pickup sensor on aircraft. Calibrated airspeed is expressed in knots and is abbreviated KCAS. Normally it doesn't differ much from IAS.

c. True airspeed (TAS) - IAS corrected for instrument installation error, compressibility error, and errors due to variations from standard air density. TAS is expressed in knots and is abbreviated KTAS. TAS is approximately equal to CAS at sea level but increases relative to CAS as altitude increases. At 35,000 ft, 250 KIAS (or KCAS) is approximately 430 KTAS.

IAS (or CAS) is important in that aircraft dynamics (such as stall speed) responds largely to this quantity. TAS is important for use in navigation (True airspeed ± wind speed = ground speed).

Figures 1 and 2 depict relations between CAS and TAS for various altitudes and non-standard temperature conditions. The first graph depicts lower speed conditions, the second depicts higher speeds.

As an example of use, consider the chart on the next page. Assume we are in the cockpit, have read our IAS from the airspeed indicator, and have applied the aircraft specific airspeed correction to obtain 370 KCAS. We start at point "A" and go horizontally to our flight altitude at point "B" (25,000 ft in this case). To find our Mach, we go down vertically to point "C" to obtain 0.86 Mach. To get our TAS at our actual environmental conditions, we go from point "B" vertically until we hit the Sea Level (S.L.) reference line at point "D", then travel horizontally until we reach our actual outside air temperature (-20°C at altitude) at point "E", then go up vertically to read our actual TAS from the scale at point "F" (535 KTAS). If we wanted our TAS at "standard" temperature and pressure conditions, we would follow the dashed lines slanting upward from point "B" to point "G" and read 515 KTAS from the scale. Naturally, we could go into the graph at any point and go "backwards" to find CAS from true Mach or TAS.

Figure 3 shows a much wider range of Mach numbers. It contains only TAS and Mach, since aircraft generally do not fly above Mach 2, but missiles (which don't have airspeed indicators) do. The data on this graph can be obtained directly from the following formula for use at <u>altitudes of 36,000 ft and below</u>:

$$Speed\ of\ Sound\ (KTAS) = 29.06 \sqrt{518.7 - 3.57\ A} \quad \textit{Where A = altitude (K\ ft)} \quad [2]$$

The speed of sound calculated from this formula can be used with the equation on the first page to obtain Mach number. This equation uses the standard sea level temperature of 59° F and a lapse rate of -3.57°/1000 ft altitude. Temperature stabilizes at -69.7° F at 36,000 ft so the speed of sound stabilizes there at 573 knots. See the last page of this section for a derivation of equation [2].

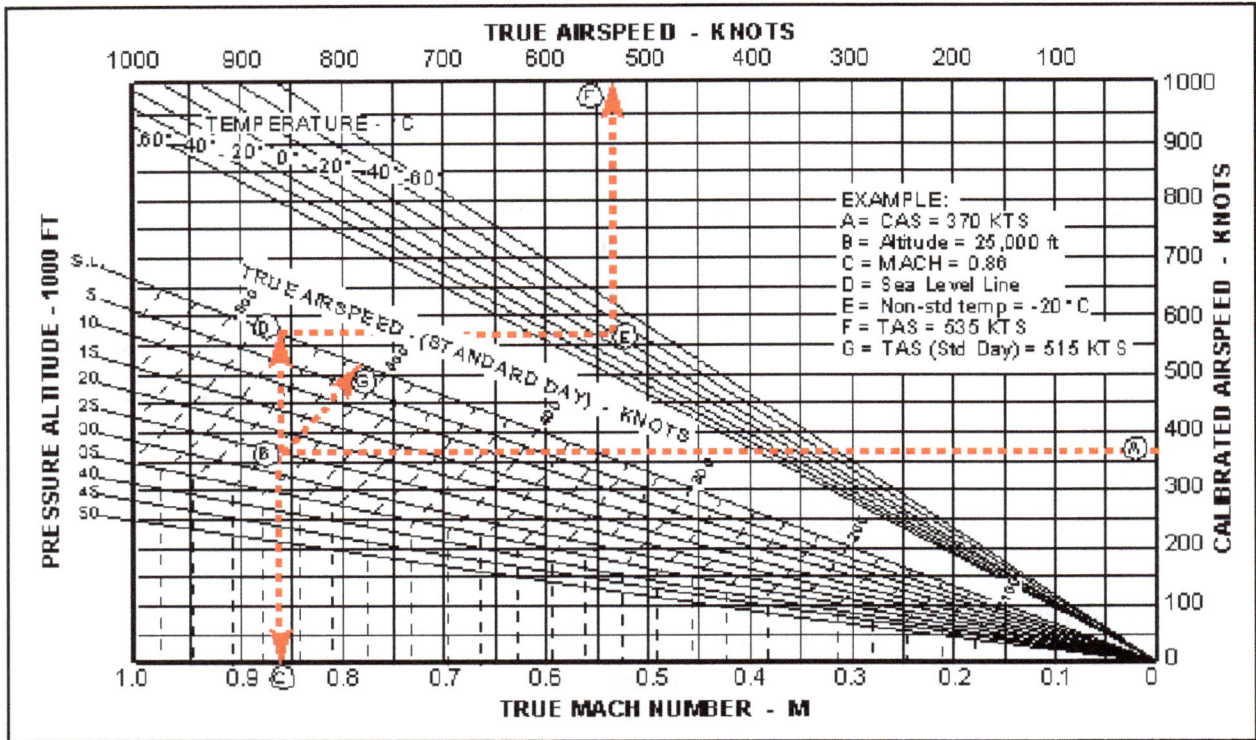

Figure 1. TAS and CAS Relationship with Varying Altitude and Temperature

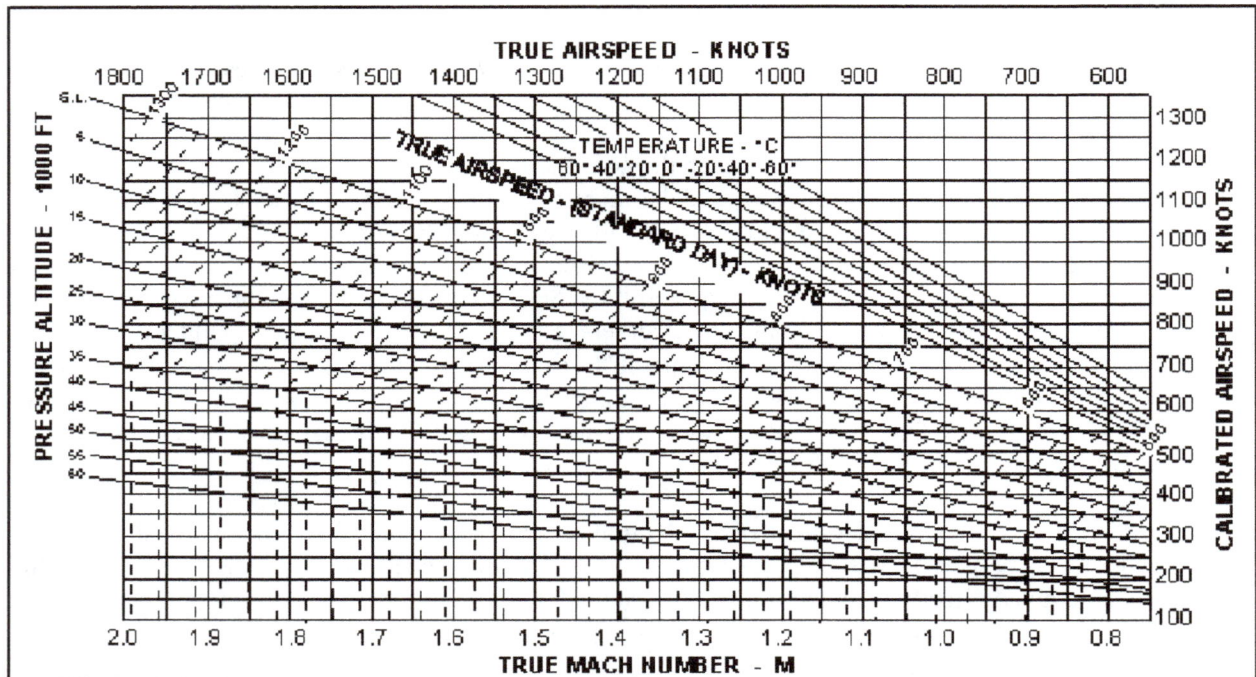

Figure 2. TAS and CAS Relationship with Varying Altitude and Temperature (Continued)

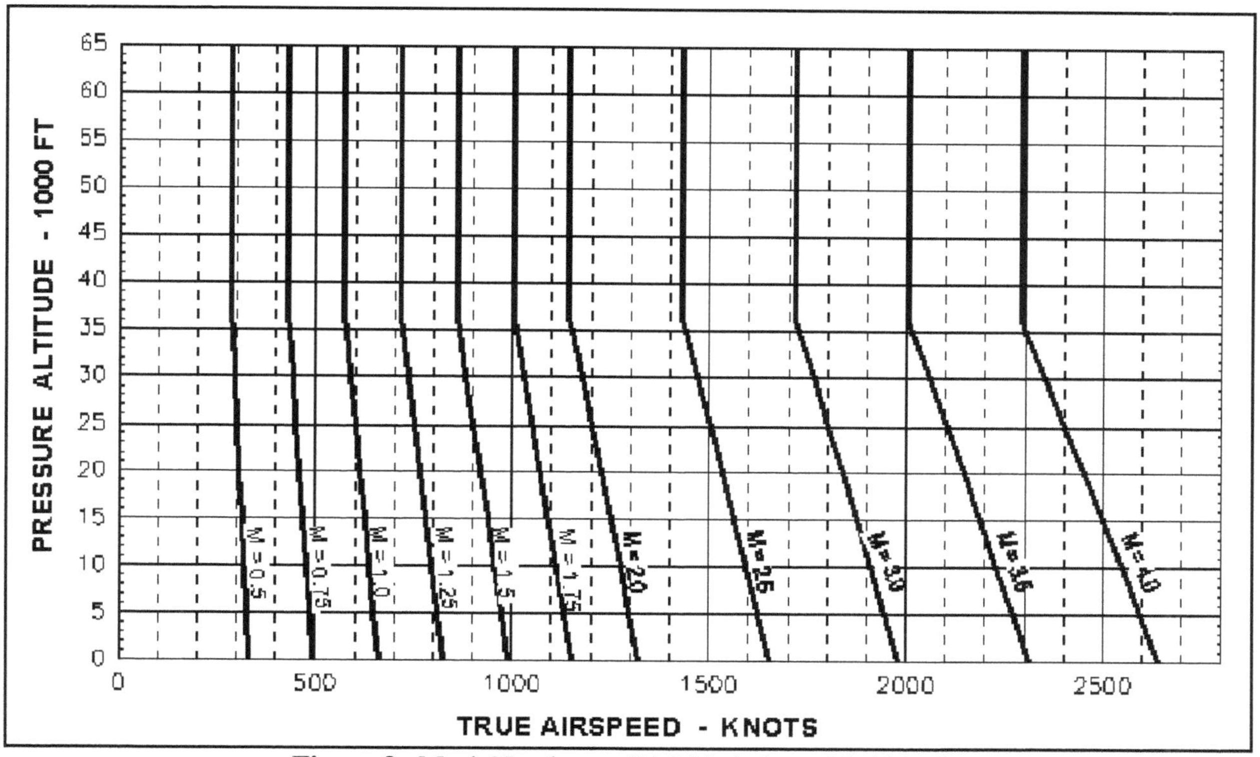

Figure 3. Mach Number vs TAS Variation with Altitude

The following is a derivation of equation [2] for the speed of sound:

Given: p = pressure (lb/ft^2) T = absolute temperature (°Rankine) = °F + 459.7
  v = specific volume (ft^3/lb) w = specific weight (lb/ft^3) = $1/v$
  R = a constant (for air: R = 53.3) ρ = density = w/g = $1/gv$ ∴ $v = 1/g\rho$

From Boyle's law of gasses: $pv = RT$, therefore we have: $p/\rho = gRT = (32.2)(53.3)T = 1718\,T$ [3]

It can also be shown that: p/ρ^{γ} = constant; for air $\gamma = 1.4$ [4]

From the continuity equation applied to a sound wave: $\rho AV_a = (\rho + dp)A(V_a + dV_a)$ [5]

Expanding and dropping insignificant terms gives: $dV_a = -V_a\,d\rho/\rho$ [6]

Using Newton's second law ($p + \rho V_a/2$ = a constant) and taking derivatives: $dp = -\rho V_a dV_a$

Substituting into [6] gives: $V_a{}^2 = dp/d\rho$ [7]

Then taking derivatives of [4] and substituting in [7] gives: $V_a = \sqrt{\dfrac{\gamma p}{\rho}}$ [8]

Then using [3] gives: $V_a = \sqrt{\gamma gRT} = \sqrt{1.4(1718)T} = 49\sqrt{T}$ [9]

Using a "Standard" atmosphere of 59° F @ Sea Level (S.L.) and a lapse rate of -3.57°/1000 ft altitude:

$$V_a = 49\sqrt{459.7 + 59 - 3.57\,A}\ \frac{ft}{sec}\left[\frac{3600\,sec}{hr}\ \frac{nm}{6076\,ft}\right] = 29.06\sqrt{518.7 - 3.57A}\quad \text{which is equation [2]}$$

This Page Blank

MANEUVERABILITY

A useful function is to determine how many "G's" an aircraft might require to make a given turn without altitude loss. From Newton's laws, F cos φ = W , where: F = force applied to an aircraft, W = weight, and φ = bank angle. By definition "G's" is the ratio of the force on an object to it's weight, i.e., G = F/W = 1/cos φ

Simple calculations will show the results presented in table 1, to the right.

Given that the average structural limit of an aircraft is about 7 G's, the maximum bank angle that can be achieved <u>in level (non-descending) flight</u> is 81.8°.

Table 1. G vs Angle of Bank (No altitude loss)

G	φ
1.0	0
1.4	45
2.0	60
3.9	75
7.2	82
11.5	85

Figure 1 can be used to determine the turn radius and rate-of-turn for any aircraft, given speed and angle of bank (assuming the aircraft maintains level flight). It may also be used in the reverse context. It should be noted that not all aircraft can fly at the speeds depicted - they may stall beforehand or may be incapable of attaining such speeds due to power/structural limitations.

In the example shown on Figure 1, we assume an aircraft is traveling at 300 kts, and decides to make a 30° angle of bank turn. We wonder what his turn radius is so we can approximate his flight path over the ground, and what his rate of turn will be. We enter the chart at the side at 300 kts and follow the line horizontally until we intercept the 30° "bank angle for rate of turn" line. We then go down vertically to determine the 2.10°/sec rate of turn. To get radius, we continue horizontally to the 30° "bank angle for turn radius" line . We can then go down vertically to determine the radius of 13,800 ft.

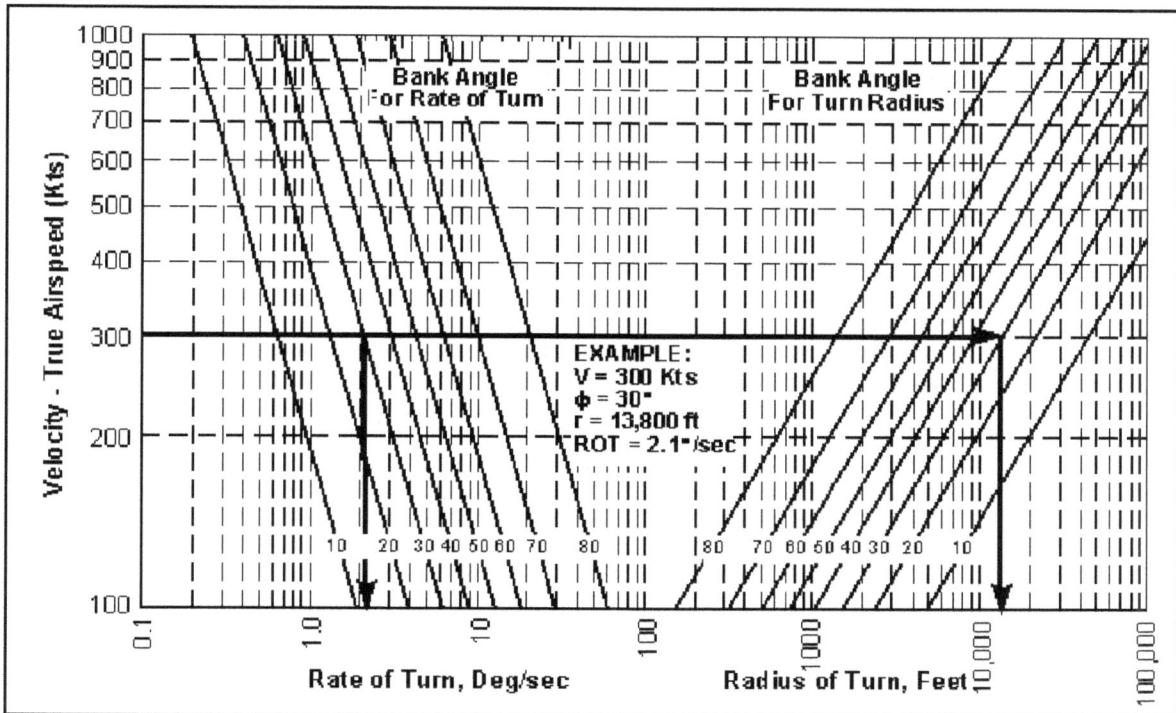

Figure 1. Aircraft Turn Rate / Radius vs Speed

The exact formulas to use are:

$$Rate\ of\ Turn = \frac{1091\tan(\phi)}{V} \qquad Radius\ of\ Turn = \frac{V^2}{11.26\tan(\phi)} \qquad Where: \begin{array}{l} V = Velocity\ (Knots) \\ and\ \phi = Angle\ of\ Bank \end{array}$$

Another interesting piece of information might be to determine the distance a typical aircraft might travel during a maneuver to avoid a missile.

Figure 2 shows a birds-eye view of such a typical aircraft in a level (constant altitude) turn.

To counter many air-to-air missiles the pilot might make a level turn, however in countering a SAM, altitude is usually

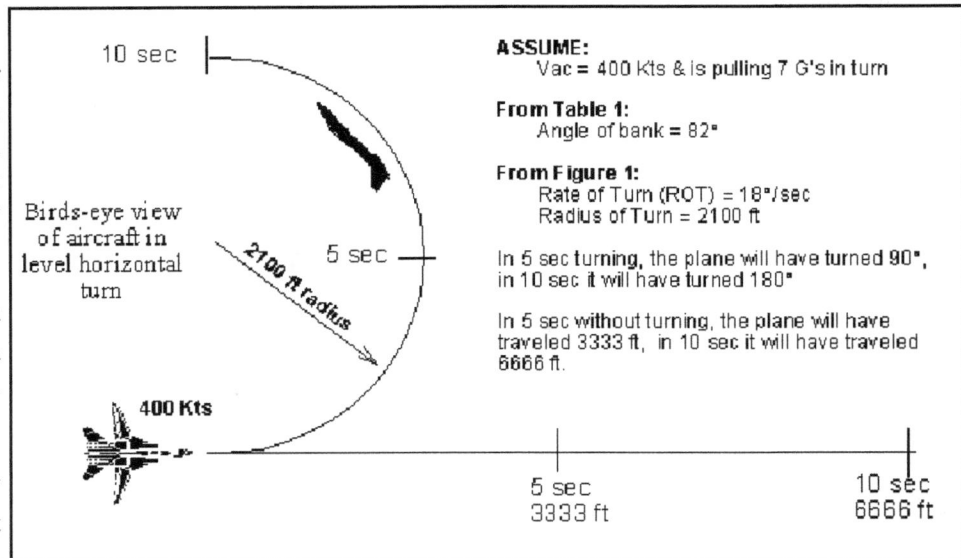

Figure 2. Maneuvering Aircraft

lost for two reasons: (1) the direction of maneuvering against the missile may be downward, and (2) many aircraft are unable to maintain altitude without also losing speed. These aircraft may have insufficient thrust for their given weight or may be at too high an altitude. The lighter an aircraft is (after dropping bombs/burning fuel), the better the performance. Likewise, the higher the altitude, the poorer the thrust-to-weight ratio. Maximum afterburner is frequently required to maintain altitude at maximum "G" level.

REFERENCE AXES (Roll, Pitch, Yaw):

The rotational or oscillating movement of an aircraft, missile, or other object about a longitudinal axis is called roll, about a lateral axis is called pitch, and about a vertical axis is called yaw as shown in Figure 3.

SAMPLE CALCULATIONS:

If we want to determine the rate of turn or turn radius more precisely than can be interpolated from the chart in Figure 1, we use the formulas. For our initial sample problem with an aircraft traveling 300 kts, in a 30° angle of bank turn, we have:

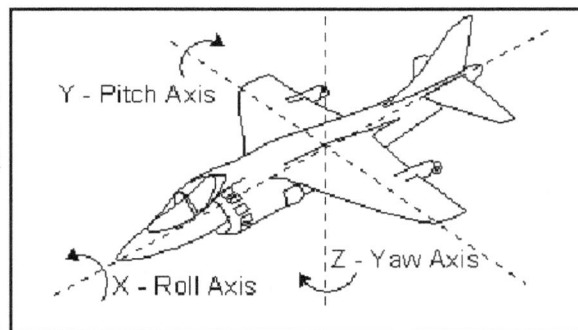

Figure 3. Reference Axes

$$Rate\ of\ Turn = \frac{1091\tan(\phi)}{V} = \frac{1091\tan(30)}{300} = 2.1°/\sec$$

$$Radius\ of\ Turn = \frac{V^2}{11.26\tan(\phi)} = \frac{300^2}{11.26\tan(30)} = 13,844\ ft$$

These are the same results as we determined using Figure 1.

EMP / AIRCRAFT DIMENSIONS

An aircraft flying in the vicinity of an electromagnetic pulse (EMP) acts like a receiving antenna and picks up EMP radiation in relation to size like a dipole (or half-wavelength dipole). The electromagnetic pulse spectrum decreases above 1 MHz as shown in Figure 1, so an F-14 aircraft that is an optimum ½ wavelength antenna at ≈8 MHz will pick up less EMP voltage than a B-52 or an aircraft with a trailing wire antenna. A rule of thumb for the voltage picked up is :

V_{EMP} = 8.1 volts/ft times the maximum dimension of the aircraft in feet

This rule of thumb was generated because a single linear relationship between voltage and aperture seemed to exist and compared favorably with more complex calculations for voltage picked up by various aircraft when subjected to EMP.

Table 1 shows various aircraft and the frequencies they would be most susceptible to, using $f = c/\lambda$, where λ matches the selected aircraft dimension for maximum "antenna reception effect". This should be a design consideration when trying to screen onboard avionics from the effects of EMP.

The following is a partial listing of aircraft types vs identifying prefix letters (several are used in Table 1):

A	Attack	K	Tanker	T	Trainer
B	Bomber	O	Observation	U	Utility
C	Cargo	P	Patrol	V	Vertical or Short Takeoff
E	Electronic Surveillance	Q	Special mission		and Landing (V/STOL)
F	Fighter	R	Reconnaissance	X	Experimental
H	Helicopter	S	Anti Sub/Ship	Y	Prototype

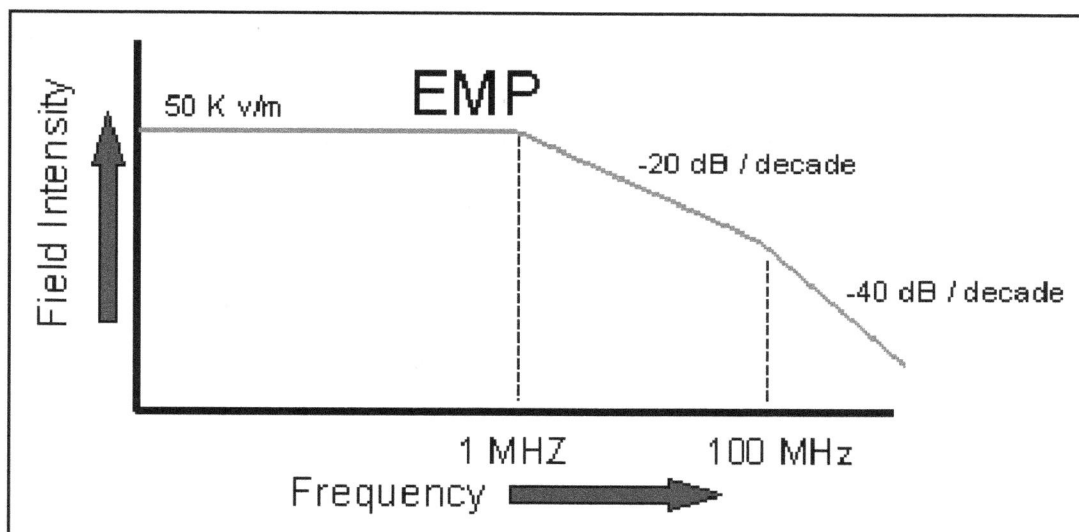

Figure 1. EMP as a Function of Frequency

Table 1. AIRCRAFT DIMENSIONS AND EQUIVALENT ANTENNA APERTURE

MISSION	AIRCRAFT TYPE	HEIGHT (ft.) A	FREQUENCY (MHz) f	f/2	LENGTH (ft.) A	FREQUENCY (MHz) f	f/2	WING SPAN (ft.) A	FREQUENCY (MHz) f	f/2
ATTACK	A-10	14.66	67.05	33.52	53.33	18.43	9.21	57.5	17.1	8.55
ELECTRONIC WARFARE	EA-6B	16.50	59.64	29.82	59.34	16.58	8.29	53.0	18.57	9.29
	EA-18	16.0	61.47	30.74	60.17	16.35	8.17	44.67	22.02	11.01
FIGHTER	F-15	18.4	53.42	26.71	63.75	15.42	7.71	42.8	22.97	11.48
	F-16	16.66	59.00	29.5	49.25	19.96	9.98	31.0	31.71	15.85
	F/A-18C/D	15.3	64.31	32.16	56.0	17.57	8.79	40.70	24.18	12.09
	F/A-18E/F	16	61.47	30.7	60.17	16.34	8.17	44.67	22.02	11.01
	F-22	17.75	55.41	27.7	64.17	15.33	7.66	43	22.87	11.43
	F-35A	14.2	69.27	34.63	51.4	19.1	9.57	35	28.1	14.05
	F-35C	15.5	63.45	31.73	51.4	19.1	9.57	43	22.8	11.43
	F-117	12.42	79.15	39.57	65.92	14.91	7.46	43.33	22.69	11.34
ASW	P-3C	33.75	29.16	14.58	116.42	8.45	4.23	99.67	9.87	4.94
	S-3A	22.75	43.25	21.63	54.34	18.45	9.23	68.67	14.33	7.17
	P-8	42.1	23.36	11.68	129.5	7.6	3.8	123.6	7.96	3.98
AEW	E-2C	18.4	53.48	26.74	56.50	17.42	8.71	80.58	12.21	6.11
V/STOL	AV-8B	11.64	84.45	42.23	46.3	21.23	10.62	30.3	32.44	16.22
	V-22	18.1	54.3	27.2	57.3	17.17	8.58	84.5	11.64	5.82
	F-35B	15.0	65.6	32.78	51.2	19.2	9.61	35	28.1	14.05
HELICOPTERS	AH-1W	14.16	69.46	34.73	58.25	16.89	8.44	48.0	20.49	10.24
	UH-1E	12.75	71.18	35.59	52.91	18.60	9.30	44.0	22.36	11.18
	U/SH-2	15.41	63.85	31.93	52.5	18.74	9.37	44.0	22.36	11.18
	SH-3D	16.42	59.93	29.97	72.67	13.54	6.77	62.00	15.87	7.84
	CH-46D	16.75	58.75	29.38	84.34	11.67	5.84	50.0	19.68	9.84
	CH-47	18.75	52.45	26.23	99.0	9.94	4.98	60.0	16.39	8.19
	CH-53A	24.91	39.50	19.75	88.16	11.16	5.58	72.25	13.62	6.81
	SH-60	17.0	57.85	28.93	64.83	15.17	7.59	53.83	18.27	9.14
	AH-64	15.25	64.5	32.25	58.25	16.89	8.44	48.0	20.49	10.24
TRANSPORT	C-2A	15.92	61.81	30.91	56.6	17.39	8.70	80.58	12.21	6.11
TANKERS	KC-130F	38.1	25.83	12.92	97.8	10.06	5.03	132.5	7.43	3.72
SPECIAL ELECTRONICS	EC-13OQ	38.5	25.56	12.78	99.34	9.91	4.96	132.58	7.42	3.71
TRAINER	T-2B	14.8	66.49	33.25	38.7	25.43	12.72	37.85	26.00	13.0
	T-39D	16.0	61.50	30.75	43.75	22.49	11.25	44.34	22.19	11.10
	T-45	13.5	72.86	36.43	39.33	25.0	12.5	30.8	31.93	15.97

DATA TRANSFER BUSSES

DATA BUSSES

INTRODUCTION

The avionics systems on aircraft frequently contain general purpose computer components which perform certain processing functions, then relay this information to other systems. Some common examples are the mission computers, the radar processors, RWRs, and jammers. Each system is frequently laid out as shown in Figure 1.

The Input/Output (I/O) modules will vary in function, but all serve the same purpose - to translate the electrical signals

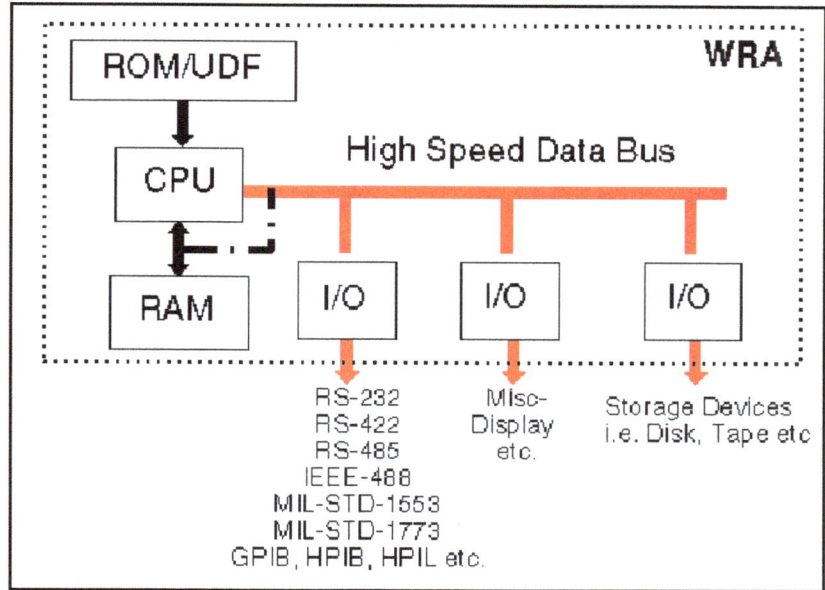

Figure 1. Avionics Block Diagram

from one protocol to one of another in order to exchange information. I/O modules are used similarly in general purpose computers in laboratories to test equipment and/or tie computers together via a local area network (LAN) to exchange information. Some of the methodologies include a star, ring, or bus type network (see terminology at the end of this section).

A number of network "models" exist for describing the functions, interfaces and protocols involved in network data interchange. Regardless of the descriptive model used, all strive toward the same end and none actually changes the detailed implementation. Table 1 shows the layer names and the number of layers of networking models presented in Internet Engineering Task Forces (IETF) Request For Comments (RFCs) and in common use by textbooks.

Table 1. Some Common Network Architecture Models

IETF-RFC 1122	Kurose, Forouzan	Comer, Kozierok	Stallings	Cisco Academy
Four layer Internet model	*Five-layer Internet model or TCP/IP protocol suite*	*TCP/IP 5-layer reference model*	*Five layer TCP/IP model*	*Four layer Internet model*
Application	Application	Application	Application	Application
Transport	Transport	Transport	Host-to-host or transport	Transport
Internet	Network	Internet	Internet	Internetwork
Link	Data link	Data link (Network interface)	Network access	Network interface
	Physical	(Hardware)	Physical	

These textbooks are secondary sources that may be contrary to the intent of RFC 1122 and other primary sources such as the Open Systems Interconnection (OSI) Reference Model developed by the International Organization for Standardization (ISO).

The OSI Reference Model is a more general description for layered communications and computer network protocol design. The IETF makes no effort to follow the OSI model although RFCs sometimes refer to it. The description of the OSI layers is shown in Table 2.

Table 2. OSI Reference Model

	Name	Use	Misc Example	IP Suitte
7	APPLICATION	Meaning of data	HL7, Modbus	DHCP, DNS, FTP, Gopher, HTTP, NTP, SMTP, SNMP, Telnet
6	PRESENTATION	Building blocks of data and encryption	ASCII, EBCDIC, MIDI	MIME, XDR, SSL
5	SESSION	Opening and closing of specific communication paths	NetBIOS, SAP, Half Duplex, Simplex, SDP	NetBIOS, RTP, SAP
4	TRANSPORT	Error checking	NBF	PPTP, SCTP, TCP, UDP
3	NETWORK	Determination of data paths within the network	NBF, Q.931, IS-IS	IP, ICMP
2	DATA LINK	Data transmission, source, destination, and checksum	802.3 (Ethernet), PPP 802.11a/b/g/n MAC, FDDI	ARP, PPP, PPTP, SLIP
1	PHYSICAL	Voltage levels, signal connections, wire, or fiber	RS-232, 10Base-T, 802.11a/b/g/n Physical	

A layer is a collection of conceptually similar functions that provide services to the layer above it and receives service from the layer below it. On each layer an instance provides services to the instances at the layer above and requests service from the layer below. For example, a layer that provides error-free communications across a network provides the path needed by applications above it, while it calls the next lower layer to send and receive packets that make up the contents of the path. Conceptually two instances at one layer are connected by a horizontal protocol connection on that layer.

Most networks do not use all layers. For example, RS-232 is only a physical layer. Ethernet is only layers 1 and 2. TCP/IP is a protocol, not a network, and uses layers 3 and 4 regardless of whether layers 1 and 2 are a telephone line, wireless connection, or 10Base-T Ethernet cable.

Most of the sections in this division discuss the lowest (physical) layer of communication. There are, however, several more areas of general interest which are included in later sections such as Ethernet and TCP/IP. These are used in general purpose computers like the desktop PC or lab networks, and are not commonly used in aircraft.

The typical high-speed data busses on avionics/computers do not operate as fast as the CPU clock speed, but they are much faster than the interface busses they connect to. There are a number of interface busses (physical layer in network model) which are widely used by aircraft, avionics systems and test equipment. The most common include the RS-232, the RS-422, the RS-485, the IEEE-488 (GP-IB/HP-IB) and the MIL-STD-1553A/B. The MIL-STD-1773 bus is a fiber optic implementation of the 1553 bus and may be used in the future when technology requires it to reduce susceptibility to emissions or other reasons.

A summary of these more common types follows in Table 3, which includes a brief descriptive comparison, while a section covering each in more detail is provided later.

Table 3. Summary of Physical Bus Characteristics

Bus	Max Length	Max Number of Terminals[1]	Type	# of Lines[2]	Data Rate	Rise Time[3]	Data Format
RS-232C	100 feet max 50 ft at 20k bps	1	Serial	3-20	150 - 19,200 baud per sec		5- to 8- bit serial
RS-422	1.2 km[4]	10[5]	Serial	3	see figure in RS-232 section	$<0.1\ T_b$	unspecified
RS-485	unspecified	32	Serial	3	10 MHz	$<0.3\ T_b$	unspecified
IEEE-488 (GP-IB/HP-IB)	20 meters	14	Parallel	16	500 kHz[6]		8-bit parallel
HP-IL	100 meters	30	Serial	2	20 k BPS		serial
MIL-STD-1553B MIL-STD-1773	300 feet N/A	32[7]	Serial	3	1 MHz	100- 300 ns	20-bit serial

NOTES FROM TABLE:

(1) Max Number of Terminals does not include the bus controller.

(2) Including ground/shield

(3) T_b = time duration of the unit interval at the applicable data signaling rate (pulse width)

(4) Length is function of data signaling rate influenced by the tolerable signal distortion, amount of longitudinally coupled noise and ground potential difference introduced between the controller and terminal circuit grounds as well as by cable balance. See RS-422 section for graph.

(5) Physical arrangement of multiple receivers involves consideration of stub line lengths, fail-safe networks, location of termination resistors, data rate, grounding, etc.

(6) Rate can go up to 1 MHz if special conventions are followed.

(7) Max Number of Terminals includes terminal reserved for broadcast commands.

BUS TERMINOLOGY

10BASE-T: Standard "Plain Vanilla" Ethernet based on Unshielded Twisted Pair wire

10BASE-F: 10Mbps fiber optic Ethernet

100BASE-T: Standard "Fast Ethernet" based on twisted pair copper wire

ADDRESS: A unique designation for the location of data or the identity of an intelligent device; allows each device on a single communications line to respond to its own message.

ASCII (American Standard Code for Information Interchange): Pronounced asky. A seven-bit-plus-parity code established by ANSI to achieve compatibility between data services.

ASYNCHRONOUS OPERATION: Asynchronous operation is the use of an independent clock source in each terminal for message transmission. Decoding is achieved in receiving terminals using clock information derived from the message.

BAUD: Unit of signaling speed. The speed in baud is the number of discrete events per second. If each event represents one bit condition, baud rate equals bits per second (BPS). When each event represents more than one bit, baud rate does not equal BPS.

BIT: Contraction of binary digit: may be either zero or one. A binary digit is equal to one binary decision or the designation of one or two possible values of states of anything used to store or convey information.

BIT RATE: The number of bits transmitted per second.

BRIDGE: A network bridge connects multiple network segments at the data link layer (layer 2) of the OSI model, and the term layer 2 switch is very often used interchangeably with bridge. Bridges are similar to repeaters or network hubs, devices that connect network segments at the physical layer; however, with bridging, traffic from one network is managed rather than simply rebroadcast to adjacent network segments.

BROADCAST: Operation of a data bus system such that information transmitted by the bus controller or a remote terminal is addressed to more than one of the remote terminals connected to the data bus.

BUS CONTROLLER: The terminal assigned the task of initiating information transfers on the data bus.

BUS MONITOR: The terminal assigned the task of receiving bus traffic and extracting selected information to be used at a later time.

BYTE: A binary element string functioning as a unit, usually shorter than a computer "word." Eight-bits per byte are most common. Also called a "character".

COMMAND/RESPONSE: Operation of a data bus system such that remote terminals receive and transmit data only when commanded to do so by the bus controller.

CRC: Cyclic Redundancy Check; a basic error-checking mechanism for link-level data transmissions; a characteristic link-level feature of (typically) bit-oriented data communications protocols. The data integrity of a received frame or packet is checked by an algorithm based on the content of the frame and then matched with the result that is performed by a sender and included in a (most often, 16-bit) field appended to the frame.

CROSSOVER CABLE: Cable with transmit/receive pairs reversed so one computer or hub or switch can link directly to another.

DATA BUS: Whenever a data bus or bus is referred to in MIL-STD-1553B, it shall imply all the hardware including twisted shielded pair cables, isolation resistors, transformers, etc., required to provide a single data path between the bus controller and all the associated remote terminals.

DCE (Data Communications Equipment): Devices that provide the functions required to establish, maintain, and terminate a data-transmission connection; e.g., a modem.

DHCP: Dynamic Host Configuration Protocol - permits auto-assignment of temporary IP addresses for new devices logging in

DNS: Domain Name Server - associates names with IP addresses

DTE (Data Terminal Equipment): Devices acting as data source, data sink, or both.

DUPLEX: Communication traveling between two nodes in both directions

DYNAMIC BUS CONTROL: The operation of a data bus system in which designated terminals are offered control of the data bus.

EIA (Electronic Industries Association): A standards organization in the U.S.A. specializing in the electrical and functional characteristics of interface equipment.

FDM (Frequency-Division Multiplexer: A device that divides the available transmission frequency range into narrower banks, each of which is used for a separate channel.

FDX (Full Duplex): Simultaneous, two-way, independent transmission in both directions (4-wire).

FTP: File Transfer Protocol - the most popular mechanism for bulk movement of files on TCP/IP.

GATEWAY: Device which links Ethernet to dissimilar networks and transfers data at the application layer level. Interface cards link the PC to Ethernet via the PCI, ISA, PCMCIA, PC/104, or other buses.

GPIB: General Purpose Interface Bus (see section 9-5),

HALF DUPLEX: Operation of a data transfer system in either direction over a single line, but not in both directions on that line simultaneously.

HANDSHAKING: Exchange of predetermined signals between two devices establishing a connection. Usually part of a communications protocol.

HPIB / HPIL: Hewlett-Packard Interface Bus / Hewlett-Packard Interface Loop

HUB: The simplest method of redistributing data, are "dumb," not interpreting or sorting messages that pass through them. A hub can be as simple as an electrical buffer with simple noise filtering. It isolates the impedances of multiple spokes in a star topology. Some hubs also have limited store-and-forward capability.

They indiscriminately transmit data to all other devices, which are still on the same collision domain, connected to the hubs. They are not assigned MAC addresses or IP addresses.

IEEE (Institute of Electrical and Electronic Engineers): An international professional society that issues its own standards and is a member of ANSI and ISO.

IP: Internet Protocol portion of TCP/IP. It is a is a protocol used for communicating data across a packet-switched network

IP ADDRESS: Address of a TCP/IP enabled device on an Intranet or Internet – in the form xxx.xxx.xxx.xxx, where xxx is an interger between 0 and 255

LAN: Local Area Network

MAC: Media Access Control - the physical components which dissasemble Ethernet message fames

MANCHESTER ENCODING: Digital encoding technique (specified for the IEEE 802.3 Ethernet baseband network standard) in which each bit period is divided into two complementary halves; a negative-to-positive (voltage) transition in the middle of the bit period designates a binary "1," while a positive-to-negative transition represents a "0". The encoding technique also allows the receiving device to recover the transmitted clock from the incoming data stream (self-clocking).

MESSAGE: A single message is the transmission of a command word, status word, and data words if they are specified. For the case of a remote terminal to remote terminal (RT to RT) transmission, the message shall include the two command words, the two status words, and data words.

MODE CODE: A means by which the bus controller can communicate with the multiplex bus related hardware, in order to assist in the management of information flow.

MODEM (Modulator-Demodulator): A device used to convert serial digital data from a transmitting terminal to a signal suitable for transmission over a telephone channel, or to reconvert the transmitted signal to serial digital data for acceptance by a receiving terminal.

MULTIPLEXOR (also Multiplexer): A device used for division of a transmission into two or more subchannels, either by splitting the frequency band into narrower bands (frequency division) or by allotting a common channel to several different transmitting devices one at a time (time division).

NETWORK: An interconnected group of nodes; a series of points, nodes, or stations connected by communications channels; the assembly of equipment through which connections are made between data stations.

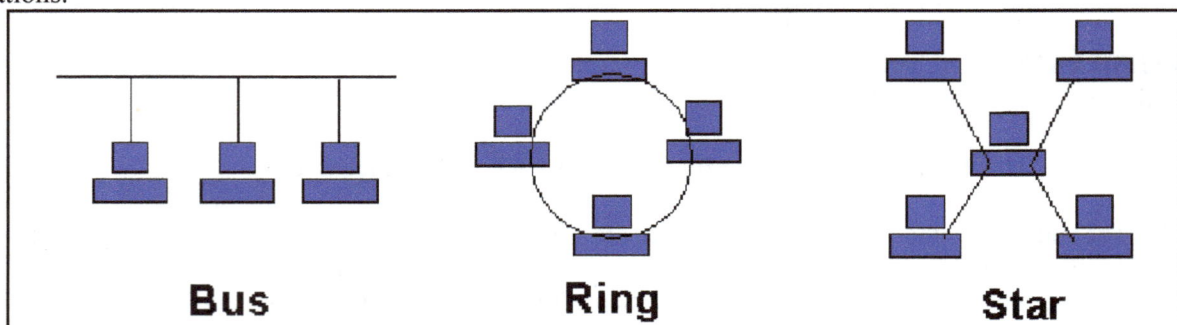

Bus   Ring   Star

NODE: A point of interconnection to a network. Normally, a point at which a number of terminals or tail circuits attach to the network.

PARALLEL TRANSMISSION: Transmission mode that sends a number of bits simultaneously over separate lines (e.g., eight bits over eight lines) to a printer. Usually unidirectional.

PHASE MODULATION: One of three ways of modifying a sine wave signal to make it "carry" information. The sine wave or "carrier" has its phase changed in accordance with the information to be transmitted.

PING: Packet Inter Net Groper - very useful utility which probes for the existence of a TCP/IP host

POLLING: A means of controlling devices on a multipoint line.

PORT: A number in TCP/IP to which services are assigned; e.g. FTP is port 21; SMTP is port 25; HTTP is port 80.

PROTOCOL: A formal set of conventions governing the formatting and relative timing of message exchange between two communicating systems.

PULSE CODE MODULATION (PCM): The form of modulation in which the modulation signal is sampled, quantized, and coded so that each element of information consists of different types or numbers of pulses and spaces.

REMOTE TERMINAL (RT): All terminals not operating as the bus controller or as a bus monitor.

REPEATER: Buffer which cleans up, strengthens and re-transmits a signal.

ROUTER: Repeater which selectively re-distributes messages based on IP address

SERIAL TRANSMISSION: The most common transmission mode; in serial, information bits are sent sequentially on a single data channel.

SNMP: Simple Network Management Protocol; allows monitoring and management of a network

SOCKET: Specific instance of an IP address and Port number that represents a single connection between two applications

STAR TOPOLOGY: Topology which allows only one device at each end of a wire and requires repeaters for more than two devices

STUBBING: Stubbing is the method wherein a separate line is connected between the primary data bus line and a terminal. The direct connection of stub line causes a mismatch which appears on the waveforms. This mismatch can be reduced by filtering at the receiver and by using bi-phase modulation. Stubs are often employed not only as a convenience in bus layout but as a means of coupling a unit to the line in such a manner that a fault on the stub or terminal will not greatly affect the transmission line operation. In this case, a network is employed in the stub line to provide isolation from the fault. These networks are also used for stubs that are of such length that the mismatch and reflection degrades bus operation. The preferred method of stubbing is to use transformer coupled stubs. The method provides the benefits of DC isolation, increased common mode protection, a doubling of effective stub impedance, and fault isolation for the entire stub and terminal. Direct coupled stubs should be avoided if at all possible. Direct coupled stubs provide no DC isolation or common mode rejection for the terminal external to its subsystem. Further, any shorting fault between the subsystems' internal isolation resistors (usually on the circuit board) and the main bus junction will cause failure of that entire bus. It can be expected that when the direct stub length exceeds 1.6 feet, that it will begin to distort the main bus waveforms. Note that this length includes the cable runs internal to a given subsystem.

SUBSYSTEM: The device or functional unit receiving data transfer service from the data bus.

SWITCH: Repeater which selectively re-distributes messages based on hardware MAC address

SYNCHRONOUS TRANSMISSION: Transmission in which data bits are sent at a fixed rate, with the transmitter and receiver synchronized. Synchronized transmission eliminates the need for start and stop bits.

TCP: Transmission Control Protocol - mechanism in TCP/IP that ensures that data arrives intact and in correct order

TELNET: Standard interface through which a client may access a host as though it were local

TERMINAL: The electronic module necessary to interface the data bus with the subsystem and the subsystem with the data bus. Terminals may exist as separate units or be contained within the elements of the subsystem.

TIME DIVISION MULTIPLEXING (TDM): The transmission of information from several signal sources through one communication system with different signal samples staggered in time to form a composite pulse train.

UDP: User Datagram Protocol - lower overhead alternative to TCP protocol which does not guarantee message delivery

WORD: A set of bits or bytes comprising the smallest unit of addressable memory. In MIL-STD-1553B, a word is a sequence of 16 bits plus sync and parity.

RS-232 INTERFACE

Introduction:

The RS-232 interface is the Electronic Industries Association (EIA) standard for the interchange of serial binary data between two devices. It was initially developed by the EIA to standardize the connection of computers with telephone line modems. The standard allows as many as 20 signals to be defined, but gives complete freedom to the user. Three wires are sufficient: send data, receive data, and signal ground. The remaining lines can be hardwired on or off permanently. The signal transmission is bipolar, requiring two voltages, from 5 to 25 volts, of opposite polarity.

Communication Standards:

The industry custom is to use an asynchronous word consisting of: a start bit, seven or eight data bits, an optional parity bit and one or two stop bits. The baud rate at which the word sent is device-dependent. The baud rate is usually 150 times an integer power of 2, ranging from 0 to 7 (150, 300, 600 ,...., 19,200). Below 150 baud, many system-unique rates are used. The standard RS-232-C connector has 25 pins, 21 pins which are used in the complete standard. Many of the modem signals are not needed when a computer terminal is connected directly to a computer, and Figure 1 illustrates how some of the "spare" pins should be linked if not needed. Figure 1 also illustrates the pin numbering used in the original DB-25 connector and that now commonly used with a DB-9 connector normally used in modern computers

Specifying compliance to RS-232 only establishes that the signal levels in two devices will be compatible and that if both devices use the suggested connector, they may be able to be connected. Compliance to RS-232 does not imply that the devices will be able to communicate or even acknowledge each other's presence.

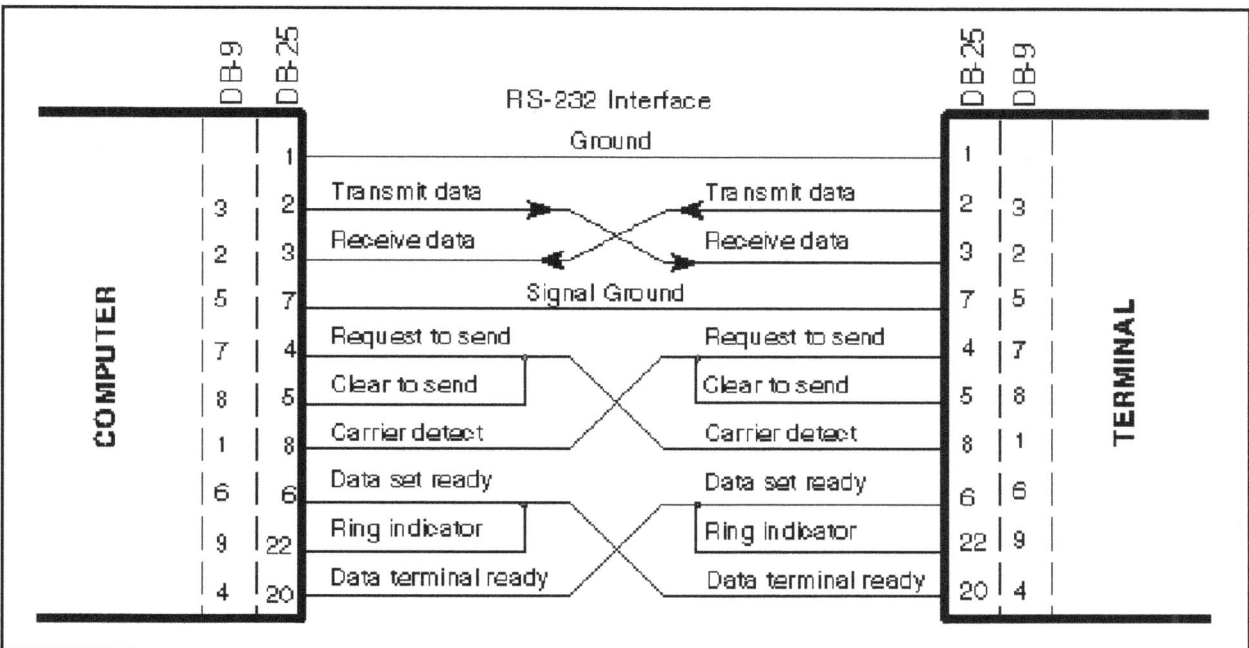

Figure 1. Direct-to-computer RS-232 Interface

Table 1 shows the signal names, and functions of the RS-232 serial port pinout. Table 2 shows a complete pin description

Table 1. RS-232 Serial Port Pinout

Name	Pin	Signal Name	Function
AA	1	PG Protective Ground	This line is connected to the chassis ground of the GPIB-232CV. Since the GPIB-232CV chassis ground is not connected to earth ground, pin 1 should be connected on both serial devices.
BA	2	TxD Transmit Data	This line carries serial data from the GPIB-232CV to the serial host.
BB	3	RxD Receive Data	This line carries serial data from the serial host to the GPIB-232CV.
CA	4	RTS Request to Send	This signal line is driven by the GPIB-232CV and when asserted indicates that the GPIB-232CV is ready to accept serial data. The GPIB-232CV un-asserts RTS when it is no longer ready to accept serial data because of a buffer full condition.
CB	5	CTS Clear to Send	This signal line is asserted by the serial host and sensed by the GPIB-232CV. When asserted, it indicates that the serial host is ready to accept serial data. When unasserted, it indicates that data transmission should be disabled.
AB	7	SG Signal Ground	This line establishes a reference point for all interface voltages.
CD	20	DTR Data Terminal Ready	This signal line is asserted by the GPIB-232CV to signal that it has been powered on, and is ready to operate.

Table 2. RS-232C Interface Signals.

Pin	Description	Pin	Description	Pin	Description
1	Protective Ground	10	(Reserved for Data Set Testing)	19	Secondary Request to Send
2	Transmitted Data	11	Unassigned	20	Data Terminal Ready
3	Received Data	12	Sec. Rec'd. Line Sig. Detector	21	Signal Quality Detector
4	Request to Send	13	Sec. Clear to Send	22	Ring Indicator
5	Clear to Send	14	Secondary Transmitted Data	23	Data Signal Rate Selector (DTE/DCE Source)
6	Data Set Ready	15	Transmission Signal Element Timing (DCE Source)	24	Transmit Signal Element Timing (DTE Source)
7	Signal Ground (Common Return)	16	Secondary Received Data	25	Unassigned
8	Received Line Signal Detector	17	Receiver Signal Element Timing (DCE Source)		
9	(Reserved for Data Set Testing)	18	Unassigned		

Electrical Characteristics: The RS-232-C specifies the signaling rate between the DTE and DCE, and a digital signal is used on all interchange circuits. The RS-232 standard specifies that logic "1" is to be sent as a voltage in the range -15 to -5 V and that logic "0" is to sent as a voltage in the range +5 to +15 V. The standard specifies that voltages of at least 3 V in amplitude will always be recognized correctly at the receiver according to their polarity, so that appreciable attenuation along the line can be tolerated. The transfer rate is rated > 20 kbps and a distance of < 15m. Greater distance and data rates are possible with good design, but it is reasonable to assume that these limits apply in practice as well as in theory. The load impedance of the terminator side of the interface must be between 3000 and 7000 ohms, and not more than 2500pF.

Table 3, summarizes the functional specifications of the most important circuits.

Table 3. RS-232-C Circuit Definitions

Name	Direction to:	Function
Data Signals		
Transmitted Data (BA)	DCE	Data generated by DTE
Received Data (BB)	DTE	Data Received by DTE
Timing signals		
Transmitter Signal Element Timing (DA)	DCE	Clocking signal, transitions to ON and OFF occur at center of each signal element
Transmitter Signal Element Timing (DB)	DTE	Clocking signal, as above; both leads relate to signals on BA
Receiver Signal Element Timing (DD)	DTE	Clocking signal, as above, for circuit BB
Control Signals		
Request to Send (CA)	DCE	DTE wishes to transmit
Clear to Send (CB)	DTE	DCE is ready to transmit; response to request to send
Data Set Ready (CC)	DTE	DCE is ready to operate
Data Terminal Ready (CD)	DCE	DTE is ready to operate
Ring Indicator (CE)	DTE	Indicates that DCE is receiving a ringing signal on the communication channel
Carrier Detect (CF)	DTE	Indicates that DCE is receiving a carrier signal
Signal Quality Detector (CG)	DTE	Asserted when there is reason to believe there is an error in the received data
Data Signal Rate Selector (CH)	DCE	Asserted to select the higher of two possible data rates
Data Signal Rate Selector (CI)	DTE	Asserted to select the higher of two possible data rates
Ground		
Protective Ground (AA)	NA	Attached to machine frame and possibly external grounds
Signal Ground (AB)	NA	Establishes common ground reference for all circuits

Range: The RS-232-C standard specifies that the maximum length of cable between the transmitter and receiver should not exceed 100 feet, Although in practice many systems are used in which the distance between transmitter and receiver exceeds this rather low figure. The limited range of the RS-232C standard is one of its major shortcomings compared with other standards which offer greater ranges within their specifications. One reason why the range of the RS-232C standard is limited is the need to charge and discharge the capacitance of the cable connecting the transmitter and receiver.

Mechanical Characteristics: The connector for the RS-232-C is a 25 pin connector with a specific arrangement of wires. In theory, a 25 wire cable could be used to connect the Data Terminal Equipment (DTE) to the Data Communication Equipment (DCE). The DTE is a device that is acting as a data source , data sink, or both, e.g. a terminal, peripheral or computer. The DCE is a device that provides the functions required to establish, maintain, and terminate a data-transmission connecting, as well as the signal conversion, and coding required for communication between data terminal equipment and data circuit; e.g. a modem. Table 4, shows the complete summary of the RS-232-C, e.g., descriptor, sponsor, data format, etc.

Table 4. Summary of the RS-232-C

Data Format	5- to 8- bit serial
Transfer Type	Asynchronous
Error Handling	Optional Parity Bit
Connector	25-pin female connector on DCE; 25-pin male connector on DTE
Length	20 meters
Speed	20 kb/s
Remarks	RS-232 is used in the microcomputer world for communications between two DTEs. The null-modem is included into one or both connecting devices, and/or cable and is seldom documented. As a result, establishing an RS-232 connection between two DTEs is frequently a difficult task.

RS-422 BALANCED VOLTAGE INTERFACE

Specifying compliance to RS-422 only establishes that the signal between the specified devices will be compatible. It does not indicate that the signal functions or operations between the two devices are compatible. The RS-422 standard only defines the characteristic requirements for the balanced line drivers and receivers. It does not specify one specific connector, signal names or operations. RS-422 interfaces are typically used when the data rate or distance criteria cannot be met with RS-232. The RS-422 standard allows for operation of up to 10 receivers from a single transmitter. The standard does not define operations of multiple tri-stated transmitters on a link.

The RS-422-A interfaces between the Data Terminal Equipment (DTE) and Data Communication Equipment (DCE) or in any point-to-point interconnection of signals between digital equipment. It employs the electrical characteristics of balanced-voltage digital interface circuits.

The balanced voltage digital interface circuit will normally be utilized on data, timing, or control circuits where the data signaling rate is up to 10 Mbit/s. While the balanced interface is intended for use at the higher data signaling rate, it may (in preference to the unbalanced interface circuit) generally be required if any of the following conditions prevail:

- The interconnecting cable is too long for effective unbalanced operation.
- The interconnecting cable is exposed to an extraneous noise source that may cause an unwanted voltage in excess of + 1 volt measured differentially between the signal conductor and circuit common at the load end of the cable with a 50 ohm resistor substituted for the generator.
- It is necessary to minimize interference with other signals.
- Inversion of signals may be required, i.e. plus to minus MARK may be obtained by inverting the cable pair.

Applications of the balanced voltage digital interface circuit are shown in Figure 1.

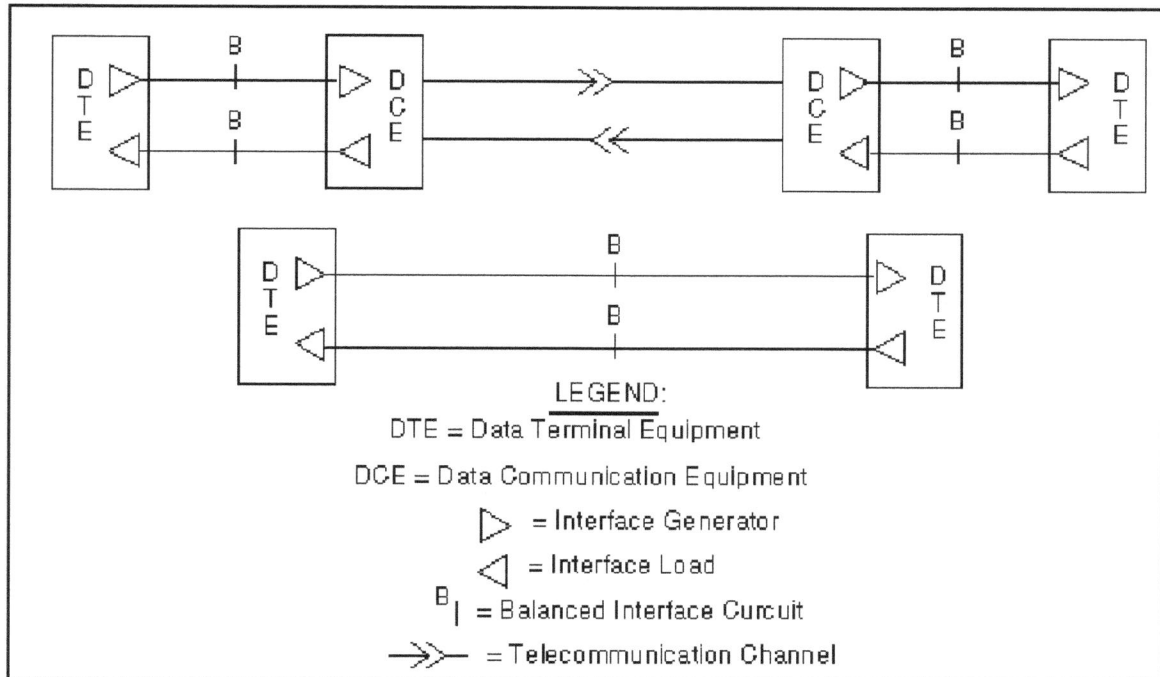

Figure 1. Application of a RS-422 Circuit

While a restriction on maximum cable length is not specified, guidelines are given later with respect to conservative operating distances as function of data signaling rate.

For a binary system in which the RS-422-A is designed, the data signaling rate in bit/s and the modulation in bauds are numerically equal when the unit interval used in each determination is the minimum interval.

Electrical Characteristics:

The balanced voltage digital interface circuit consists of three parts: the generator (G), the balanced interconnecting cable, and the load. The load is comprised of one or more receivers (R) and an optional cable termination resistance (RT). The balanced voltage interface circuit is shown in Figure 2.

Environmental Constraints:

Balanced voltage digital interface conforming to this standard will perform satisfactorily at data signaling rates up to 10 Mbit/s providing that the following operational constraints are satisfied:
- The interconnecting cable length is within that recommended for the applicable data signaling rate (see Figure 3) and the cable is appropriately terminated.
- The common mode voltage at the receiver is less than 7 volts (peak). The common mode voltage is defined to be any uncompensated combination of generator-receiver ground potential difference, the generator offset voltage (Vos), and longitudinally coupled peak noise voltage measured between the received circuit ground and cable within the generator ends of the cable short-circuited to ground.

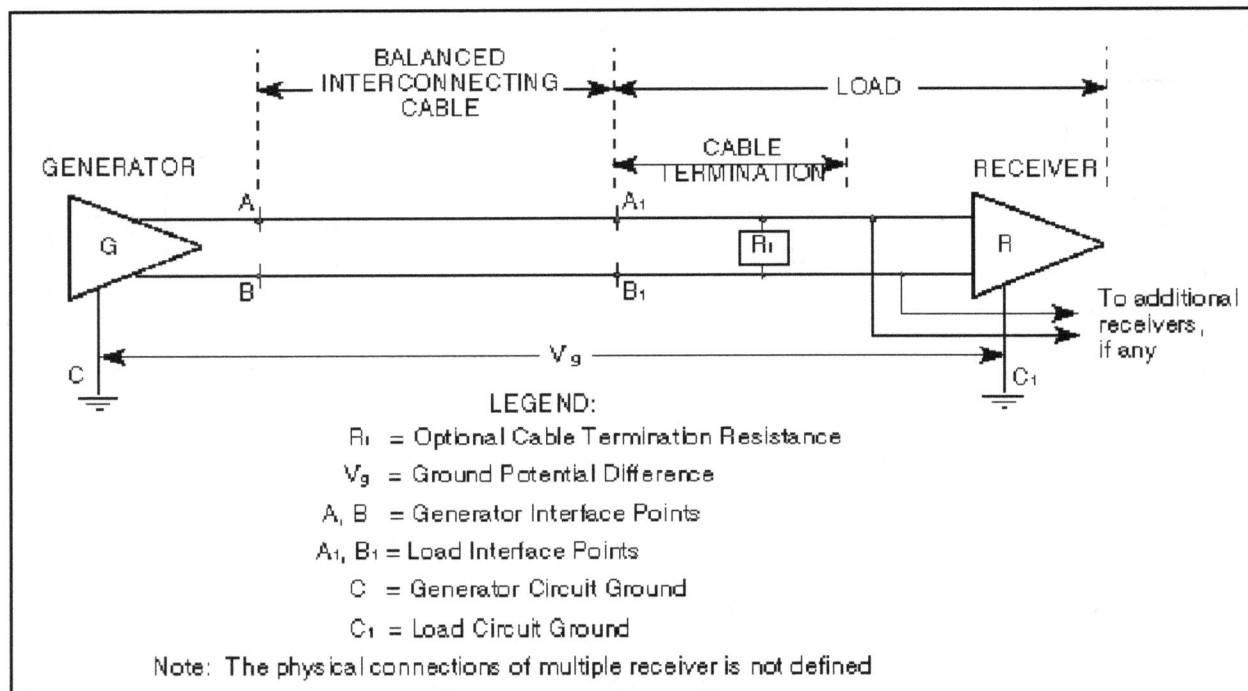

Figure 2. Balanced Digital Interface Circuit

Interconnecting Cable Guidelines:

The maximum permissible length of cable separating the generator and the load is a function of data signaling rate and is influenced by the tolerable signal distortion, the amount of coupled noise and ground potential difference introduced between the generator and load circuit as well as by cable balance. The curve of cable length versus signaling rate is given in Figure 3. This curve is based upon using 24 AWG copper, twisted-pair cable with a capacitance of 52.5 pF/meter terminated in a 100 ohm load. As data signaling rate is reduced below 90 kbit/s, the cable length has been limited at 1200 meters by the assumed maximum allowable 6 dBV signal loss.

Industry customs are not nearly as well established for RS-422 interfaces as they are for RS-232. The standard specifies use of the 37-pin "D"; the 9-pin "D" is specified for use with the secondary channel. Most data communications equipment uses the 37-pin "D"; many computer applications use a 9-pin "D" only. Some equipment applications use the 25-pin "D" defined for RS-232.

Compatibility With Other Interfaces:

Since the basic differential receivers of RS-423-A and RS-422-A are electrically identical, it is possible to interconnect an equipment using RS-423-A receivers and generators on one side of the interface with an equipment using RS-422-A generators and receivers on the other side of the interface, if the leads of the receivers and generators are properly configured to accommodate such an arrangement and the cable is not terminated.

This circuit is not intended for interoperation with other interface electrical circuits such as RS-232-C, MIL-STD-188C, or CCITT (Comite Consultatif Internationale Telegraphique et Telephonique), recommendations V.28 and V.35. Under certain conditions, the above interfaces may be possible but may require modification of the interface or equipment; therefore satisfactory operation is not assured and additional provisions not specified herein may be required.

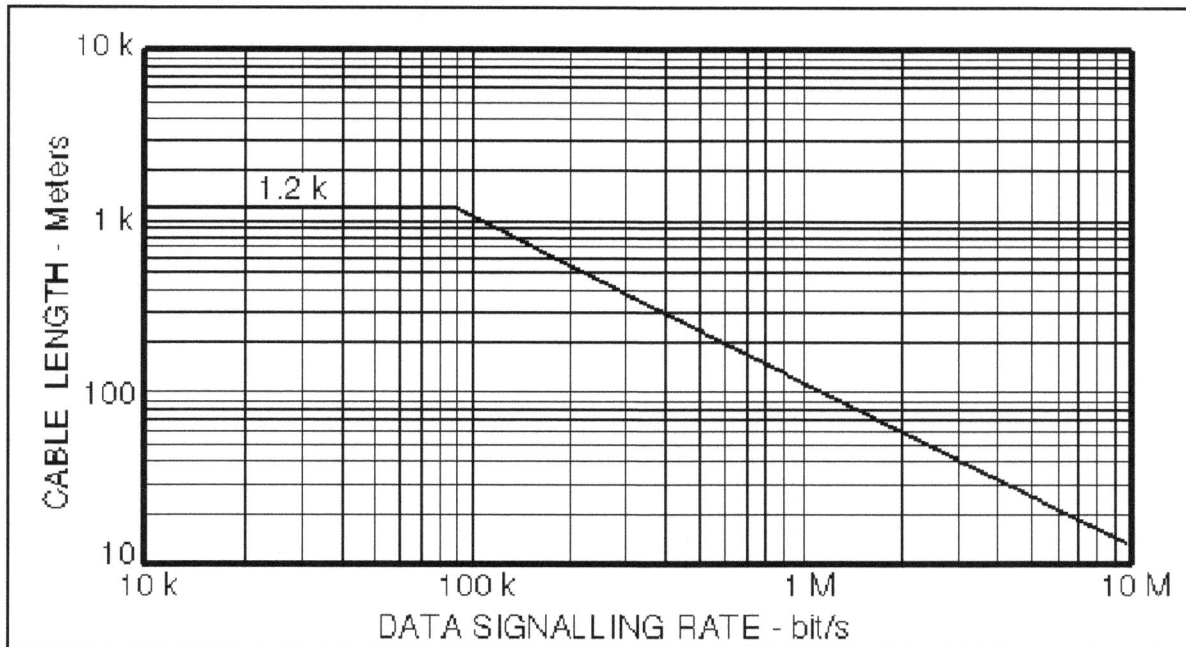

Figure 3. Data Signaling Rate vs Cable Length

This Page Blank

RS-485 INTERFACE

STANDARD FOR ELECTRICAL CHARACTERISTICS OF GENERATORS AND RECEIVERS FOR USE IN BALANCED DIGITAL MULTIPOINT SYSTEMS

Introduction: The RS-485 is the recommend standard by the Electronic Industries Association (EIA) that specifies the electrical characteristics of generators and receivers that may be employed for the interchange of binary signals in multipoint interconnection of digital equipments. When implemented within the guidelines, multiple generators and receivers may be attached to a common interconnecting cable. An interchange system includes one or more generators connected by a balanced interconnecting cable to one or more receivers and terminating resistors.

Electrical Characteristics:
The electrical characteristics that are specified are measured at an interconnect point supplied by the devices manufacturer. Figure 1 shows an interconnection application of generators and receivers having the electrical parameters specified. The elements in the application are: generators, receivers, transmission cables, and termination resistances (Rt). The loads on the system caused by each

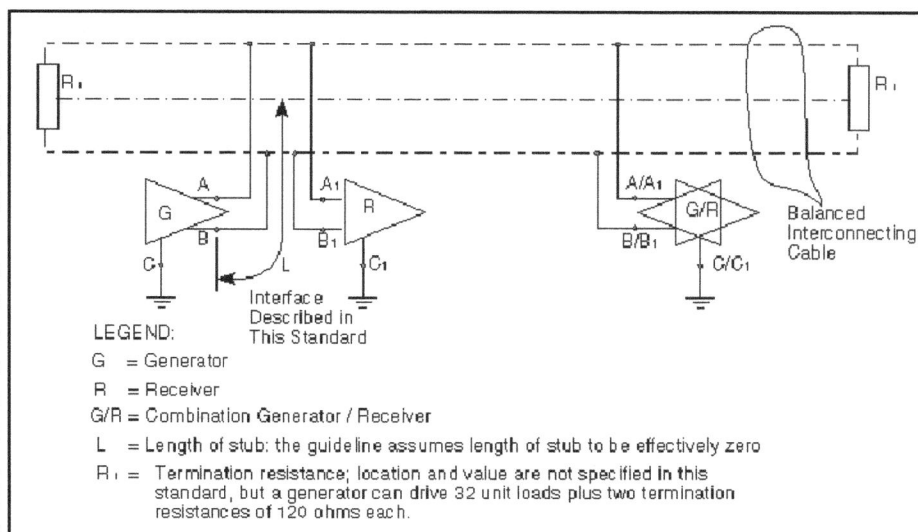

Figure 1. Multipoint Interconnect Application

receiver and passive generator shall be defined in terms of unit loads. Each generator can drive up to 32 unit loads consisting of both receivers and generators in the passive state. The loading caused by receivers and passive generators on the interconnect must be considered in defining the device electrical characteristics. Two areas are of concern: the DC load and the AC load characteristics. The DC load is defined as a number or fractions of "unit loads". The AC loading is not standardized but must be considered in the design of a system using the devices meeting this standard.

General System Configuration: The generators and receivers conforming to the RS-485 standard can operate with a common mode voltage between -7 volts and +7 volts (instantaneous). The common mode voltage is defined to be any uncompensated combination of generator-receiver ground potential difference and longitudinally coupled peak noise voltage measured between the receiver circuit ground and cable with the generator ends of the cable short circuited to ground, plus the generator offset voltage (Vos).

Grounding Arrangements: Proper operation of the generator and receiver circuits requires the presence of a signal return path between the circuit grounds of the equipment at each end of the interconnection. The grounding arrangements are shown in Figure 2. Where the circuit reference is provided by a third conductor, the connection between circuit common and the third conductor must contain some resistance (e.g., 100 ohms) to limit circulating currents when other ground connections are provided for safety. Some applications may require the use of shielded interconnecting cable for EMI or other purposes. The shield shall be connected to frame ground at either or both ends, depending on the application.

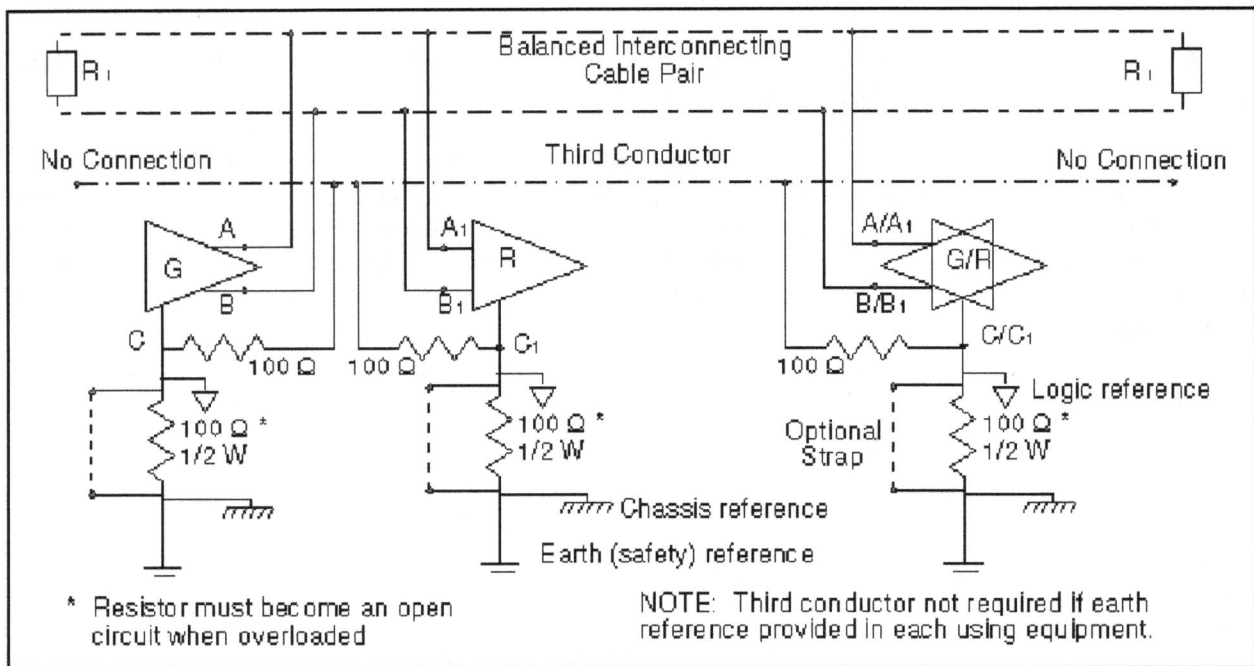

Figure 2. Grounding Arrangements

Similarity with RS-422-A:

In certain instances, it may be possible to produce generators and receivers that meet the requirements of both RS-422-A and of RS-485. Table 1 depicts the differences in parameter specifications which exist between the two documents.

Table 1. Comparison of RS-422-A and RS-485 Characteristics

Characteristic	RS-422-A	RS-485
Min. output voltage	2V into 100 ohm > 1/2 open circuit V	1.5 V into 54 ohms
I_{short} to ground	150 mA maximum	
I_{short} to -7, +12 volts		250 mA peak
t_{rise} time	< 0.1 t_b , 100 ohm load	< 0.3 t_b , 54 ohm, 50 pF load

Where t_b = time duration of the unit interval at the applicable data signaling rate (pulse width).

IEEE-488 INTERFACE BUS (HP-IB/GP-IB)

In the early 1970's, Hewlett-Packard came out with a standard bus (HP-IB) to help support their own laboratory measurement equipment product lines, which later was adopted by the IEEE in 1975. This is known as the IEEE Std. 488-1975. The IEEE-488 Interface Bus (HP-IB) or general purpose interface bus (GP-IB) was developed to provide a means for various instruments and devices to communicate with each other under the direction of one or more master controllers. The HP-IB was originally intended to support a wide range of instruments and devices, from the very fast to the very slow.

Description:

The HP-IB specification permits up to 15 devices to be connected together in any given setup, including the controller if it is part of the system. A device may be capable of any other three types of functions: controller, listener, or talker. A device on the bus may have only one of the three functions active at a given time. A controller directs which devices will be talkers and listeners. The bus will allow multiple controllers, but only one may be active at a given time. Each device on the bus should have a unique address in the range of 0-30. The maximum length of the bus network is limited to 20 meters total transmission path length. It is recommended that the bus be loaded with at least one instrument or device every 2 meter length of cable (4 meters is maximum). The use of GP-IB extenders may be used to exceed the maximum permitted length of 20 meters.

Electrical Interface:

The GP-IB is a bus to which many similar modules can be directly connected, as is shown in Figure 1. A total of 16 wires are shown in the figure - eight data lines and eight control lines. The bus cables actually have 24 wires, providing eight additional for shielding and grounds.

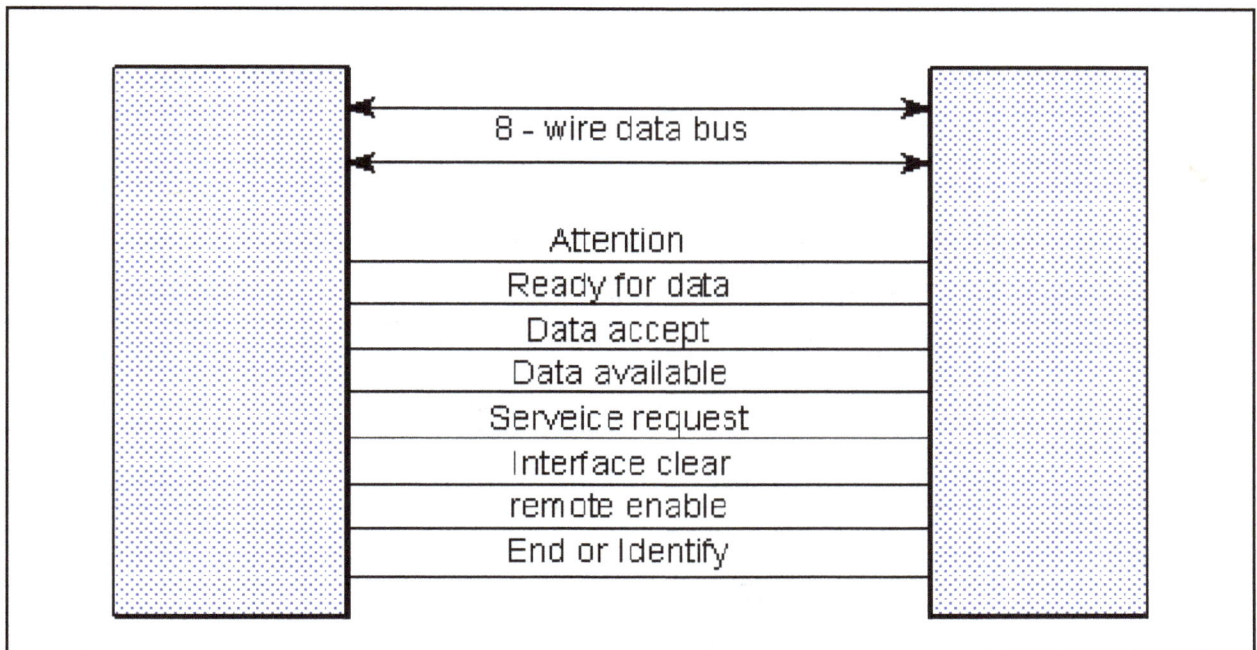

Figure 1. IEEE-488 (HP-IB / GP-IB) Bus Configuration

The GP-IB defines operation of a three-wire handshake that is used for all data transfers on the bus. The bus operation is asynchronous in nature. The data-transfer rate of the GP-IB is 500 kHz for standard applications and can go up 1 MHz if special conventions are followed. Each transaction carries 8 bits, the maximum data bandwidth is on the order of 4 to 8 megabits (1 M byte) per second. The bus is a two way communications channel and data flows in both directions. Figure 2 illustrates the structure of the GP-IB bus and identifies the 16 connections of the interconnecting cable.

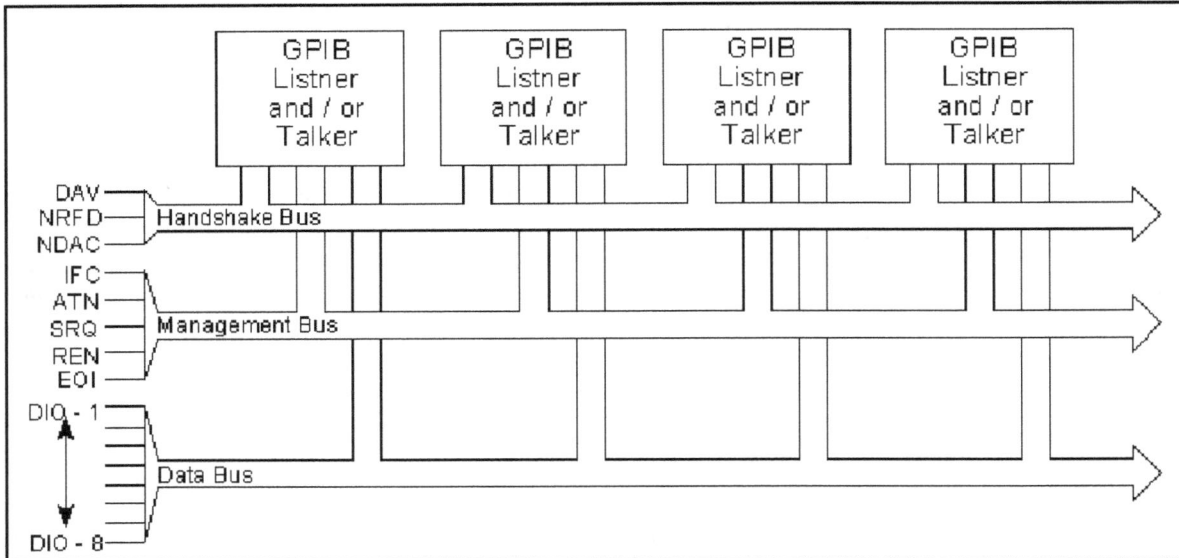

Figure 2. GP-IB Instrumentation Bus Structure

The cabling limitations make it a less-than-ideal choice for large separation between devices. These limitations can be overcome with bus extenders. Those attempting to use bus extenders should be aware that few extenders are as transparent as claimed. This is especially true in handling of continuous data and interrupts. In nonextended environments, it provides an excellent means for high-speed computer control of multiple devices.

The following table shows the various interface functions, the mnemonics and the descriptions.

Table 1. GP-IB Interface Functions

Interface Function	Mnemonic	Description
Talker (extended talker)	T (TE)	Device must be able to transmit
Listener (Extended listener)	L (LE)	Device must receive commands and data
Source Handshake	SH	Device must properly transfer a multiline message
Acceptor Handshake	AH	Device must properly receive remote multiline messages
Remote/Local	RL	Device must be able to operate from front panel and remote information from bus
Service Request	SR	Device can asynchronously request service from the controller
Parallel Poll	PP	Upon controller request, device must uniquely identify itself if it requires service
Device Clear	DC	Device can be initialized to a predetermined state
Device Trigger	DT	A device function can be initiated by the talker on the bus
Controller	C	Device can send addresses, universal commands, address commands, and conduct polls
Drivers	E	This code describes the type of electrical drivers in a device

The cabling specifications of the GP-IB interface system permit interconnecting all devices together in a star or linear configuration. The GP-IB connector is a 24-pin ribbon-type connector.

In summary, Table 2 on this page and the next shows the complete description of the GP-IB data bus.

Table 2. GP-IB Data Bus Description

IEEE-488, GP-IB, HP-IB, or IEC-625					
Descriptor	8-bit parallel, monodirectional, multi-master (token passing) One controller, one talker, several listeners	Arbitration	Token passing: the controller addresses the next controller SRQ Service request when the controller assigns modes	Connector	24-pin Amphenol Female connector on equipment chassis. DIO1 1 ■ ■ 13 DIO5 DIO2 2 ■ ■ 14 DIO6 DIO3 3 ■ ■ 15 DIO7
Sponsor	Hewlett-Packard	Error handling	Parity bit DI07 when 7-bit ACSII characters		DIO4 4 ■ ■ 16 DIO8 EOI 5 ■ ■ 17 REN DAV 6 ■ ■ 18 Gnd
Standard	IEEE 488, IEC 625	Bus length	15 m		NRFD 7 ■ ■ 19 Gnd
Address space	31 devices	Driver	Special 24 mA drivers		NDAC 8 ■ ■ 20 Gnd IFC 9 ■ ■ 21 Gnd
Data format	8-bit parallel	Speed	1 MByte/s		SRQ 10 ■ ■ 22 Gnd ATN 11 ■ ■ 23 Gnd Shld 2 ■ ■ 24 Gnd
Transfer type	Write only, talker toward listener(s) or commander toward all others			Remarks	The 488 is most commonly used for data acquisition of H-P peripherals. Programmable interfaces and drivers exist and simplify the development of microprocessor interfaces.
Timing	Handshaken 3-wire broadcast transfer: DAV data valid NDAC Not data accepted NRFD Not ready for data	References	IEEE Computer Society		

HP-IL Variation:

Since introduction of the IEEE-488, technology produced a generation of medium-speed, low-power, instrumentation which had a need to operate in an automatic test system such as the GP-IB. The HP-IL (Hewlett-Packard Interface Loop), was introduced to meet this need. The HP-IL is a low-cost, low-power alternative to the GP-IB system. The HP-IL and GP-IB provide the same basic functions in interfacing controllers, instruments, and peripherals, but they differ in many other respects. HP-IL is suitable for use in low-power, portable applications (typically used for interface of battery-power systems). The GP-IB is not practical to operate from battery power. The HP-IL maximum data rate is 20K bytes per second. This is a high rate compared to the RS-232C, but much slower than GP-IB. The HP-IL can operate over distances of up to 100 meters between any two devices. Since it is a loop environment, there is no maximum system cable restriction. The basic device-addressing scheme allows for up to 30 devices on a loop.

This Page Blank

MIL-STD-1553 & 1773 DATA BUS

PURPOSE

In recent years, the use of digital techniques in aircraft equipment has greatly increased, as have the number of avionics subsystems and the volume of data processed by them.

Because analog point-to-point wire bundles are inefficient and cumbersome means of interconnecting the sensors, computers, actuators, indicators, and other equipment onboard the modern military vehicle, a serial digital multiplex data bus was developed. MIL-STD-1553 defines all aspects of the bus, therefore, many groups working with the military tri-services have chosen to adopt it.

The 1553 multiplex data bus provides integrated, centralized system control and a standard interface for all equipment connected to the bus. The bus concept provides a means by which all bus traffic is available to be accessed with a single connection for testing and interfacing with the system. The standard defines operation of a serial data bus that interconnects multiple devices via a twisted, shielded pair of wires. The system implements a command-response format.

MIL-STD-1553, "Aircraft Internal Time-Division Command/Response Multiplex Data Bus," has been in use since 1973 and is widely applied. MIL-STD-1553 is referred to as "1553" with the appropriate revision letter (A or B) as a suffix. The basic difference between the 1553A and the 1553B is that in the 1553B, the options are defined rather than being left for the user to define as required. It was found that when the standard did not define an item, there was no coordination in its use. Hardware and software had to be redesigned for each new application. The primary goal of the 1553B was to provide flexibility without creating new designs for each new user. This was accomplished by specifying the electrical interfaces explicitly so that compatibility between designs by different manufacturers could be electrically interchangeable.

The Department of Defense chose multiplexing because of the following advantages:
- Weight reduction
- Simplicity
- Standardization
- Flexibility

Some 1553 applications utilize more than one data bus on a vehicle. This is often done, for example, to isolate a Stores bus from a Communications bus or to construct a bus system capable of interconnecting more terminals than a single bus could accommodate. When multiple buses are used, some terminals may connect to both buses, allowing for communication between them.

MULTIPLEXING

Multiplexing facilitates the transmission of information along the data flow. It permits the transmission of several signal sources through one communications system.

BUS

The bus is made up of twisted-shielded pairs of wires to maintain message integrity. MIL-STD-1553 specifies that all devices in the system will connect to a redundant pair of buses. This provides a second path for bus traffic should one of the buses be damaged. Signals are only allowed to appear on one of the two buses at a time. If a message cannot be completed on one bus, the bus controller may switch to the other bus. In some applications more than one 1553 bus may be implemented on a given vehicle. Some terminals on the bus may actually connect to both buses.

BUS COMPONENTS

There are only three functional modes of terminals allowed on the data bus: the bus controller, the bus monitor, and the remote terminal. Devices may be capable of more than one function. Figure 1 illustrates a typical bus configuration.

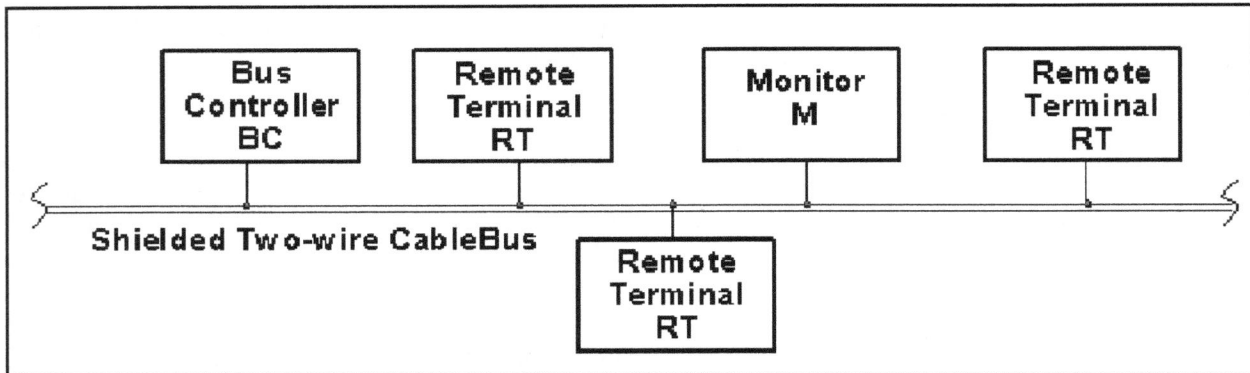

Figure 1. 1553 Bus Structure

- Bus Controller - The bus controller (BC) is the terminal that initiates information transfers on the data bus. It sends commands to the remote terminals which reply with a response. The bus will support multiple controllers, but only one may be active at a time. Other requirements, according to 1553, are: (1) it is "the key part of the data bus system," and (2) "the sole control of information transmission on the bus shall reside with the bus controller."

- Bus Monitor - 1553 defines the bus monitor as "the terminal assigned the task of receiving bus traffic and extracting selected information to be used at a later time." Bus monitors are frequently used for instrumentation.

- Remote Terminal - Any terminal not operating in either the bus controller or bus monitor mode is operating in the remote terminal (RT) mode. Remote terminals are the largest group of bus components.

MODULATION

The signal is transferred over the data bus using serial digital pulse code modulation.

DATA ENCODING

The type of data encoding used by 1553 is Manchester II biphase.
- A logic one (1) is transmitted as a bipolar coded signal 1/0 (in other words, a positive pulse followed by a negative pulse).
- A logic zero (0) is a bipolar coded signal 0/1 (i.e., a negative pulse followed by a positive pulse).

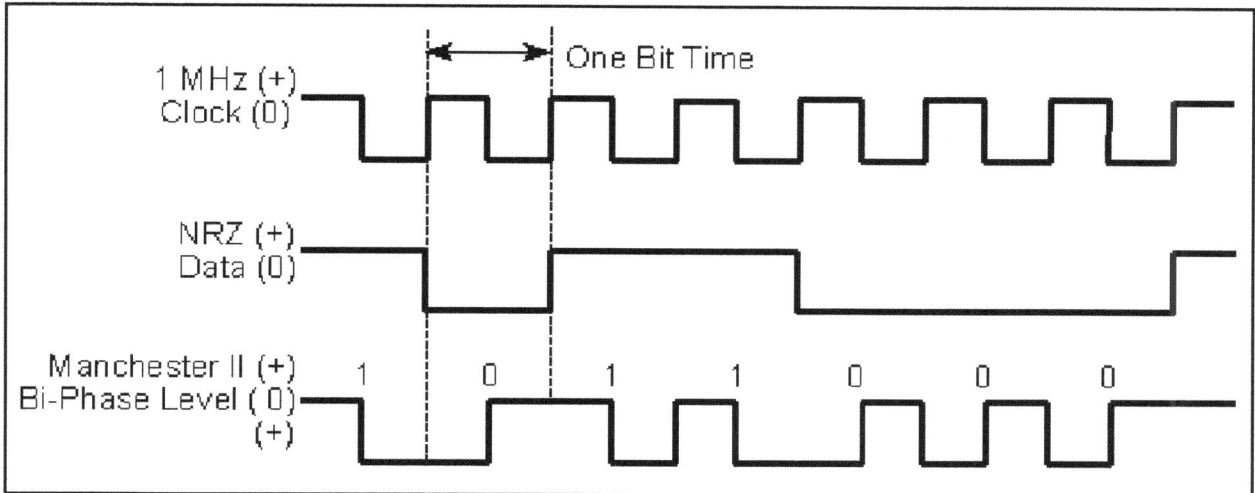

Figure 2. Data Encoding

A transition through zero occurs at the midpoint of each bit, whether the rate is a logic one or a logic zero. Figure 2 compares a commonly used Non Return to Zero (NRZ) code with the Manchester II biphase level code, in conjunction with a 1 MHz clock.

BIT TRANSMISSION RATE

The bit transmission rate on the bus is 1.0 megabit per second with a combined accuracy and long-term stability of +/- 0.1%. The short-term stability is less than 0.01%.

There are 20 1.0-microsecond bit times allocated for each word. All words include a 3 bit-time sync pattern, a 16-bit data field that is specified differently for each word type, and 1 parity check bit.

WORD FORMATS

Bus traffic or communications travels along the bus in words. A word in MIL-STD-1553 is a sequence of 20 bit times consisting of a 3 bit-time sync wave form, 16 bits of data, and 1 parity check bit. This is the word as it is transmitted on the bus; 1553 terminals add the sync and parity before transmission and remove them during reception. Therefore, the nominal word size is 16 bits, with the most significant bit (MSB) first.

There are three types of words: command, status, and data. A packet is defined to have no inter-message gaps. The time between the last word of a controller message and the return of the terminal status byte is 4-12 microseconds. The time between status byte and the next controller message is undefined. Figure 3 illustrates these three formats.

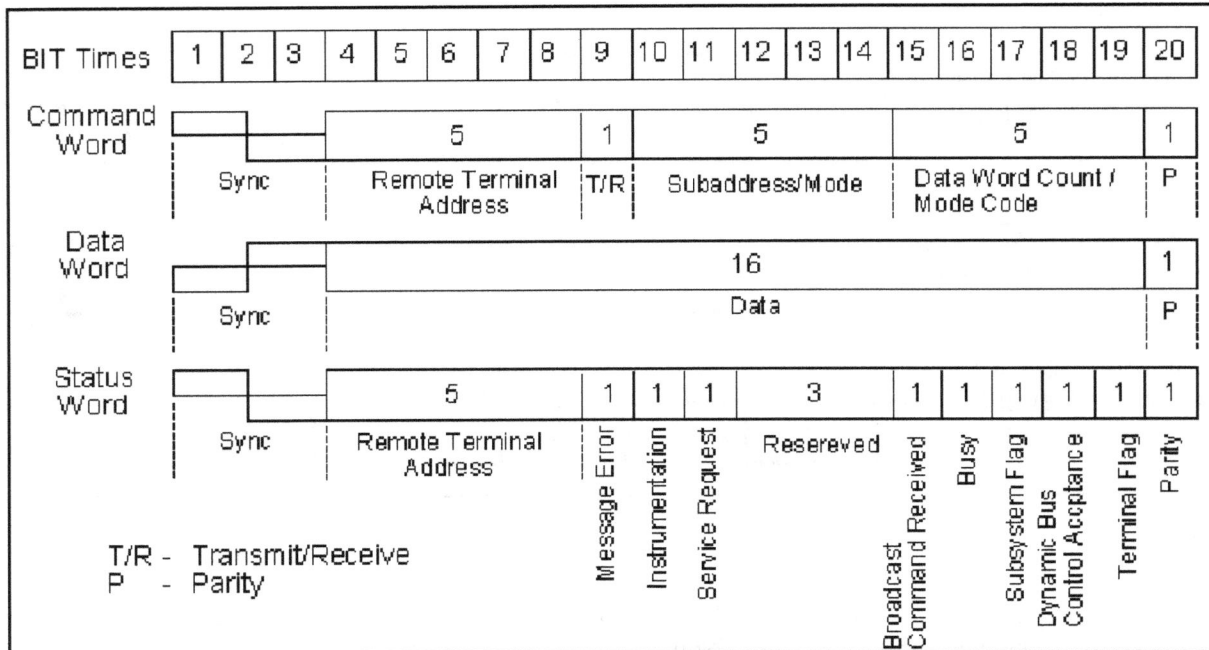

Figure 3. 1553 Word Formats

COMMAND WORD

      Command words are transmitted only by the bus controller and always consist of:
- 3 bit-time sync pattern
- 5 bit RT address field
- 1 Transmit/Receive (T/R) field
- 5 bit subaddress/mode field
- 5 bit word count/mode code field
- 1 parity check bit.

DATA WORD

      Data words are transmitted either by the BC or by the RT in response to a BC request. The standard allows a maximum of 32 data words to be sent in a packet with a command word before a status response must be returned. Data words always consist of:
- 3 bit-time sync pattern (opposite in polarity from command and status words)
- 16 bit data field
- 1 parity check bit.

STATUS WORD

      Status words are transmitted by the RT in response to command messages from the BC and consist of:
- 3 bit-time sync pattern (same as for a command word)
- 5 bit address of the responding RT
- 11 bit status field
- 1 parity check bit.

The 11 bits in the status field are used to notify the BC of the operating condition of the RT and subsystem.

INFORMATION TRANSFERS

Three basic types of information transfers are defined by 1553:
- Bus Controller to Remote Terminal transfers
- Remote Terminal to Bus Controller transfers
- Remote Terminal to Remote Terminal transfers

These transfers are related to the data flow and are referred to as messages. The basic formats of these messages are shown in Figure 4.

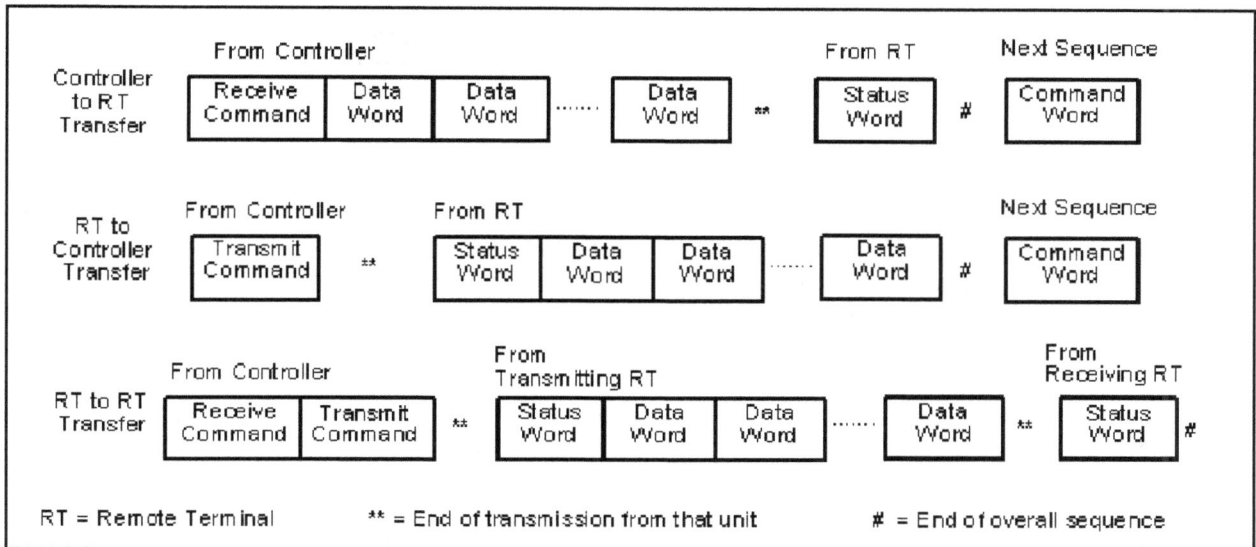

Figure 4. 1553 Data Message Formats

The normal command/response operation involves the transmission of a command from the BC to a selected RT address. The RT either accepts or transmits data depending on the type (receive/transmit) of command issued by the BC. A status word is transmitted by the RT in response to the BC command if the transmission is received without error and is not illegal.

Figure 5 illustrates the 1553B Bus Architecture in a typical aircraft.

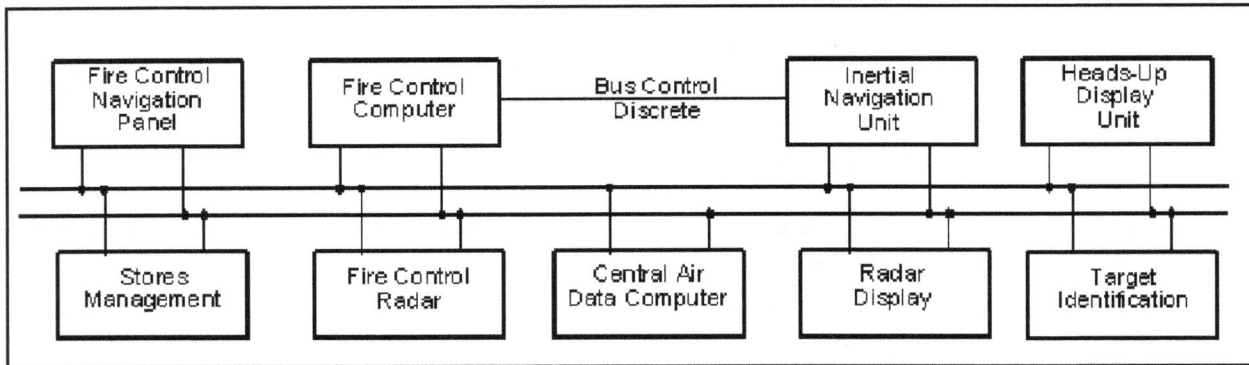

Figure 5. Typical Bus Architecture

MIL-STD-1773

MIL-STD-1773 contains the requirements for utilizing a fiber optic "cabling" system as a transmission medium for the MIL-STD-1553B bus protocol. As such, the standard repeats MIL-STD-1553 nearly word-for-word. The standard <u>does not</u> specify power levels, noise levels, spectral characteristics, optical wavelength, electrical/optical isolation or means of distributing optical power. These must be contained in separate specifications for each intended use.

Data encoding and word format are identical to MIL-STD-1553, with the exception that pulses are defined as transitions between 0 (off) and 1 (on) rather than between + and - voltage transitions since light cannot have a negative value.

Since the standard applies to cabling only, the bus operates at the same speed as it would utilizing wire. Additionally, data error rate requirements are unchanged.

Different environmental considerations must be given to fiber optic systems. Altitude, humidity, temperature, and age affects fiber optics differently than wire conductors. Power is divided evenly at junctions which branch and connectors have losses just as wire connectors do.

ETHERNET

HISTORY

Ethernet was originally developed by Digital, Intel, and Xerox (DIX) in the early 1970's at the Xerox Palo Alto Research Center (PARC). The two primary inventors were Robert Metcalf and David Boggs.

IEEE Project 802 was set up in February 1980 to provide standards for Local Area Network (LAN) and Metropolitan Area Network (MAN) Architecture. The following committees were established:

- IEEE 802.1: Standards related to network management.
- IEEE 802.2: General standard for the data link layer in the OSI Reference Model. The IEEE divides this layer into two sub-layers -- the logical link control (LLC) layer and the media access control (MAC) layer. The MAC layer varies for different network types and is defined by standards IEEE 802.3 through IEEE 802.5.
- IEEE 802.3: Defines the MAC layer for bus networks that use CSMA/CD. **This is the basis of what is most commonly referred to as the Ethernet standard.**
- IEEE 802.4: Defines the MAC layer for token-bus networks.
- IEEE 802.5: Defines the MAC layer for token-ring networks.
- IEEE 802.6: Standard for Metropolitan Area Networks (MANs).
- IEEE 802.7-9 Inactive Standards:
  - IEEE 802.7: Broadband local area networks. The working group is currently inactive.
  - IEEE 802.8: The Fibre Optic Technical Advisory Group was to create a LAN standard for fiber optic media used in token passing computer networks like FDDI.
  - IEEE 802.9: Integrated Services (IS) LAN Interface at the MAC and Physical Layers.
- IEEE 802.10: A standard for security functions that can be used in both LAN and MAN.
- IEEE 802.11: Wireless LAN MAC and physical layer
- IEEE 802.12: Demand Priority Access Method, Physical Layer and Repeater used in star topology, 100VG-AnyLAN, etc
- IEEE 802.16: Fixed broadband wireless LAN

Ethernet versions 1.0 and 2.0 followed after the original development until the IEEE 802.3 committee altered the structure of the Ethernet II packet to form the Ethernet 802.3 packet. You will currently see either Ethernet II (DIX) format or Ethernet 802.3 format being used.

IEEE 802.3 Ethernet uses Manchester Phase Encoding (MPE) for coding the data bits on the outgoing signal.

INTRODUCTION

Ethernet is a family of frame-based computer networking technologies for local area networks (LANs). The 'Ether' part of Ethernet denotes that the system is not meant to be restricted for use on only one medium type. Copper cables, fiber cables and even radio waves can be used.

Ethernet is standardized as IEEE 802.3. The combination of the twisted pair versions of Ethernet for connecting end systems to the network, along with the fiber optic versions for site backbones, is the most widespread wired LAN technology. It has largely replaced competing LAN standards such as token ring, FDDI, and ARCNET.

Ethernet defines a number of wiring and signaling standards for the Physical Layer of the OSI networking model, through means of network access at the Media Access Control (MAC) / Data Link Layer, and a common addressing format. Therefore, Ethernet defines the lower two layers of the OSI Reference Model (see Data Busses Section intro). Ethernet sometimes implies an attached protocol—such as TCP/IP, however TCP/IP really defines the transport and network layers, respectively, of the OSI model.

Ethernet was designed as a 'broadcast' system, i.e. stations on the network can send messages whenever and wherever they want. All stations may receive the messages, however only the specific station to which the message is directed responds. To handle simultaneous demands, the system uses Carrier Sense Multiple Access with Collision Detection (CSMA/CD).

CARRIER SENSE MULTIPLE ACCESS WITH COLLISION DETECTION (CSMA/CD)

When an Ethernet station is ready to transmit, it checks for the presence of a signal on the cable. If no signal is present then the station begins transmission. However, if a signal is present then the station delays transmission until the cable is not in use. If two stations detect an idle cable and at the same time transmit data, then a collision occurs. On a star-wired, unshielded twisted pair (UTP) network, if the transceiver of a sending station detects activity on both its receive and transmit pairs before it has completed transmitting, then it decides that a collision has occurred. On a coaxial system, a collision is detected when the DC signal level on the cable is the same or greater than the combined signal level of the two transmitters. Line voltage drops dramatically if two stations transmit at the same time and the first station to notice this sends a high voltage jamming signal around the network as a signal. The two stations involved with the collision quit transmitting again for a random time interval.

A Collision Domain is that part of the network where each station can 'see' other stations' traffic both unicast and broadcasts. The Collision Domain is made up of one segment of Ethernet coax (with or without repeaters) or a number of UTP shared hubs. A network is segmented with bridges that create two segments, or two Collision Domains where a station on one segment cannot see traffic between stations on the other segment unless the packets are destined for itself. If it can still see all broadcasts on the segmented network(s), no matter the number of segments, it is a Broadcast Domain.

ETHERNET FRAME

Figure 1 depicts the structure of the older DIX (Ethernet II) and the now standard 802.3 Ethernet frames. The numbers above each field represent the number of bytes.

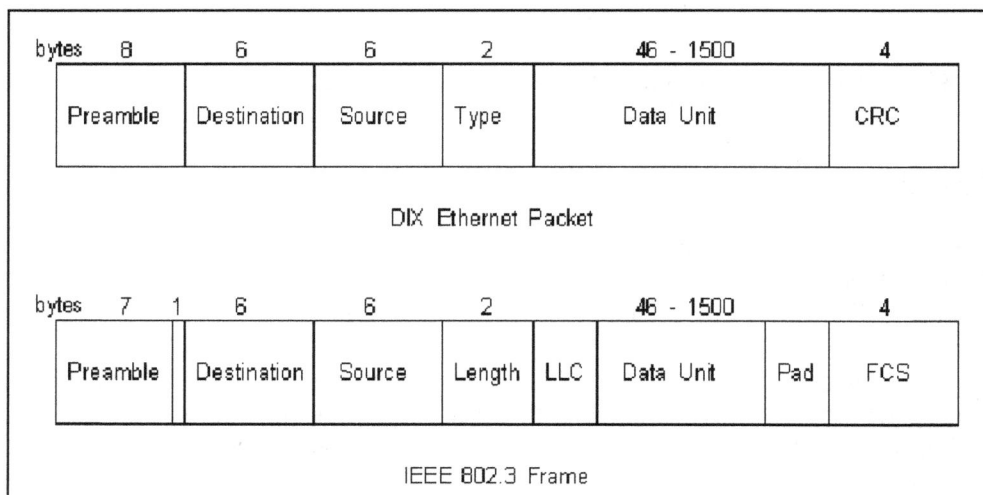

bytes	8	6	6	2	46 - 1500	4
	Preamble	Destination	Source	Type	Data Unit	CRC

DIX Ethernet Packet

bytes	7	1	6	6	2		46 - 1500		4
	Preamble		Destination	Source	Length	LLC	Data Unit	Pad	FCS

IEEE 802.3 Frame

Figure 1. Ethernet Frame

- **Preamble field:** Establishes bit synchronization and transceiver conditions so that the circuitry synchs with the received frame timing. The DIX frame has 8 bytes for the preamble rather than 7, as it does not have a Start Frame Delimiter (or Start of Frame).
- **Start Frame Delimiter:** Sequence 10101011 in a separate field, only in the 802.3 frame.
- **Destination address:** Hardware address (MAC address) of the destination station (usually 48 bits i.e. 6 bytes).
- **Source address:** Hardware address of the source station (must be of the same length as the destination address, the 802.3 standard allows for 2 or 6 byte addresses, although 2 byte addresses are never used, N.B. Ethernet II *only* uses 6 byte addresses).
- **Type:** Specifies the protocol sending the packet such as IP or IPX (only applies to DIX frame).
- **Length:** Specifies the length of the data segment, actually the number of LLC data bytes, (only applies to 802.3 frame and replaces the Type field).
- **LLC:** Logical Length Control is a data communication protocol layer which provides multiplexing and flow control mechanisms that make it possible for several network protocols (IP, IPX) to coexist within a multipoint network and to be transported over the same network media.
- **Data:** Actual data which is allowed anywhere between 46 to 1500 bytes within one frame.
- **Pad:** Zeros added to the data field to 'Pad' a short data field to 46 bytes (applies to 802.3 frame).
- **CRC:** Cyclic Redundancy Check to detect errors during transmission (DIX version of FCS).
- **FCS:** Frame Check Sequence to detect errors that occur during transmission (802.3 version of CRC). This 32-bit code has an algorithm applied to it, which will give the same result as the other end of the link, provided that the frame was transmitted successfully.

From the above we can see that the maximum 802.3 frame size is 1518 bytes and the minimum size is 64 bytes. Packets that have correct CRC's (or FCS's) but are smaller than 64 bytes are known as 'Runts'.

The hardware address, or MAC address is transmitted and stored in Ethernet network devices in **Canonical** format i.e. Least Significant Bit (LSB) first. The expression **Little-Endian** is used to describe the LSB format in which Ethernet is transmitted. On the other hand, Token Ring networks and the subset, Fiber Distributed Data Interface (FDDI) networks, transmit the MAC address with the Most Significant Bit (MSB) first, or **Big-Endian**, This is known as **Non-Canonical** format. Note that this applies on a byte-by-byte basis i.e. the bytes are transmitted in the same order, it is just the bits in each of those bytes that are reversed. The storage of the MAC addresses in Token Ring and FDDI devices however, may sometimes still be in Canonical format so this can sometimes cause confusion. The reference to, the distribution of MAC addresses and the Organizationally Unique Identifier (OUI) destinations are always in Canonical format.

The Logical Link Control (LLC) protocol data unit (PDU) is defined in IEEE 802.2 and operates with 802.3 Ethernet as seen in Figure 2.

DSAP address	SSAP address	Control	Information
8 bits	8 bits	8 or 16 bits	M*8 bits

DSAP address	=	Destination service access point address field
SSAP address	=	Source service access point address field
Control	=	Control field [16 bits for formats that include sequence numbering, and 8 bits for formats that do not (see 5.2)]
Information	=	Information field
*	=	Multiplication
M	=	An integer value equal to or greater than 0. (Upper bound of M is a function of the medium access control methodology used.)

Figure 2. LLC Format

LLC is based on the High-Level Data Link Control (HDLC) data protocol format. Whereas Ethernet II (2.0) combines the MAC and the Data link layers restricting itself to connectionless service in the process, IEEE 802.3 separates out the MAC and Data Link layers. IEEE 802.2 is also required by Token Ring and FDDI but cannot be used with the Novell 'Raw' format. There are three types of LLC; Type 1, which is connectionless, Type 2 which is connection-oriented, and Type 3 for Acknowledged Connections.

The **Service Access Point (SAP)** is used to distinguish between different data exchanges on the same end station and basically replaces the Type field for the older Ethernet II frame. The **Source Service Access Point (SSAP)** indicates the service from which the LLC data unit is sent, and the **Destination Service Access Point (DSAP)** indicates the service to which the LLC data unit is being sent. As examples, NetBIOS uses the SAP address of **F0** whilst IP uses the SAP address of **06**. The following lists some common SAPs:

00 - Null LSAP
02 - Individual LLC Sublayer Management Function
03 - Group LLC Sublayer Management Function
05 - IBM SNA Path Control (group)
06 - **Internet Protocol (IP)**
18 - Texas Instruments
42 - IEEE 802.1 Bridge Spanning Tree Protocol
80 - Xerox Network Systems (XNS)
86 - Nestar
AA - SubNetwork Access Protocol (SNAP)
E0 - Novell NetWare
F0 - **IBM NetBIOS**
F5 - IBM LAN Management (group)
FE - ISO Network Layer Protocol

The **Control Field** identifies the type of LLC, of which there are three:

- **Type 1** - Data (PDUs) shall be exchanged between LLCs without the need for the establishment of a data link connection. These PDUs shall not be acknowledged, nor shall there be any flow control or error recovery. This is called **Unsequenced Information (UI)**.
- **Type 2** – Needs data link connection. Uses **Information (I)** frames and maintains the sequence numbers during an acknowledged connection-oriented transmission.
- **Type 3** – No data link connection required. Uses **Acknowledged Connection (AC)** frames in an acknowledged connectionless service.

MAC ADDRESS

With an Ethernet MAC address, the first octet uses <u>only</u> the lowest significant bit as the Individual/Group (I/G) bit address identifier. If it is zero it indicates the address is individual. If it is one, it indicates a group address. The Universally or Locally administered (U/L) address bit is the bit of octet 0 adjacent to the I/G address bit. This bit indicates whether the address has been assigned by a local or universal administrator. Universally administered addresses have this bit set to 0. If this bit is set to 1, the entire address (i.e., 48 bits) has been locally administered.

The first 3 octets of the MAC address form the Organizational Unique Identifier (OUI) assigned by IEEE to organizations that requires their own group of MAC addresses.

SUBNETWORK ACCESS PROTOCOL (SNAP)

The SNAP protocol was introduced to allow an easy transition to the new LLC frame format for vendors. SNAP allows older frames and protocols to be encapsulated in a Type 1 LLC header making any protocol 'pseudo-IEEE compliant'. SNAP is described in RFC 1042. Figure 3 shows how it looks:

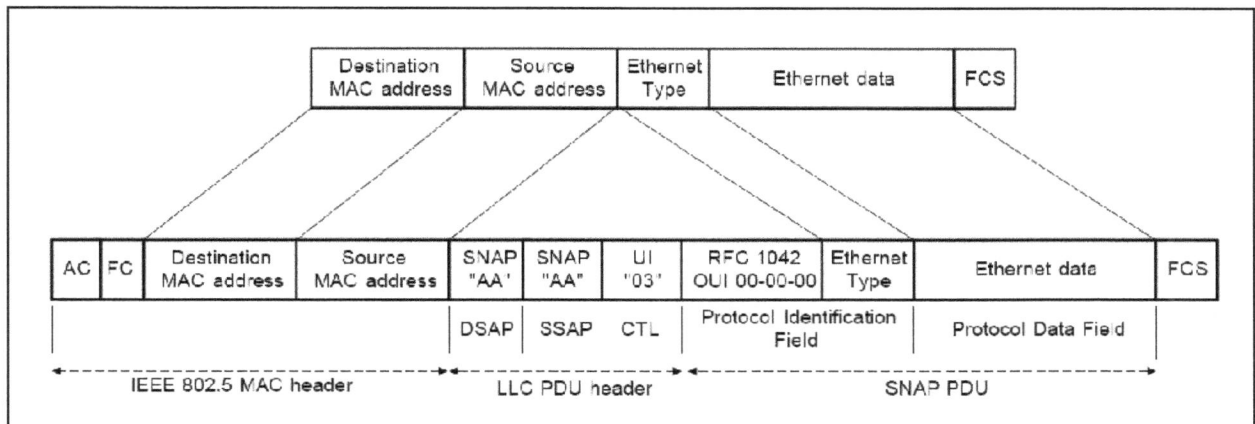

Figure 3. Ethernet frame SNAP-encapsulated in IEEE 802.5 frame

It can be seen that it is an LLC data unit (sometimes called a Logical Protocol Data Unit (LPDU)) of Type 1 (indicated by 03). The DSAP and SSAP are set to AA to indicate that this is a SNAP header coming up. The SNAP header then indicates the vender via the Organizational Unique Identifier (OUI) and the protocol type via the Ethertype field. In the example above the OUI is 00-00-00 which means that there is an Ethernet frame. An ethernet type (Ethertype) of 08-00 would indicate Internet IP is the protocol. The official list of types can be found at IEEE. Although vendors tend to be are moving to LLC1 on the LAN, SNAP still remains.

TRANSMISSION MEDIA

<u>10Base-5</u> - largely obsolete, but may still be in use.

Traditionally, Ethernet is used over 'thick' coaxial cable called 10Base-5 (the '10' denotes 10Mbps, base means that the signal is baseband i.e. takes the whole bandwidth of the cable (so that only one device can transmit at one time on the same cable), and the '5' denotes 500m maximum length). The minimum length between stations is 2.5m.

The cable is run in one long length forming a 'Bus Topology'. Stations attach to it by way of inline N-type connections or a transceiver, which is literally screwed into the cable (by way of a 'Vampire Tap') providing a 15-pin Attachment Unit Interface (AUI) connection (also known as a DIX connector or a DB-15 connector) for a drop lead connection (maximum of 50m length) to the station. The segments are terminated with 50 ohm resistors and the shield grounded at one end only.

The segment could be appended with up to a maximum of 4 repeaters; therefore 5 segments (total length of 2,460m) can be connected together. Of the 5 segments only 3 can have devices attached (100 per segment). This is known as the 5-4-3 rule. A total of 300 devices can be attached on a Thicknet broadcast domain.

<u>10Base-2</u> – dominant for many years

It was common to see the Thick coax used in Risers to connect Repeaters, which in turn provide 'Thin Ethernet' coaxial connections for runs up to 30 workstations. Thin Ethernet (Thinnet) uses RG-58 cable and is called 10Base-2 (The '2' now denoting 200m maximum length - strictly speaking this is 185m). The minimum length between stations is 0.5m.

Each station connects to the thinnet by way of a Network Interface Card (NIC), which provides a BNC (British Naval Connector). At each station the thinnet terminates at a T-piece and at each end of the thinnet run (or 'Segment') a 50-ohm terminator is required to absorb stray signals, thereby preventing signal bounce. The shield is grounded at one end only.

A segment can be appended with other segments using up to 4 repeaters, i.e. 5 segments in total. Two of these segments however, cannot be tapped; they can only be used for extending the length of the broadcast domain (to 925m). What this means is that 3 segments with a maximum of 30 stations on each can give you 90 devices on a Thinnet broadcast domain.

<u>10Base-T</u> – Most widespread today

The most common media used in of Ethernet today involves use of Unshielded Twisted Pair (UTP) or Shielded Twisted Pair (STP). For instance, Category 5 UTP can be installed in a 'Star-wired' format, with runs recommended at no greater than 100m (including patch leads, cable run and flyleads) and Ethernet Hubs with UTP ports (RJ45) centrally located. It has been found though that runs of up to 150m are feasible, the limitations being signal strength. Also, there should be no more than a 11.5dB signal loss and the minimum distance between devices is 2.5m. The maximum delay for the signal in a 10Mbps network is 51.2 microseconds. This comes from the fact that the bit time (time to transmit one bit) is 0.1 microseconds and that the slot time for a frame is 512 bit times.

In order to connect to Ethernet in this 'Star Topology', each station has a NIC, which contains an RJ45 socket which is used by a 4-pair droplead to connect to another nearby RJ45 socket.

Each port on the hub sends a 'Link Beat Signal' which checks the integrity of the cable and devices attached, a flickering LED on the front of the port of the hub tells you that the link is running fine. The maximum number of hubs (or, more strictly speaking, repeater counts) that you can have in one segment is four and the maximum number of stations on one broadcast domain is 1024.

The advantages of the UTP/STP technology are gained from the flexibility of the system, with respect to moves, changes, fault finding, reliability and security. The following table shows the RJ45 pin-outs for 10Base-T:

RJ45 Pin	Function	Color
1	Transmit	White/Orange stripe
2	Transmit	Orange
3	Receive	White/Green stripe
4		Blue
5		White/Blue stripe
6	Receive	Green
7		White/Brown stripe
8		Brown

If you wish to connect hub to hub, or a NIC directly to another NIC, then the following 10Base-T crossover cable should be used:

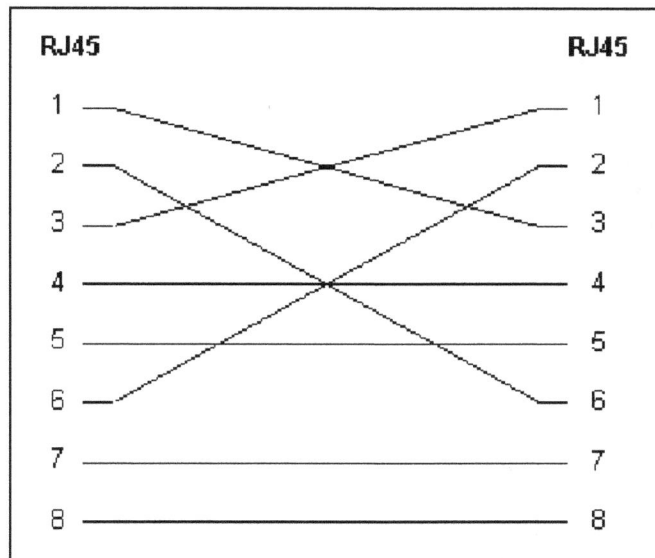

Figure 4. 10Base-T Crossover

The 4-repeater limit manifests itself in 10/100Base-T environments where the active hub/switch port is in fact a repeater, hence the name multi-port repeater. Generally, the hub would only have one station per port but you can cascade hubs from one another up to the 4-repeater limit. The danger here is that you will have all the traffic from a particular hub being fed into one port so care would need to be taken on noting the applications being used by the stations involved, and the likely bandwidth that the applications will use.

There is a pre-standard called Lattisnet (developed by Synoptics), which runs 10MHz Ethernet over twisted pair but instead of bit synchronization occurring at the sending (as in 10Base-T) the synchronization occurs at the receiving end.

10Base-F

The 10Base-F standard defines the use of fiber for Ethernet. 10Base-FL describes the standards for the fiber optic links between stations and repeaters, allowing up to 2km per segment on multi-mode fiber. It replaces the 1987 FOIRL (Fiber Optic Inter-Repeater Link), which provides the specification for a fiber optic Media Attachment Unit (MAU) and other interconnecting components. Two related standards are essentially "dead": 10Base-FB allows up to 2km per segment (on multi-mode fiber) and is designed for backbone applications such as cascading repeaters. In addition, there is the 10Base-FP (Passive components) standard

The 10Base-F standard allows for 1024 devices per network.

Fast Ethernet - 100Base-T

Fast Ethernet is the most popular of the newer standards and is an extension to 10Base-T. The '100' denotes 100Mbps data speed and it uses the same two pairs as 10Base-T and must only be used on Category 5 UTP cable installations with provision for it to be used on Type 1 STP. The actual data throughput increases by between 3 to 4 times that of 10Base-T.

Fast Ethernet is specified in 802.3u and uses the same frame formats and CSMA/CD technology as normal 10Mbps Ethernet. The difference is that the maximum delay for the signal across the segment is 5.12 microseconds instead of 51.2 microseconds. This stems from the fact that the bit time (time to transmit one bit) is 0.01 microseconds and that the slot time for a frame is 512 bit times. The Inter-Packet Gap (IPG) for 802.3u is 0.96 microseconds as opposed to 9.6 microseconds for 10Mbps Ethernet.

Whereas 10Base-T uses Normal Link Pulses (NLP) for testing the integrity of the connection, 100Base-T uses Fast Link Pulses (FLP) which are backwardly compatible with NLPs but contain more information. FLPs are used to detect the speed of the network (e.g. in 10/100 switchable cards and ports). The distance must not exceed 100m.

The IEEE uses the term 100Base-X to refer to both 100Base-Tx and 100Base-Fx and the Media-Independent Interface (MII) allows a generic connector for transceivers to connect to 100Base-Tx, 100Base-Fx and 100Base-T4 LANs.

There is no such thing as the 5-4-3 rule in Fast Ethernet. All 10Base-T repeaters are considered to be functionally identical. Fast Ethernet repeaters are divided into two classes of repeater: Class I and Class II. A Class I repeater has a repeater propagation delay value of 140 bit times, whilst a Class II repeater is 92 bit times. The Class I repeater (or Translational Repeater) can support different signaling types such as 100Base-Tx and 100Base-T4. A Class I repeater transmits or repeats the incoming line signals on one port to the other ports by first translating them to digital signals and then retranslating them to line signals. The translations are necessary when connecting different physical media (media conforming to more than one physical layer specification) to the same collision domain. Any repeater with an MII port would be a Class I device. Only one Class I repeater can exist within a single collision domain, so this type of repeater cannot be cascaded.

A Class II repeater immediately transmits or repeats the incoming line signals on one port to the other ports: it does not perform any translations. This repeater type connects identical media to the same collision domain (for example, TX to TX). At most, two Class II repeaters can exist within a single collision domain. The cable used to cascade the two devices is called and unpopulated segment or IRL (Inter-Repeater Link). The Class II repeater (or **Transparent Repeater**) can only support one type of physical signaling.

100Base-T4

100Base-T4 was an early implementation of Fast Ethernet. It requires four twisted copper pairs, but those pairs were only required to be category 3 rather than the category 5 required by TX. One pair is reserved for transmit, one for receive, and the remaining two will switch direction as negotiated. A very unusual 8B6T code is used to convert 8 data bits into 6 base-3 digits (the signal shaping is possible as there are three times as many 6-digit base-3 numbers as there are 8-digit base-2 numbers). The two resulting 3-digit base-3 symbols are sent in parallel over 3 pairs using 3-level pulse-amplitude modulation (PAM-3).

100Base-T2

This little known version of Fast Ethernet is for use over two pairs of Category 3 cable and uses PAM-5 for encoding. There is simultaneous transmission and reception of data in both pairs and the electronics uses DSP technology to handle alien signals in adjacent pairs. 100Base-T2 can run up to 100m on Category 3 UTP.

100Base-Fx

100Base-Fx is a version of Fast Ethernet over optical fiber. It uses a 1300 nm near-infrared (NIR) light wavelength transmitted via two strands of optical fiber, one for receive (RX) and the other for transmit (TX). Maximum length is 400 metres (1,310 ft) for half-duplex connections (to ensure collisions are detected) or 2 kilometres (6,600 ft) for full-duplex over multimode optical fiber. Longer distances are possible when using single-mode optical fiber. 100Base-Fx uses the same 4B5B encoding and NRZI line code that 100Base-TX does. 100Base-Fx should use the following connectors: SC, straight tip (ST), or media independent connectors (MIC), with SC being the preferred option. 100Base-Fx is not compatible with 10Base-Fl, the 10 MBit/s version over optical fiber.

100Base-Sx

100Base-Sx is a version of Fast Ethernet over optical fiber. It uses two strands of multi-mode optical fiber for receive and transmit. It is a lower cost alternative to using 100Base-Fx, because it uses short wavelength optics (850 nm) which are significantly less expensive than the long wavelength optics used in 100Base-Fx. 100Base-Sx can operate at distances up to 300 metres (980 ft).

100Base-Sx uses the same wavelength as 10Base-Fl, the 10 MBit/s version over optical fiber. Unlike 100Base-Fx, this allows 100Base-Sx to be backwards-compatible with 10Base-Fl. Because of the shorter wavelength used and the shorter distance it can support, 100Base-Sx uses less expensive optical components (LEDs instead of lasers) which makes it an attractive option for those upgrading from 10Base-Fl and those who do not require long distances.

100Base-Bx

100Base-Bx is a version of Fast Ethernet over a single strand of optical fiber (unlike 100Base-Fx, which uses a pair of fibers). Single-mode fiber is used, along with a special multiplexer which splits the signal into transmit and receive wavelengths. The two wavelengths used for transmit and receive are either 1310/1550nm or 1310/1490nm. Distances can be 10, 20 or 40 km.

100VG-AnyLAN

Based on 802.12 (Hewlett Packard), 100VG-AnyLAN uses an access method called Demand Priority. The 'VG' stands for 'Voice Grade' as it is designed to be used with Category 3 cable. This is where the repeaters (hubs) carry out continuous searches around all of the nodes for those that wish to send data. If two devices cause a 'contention' by wanting to send at the same time, the highest priority request is dealt with first, unless the priorities are the same, in which case both requests are dealt with at the same time (by alternating frames). The hub only knows about connected devices and other repeaters so communication is only directed at them rather than broadcast to every device in the broadcast domain (which could mean 100's of devices!). This is a more efficient use of the bandwidth. This is the reason why the new standard was developed as it is not strictly Ethernet. Standard 802.12 is designed to better support both Ethernet and Token Ring.

The encoding techniques used are 5B/6B and NRZ. All four pairs of UTP are used. On Cat3 the longest cable run is 100m but this increases to 200m on Cat5. The clock rate on each wire is 30MHz; therefore 30Mbits per second are transmitted on each pair giving a total bit rate of 120Mbps. Since each 6-bits of data on the line represents 5 bits of real data due to the 5B/6B encoding, the rate of real data being transmitted is 25Mbps on each pair, giving a total real data rate of 100Mbps. For 2-pair STP and fiber, the bit rate is 120Mbps on the transmitting pair, for a real data transmission rate of 100Mbps.

Gigabit Ethernet

Although the functional principles of Gigabit Ethernet are the same as Ethernet and Fast Ethernet i.e. CSMA/CD and the Framing format, the physical setup is very different. One difference is the slot time. The standard Ethernet slot time required in CSMA/CD half-duplex mode is not long enough for running over 100m of copper, so Carrier Extension is used to guarantee a 512-bit slot time.

1000Base-X (802.3z)

The 802.3z committee is responsible for formalizing the standard for Gigabit Ethernet. The 1000 refers to 1Gbps data speed. The existing Fiber Channel interface standard (ANSI X3T11) is used and allows up to 4.268Gbps speeds. The Fiber Channel encoding scheme is 8B/10B.

Gigabit Ethernet can operate in half or full duplex modes and there is also a standard 802.3x which manages XON/XOFF flow control in full duplex mode. With 802.3x, a receiving station can send a packet to a sending station to stop it sending data until a specified time interval has passed.

There are three media types for 1000Base-X. 1000Base-Sx, 1000Base-Lx and 1000Base-Cx.

1000Base-Sx ('S' is for Short Haul) uses short-wavelength laser (850nm) over multi-mode fiber. 1000Base-Sx can run up to 300m on 62.5/125µm multimode fiber and up to 550m on 50/125µm multimode fiber.

1000Base-Lx ('L' is for Long Haul) uses long wavelength laser (1300nm) and can run up to 550m on 62.5/125µm multi-mode fiber or 50/125µm multi-mode fiber. In addition, 1000Base-Lx can run up to 5km (originally 3km) on single-mode fiber using 1310nm wavelength laser.

1000Base-Cx is a standard for STP copper cable and allows operation up to 25m over STP cable.

1000Base-T (802.3ab)

Many cable manufacturers are enhancing their cable systems to 'enhanced Category 5' (Cat 5e) standards in order to allow Gigabit Ethernet to run at up to 100m on copper. The Category 6 standard has yet to be ratified, and is not being persued very strongly.

In order to obtain the 1000Mbps data bit rate across the UTP cable without breaking the FCC rules for emission, all 4 pairs of the cable are used. Hybrid circuits at each end of each pair are used to allow simultaneous transmission and reception of data (full-duplex) by separating the transmission signal from the receiving signal. Because some transmission signal still manages to couple itself to the receiving side there is an additional echo canceller built in, this is called a Near End Crosstalk (NEXT) canceller.

Encoding is carried out with PAM-5.

This Page Blank

TRANSMISSION CONTROL PROTOCOL / INTERNET PROTOCOL

Transmission Control Protocol / Internet Protocol (TCP/IP) is the communication protocol for the Internet.

Whereas IP handles lower-level transmissions from computer to computer as a message makes its way across the Internet, TCP operates at a higher level, concerned only with the two end systems, for example a Web browser and a Web server. In particular, TCP provides ordered delivery of a stream of bytes from a program on one computer to another program on another computer. Besides the Web, other common applications of TCP include e-mail and Internet file transfer like FTP. Among its other management tasks, TCP controls message size, the rate at which messages are exchanged, and network traffic congestion. Your computer Internet address is a part of the standard TCP/IP protocol and so is your domain name.

Inside the TCP/IP standard there are several protocols for handling data communication:
- TCP (Transmission Control Protocol) communication between applications
- UDP (User Datagram Protocol) simple communication between applications
- IP (Internet Protocol) communication between computers
- ICMP (Internet Control Message Protocol) for errors and statistics
- DHCP (Dynamic Host Configuration Protocol) for dynamic addressing

TCP: If one application wants to communicate with another via TCP, it sends a communication request. This request must be sent to an exact address. TCP is responsible for breaking data down into IP packets before they are sent, and for assembling the packets when they arrive. After a "handshake" between the two applications, TCP will set up a "full-duplex" communication between the two applications. The "full-duplex" communication will occupy the communication line between the two computers until it is closed by one of the two applications. UDP is very similar to TCP, but simpler and less reliable.

IP: IP is responsible for sending the packets to the correct destination. IP does not occupy the communication line between two computers and hence reduces the need for network lines. Each line can be used for communication between many different computers at the same time. With IP, messages (or other data) are broken up into small independent "packets" and sent between computers via the Internet. IP is responsible for "routing" each packet to the correct destination.

When an IP packet is sent from a computer, it arrives at an IP router. The IP router is responsible for "routing" the packet to the correct destination, directly or via another router. The path the packet will follow might be different from other packets in the same communication chain. The router is responsible for the right addressing, depending on traffic volume, errors in the network, or other parameters.

Each computer must have a unique IP address before it can connect to the Internet. Each IP packet must also have an address before it can be sent to another computer. TCP/IP uses 32 bits (4 bytes), of four numbers between 0 and 255, separated by a period, to address a computer. A typical IP address is of the format: 192.255.30.50

Domain Names: A name is much easier to remember than a 12 digit number. The names used for TCP/IP addresses are called domain names. "**navy.mil**" is a domain name. Subdomains are identified by prefixes to the main domain name (separated by a period). When you address a web site, like **http://www.navair.navy.mil** the name is translated to a number by a Domain Name System (DNS) Server. DNS servers are connected to the Internet all over the world. When a new domain name is registered together with a TCP/IP address, all DNS servers are updated with this information.

BROWSER AND OTHER APPLICATION LAYER PROTOCOLS

DHCP - Dynamic Host Configuration Protocol. DHCP is used for allocation of dynamic IP addresses to computers in a network.

FTP - File Transfer Protocol. FTP takes care of transmission of data files between computers.

HTTP - Hyper Text Transfer Protocol - takes care of the communication between a web server and a web browser. It is used for sending requests from a web client (a browser) to a web server, returning web content (web pages) from the server back to the client.

HTTPS - Secure HTTP. HTTPS typically handles credit card transactions and other sensitive data.

IMAP - Internet Message Access Protocol. The IMAP protocol is used by email programs (like Microsoft Outlook) just like the POP protocol (discussed below). The main difference between the IMAP protocol and the POP protocol is that the IMAP protocol will not automatically download all your emails each time your email program connects to your email server. The IMAP protocol allows you to look through your email messages at the email server before you download them. With IMAP you can choose to download your messages or just delete them. This way IMAP is useful if you need to connect to your email server from different locations, but only want to download your messages when you are back in your office.

NTP - Network Time Protocol. NTP is used to synchronize the time (the clock) between computers.

POP - Post Office Protocol. The POP protocol is used by email programs (like Microsoft Outlook) to retrieve emails from an email server. If your email program (also called email client), uses POP, all your emails are downloaded to your computer each time it connects to your email server.

SMTP - Simple Mail Transfer Protocol. The SMTP protocol is used for the transmission of e-mails to another computer. Normally your email is sent to an email server (SMTP server), and then to another server or servers, and finally to its destination. SMTP can only transmit pure text. It cannot transmit binary data like pictures, sounds or movies directly. SMTP uses the MIME protocol to send binary data across TCP/IP networks.

SNMP - Simple Network Management Protocol. SNMP is used for administration of computer networks.

PRESENTATION LAYER PROTOCOLS:

MIME - Multi-purpose Internet Mail Extensions protocol. Converts binary data to pure text. The MIME protocol lets SMTP transmit multimedia files including voice, audio, and binary data.

SSL - Secure Sockets Layer. The SSL protocol is used for encryption of data for secure data transmission.

LOWER LAYER PROTOCOLS:

ARP - Address Resolution Protocol. ARP is used by IP to find the hardware address of a computer network card based on the IP address.

BOOTP - Boot Protocol. BOOTP is used for booting (starting) computers from the network.

ICMP - Internet Control Message Protocol. ICMP takes care of error-handling in the network.

LDAP - Lightweight Directory Access Protocol. LDAP is used for collecting information about users and e-mail addresses from the internet.

PPTP - Point to Point Tunneling Protocol. PPTP is used for setting up a connection (tunnel) between private networks.

RARP - Reverse Address Resolution Protocol. RARP is used by IP to find the IP address based on the hardware address of a computer network card.

This Page Blank

GLOSSARY

ACCEPTABLE DEGRADATION - The allowable reduction in system performance. For a fire control radar, the acceptable degradation is usually expressed as a reduction in range; for example, the maximum lock-on range might be degraded by 25 percent without loss of essential defense capability.

ACQUISITION - A procedure by which a fire control tracking radar attains initial lock-on. Usually, the approximate target coordinates are supplied to the tracking radar and it searches a predetermined volume of space to locate the target.

AEROSOLS - Solid particles dispersed in the atmosphere having resonant size particles with a high index of refraction. The particles both scatter and absorb visual and laser directed energy so as to cut down on weapon systems directed by these techniques.

AFC (AUTOMATIC FREQUENCY CONTROL) - An arrangement whereby the frequency of an oscillator or the tuning of a circuit is automatically maintained within specified limits with respect to a reference frequency. A magnetron drifts in frequency over a period of time. The AFC of a radar makes the local oscillator shift by an equal amount so the IF frequency will remain constant.

AGC (AUTOMATIC GAIN CONTROL) - A method for automatically obtaining an essentially constant receiver output amplitude. The amplitude of the received signal in the range gate determines the AGC bias (a DC voltage) which controls the receiver gain so as to maintain a nearly constant output even though the amplitude of the input signal changes.

AMPLIFIER - An electronic device used to increase signal magnitude or power. See also GaAs FET Amplifier, Klystron Amplifier, Traveling-Wave Tube Amplifier.

AMPLITUDE MODULATION (AM) - A method of impressing a message upon a carrier signal by causing the carrier amplitude to vary proportionally to the message waveform.

AMPLITUDE SHIFT KEYING (ASK) - A method of impressing a digital signal upon a carrier signal by causing the carrier amplitude to take different values corresponding to the different values of the digital signal.

ANGLE JAMMING - ECM technique, when azimuth and elevation information from a scanning fire control radar is jammed by transmitting a jamming pulse similar to the radar pulse, but with modulation information out of phase with the returning target angle modulation information.

ANGULAR SEPARATION - This term is frequently used to indicate a protective (from EMI) zone for a missile. The interfering antenna axis must be separated, throughout the critical portion of the missile flight, from the missile by the specified angle. The vertex of the angle is at the interference source antenna.

ANTENNA BEAMWIDTH - The angle, in degrees, between the half-power points (-3 dB) of an antenna beam. This angle is also nearly that between the center of the mainlobe and the first null. The angle is given for both horizontal and vertical planes unless the beam is circular. When so indicated, the term may refer to the angular width of the mainlobe between first nulls [beamwidth between first nulls (BWFN)]. See also Antenna Pattern. The figure illustrates vertical profile for antenna displaying a 10-degree beamwidth characteristic. The values can vary dramatically with frequency.

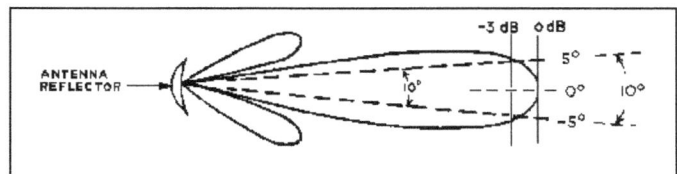

ANTENNA CROSS TALK - A measure of undesired power transfer through space from one antenna to another. Ratio of power received by one antenna to power transmitted by the other, usually expressed in decibels.

ANTENNA ISOLATION - The ratio of the power input to one antenna to the power received by the other. It can also be viewed as the insertion loss from transmit antenna input to receive antenna output to circuitry.

ANTENNA LOBING - Two lobes are created that overlap and intercept at -1 to -3dB. The difference between the two lobes produces much greater spatial selectivity than provided by either lobe alone. (See also "Lobe, Antenna").

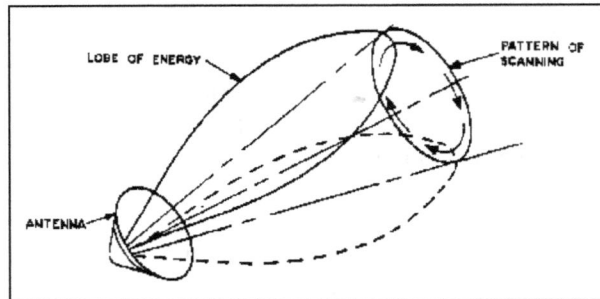

ANTENNA NUTATING - An antenna, as used in automatic-tracking radar systems, consisting of a parabolic reflector combined with a radiating element which is caused to move in a small circular orbit about the focus of the antenna with or without change of polarization. The radiation pattern is in the form of a beam that traces out a cone centered on the reflector axis. The process is also known as nutating conical scanning.

ANTENNA PATTERN - A cross section of the radiating pattern (representing antenna gain or loss) in any plane that includes the origin (source reference point) of the pattern. Both horizontal and vertical polar plots are normally used to describe the pattern. Also, termed "polar diagram" and "radiation pattern."

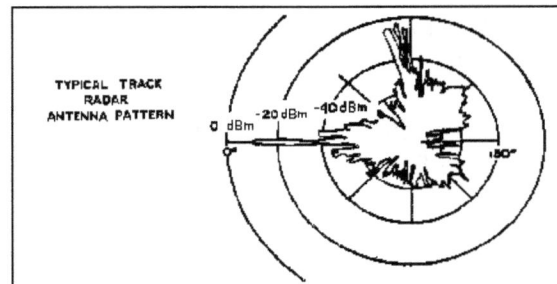

ANTENNA, PENCIL-BEAM - A highly directional antenna designed that cross sections of the major lobe are approximately circular, with a narrow beamwidth.

ANTI-CLUTTER CIRCUITS (IN RADAR) - Circuits which attenuate undesired reflections to permit detection of targets otherwise obscured by such reflections.

APERTURE - In an antenna, that portion of the plane surface area near the antenna perpendicular to the direction of maximum radiation through which the major portion of the radiation passes. The effective and/or scattering aperture area can be computed for wire antennas which have no obvious physical area.

A-SCOPE - A cathode-ray oscilloscope used in radar systems to display vertically the signal amplitude as a function of time (range) or range rate. Sometimes referred to as Range (R)-Scope.

ASYNCHRONOUS PULSED JAMMING - An effective form of pulsed jamming. The jammer nearly matches the pulse repetition frequency (PRF) of the radar; then it transmits multiples of the PRF. It is more effective if the jammer pulsewidth is greater than that of the radar. Asynchronous pulsed jamming is similar to synchronous jamming except that the target lines tend to curve inward or outward slightly and appear fuzzy in the jammed sector of a radar scope.

ATTENUATION - Decrease in magnitude of current, voltage, or power of a signal in transmission between two points. May be expressed in decibels.

AUTOMATIC FREQUENCY CONTROL - See AFC.

AUTOMATIC GAIN CONTROL - See AGC.

BACKWARD WAVE OSCILLATOR (BWO) - A cross-field device in which an electron stream interacts with a backward wave on a nonreentrant circuit. This oscillator may be electronically tuned over a wide range of frequencies, is relatively unaffected by load variations and is stable. BWO is commonly pronounced "be woe".

BALANCED MIXERS - The two most frequently encountered mixer types are single-balanced and double-balanced. In a double-balanced mixer, four Schottky diodes and two wideband transformers are employed to provide isolation of all three ports.

BALLISTIC MISSILE - Any missile which does not rely upon aerodynamic surfaces to produce lift and consequently follows a ballistic trajectory when thrust is terminated.

BANDPASS FILTER - A type of frequency discrimination network designed to pass a band or range of frequencies and produce attenuation to all other frequencies outside of the pass region. The figure illustrates a typical bandpass filter, incorporating a band-pass region of $(F_h)-(F_l)$, offering no rejection (0 dB) to desired signal (F_m) and much higher rejection to the adjacent undesired signals F_h, and F_l.

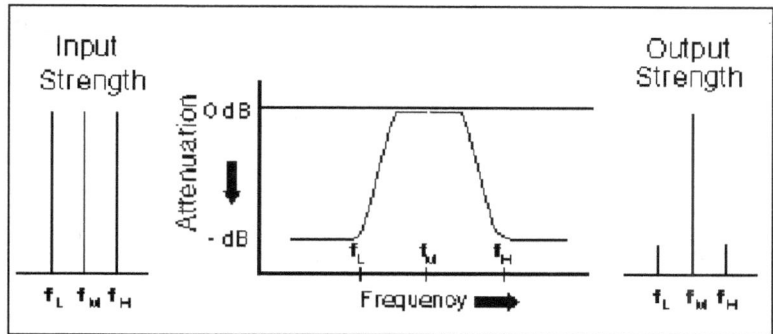

The upper and lower frequencies are usually specified to be the half power (-3dB) or half voltage points (-6dB).

BANDWIDTH - An expression used to define the actual operational frequency range of a receiver when it is tuned to a certain frequency. For a radar receiver, it is the difference between the two frequencies at which the receiver response is reduced to some fraction of its maximum response (such as 3 dB, 6 dB, or some other specified level). The frequencies between which "satisfactory" performance is achieved. Two equations are used:

$$Narrowband\ by\ \%\ (\frac{F_u - F_l}{F_c})(100)\ \ ;\ \ Broadband\ by\ ratio\ \frac{F_u}{F_l}$$

$$Where\ F_u = Upper\ ;\ F_l = lower\ ;\ F_c = center = (\ F_u + F_l\ ) \div 2$$

See also Receiver Bandwidth and Spectrum Width.

BARRAGE NOISE JAMMING - Noise jamming spread in frequency to deny the use of multiple radar frequencies to effectively deny range information. Although this is attractive because it enables one jammer to simultaneously jam several radars of different frequencies, it does have the inherent problem that the wider the jamming spread, the less jamming power available per radar, i.e. the watts per MHz bandwidth is low.

BATTERY, MISSILE - A missile battery consists of a missile launcher and its associated missile fire control systems (such as a MK 11 MOD 0 Missile Launcher and two MK 74 MOD 0 Missile Fire Control Systems).

BEACON - A system wherein a transponder in a missile receives coded signals from a shipboard radar guidance transmitter and transmits reply signals to a shipboard radar beacon receiver to enable a computer to determine missile position. The missile beacon transmitter and shipboard radar beacon receiver are tuned to a frequency different from that of the guidance transmitter.

BEAM - See Lobe, antenna. The beam is to the side of an aircraft or ship.

BEAM, CAPTURE - See Capture Beam.

BEAM-TO-BEAM CORRELATION (BBC) - BBC is used by frequency scan radars to reject pulse jamming and jamming at a swept frequency. Correlation is made from two adjacent beams (pulses). The receiver rejects those targets (signals) that do not occur at the same place in two adjacent beams.

BEAMWIDTH - See Antenna Beamwidth.

BEAT FREQUENCY OSCILLATOR (BFO) - Any oscillator whose output is intended to be mixed with another signal to produce a sum or difference beat frequency. Used particularly in reception of CW transmissions.

BINGO - The fuel state at which an aircraft must leave the area in order to return and land safely. Also used when chaff/flares reach a preset low quantity and automatic dispensing is inhibited.

BIPOLAR VIDEO - Unrectified (pre-detection) IF (both positive and negative portions of the RF envelope) signals that arise from the type of detection and console display employed in pulse Doppler and MTI receivers.

BISTATIC RADAR - A radar using antennas at different locations for transmission and reception.

BLANKING - The process of making a channel, or device non-effective for a certain interval. Used for retrace sweeps on CRTs or to mask unwanted signals such as blanking ones own radar from the onboard RWR.

BOGEY - Unknown air target

BURN-THROUGH RANGE - The ability of a radar to see through jamming. Usually, described as the point when the radar's target return is a specified amount stronger than the jamming signal. (typical values are 6dB manual and 20 dB automatic). See Section 4-8.

BUTT LINE - Line used for reference in measurement of left/right location. One of several aircraft references. See also fuselage station and water line.

CAPTURE BEAM - A wide beam incorporated in capture transmitters of beam rider (command guided) missile systems to facilitate gaining initial control of a missile immediately after launch. Upon capture, the system then centers the missile in the narrow guidance beam. The figure illustrates a launched missile at point of capture.

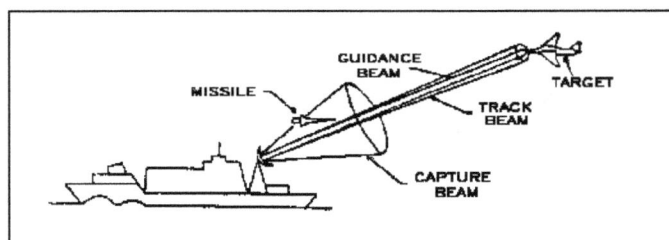

CAPTURE TRANSMITTER - A transmitter employing a wide beam antenna to gain initial control of in-flight missile for the purpose of centering the missile in the guidance transmitter antenna beam. See also Capture Beam.

CARRIER FREQUENCY - The basic radio frequency of the wave upon which modulations are impressed. Also called "Carrier" or f_c . See figure at right.

CATCH-22 - A lose-lose situation, from the book of the same name.

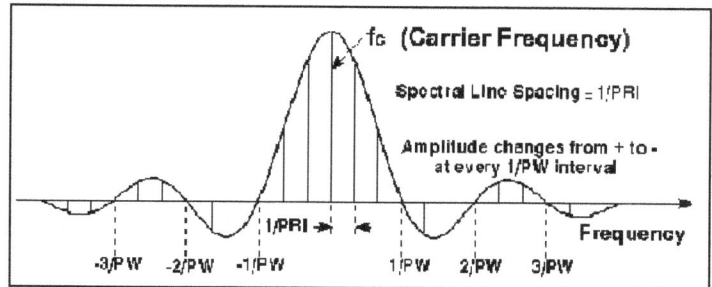

CAVITY - A space enclosed by a conducting surface used as a resonant circuit at microwave frequencies. Cavity space geometry determines the resonant frequency. A storage area for oscillating electromagnetic energy.

CENTER FREQUENCY - The tuned or operating frequency. Also referred to as center operating frequency. In frequency diversity systems, the midband frequency of the operating range. See also Carrier Frequency.

CHAFF - Ribbon-like pieces of metallic materials or metallized plastic which are dispensed by aircraft or ships to mask or screen other "targets". The radar reflections off the chaff may cause a tracking radar to break lock on the target. The foil materials are generally cut into small pieces for which the size is dependent upon the radar interrogation frequency (approximately 1/2 wave length of the victim radar frequency). Being this length, chaff acts as a resonant dipole and reflects much of the energy back to the radar. Also see rainbow, rope, stream chaff, and window.

CHANNEL - A frequency or band of frequencies. In guided missile systems, an assigned center frequency and a fixed bandwidth around it. Designates operating frequency of track radars and frequency/code assignments of X-band CW illuminators.

CHIRP - A pulse compression technique which uses frequency modulation (usually linear) on pulse transmission.

CHIRP RADAR - See PC.

CIRCULARLY POLARIZED JAMMING - The techniques of radiating jamming energy in both planes of polarization simultaneously. With this method, there is a loss of 3 dB of effective power in either linear plane, and substantial loss if the opposite sense of circular polarization is used (i.e. left vs right). See Section 3-2.

CLUTTER, RADAR - Undesired radar returns or echoes resulting from man-made or natural objects including chaff, sea, ground, and rain, which interfere with normal radar system observations. The figure illustrates a target being masked by ground clutter

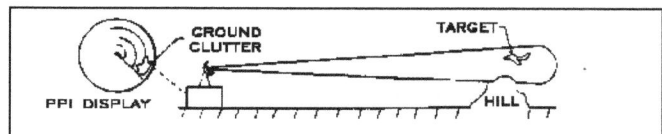

CO-CHANNEL - This term is used to indicate that two (or more) equipments are operating on the same frequency.

COHERENT - Two signals that have a set (usually fixed) phase relationship.

COINCIDENCE DETECTOR - This radar video process requires more than one hit in a range cell before a target is displayed. This prevents video interference from pulses coming from another radar, because such interference is unlikely to occur twice in the same range cell.

COLLIMATION - The procedure of aligning fire control radar system antenna axes with optical line of sight, thereby ensuring that the radars will provide for correct target illumination and guidance beam positioning.

COMMAND CODE - Modulations superimposed upon transmitter carrier signals to provide electronic instructions to an airborne guided missile or pilotless aircraft. The receiver of the remotely guided vehicle is preset to accept only a selected transmitter code to eliminate the possibility of the vehicle responding to commands of extraneous signals. Missile command codes include instructions such as arm, warhead detonate, and self destruct.

COMMAND GUIDANCE - A guidance system wherein intelligence transmitted to the missile from an outside source causes the missile to traverse a directed flight path.

CONICAL SCAN - See Antenna, Nutating.

CONTINUOUS WAVE and CONTINUOUS WAVE ILLUMINATOR - See CW and CWI.

COOPERATIVE COUNTERMEASURES - (CO-OP) Generic term for jamming the same threat radar from two or more separate platforms that are in the same radar resolution cell.

COUPLING FACTOR - A multiplying factor, expressed in dB, used to express the change in EM energy intensity from a radar transmitter to a receiver. The factor includes the antenna gains and the loss (basic transmission loss) caused by the distance between the antennas. The factor will usually be a negative dB figure (a reduction in intensity) because basic transmission loss is always a large negative value. The antenna gains may be positive (pointed toward each other) or negative (no main beam interactions).

CROSS MODULATION - Intermodulation caused by modulation of the carrier by an undesired signal wave.

CROSS POLARIZATION - or "Cross Pole", is a monopulse jamming technique where a cross-polarized signal is transmitted to give erroneous angle data to the radar. The component of the jamming signal with the same polarization as the radar must be very small.

CW (CONTINUOUS WAVE) - In radar and EW systems this term means that the transmitter is on constantly; i.e., not pulsed (100% duty cycle). These systems may frequency or phase modulate the transmitter output. A CW radar has the ability to distinguish moving targets against a stationary background while conserving spectrum bandwidth compared to pulsed radar requirements. A CW radar extracts accurate target range-rate data but cannot determine target range.

CWI (CONTINUOUS WAVE ILLUMINATOR) - A surface or aircraft-based CW transmitter employed in semiactive homing missile systems where the transmitter illuminates the target and the missile senses the reflected energy. The transmitter also provides a reference signal to the missile rear receiver to allow determination of range-rate data and target identification. CW transmitter emissions are sometimes coded corresponding to the associated missile receiver codes to reduce the possibility of the "missile accepting commands of extraneous signals.

DECIBEL (dB) - A dimensionless unit for expressing the ratio of two values of power, current, or voltage. The number of decibels being equal to: $dB = 10 \log P_2/P_1 = 20 \log V_2/V_1 = 20 \log I_2/I_1$
Normally, used for expressing transmission gains, losses, levels, and similar quantities. See Section 2-4.

DECEPTION - The deliberate radiation, reradiation, alteration, absorption or reflection of electromagnetic energy in a manner intended to mislead the enemy interpretation or use of information received by his electronic systems.

dB - See Decibel, or Decibel Section 2-4.

dBc - Decibels referenced to the carrier signal.

dBi - Decibels referenced to an isotropic radiator. (dBLi indicating linear isotropic radiator is sometimes used).

dBm - Decibels relative to 1 mW. dBm is calculated by using the ratio of some power (expressed in mW) to 1 mW. For example, 1 mW is 0 dBm and 10 mW is +10 dBm.

dBsm - Decibel referenced to one square meter.

dBv / dBμv - Decibels referenced to one volt or microvolt, i.e. 0 dBv is 1 volt or 120 dBμv.

dBW / dBμW - Decibels referenced to 1 watt or one microwatt, i.e. 0 dBW is 1 watt or 30 dBm or 60 dBμW.

DEMODULATOR - A device employed to separate the modulation signal from its associated carrier, also called Second Detector. See also Detection.

DESIGNATION - The assignment of a fire control radar to a specific target by supplying target coordinate data to the radar system.

DETECTION - Usually refers to the technique of recovering the amplitude modulation signal (envelope) superimposed on a carrier. See figure at right.

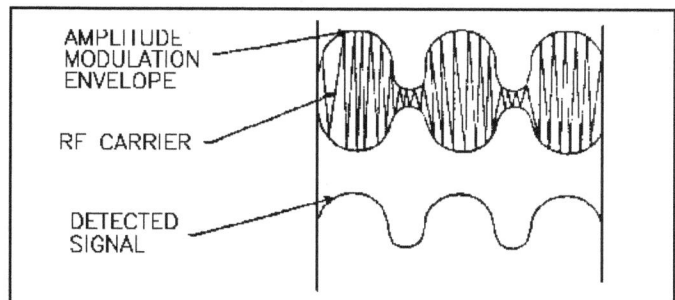

DICKE FIX - This type of radar processing occurs in the IF amplifier. A limiter follows a wideband amplifier, and then the signal goes to a matched filter amplifier. This discriminates against pulses that are too long (clutter) or too short (interference). The "DICKE FIX" is a technique that is specifically designed to protect the receiver from ringing caused by noise, fast-sweep, or narrow pulse jamming. The basic configuration consists of a broadband limiting IF amplifier, followed by an IF amplifier of optimum bandwidth. The limit level is preset at approximately the peak amplitude of receiver noise, the bandwidth may vary from 10 to 20 MHz, depending on the jamming environment. This device provides excellent discrimination against fast sweep jamming (10-500 MHz), usually something on the order of 20 to 40 dB, without appreciable loss in sensitivity. However, strong CW jamming will seriously degrade the performance of the DICKE FIX because the CW signal captures the radar signal in the limiter.

DIELECTRICALLY STABILIZED OSCILLATOR - The DSO uses a dielectric resonator as the frequency determining element. When the dielectric material is properly selected and used, the variations in dielectric constant vs temperature and the dimensions of the resonant structure vs temperature tend to cancel out, providing relatively good frequency vs temperature stability. The DSO offers frequency accuracy and stability, low power consumption and high reliability. Some of the commonly used materials are barium, zirconium, or tin tinates. The composition of these materials may be controlled to achieve any frequency variation with temperature with close tolerances.

DIODE - An electronic device which restricts current flow chiefly to one direction. See also Gunn diode, IMPATT diode, PIN diode, point contact diode, Schottky barrier diode, step recovery diode, tunnel diode, varactor diode.

DIODE SWITCH - PIN-diode switches provide state-of-the-art switching in most present-day microwave receivers. These switches are either reflective or nonreflective in that the former reflect incident power back to the source when in the isolated state. While both types of switches can provide high isolation and short transition times, the reflective switch offers multi octave bandwidth in the all shunt diode configuration, while the non-reflective switch offers an octave bandwidth.

DIPLEX - The simultaneous transmission or reception of two signals using a common feature such as a single antenna or carrier. Typically, two transmitters operate alternately at approximately the same RF and using a common antenna. See Section 6-7 for a discussion of diplexers.

DIRECTIONAL COUPLER - A 4-port transmission coupling device used to sample the power traveling in one direction through the main line of the device. There is considerable isolation (typically 20 dB) to signals traveling in the reverse direction. Because they are reciprocal, the directional coupler can also be used to directively combine signals with good reverse isolation. The directional coupler is implemented in waveguide and coaxial configurations. See Section 6-4.

DIRECTIVITY - For antennas, directivity is the maximum value of gain in a particular direction. (Isotropic point source has directivity = 1). For directional couplers, directivity is a measure (in dB) of how good the directional coupling is and is equal to the isolation minus the coupling. See Section 6-4.

DISH - A microwave reflector used as part of a radar antenna system. The surface is concave and is usually parabolic shaped. Also called a parabolic reflector.

DOPPLER EFFECT - The apparent change in frequency of an electromagnetic wave caused by a change in distance between the transmitter and the receiver during transmission/reception. The figure illustrates the Doppler increase that would be realized by comparing the signal received from a target approaching the radar site to the transmitted reference signal. An apparent frequency decrease would be noted for targets departing the radar location. Differences can be calibrated to provide target range-rate data.

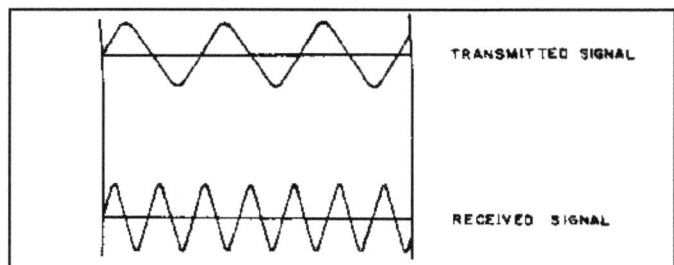

DRY RUN - A test run with aircraft/ship armament and/or EW switches off.

DUCTING - The increase in range that an electromagnetic wave will travel due to a temperature inversion of the atmosphere. The temperature inversion forms a channel or waveguide (duct) for the waves to travel in, and they can be trapped, not attenuating as would be expected from the radar equation.

DUMMY LOAD (Radio Transmission) - A dissipative but essentially nonradiating substitute device having impedance characteristics simulating those of the antenna. This allows power to be applied to the radar transmitter without radiating into free space. Dummy loads are commonly used during EMCON conditions or when troubleshooting a transmitter at a workbench away from it's normal environment.

DUPLEXER - A switching device used in radar to permit alternate use of the same antenna for both transmitting and receiving.

DUTY CYCLE - The ratio of average power to peak power, or ratio of pulse length to interpulse period for pulsed transmitter systems. Interpulse period is equal to the reciprocal of the pulse repetition rate.
See Section 2-5.

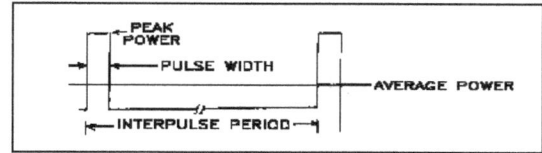

The duty cycle of a radar having a pulse length of 0.3 μsec and a PRF of 2000 pulses/sec is computed as follows:

$$\text{Interpulse Period, } T = PRI = 1/PRF = 500 \text{ μsec}$$

$$Duty\ Cycle = \frac{Pulse\ length}{Interpulse\ Period} = \frac{0.3\ \mu sec}{500\ musec} = 0.0006\ (or\ 0.06\%) \text{ or Duty Cycle (dB)} = 10\log(\text{Duty cycle}) = -32.2 \text{ dB}$$

An output tube providing an average power of only 90 watts for such a system would, therefore, provide a peak power of:

$$Peak\ Power = \frac{Average\ Power}{Duty\ Cycle} = \frac{90}{0.0006} = 150,000\ W\ or\ 52\ dBW\ or\ 82\ dBm$$

EFFECTIVE RADIATED POWER (ERP) - Input power to the antenna in watts times the gain ratio of the antenna. When expressed in dB, ERP is the transmitter power (P_T), in dBm (or dBW) plus the antenna gain (G_T) in dB. The term EIRP is used sometimes and reiterates that the gain is relative to an isotropic radiator.

EGRESS - Exit the target area.

ELECTROMAGNETIC COUPLING - The transfer of electromagnetic energy from one circuit or system to another circuit or system. An undesired transfer is termed EMI (electromagnetic interference).

EMC (ELECTROMAGNETIC COMPATIBILITY) - That condition in which electrical/electronic systems can perform their intended function without experiencing degradation from, or causing degradation to other electrical/electronic systems. More simply stated, EMC is that condition which exists in the absence of EMI. See also Intersystem and Intrasystem EMC tests.

EME (ELECTROMAGNETIC ENVIRONMENT) - The total electromagnetic energy in the RF spectrum that exists at any given location.

EMI (ELECTROMAGNETIC INTERFERENCE) - Any induced, radiated, or conducted electrical emission, disturbance, or transient that causes undesirable responses, degradation in performance, or malfunctions of any electrical or electronic equipment, device, or system. Also synonymously referred to as RFI (Radio Frequency Interference).

EMI MODEL - Usually a set of equations or logical concepts designed to illustrate the interactions, the detailed parameters considerations, and mathematical procedures necessary for proper analysis of a given EMI situation.

EMITTER - Any device or apparatus which emits electromagnetic energy.

EMP (ELECTROMAGNETIC PULSE) - The generation and radiation of a very narrow and very high-amplitude pulse of electromagnetic noise. It is associated with the high level pulse as a result of a nuclear detonation and with intentionally generated narrow, high-amplitude pulse for ECM applications. In the case of nuclear detonations, EMP consists of a continuous spectrum with most of its energy distributed through the low frequency band of 3 KHz to 1 MHz.

ERROR SIGNAL - In servomechanisms, the signal applied to the control circuit that indicates the degree of misalignment between the controlling and the controlled members. In tracking radar systems, a voltage dependent upon the signal received from a target whose polarity and magnitude depend on the angle between the target and the center axis of the scanning beam.

FAST TIME CONSTANT - See FTC.

FEET DRY / WET - Aircraft has crossed from water to shore / aircraft has crossed from shore to water.

FERRET - An aircraft, ship, or vehicle especially equipped for the detection, location, recording, and analyzing of electromagnetic radiations.

FIELD STRENGTH - The magnitude of a magnetic or electric field at any point, usually expressed in terms of ampere turns per meter or volts per meter. Sometimes called field intensity and is expressed in volts/meter or dBμv/meter. Above 100 MHz, power density terminology is used more often. See Sect 4-1.

FIRST HARMONIC - The fundamental (original) frequency.

FREQUENCY AGILITY - A radar's ability to change frequency within its operating band, usually on a pulse-to-pulse basis. This is an ECCM technique employed to avoid spot jamming and to force the jammer to go into a less effective barrage mode.

FREQUENCY AGILITY RADAR - A radar that automatically or semiautomatically tunes through a discrete set of operating frequencies in its normal mode of operation.

FREQUENCY DIVERSITY RADAR - A radar system technique, employed primarily as an antijamming feature, where the transmitter output frequency varies randomly from pulse to pulse over a wide frequency range.

FREQUENCY RANGE - (1) A specifically designated portion of the frequency spectrum; (2) of a device, the band of frequencies over which the device may be considered useful with various circuit and operating conditions; (3) of a transmission system, the frequency band in which the system is able to transmit power without attenuating or distorting it more than a specified amount.

FREQUENCY SHIFT KEYING (FSK) - A form of FM where the carrier is shifted between two frequencies in accordance with a predetermined code. In multiple FSK, the carrier is shifted to more than two frequencies. FSK is used principally with teletype communications.

"FRUIT" - In a radar beacon system, there is a type of interference called "FRUIT", caused by beacon replies to interrogation asynchronous with the observer's interrogator. The largest amount of this interference is received through the sidelobes of the interrogating antenna, but it can become dense enough to cause false target indications.

FTC (FAST TIME CONSTANT) - An antijam feature employed in radar systems where receiver circuits may be selected to provide a short time constant to emphasize signals of short duration to produce discrimination against the low frequency components of clutter.

FUNDAMENTAL FREQUENCY - Used synonymously for tuned frequency, carrier frequency, center frequency, output frequency, or operating frequency.

FUSELAGE STATION or just STATION - A reference point (usually the nose of an aircraft) used to measure or identify fore and aft locations. One of several aircraft location designations - also see butt line and water line.

GaAs FET AMPLIFIER - Because of their low noise, field-effect transistors are often used as the input stage of wideband amplifiers. Their high input resistance makes this device particularly useful in a variety of applications. Since the FET does not employ minority current carriers, carrier storage effects are eliminated giving the device faster operating characteristics and improved radiation resistant qualities.

GAIN: - For antennas, the value of power gain in a given direction relative to an isotropic point source radiating equally in all directions. Frequently expressed in dB (gain of an isotropic source = 0 dB). The formula for calculating gain is:

$$G = \frac{4\pi \, P(\theta,\phi)}{P_{in}} \quad ; where \qquad \begin{array}{l} P(\theta,\phi) = Radiation\,intensity\,in\,given\,direction \\ P_{in} = Power\,into\,lossless\,antenna\,radiating\,uniformly\,in\,all\,directions \end{array}$$

Note: (1) If radiation efficiency is unity, then gain = directivity i.e. if directivity = 2, then gain = 3 dB, etc.
        (2) Interference losses within an array also affect gain
        (3) See Section 3-1 for further details

For amplifiers, gain is the ratio of the output power to input power (usually in dB).

GATE (RANGE) - A signal used to select radar echoes corresponding to a very short range increment. Range is computed by moving the range gate or marker to the target echo; an arrangement which permits radar signals to be received in a small selected fraction of the time period between radar transmitter pulses.

GATING - (1) The process of selecting those portions of a wave which exist during one or more selected time intervals; (2) the application of a square waveform of desired duration and timing to perform electronic switching; (3) the application of receiver operating voltages to one or more stages only during that portion of a cycle of operation when reception is desired. See also Gate (Range).

GCI (GROUND-CONTROLLED INTERCEPT) - vectoring an interceptor aircraft to an airborne target by means of information relayed from a ground-based radar site which observes both the interceptor and the target.

GIGA - A prefix meaning 10^9 (times a billion). For example, gigahertz (GHz).

GLINT (In Radar) - 1. The random component of target location error caused by variations in the phase front of the target signal (as contrasted with Scintillation Error). Glint may affect angle, range of Doppler measurements, and may have peak values corresponding to locations beyond the true target extent in the measured coordinate. 2. Electronic countermeasures that uses the scintillating, or flashing effect of shuttered or rotating reflectors to degrade tracking or seeking functions of an enemy weapons system.

GUARDBAND - A frequency band to which no other emitters are assigned as a precaution against interference to equipments susceptible to EMI in that band.

GUIDANCE, BEAM RIDER - A missile guidance technique which is dependent on the missile's ability to determine its positions with reference to the center of scan of the guidance radar beam and thus correct its trajectory on the basis of detected errors.

GUIDANCE CODE - A technique of modulating guidance transmitter carriers with coded pulses compatible with the receiver code of the missile assigned that system, thus reducing the possibility of the missile accepting erroneous commands of other transmissions.

GUIDANCE, COMMAND - A guidance system wherein intelligence transmitted to the missile from an outside source causes the missile to traverse a directed path in space.

GUIDANCE, HOMING, ACTIVE - A system of homing guidance wherein both the transmitter and receiver are carried within the missile.

GUIDANCE, HOMING, PASSIVE - A form of homing guidance, which is dependent on a missile's ability to detect energy emitted by the target. Frequently termed Home-On-Jam (HOJ).

GUIDANCE, HOMING, SEMIACTIVE - A system of homing guidance wherein the missile uses reflected signals from the target which has been illuminated by a source other than within the missile. See also CWI.

GUIDANCE, INERTIAL - A self-contained system independent of information obtained from outside the missile, usually using Newton's second law of motion.

GUNN DIODE - The Gunn diode is a transferred electron device which because of its negative resistance can be used in microwave oscillators or amplifiers. When the applied voltage exceeds a certain critical value, periodic fluctuations in current occur. The frequency of oscillation depends primarily upon the drift velocity of electrons through the effective length of the device. This frequency may be varied over a small range by means of mechanical tuning.

HARMONIC - A sinusoidal component of a periodic wave or quantity having a frequency that is an integral multiple of the fundamental frequency. For example, a component which is twice the fundamental frequency is called the second harmonic. (the fundamental is the first harmonic, which is frequently misunderstood).

HERTZ - The unit of frequency equal to one cycle per second.

HOME-ON-JAM (HOJ) - See Guidance, Homing, Passive.

HORN ANTENNA - A flared, open-ended section of waveguide used to radiate the energy from a waveguide into space. Also termed "horn" or "horn radiator." Usually linearly polarized, it will be vertically polarized when the feed probe is vertical, or horizontally polarized if the feed is horizontal. Circular polarization can be obtained by feeding a square horn at a 45° angle and phase shifting the vertical or horizontal excitation by 90°.

HYPERABRUPT VARACTOR OSCILLATOR - Due to a non-uniform concentration of N-type material (excess electrons) in the depletion region, this varactor produces a greater capacitance change in tuning voltage and a far more linear voltage-vs-frequency tuning curve. As a result, this device has an improved tuning linearity and low tuning voltage.

IF (INTERMEDIATE FREQUENCY) - The difference frequency resulting from mixing (beating) the received signal in a superheterodyne receiver with the signal from the local oscillator. The difference frequency product provides the advantages inherent to the processing (amplification, detection, filtering, and such) of low frequency signals. The receiver local oscillator may operate either below or above the receiver tuned frequency. A single receiver may incorporate multiple IF detection.

$IF = F_{LO} - F_O$. (for a local oscillator operating above the fundamental)

where:

F_O = Received fundamental frequency

F_{LO} = Local oscillator frequency

The simplified block diagram illustrates a typical mixing procedure employed in radar systems to obtain desired IF frequencies. The local oscillator is tuned above the fundamental frequency in this example. It should be noted that an undesired signal received at the receiver image frequency of 1170 MHz will also produce the desired 60 MHz IF frequency; this relationship provides the receiver image. See also Image Frequency.

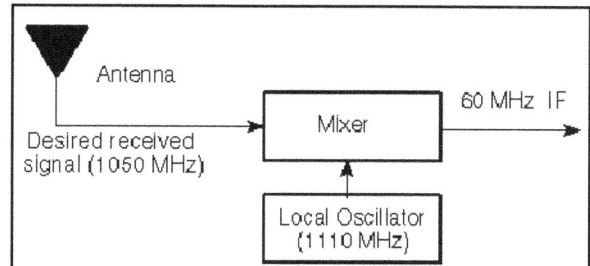

IFF (IDENTIFICATION FRIEND OR FOE) - A system using radar transmission to which equipment carried by friendly forces automatically responds by emitting a unique characteristic series of pulses thereby distinguishing themselves from enemy forces. It is the "Mode IV" for the aircraft transponder. See also transponder.

IMAGE FREQUENCY - That frequency to which a given superheterodyne receiver is inherently susceptible, thereby rendering such a receiver extremely vulnerable to EMI at that frequency. The image frequency is located at the same frequency difference (Δf) to one side of the local oscillator as the tuned (desired) frequency is to the other side. An undesired signal received at the image frequency by a superheterodyne receiver not having preselection would, therefore, mix (beat) with the oscillator, produce the proper receiver IF, and be processed in the same manner as a signal at the desired frequency. See also receiver selectivity.

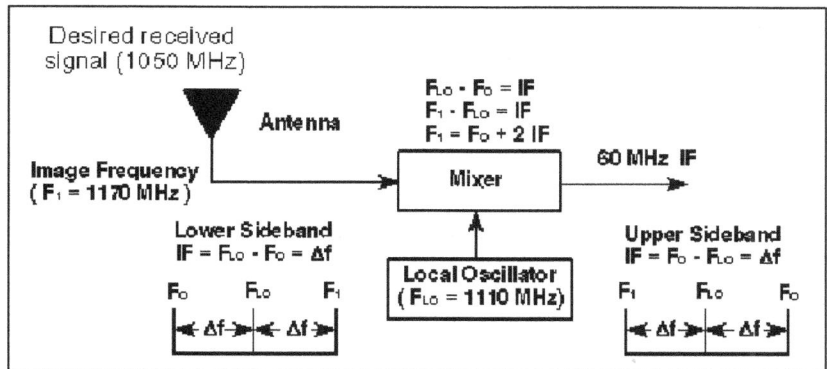

IMAGE JAMMING - Jamming at the image frequency of the radar receiver. Barrage jamming is made most effective by generating energy at both the normal operating and image frequency of the radar. Image jamming inverts the phase of the response and is thereby useful as an angle deception technique. Not effective if the radar uses image rejection.

IMPATT DIODE - The IMPATT (IMPact Avalanche and Transit Time) diode acts like a negative resistance at microwave frequencies. Because of this property, Impatt diodes are used in oscillators and amplifiers. Usually the frequency range is in the millimeter wave region where other solid state devices cannot compete.

INGRESS - Go into the target area.

INSERTION LOSS - The loss incurred by inserting an element, device, or apparatus in an electrical/electronic circuit. Normally expressed in decibels determined as 10 log of the ratio of power measured at the point of insertion prior to inserting the device (P_1) to the power measured after inserting the device (P_2). Insertion loss (dB) = 10 log P_1/P_2.

INTEGRATION EFFECT - Pulse radars usually obtain several echoes from a target. If these echoes are added to each other, they enhance the S/N ratio, making a weak target easier to detect. The noise and interference do not directly add from pulse to pulse, so the ratio of target strength to undesired signal strength improves making the target more detectable. Random noise increases by the square root of the number of integrations, whereas the signal totally correlates and increases directly by the number of integrations, therefore the S/N enhancement is equal to the square root of the number of integrations.

INTERFERENCE - See EMI.

INTERFERENCE PARAMETERS - Equipment and propagation characteristics necessary for the proper evaluation of a given EMI situation.

INTERFERENCE/SIGNAL RATIO = See I/S Ratio.

INTERFERENCE THRESHOLD - The level of interference normally expressed in terms of the I/S (interference/signal) ratio at which performance degradation in a system first occurs as a result of EMI.

INTERFEROMETER - When two widely spaced antennas are arrayed together, they form an interferometer. The radiation pattern consists of many lobes, each having a narrow beamwidth. This antenna can provide good spatial selectivity if the lobe-to-lobe ambiguity can be solved such as using amplitude comparison between the two elements.

INTERMODULATION - The production, in a nonlinear element (such as a receiver mixer), of frequencies corresponding to the sums and differences of the fundamentals and harmonics of two or more frequencies which are transmitted through the element; or, the modulation of the components of a complex wave by each other, producing frequencies equal to the sums and differences of integral multiples of the component frequencies of the complex wave.

INTERSYSTEM EMC - EMC between the external electromagnetic environment (EME) and an aircraft with it's installed systems. Generally, only system BIT must operate properly on the carrier deck while all system functions must operate properly in the operational EME.

INTRASYSTEM EMC - EMC between systems installed on an aircraft, exclusive of an external environment.

INVERSE CON SCAN - One method of confusing a radar operator or fire control radar system is to provide erroneous target bearings. This is accomplished by first sensing the radar antenna scan rate and then modulating repeater amplifier gain so the weak portion of the radar signal is amplified by the jammer, while the strong portion is not, so the weapons systems will fire at some bearing other than the true target bearing. The angle deception technique is used to break lock on CONSCAN radars.

INVERSE GAIN - Amplification, inverse modulation, and re-radiation of a radar's pulse train at the rotation rate of the radar scan. Deceives a conical scanning radar in angle.

ISOTROPIC ANTENNA - A hypothetical antenna which radiates or receives energy equally in all directions.

I/S RATIO (INTERFERENCE-TO-SIGNAL RATIO) (ISR) - The ratio of electromagnetic interference level to desired signal level that exists at a specified point in a receiving system. The ratio, normally expressed in dB, is employed as a tool in prediction of electronic receiving system performance degradation for a wide range of interference receiver input levels. Performance evaluations compare actual I/S ratios to minimum acceptable criteria.

JAFF - Expression for the combination of electronic and chaff jamming.

JAMMING - The deliberate radiation, reradiation, or reflection of electromagnetic energy with the object of impairing the use of electronic devices, equipment, or systems by an enemy.

JATO - JAmmer Technique Optimization group -- Organization coordinates EA technique development and evaluation efforts in support of naval airborne EA (ALQ-99, USQ-113, ALQ-227, ALQ-231, ALE-43). The group is run out of PMA-234 and comprised of elements at NAWCWD-Point Mugu, Johns Hopkins University/ Applied Physics Lab, Naval Research Lab, and NSWC-Crane. Efforts include M&S, Lab test, ground test, flight test, direct fleet support, and requirements and test support to development programs.

JINK - An aircraft maneuver which sharply changes the instantaneous flight path but maintains the overall route of flight. More violent than a weave.

JITTERED PRF - An antijam feature of certain radar systems which varies the PRF consecutively, and randomly, from pulse to pulse to prevent enemy ECM equipment from locking on, and synchronizing with, the transmitted PRF. PRF is synonymous with pulse repetition rate (PRR).

JTAT - JATO Techniques Analysis and Tactics -- Analysis documents produced by JATO to advise EA systems operators on the best ways to employ their systems by region, mission and equipment configuration. Currently covering ALQ-99 and USQ-113 on EA-6B; ALQ-99 and ALQ-227 on EA-18G, and ALQ-231 on AV-8B.

KILO - A prefix meaning 10^3 (times one thousand). For example, kilohertz.

KLYSTRON AMPLIFIER - An electron beam device which achieves amplification by the conversion of periodic velocity variations into conduction-current modulation in a field-free drift region. Velocity variations are launched by the interaction of an RF signal in an input resonant cavity and are coupled out through an RF output cavity. Several variations including reflex and multi cavity klystrons are used.

KLYSTRON, MULTICAVITY - An electron tube which employs velocity modulation to generate or amplify electromagnetic energy in the microwave region. Since velocity modulation employs transit time of the electron to aid in the bunching of electrons, transient time is not a deterrent to high frequency operations as is the case in conventional electron tubes. See also Velocity Modulation.

KLYSTRON, REFLEX - A klystron which employs a reflector (repeller) electrode in place of a second resonant cavity to redirect the velocity-modulated electrons through the resonant cavity. The repeller causes one resonant circuit to serve as both input and output, which simplifies the tuning operation. This type of klystron is well adapted for use as an oscillator because the frequency is easily controlled by varying the position of the repeller. See also Velocity Modulation.

LEAKAGE - Undesired radiation or conduction of RF energy through the shielding of an enclosed area or of an electronic device.

LENS, RADAR (MICROWAVE) - The purpose of any such lens is to refract (focus) the diverging beam from an RF feed into a parallel beam (transmitting) or vice versa (receiving). The polarization is feed dependent.

LIGHT AMPLIFICATION BY STIMULATED EMISSION OF RADIATION (LASER) - A process of generating coherent light. The process utilizes a natural molecular (and atomic) phenomenon whereby molecules absorb incident electromagnetic energy at specific frequencies, store this energy for short but usable periods, and then release the stored energy in the form of light at particular frequencies in an extremely narrow frequency-band.

LIMITING - A term to describe that an amplifier has reached its point of saturation or maximum output voltage swing. Deliberate limiting of the signal is used in FM demodulation so that AM will not also be demodulated.

LITTORAL - Near a shore.

LOBE, ANTENNA - Various parts of the antenna's radiation pattern are referred to as lobes, which may be subclassified into major and minor lobes. The major lobe is the lobe of greatest gain and is also referred to as the main lobe or main beam. The minor lobes are further subclassified into side and back lobes as indicated in the figure to the right. The numbering of the side lobes are from the main lobe to the back lobe.

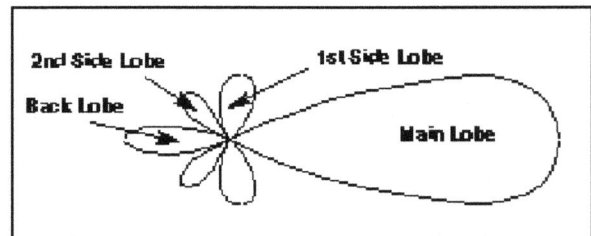

LOCAL OSCILLATOR FREQUENCY - An internally generated frequency in a superheterodyne receiver. This frequency differs from the receiver operating frequency by an amount equal to the IF of the receiver. The local oscillator frequency may be designed to be either above or below the incoming signal frequency.

LOG VIDEO - This receiver process, generally implemented in the IF, compresses the dynamic range of the signal so both weak and strong signals are displayed without changing the gain setting. Output voltage can be calibrated in volts/dB of input power.

LONG PULSE MODE - Many pulsed radars are capable of transmitting either long or short pulses of RF energy. When the long pulses of RF energy are selected manually (or sometimes automatically), the radar is said to be operating in the long pulse mode. In general, "long pulse mode" is used to obtain high average power for long-range search or tracking, and "short pulse mode" gives low average power for short-range, high-definition, tracking or search.

LOOSE DEUCE - General term for two aircraft working in mutual support of each other.

LORO (LOBE-ON-RECEIVE-ONLY) - A mode of operation generally consisting of transmitting on one non-scanning antenna system and receiving the reflected energy on another scanning system (The receiver could be TWS, Conical, or monopulse).

MACH NUMBER - The ratio of the velocity of a body to the speed of sound in the medium that is being considered. In the atmosphere, the speed of sound varies with temperature and atmospheric pressure, hence, so does mach number.

MAGNETIC ANOMALY DETECTOR - A means of detecting changes in the earth's magnetic field caused by the presence of metal in ships and submarines.

MAGNETRON - A magnetron is a thermionic vacuum tube which is constructed with a permanent magnet forming a part of the tube and which generates microwave power. These devices are commonly used as the power output stage of radar transmitters operating in the frequency range above 1000 MHz and are used less commonly down to about 400 MHz. A magnetron has two concentric cylindrical electrodes. On a conventional magnetron, the inner one is the cathode and the outer one is the anode. The opposite is true for a coaxial magnetron.

MAGNETRON OSCILLATOR - A high-vacuum tube in which the interaction of an electronic space charge and a resonant system converts direct current power into ac power, usually at microwave frequencies. The magnetron has good efficiency, is capable of high power outputs, and is stable.

MATCHED FILTER - This describes the bandwidth of an IF amplifier that maximizes the signal-to-noise ratio in the receiver output. This bandwidth is a function of the pulsewidth of the signal.

MDS (MINIMUM DETECTABLE/DISCERNIBLE SIGNAL) - The receiver input power level that is just sufficient to produce a detectable/discernible signal in the receiver output. The detectable term is interchangeable with S_{min} and the discernable term is interchangeable with MVS. See Section 5-2.

MEACONING - A system receiving radio signals and rebroadcasting them (or just transmitting) on the same frequency to confuse navigation. The meaconing station attempts to cause aircraft to receive inaccurate range or bearing information.

MEATBALL - Visual light "ball" seen in Fresnel lens optical landing system (FLOLS) by pilot during carrier or Navy field landing. Used as a reference to determine if flight path is high or low.

MEGA - A prefix meaning 10^6 (times one million). For example megahertz (MHz)

MICROVOLT PER METER - A commonly used unit of field strength at a given point. The field strength is measured by locating a standard receiving antenna at that point, and the "microvolts per meter" value is then the ratio of the antenna voltage in microvolts to the effective antenna length in meters. Usually used below 100 MHz. Above 100 MHz, power density terminology is normally used.

MICROWAVE AMPLIFICATION BY STIMULATED EMISSION OF RADIATION (MASER) - A low-noise radio-frequency amplifier. The emission of energy stored in a molecular or atomic system by a microwave power supply is stimulated by the input signal.

MISS DISTANCE - Used variously in different contexts. The distance from the missile to the geometric center of the aircraft, or the closest point of approach (CPA) of the missile to any portion of the aircraft such as the aircraft nose or telemetry pod, etc.

MISSILE SYSTEMS FUNCTIONS - Examples of missile system functions are: "acquisition" (ability to lock-on a desired target); "tracking" of a target; "guidance" of a missile toward a target; "illumination" of a target so that a homing missile can home on the reflected RF illumination; and "command" signal transmission to a missile to cause it to arm, to detonate, to commence homing, or to destroy itself.

MIXERS - See Balanced and Schottky Diode Mixers.

MODULATION - The process whereby some
characteristic of one wave is varied in accordance with
some characteristic of another wave. The basic types of
modulation are angle modulation (including the special
cases of phase and frequency modulation) and amplitude
modulation. In missile radars, it is common practice to
amplitude modulate the transmitted RF carrier wave of
tracking and guidance transmitters by using a pulsed wave
for modulating, and to frequency module the transmitted
RF carrier wave of illuminator transmitters by using a sine
wave.

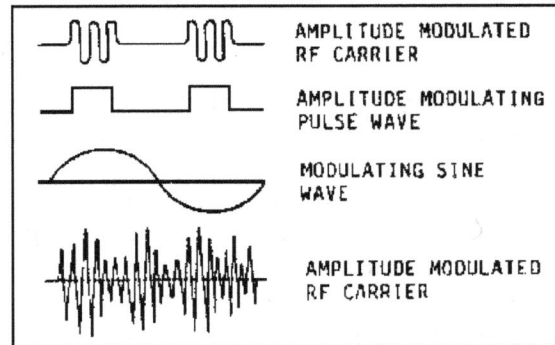

MODULATION, AMPLITUDE - This type of modulation changes the amplitude of a carrier wave in
responses to the amplitude of a modulating wave. This modulation is used in radar and EW only as a switch
to turn on or turn off the carrier wave; i.e., pulse is a special form of amplitude modulation.

MODULATION, FREQUENCY - The frequency of the modulated carrier wave is varied in
proportion to the amplitude of the modulating wave and therefore, the phase of the carrier varies with the
integral of the modulating wave. See also Modulation.

MODULATION, PHASE - The phase of the modulated carrier is varied in proportion to the
amplitude of the modulating wave. See also Modulation.

MONOPULSE - (See figure to right) A
type of tracking radar that permits the extracting
of tracking error information from each received
pulse and offers a reduction in tracking errors as
compared to a conical-scan system of similar
power and size. Multiple (commonly four)
receiving antennas or feeds are placed
symmetrically about the center axis and operate
simultaneously to receive each RF pulse
reflected from the target. A comparison of the
output signal amplitude or phase among the four
antennas indicates the location of the target with
respect to the radar beam center line. The output
of the comparison circuit controls a servo system
that reduces the tracking error to zero and
thereby causes the antenna to track the target.

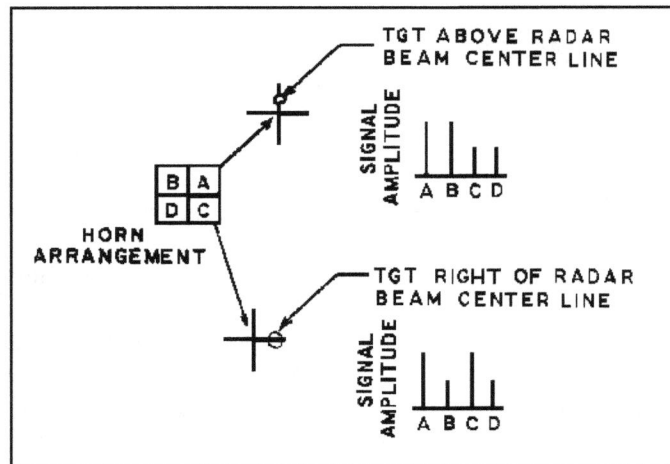

MOS (MINIMUM OPERATIONAL SENSITIVITY) - The minimum signal which can be detected
and automatically digitally processed by a radar without human discrimination.

MTI (MOVING TARGET INDICATOR) - This radar signal process shows only targets that are in
motion. Signals from stationary targets are subtracted out of the return signal by a memory circuit.

MULTIPATH - The process by which a transmitted signal arrives at the receiver by at least two
different paths. These paths are usually the main direct path, and at least one reflected path. The signals
combine either constructively or destructively depending upon phase, and the resultant signal may be either
stronger or weaker than the value computed for free space.

MULTIPLEX - Simultaneous transmission of two or more signals on a common carrier wave. The three types of multiplex are called time division, frequency division, and phase division.

MULTIBAND RADAR - A type of radar which uses simultaneous operation on more than one frequency band through a common antenna. This technique allows for many sophisticated forms of video processing and requires any jammer to jam all channels at the same time in order to be effective.

MVS (MINIMUM VISIBLE SIGNAL) - The minimum input pulse signal power level which permits visibility of the output pulse, such as on a radar A-scope display. This level is determined by initially setting the input level above the visible detection threshold, and then slowly decreasing the amplitude.

NOISE FIGURE, RECEIVER - A figure of merit (NF or F) of a system given by the ratio of the signal-to-noise ratio at the input, S_i / N_i, divided by the signal-to-noise ratio at the output, S_o / N_o. It essentially expresses the ratio of output noise power of a given receiver to that of a theoretically perfect receiver which adds no noise.

$$Noise\ Figure = \frac{S_i / N_i}{S_o / N_o} = \frac{N_o}{G\,N_i}$$

Where $S_o = GS_i$ and G is the gain of the system.

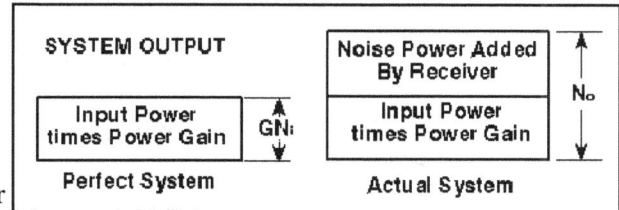

Noise figure is usually expressed in dB and given for an impedance matched condition. Impedance mismatch will increase the noise figure by the amount of mismatch loss. NF is usually given at room temperature; 17°C or 290°K.   See Section 5-2.

NOISE JAMMING - A continuous random signal radiated with the objective of concealing the aircraft echo from the enemy radar. In order for it to be effective, it must have an average amplitude at least as great as the average amplitude of the radar echo. There are three major categories of noise jamming which are grouped by how jamming power is concentrated: Spot, barrage, and swept jamming. (See individual definitions)

NONCOHERENT - Two signals that have no set phase relationship.

NOTCH - The portion of the radar velocity display where a target disappears due to being notched out by the zero Doppler filter. If not filtered (notched), ground clutter would also appear on the display. A notch filter is a narrow band-reject filter. A "notch maneuver" is used to place a tracking radar on the beam of the aircraft so it will be excluded.

NULL, ANTENNA PATTERN - The directions of minimum transmission (or reception) of a directional antenna. See also Lobe, Antenna.

NULL FILL - The nulls in an antenna pattern may be reduced (filled) by using a second ancillary (spoiler) antenna whose pattern is such that it fills in the nulls of the main antenna pattern.

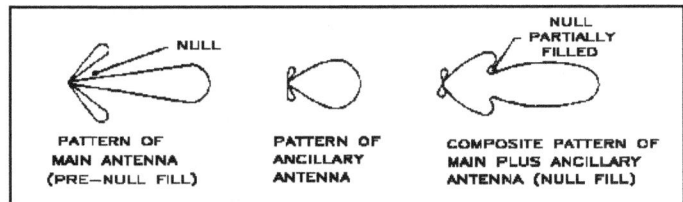

NUTATION - As applied to current missile system radars, this term refers to the mechanical motion of an antenna feed to produce a conical scan (fixed polarization) by the main beam of a tracking antenna, thus providing a means of developing tracking error signals. See also Antenna, Nutating. By analogy, "Nutation" also is used to denote the electrical switching of the quadrants of a seeker antenna. See also Interferometer. The effect is similar to that of a conical scan.

NUTATOR - A motor-driven rotating antenna feed used to produce a conical scan for a tracking radar. See also Antenna, Nutating. Also, the electrical circuits necessary to effect nonmechanical conical scans. See also Nutation.

OPERATIONAL CONSTRAINTS - Limitations on operating procedures in order to prevent interference between missile systems on a ship or between missile systems in a formation of ships under operational conditions. These limitations consist of such things as limited frequency bands or channels in which the radars may be tuned, limited sectors of space into which radar beams may be pointed, limits on minimum spacing between ships, limits on what codes may be used by radars and missiles on each ship, and limits on minimum interval between firing of certain missiles.

OSCILLATORS - Devices which generate a frequency. See also Backward Wave, Dielectrically Stabilized Oscillator, Hyperabrupt Varactor Oscillator, Magnetron Oscillator, Varactor Tuned Oscillator, and YIG tuned oscillator.

OSCILLATOR, LOCAL - See Local Oscillator Frequency.

PALMER SCAN - Conical scan superimposed on another type of scan pattern - usually a spiral pattern.

PARAMETER - A quantity which may have various values, each fixed within the limits of a stated case or discussion. In the present case, some examples of parameters; would be: radar frequency, limited by the tuning range of the radar; missile range, limited by the maximum operating range of the missile; or a missile code, limited by the number of codes available and by the codes that the ship radars are set up to operate on.

PASSIVE ANGLE TRACKING - Tracking of a target using radiation from the target (such as jamming), with no radiation from the radar itself. Only angular tracking is possible under these conditions since no measurement of time of travel of radiation to the target is possible, as is required to obtain target range.

PC (PULSE COMPRESSION) - The process used in search and tracking pulse radars whereby the transmitted pulse is long, so as to obtain high average transmitter output power, and the reflected pulse is processed in the radar receiver to compress it to a fraction of the duration of the transmitted pulse to obtain high definition and signal strength enhancement. Pulse compression may be accomplished by sweeping the transmitted frequency (carrier) during the pulse. The returned signal is then passed through a frequency-dependent delay line. The leading edge of the pulse is therefore delayed so that the trailing edge catches up to the leading edge to produce effectively a shorter received pulse than that transmitted. Pulse compression radars are also referred to as CHIRP radars. Other more sophisticated pulse compression techniques are also possible and are becoming more popular.

PENCIL BEAM - A narrow circular radar beam from a highly directional antenna (such as a parabolic reflector).

PHASED ARRAY RADAR - Radar using many antenna elements which are combined in a controlled phase relationship. The direction of the beam can be changed as rapidly as the phase relationships (usually less than 20 microseconds). Thus, the antenna typically remains stationary while the beam is electronically scanned. The use of many antenna elements allows for very rapid and high directivity of the beam(s) with a large peak and/or average power. There is also a potential for greater reliability over a conventional radar since the array will fail gracefully, one element at a time.

PIN DIODE - A diode with a large intrinsic (I) region sandwiched between the P- and N- doped semiconducting regions. The most important property of the PIN diode is the fact that it appears as an almost pure resistance at RF. The value of this resistance can be varied over a range of approximately one-10,000 ohms by direct or low frequency current control. When the control current is varied continuously, the PIN diode is useful for attenuating, leveling and amplitude modulation of an RF signal. When the control current is switched on and off or in discrete steps, the device is useful in switching, pulse modulating, and phase shifting an RF signal.

POINT CONTACT DIODE - This was one of the earliest semiconductor device to be used at microwave frequencies. Consisting of a spring-loaded metal contact on a semiconducting surface, this diode can be considered an early version of the Schottky barrier diode. Generally used as a detector or mixer, the device is somewhat fragile and limited to low powers.

POLARIZATION - The direction of the electric field (E-field) vector of an electromagnetic (EM) wave. See Section 3-2. The most general case is elliptical polarization with all others being special cases. The E-field of an EM wave radiating from a vertically mounted dipole antenna will be vertical and the wave is said to be vertically polarized. In like manner, a horizontally mounted dipole will produce a horizontal electric field and is horizontally polarized. Equal vertical and horizontal E-field components produce circular polarization.

PORT - The left side of a ship or aircraft when facing the bow (forward)

POWER (AVERAGE) FOR PULSED RADARS - Average power for a pulse radar is the average power transmitted between the start of one pulse and the start of the next pulse (because the time between pulses is many times greater than the pulse duration time, the average power will be a small fraction of peak power).

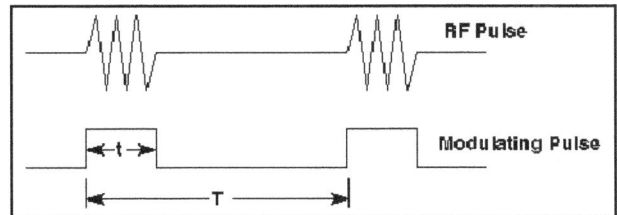

For this example: Peak Power = 1 MW, Pulse Time (t) = 0.5 micro-second, and Interval Between Pulses (T) = 1000 microseconds (1000 pps).

Peak Power = Pwr during pulse time (t) = 1 MW = 10^6 Watts = 90 dBm.

Avg Power = Average Power During Time (T) = 10^6 x t/T
= 10^6 x 0.5/1000 = 0.5 x 10^3 = 0.5 kilowatt = 57 dBm or 27 dBW

POWER OUTPUT - Power output of a transmitter or transmitting antenna is commonly expressed in dBW or dBm. One megawatt would be expressed as 60 dBW or 90 dBm:

10 log (1 megawatt / 1 watt) = 10 log ($10^6/10^0$)   10 log (1 megawatt / 1 milliwatt)   = 10 log ($10^6/10^{-3}$)
= 10 x 6 = 60 dBW   = 10 x 9 = 90 dBm

POWER (PEAK) FOR PULSED RADARS - Peak power for a pulsed radar is the power radiated during the actual pulse transmission (with zero power transmitted between pulses).

POWER FOR CW RADARS - Since the power output of CW transmitters (such as illuminator transmitters) usually have a duty cycle of one (100%), the peak and average power are the same.

POWER DENSITY - The density of power in space expressed in Watts/meter^2, dBW/m^2, etc. Generally used in measurements above 100 MHz. At lower frequencies, field intensity measurements are taken. See Section 4-1.

PPI-SCOPE - A radar display yielding range and azimuth (bearing) information via an intensity modulated display and a circular sweep of a radial line. The radar is located at the center of the display.

PRESELECTOR - A device placed ahead of the mixer in a receiver, which has bandpass characteristics such that the desired (tuned) RF signal, the target return, is allowed to pass, and other undesired signals (including the image frequency) are attenuated.

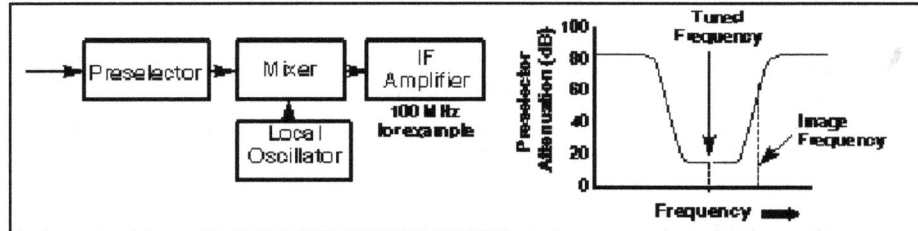

PROPAGATION - In electrical practice, the travel of waves through or along a medium. The path traveled by the wave in getting from one point to another is known as the propagation path (such as the path through the atmosphere in getting from a transmitting antenna to a receiving antenna, or the path through the waveguides and other microwave devices in getting from an antenna to a receiver).

PULSE COMPRESSION - See PC.

PULSED DOPPLER (PD) - A type of radar that combines the features of pulsed radars and CW Doppler radars. It transmits pulses (instead of CW) which permits accurate range measurement. This is an inherent advantage of pulsed radars. Also, it detects the Doppler frequency shift produced by target range rate which enables it to discriminate between targets of only slightly different range rate and also enables it to greatly reduce clutter from stationary targets. See also Doppler Effect.

PULSE LENGTH - Same meaning as Pulsewidth.

PULSE MODULATION - A special case of amplitude modulation wherein the carrier wave is varied at a pulsed rate. Pulse Modulation - The modulation of a carrier by a series of pulses generally for the purpose of transmitting data. The result is a short, powerful burst of electromagnetic radiation which can be used for measuring the distance from a radar set to a target.

PULSE REPETITION FREQUENCY (PRF) - The rate of occurrence of a series of pulses, such as 100 pulses per second. It is equal to the reciprocal of the pulse spacing (T) or PRT. (PRF = 1/T = 1/PRI). Sometimes the term pulse repetition rate (PRR) is used.

PULSE REPETITION FREQUENCY (PRF) STAGGER - The technique of switching PRF (or PRI) to different values on a pulse-to-pulse basis such that the various intervals follow a regular pattern. This is useful in compensating for blind speeds in pulsed MTI radars. Interpulse intervals which differ but follow a regular pattern.

PULSE REPETITION INTERVAL (PRI) or TIME (PRT) - Time between the beginning of one pulse and the beginning of the next.

PULSE SPACING - The interval of time between the leading edge of one pulse and the leading edge of the next pulse in a train of regularly recurring pulses. See also Pulse Repetition Frequency. Also called "the interpulse period."

PULSEWIDTH - The interval of time between the leading edge of a pulse and the trailing edge of a pulse (measured in microseconds for the short pulses used in radar). Usually measured at the 3 dB midpoint (50-percent power or 70% voltage level) of the pulse, but may be specified to be measured at any level. See Section 6-10 for measurement techniques.

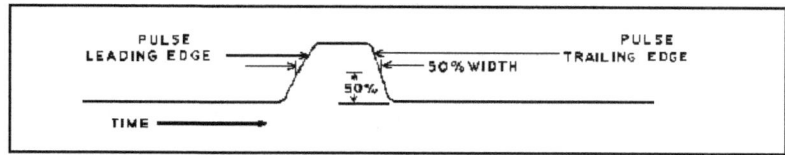

QUANTIZE - The process of restricting a variable to a number of discrete values. For example, to limit varying antenna gains to three levels.

RADAR - Radio detection and ranging.

RADAR CROSS SECTION - A measure of the radar reflection characteristics of a target. It is equal to the power reflected back to the radar divided by power density of the wave striking the target. For most targets, the radar cross section is the area of the cross section of the sphere that would reflect the same energy back to the radar if the sphere were substituted. RCS of sphere is independent of frequency if operating in the far field region. See Section 4-11.

RADAR RANGE EQUATION - The radar range equation is a basic relationship which permits the calculation of received echo signal strength, if certain parameters of the radar transmitter, antenna, propagation path, and target are known.

Given: $P_r = \dfrac{P_t G_t G_r \lambda^2 \sigma}{(4\pi)^3 R^4}$ *(freespace)*   as the basic two-way radar equation (see Sections 4-4 thru 4-6)

where:

P_r	=	Peak power at receiver input	λ	=	Wavelength of signal (length) = c/f
P_t	=	Transmitted signal level (power)	R	=	Range of target to radar (distance)
G_t	=	Gain of transmitting antenna (dimensionless ratio)	σ	=	Radar cross section of target
G_r	=	Gain of receiving antenna (dimensionless ratio)			

In practical use, the radar range equation is often written in logarithmic form, all terms expressed in decibels, so that the results can be found by simple processes of addition and subtraction. Using the above equation and λ = c/f

$10 \log P_r = 10 \log P_t + 10 \log G_t + 10 \log G_r + 10 \log \sigma - 40 \log R - 20 \log f + 20 \log c - 30 \log 4\pi$
where: f = Signal frequency (cycles {dimensionless}/time)   c = Speed of light (length/time)

$10 \log P_r = 10 \log P_t + 10 \log G_t + 10 \log G_r + G_\sigma - 2\alpha_1$

where α_1 and G_σ are factors containing the constants and conversion factors to keep the equations in consistent units.

Refer to Sections 4-4 through 4-6

RADAR TRIGGER KILL - see Trigger Kill, Radar

RADIATION EFFICIENCY - $E = P_{radiated}/P_{in}$ (ideal=1)

RADIATION PATTERN - See Antenna Pattern.

RADIO FREQUENCY - See RF.

RADIO FREQUENCY INTERFERENCE - See RFI.

RAIL KEEPING - Ability of countermeasures to keep the missile on the launch rail, i.e., prevent launch.

RAINBOW - A technique which applies pulse-to-pulse frequency changing to identifying and discriminating against decoys and chaff.

RANGE CELL - In a radar, a range cell is the smallest range increment the radar is capable of detecting. If a radar has a range resolution of 50 yards and a total range of 30 nautical miles (60,000 yds), there are: 60000/50 = 1,200 range cells.

RANGE GATE - A gate voltage used to select radar echoes from a very short range interval.

RANGE GATE PULL OFF (RGPO) - Deception technique used against pulse tracking radars using range gates. Jammer initially repeats the skin echo with minimum time delay at a high power to capture the AGC circuitry. The delay is progressively increased, forcing the tracking gates to be pulled away ("walked off") from the target echo. Frequency memory loops (FML's), or transponders provide the variable delay.

RANGE RATE - The rate at which a radar target is changing its range with respect to the radar (in feet per second for example). Note that this rate is not the same as target velocity unless the target is moving straight toward or straight away from the radar.

RANGE SCOPE - See A-Scope or PPI.

RECEIVER BANDWIDTH - The difference between the limiting frequencies within which receiver performance in respect to some characteristic falls within specified limits. (In most receivers this will be the difference between the two frequencies where the intermediate frequency (IF) amplifier gain falls off 3 dB from the gain at the center IF frequency.) See also Receiver Selectivity.

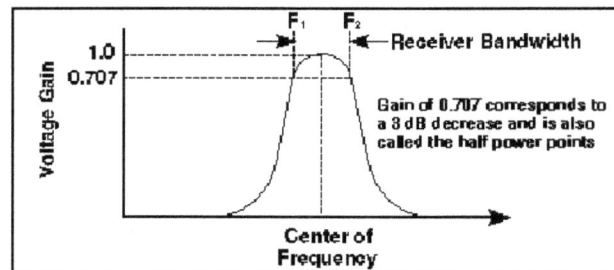

RECEIVER SELECTIVITY - The degree to which a receiver is capable of differentiating between the desired signal and signals or interference at other frequencies. (The narrower the receiver bandwidth, the greater the selectivity.)

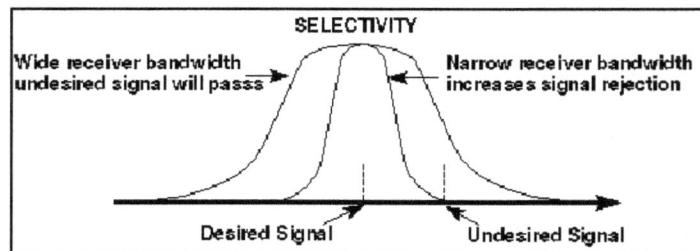

REFLECTION - The turning back (or to the side) of a radio wave as a result of impinging on any conducting surface which is at least comparable in dimension to the wavelength of the radio wave.

RESOLUTION - In radar, the minimum separation in angle or in range between two targets which the radar is capable of distinguishing.

RF (RADIO FREQUENCY) - A term indicating high frequency electromagnetic energy.

RFI (RADIO FREQUENCY INTERFERENCE) - Any induced, radiated, or conducted electrical disturbance or transient that causes undesirable responses or malfunctioning in any electrical or electronic equipment, device, or system. Same as EMI. Not to be confused with the logistic term ready for issue (also RFI).

RING AROUND - A condition in which a repeater jammer's total gain, from receiver antenna to transmitter antenna, exceeds the antenna isolation resulting in the repeater amplifying it's own internal noise. Akin to positive feedback in an amplifier that causes unwanted oscillations.

RING AROUND (RADAR-TO-MISSILE) - The condition where radio frequency interference signals from a transmitter of one missile radar enter the receiving circuits of a missile under the control of another missile radar.

RING AROUND (RADAR-TO-RADAR) - The condition where radio frequency interference signals from a transmitter of one radar enter the receiving circuits of another radar.

ROPE - An element of chaff consisting of a long roll of metallic foil or wire which is designed for broad, low-frequency response. See Chaff.

R-SCOPE - (RANGE SCOPE) See A-scope or PPI.

SAFETY OF FLIGHT (SOF) TEST - A flight test to verify that a new or modified subsystem will not cause a major problem with the aircraft, i.e., interference can occur, but will not be such that required navigational systems will fail or which might potentially cause the loss of an aircraft under all normally expected weather conditions.

SCAN - To transverse or sweep a sector or volume of airspace with a recurring pattern, by means of a controlled directional beam from a radar antenna. See also "Antenna, nutating".

SCHOTTKY BARRIER DIODE - The Schottky barrier diode is a simple metal-semiconductor boundary with no P-N junction. A depletion region between the metal contact and the doped semiconductor region offers little capacitance at microwave frequencies. This diode finds use as detectors, mixers, and switches.

SCHOTTKY DIODE MIXER - The mixer is a critical component in modern RF systems. Any nonlinear element can perform the mixing function, but parameters determining optimal mixing are noise figure, input admittance, and IF noise and impedance. The Schottky diode is particularly effective because of its low noise figure and nearly square law characteristics.

SCHOTTKY DIODE SWITCH - Standard P-N diodes are limited in switching ability at high frequencies because of capacitance provided by the minority carriers. The Schottky diode overcomes this problem by use of the metal-semiconductor junction with inherently low carrier lifetimes, typically less than 100 picoseconds.

SEARCH RADAR - A radar whose prime function is to scan (search) a specified volume of space and indicate the presence of any targets on some type of visual display, and, in some cases, to provide coordinates of the targets to a fire control system to assist in target acquisition and tracking.

SEEKER - The seeker consists of circuitry in a homing missile which detects, electronically examines, and tracks the target; provides data for controlling the flight path of the missile; and provides signals for destroying the missile or for detonating it at intercept. (The seeker function is similar to that of an interferometer.)

SELF-SCREENING JAMMING (SSJ) - Each aircraft carries it's own jamming equipment for it's own protection.

SENSITIVITY - The sensitivity of a receiver is taken as the minimum signal level required to produce an output signal having a specified signal-to-noise ratio. See also Minimum Visible Signal and Minimum Discernible Signal (MDS).

SENSITIVITY TIME CONTROL - See STC.

SENSOR - The receiver portion of a transmitter/receiver pair used to detect and process electromagnetic energy.

SHIELDING - The physical arrangement of shields for a particular component, equipment, or system, (A shield is a housing, screen, or other material, usually conducting, that substantially reduces the effect of electric or magnetic fields on one side of the shield upon devices or circuits on the other side.) Examples are tube shields, a shielded enclosure or cabinet for a radar receiver, and the screen around a screen room.

SHORT PULSE MODE - See Long Pulse Mode.

SIDEBAND - A signal either above or below the carrier frequency, produced by the modulation of the carrier wave by some other wave. See figure at right ⇒

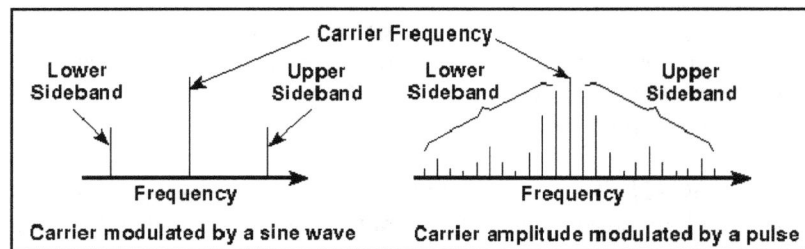

SIDELOBE - See Lobe, Antenna.

SIGNAL STRENGTH - The magnitude of a signal at a particular location. Units are volts per meter or dBV/m.

SIGNATURE - The set of parameters which describe the characteristics of a radar target or an RF emitter and distinguish one emitter from another. Signature parameters include the radio frequency of the carrier, the modulation characteristics (typically the pulse modulation code), and the scan pattern.

SILICON CONTROLLED SWITCH - A P-N-P-N device able to operate at sub-microsecond switching speeds by the application of gate signals. Because it is a four layer device, this switch is also known as a tetrode thyristor.

SLANT POLARIZATION - Technique of rotating a linear antenna 45° so it can receive or jam both horizontal and vertical polarization although there is a 3 dB loss. See Section 3.2.

SOLID STATE STAMO - A stable master oscillator constructed using transistors and other solid state devices as opposed to vacuum tubes. See also STAMO.

SPECTRUM - The distribution of power versus frequency in an electromagnetic wave. See also Spectrum Signature Analysis and illustrations under Sideband.

SPECTRUM ANALYZER - An electronic device for automatically displaying the spectrum of the electromagnetic radiation from one or more devices. A cathode ray tube display is commonly used to display this power-versus frequency spectrum. For examples of two types of displays, see illustrations under Sideband.

SPECTRUM SIGNATURE ANALYSIS - The analysis of the electromagnetic radiation from an electronic device to determine the relative power in each sideband, harmonic, and spurious emission compared to the carrier frequency. This particular distribution (or spectrum) is peculiar to the device and can identify this type of device, thereby acting as an identifying "signature."

SPECTRUM WIDTH (TRANSMITTER) - The difference between the frequency limits of the band which contains all the spectrum frequency components of significant magnitude.

SPOILER ANTENNA - An antenna used to change (spoil) the antenna pattern of a second antenna so as to reduce the nulls in the pattern of the second antenna. See also Null Fill .

SPOKING (RADAR) - Periodic flashes of the rotating radial display. Sometimes caused by mutual interference.

"SPOOFING" - A type of deception by using an electronic device to transmit a "target" echo. The spoofing transmitter must operate at the same frequency and PRF as the radar to be deceived. The radar main pulse triggers the spoofing transmitter which, after a delay, transmits a false echo.

SPOT JAMMING - Narrow frequency band jamming concentrated against a specific radar at a particular frequency. The jamming bandwidth is comparable to the radar bandpass. Can deny range and angle information.

SPURIOUS EMISSION - Electromagnetic radiation transmitted on a frequency outside the bandwidth required for satisfactory transmission of the required waveform. Spurious emissions include harmonics, parasitic emissions, and intermodulation products, but exclude necessary modulation sidebands of the fundamental carrier frequency.

SQUINT ANGLE - The angular difference between the axis of the antenna mainlobe and the geometric axis of the antenna reflector, such as the constant angle

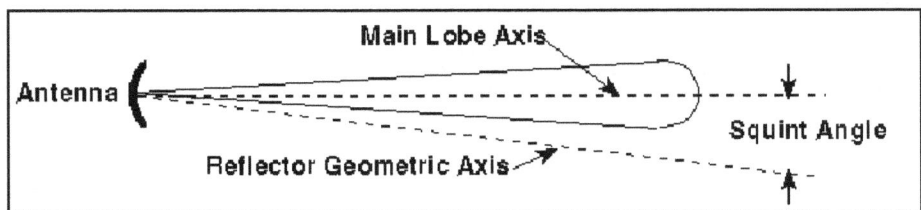

maintained during conical scan as the mainlobe rotates around the geometric axis of the reflector.

STAGGERED PRF - Staggered PRF allows an increase in MTI blind speeds such that no zeros exist in the velocity response at lower velocities. In a two-period mode, the usual "blind speed" or occurrence of a zero in the velocity response is multiplied by a factor which is a function of the ratio of the two repetition periods.

STAMO (STABLE MASTER OSCILLATOR) - A very stable (drift free) oscillatory used to provide a precise frequency for transmission and for comparison with the reflected radar signal returned to the receiver, such as in a Doppler radar where a precise difference between transmitted and received signals must be measured to determine accurately the Doppler frequency.

STAND-FORWARD JAMMING - A method which places the jamming vehicle between the enemy sensors and attack aircraft.

STAND-IN JAMMING (SIJ) - Similar to stand-forward jamming but usually using an UAV with a lower powered jammer instead of a jammer aircraft.

STAND-OFF JAMMING (SOJ) - An ECM support aircraft orbits in the vicinity of the intended target. As the fighter-bomber pilot starts his strike penetration, the ECM aircraft directs jamming against all significant radars in the area. This technique provides broad frequency band ECM without affecting performance of the strike aircraft.

STARBOARD - The right side of a ship or airplane when facing the bow (forward).

STC (SENSITIVITY TIME CONTROL) - Gain control that reduces the radar receiver gain for nearby targets as compared to more distant targets. STC prevents receiver saturation from close-in targets.

STEP RECOVERY DIODE - A charge-controlled switch which ceases current conduction so rapidly that it can be used to produce an impulse. Cyclic operation of the diode can produce a train of impulses which when used with a resonant circuit can produce a single frequency output at any harmonic of the pulse frequency.

STERADIAN - Unit of solid angle. An entire sphere has 4π steradians.

STREAM CHAFF - Operational technique of dropping large quantities of chaff for a continuous period of time. This results in a "ribbon" or "stream" of returns many miles in lengths on radarscopes. The penetrating strike force can then use the resulting chaff corridor to mask their penetration.

SUBHARMONIC - A frequency which is an integral submultiple of another frequency. For example, a sine wave whose frequency is one-third of the frequency of another sine wave is called the third subharmonic. (3 MHz is the third subharmonic of 9 MHz).

SUPERHETERODYNE RECEIVER - A receiver that mixes the incoming signal with a locally generated signa] (local oscillator) to produce a fixed, low intermediate frequency (IF) signal for amplification in the IF amplifiers.

SUPPRESSION - Elimination or reduction of any component of an emission, such as suppression of a harmonic of a transmitter frequency by band rejection filter.

SUPPRESSION OF ENEMY AIR DEFENSES (SEAD) - Activity which neutralizes, destroys, or temporarily degrades enemy air defense systems by using physical attack or electronic means (SEAD pronounced "seed" or "C add").

SUSCEPTIBILITY - The degree to which an equipment or a system is sensitive to externally generated interference.

SWEPT JAMMING - Narrowband jamming which is swept through the desired frequency band in order to maximize power output. This technique is similar to sweeping spot noise to create barrage jamming, but at a higher power.

SWITCHES - See also Diode Switch, Silicon Controlled Switch, Schottky Diode Switch.

SYNCHRODYNE - A klystron mixer amplifier stage in a transmitter, where two signal frequencies are applied as inputs and a single amplified signal is taken out.

TARGET SIZE - A measure of the ability of a radar target to reflect energy to the radar receiving antenna. The parameter used to describe this ability is the "radar cross section" of the target. The size (or radar cross section) of a target, such as an aircraft, will vary considerably as the target maneuvers and presents different views to the radar. A side view will normally result in a much larger radar cross section than a head-on view. See also Radar Cross Section.

TERMINAL IMPEDANCE: - The equivalent impedance as seen by the transmitter/receiver.

TERRAIN BOUNCE - Term for jamming that is directed at the earth's surface where it is reflected toward the threat radar. Reflected jamming creates a virtual image of the jamming source on the earth as a target for HOJ missiles.

THERMISTOR - A resistor whose resistance varies with temperature in a defined manner. The word is formed from the two words "thermal" and "resistor,"

THRESHOLD ISR - The interference to signal ratio (ISR) at which the performance of a receiver starts undergoing degradation. It must be determined by tests.

TRACKING RADAR - A radar whose prime function is to track a radar target and determine the target coordinates (in range and angular position) so that a missile may be guided to the target, or a gun aimed at the target.

TRACKING RADAR RECEIVER - These are of two primary types: conical scan and monopulse. (1) The conical scan system directs the radar signal in a circle around the target. The radar paints this circle 15 to 40 times per second. As the target moves out of the center of this circle, the radar develops aim error voltages and re-aims the antenna. (2) The monopulse system directs four beams at the target simultaneously. The target is in the middle of the four beams. If the target is not in the center, the radar return develops an aim error voltage to re-aim the antenna.

TRACK WHILE SCAN (TWS) RADAR - Although it is not really a tracking radar in the true sense of the word, it does provide complete and accurate position information for missile guidance. In one implementation it would utilize two separate beams produced by two separate antennas on two different frequencies. The system utilizes electronic computer techniques whereby raw datum is used to track an assigned target, compute target velocity, and predict its future position, while maintaining normal sector scan. Most aircraft use only a single antenna.

TRADE-OFF TABLES - A set of tables showing the various combinations of two or more variables that are related in that making one variable better will make the other variable worse. The trade-off helps find the best solution considering all combinations. (For example, how a no-interference condition can be maintained if two emitter platforms are brought close together, if at the same time the frequency separation between their radar transmitters is increased.)

TRANSIENT - A phenomenon (such as a surge of voltage or current) caused in a system by a sudden change in conditions, and which may persist for a relatively short time after the change (sometimes called ringing).

TRANSPONDER - A transmitter-receiver capable of accepting the electronic challenge of an interrogator and automatically transmitting an appropriate reply. There are four modes of operation currently in use for military aircraft. Mode 1 is a nonsecure low cost method used by ships to track aircraft and other ships. Mode 2 is used by aircraft to make carrier controlled approaches to ships during inclement weather. Mode 3 is the standard system used by commercial aircraft to relay their position to ground controllers throughout the world. Mode 4 is IFF. See also IFF.

TRAVELING-WAVE TUBE AMPLIFIER - The TWT is a microwave amplifier capable of operation over very wide bandwidths. In operation, an electron beam interacts with a microwave signal which is traveling on a slow wave helical structure. The near synchronism of the beam and RF wave velocities results in amplification. Bandwidths of 3:1 are possible. Operation at high powers or at millimeter wavelengths is possible at reduced bandwidths.

TRIGGER KILL (RADAR) - A method employed to momentarily disable certain radar system circuits to reduce or eliminate RF emissions which may cause an EMI/EMC or RADHAZ situation such as on the deck of a ship.

TUNNEL DIODE - The tunnel diode is a heavily doped P-N junction diode that displays a negative resistance over a portion of its voltage-current characteristic curve. In the tunneling process, electrons from the p-side valence bands are able to cross the energy barrier into empty states in the N-side conduction band when a small reverse bias is applied. This diode is used as a microwave amplifier or oscillator.

UPLINK - The missile guidance signal which passes midcourse correction command guidance intelligence from the guidance radar site to the missile.

VARACTOR DIODE - A P-N junction employing an external bias to create a depletion layer containing very few charge carriers. The diode effectively acts as a variable capacitor.

VARACTOR TUNED OSCILLATOR - A varactor diode serves as a voltage-controlled capacitor in a tuned circuit to control the frequency of a negative resistance oscillator. The major feature of this oscillator is its extremely fast tuning speed. A limiting factor is the ability of the external voltage driver circuit to change the voltage across the varactor diode, which is primarily controlled by the driver impedance and the bypass capacitors in the tuning circuit.

VELOCITY GATE PULL-OFF (VGPO) - Method of capturing the velocity gate of a Doppler radar and moving it away from the skin echo. Similar to the RGPO, but used against CW or Doppler velocity tracking radar systems. The CW or pulse doppler frequency, which is amplified and retransmitted, is shifted in frequency (velocity) to provide an apparent rate change or Doppler shift.

VELOCITY MODULATION - Velocity modulation is modification of the velocity of an electron beam by alternately accelerating and decelerating the electrons at a frequency equal to the input frequency. Thus, the electrons are segregated in bunches, each bunch causing a cycle or current as it passes an output electrode. The velocity of the electrons is thus a function of the modulation voltage. See also Klystron, Multicavity and Klystron, Reflex.

VICTIM - A receiver (radar or missile) that suffers degradation due to ECM or EMI effects.

VIDEO - Receiver RF signals that have been converted (post detection) into a pulse envelope that can be seen when applied to some type of radar visual display; also used to describe the actual display itself (such as the video on an A-scope).

WARM - Acronym for Wartime Reserve Mode. Any mode of operation of a radar or ECM that is held in reserve, and never used, except in actual combat.

WATER LINE - A reference line used for vertical measurements. When used with an aircraft it is usually the ground with the landing gear extended normally. One of several aircraft location designations, also see butt line and fuselage station.

WAVEGUIDE - A transmission line consisting of a hollow conducting tube of arbitrary geometry (usually rectangular, but may be circular) within which electromagnetic waves may propagate.

WAVELENGTH (λ) - The distance traveled by a wave in one period (the period is the time required to complete one cycle). $\lambda = c/f$. In the atmosphere, electromagnetic waves travel at c, the speed of light (300 million meters per second or 30 cm/nsec). At 5 GHz, one wavelength = 6 cm. At 10 GHz, one wavelength = 3 cm.

WAVEMETER - An instrument for measuring the frequency of a radio wave. The wavemeter is a mechanically tunable resonant circuit. It must be part of a reflection of transmission measurement system to measure the maximum response of a signal. Below 20 GHz, the wavemeter has been replaced by the frequency counter with much greater accuracy and ease of use.

WEAVE - An aircraft maneuver that smoothly changes the instantaneous flight path but maintains the overall route of flight. Not as violent as a jink.

WET RUN - A test run with ship / aircraft armament and/or EW switches on.

WILD WEASEL - USAF aircraft (F-4Gs during Desert Storm) used for suppression of enemy air defense (SEAD) mission.

WINDOW - WWII name for chaff

YIG TUNED OSCILLATOR - A YIG (yttrium iron garnet) sphere, when installed in the proper magnetic environment with suitable coupling will behave like a tunable microwave cavity with Q on the order of 1,000 to 8,000. Since spectral purity is related to Q, the device has excellent AM and FM noise characteristics.

ZENER DIODE - A diode that exhibits in the avalanche-breakdown region a large change in reverse current over a very narrow range in reverse voltage. This characteristic permits a highly stable reference voltage to be maintained across the diode despite a wide range of current.

www.ingramcontent.com/pod-product-compliance
Lightning Source LLC
Chambersburg PA
CBHW072010230326
41598CB00082B/6965